Global Trends in Health, Technology and Management

Anil Kumar Saxena • Sisir Nandi
Editors

Global Trends in Health, Technology and Management

Proceedings of the 1st International Symposium GTHTM-2024

Springer

Editors
Anil Kumar Saxena
Global Institute of Pharmaceutical Education and Research
Kashipur, Uttarakhand, India

Sisir Nandi
Global Institute of Pharmaceutical Education and Research
Kashipur, Uttarakhand, India

ISBN 978-3-031-75456-2 ISBN 978-3-031-75457-9 (eBook)
https://doi.org/10.1007/978-3-031-75457-9

This Springer imprint is published by the registered company Springer Nature Switzerland AG
The registered company address is: Gewerbestrasse 11, 6330 Cham, Switzerland

Tribute to Dr Nityanand and Dr V. P. Kamboj

As chairman of GIPER, GHTMF, and parent body Krishna Memorial Trust, it gives me immense pleasure to welcome you all to this international symposium. The Krishna Memorial Trust was founded in 1988 by the settler late Smt. Parwati Devi, my aunt/buaji, my father late Sri A L Saxena, and my mother late Mrs RD Saxena.

My father's principles that one should serve the place where he is born and my buaji wish to impart education to society inspired me to establish a primary school the Krishna Public Collegiate in the year 1987 which we started in huts. Now, after almost 37 years the small primary school has grown into a 10+2 school and a pharmaceutical research institute. Next to my father, Dr Nityanand, my mentor (guru) has been a key person who has supported me in my career and also beyond it.

Dr Nityanand was born on 1st January 1925 at Lyalpur, now in Pakistan. He got his B.Sc. from Lahore in 1943 and moved to St. Stephen's College, Delhi, for his M.Sc. in chemistry which he completed in 1945. He did his first Ph.D. from UDCT Mumbai with Prof K. Venkatraman and his second Ph.D. from Cambridge under Nobel laureate Lord Todd in 1949. He joined CSIR-CDRI in 1951 as a junior scientific officer and continued to work there till he retired on 31st Dec 1984 as Director CDRI. He made an immense contribution to Medicinal Chemistry in terms of more than 400 research publications, 130 patents, and 30 book chapters. During his tenure, he guided 90 Ph.D. students and above all invented drugs like Centbutindole, which was the first drug to be released from CDRI in the year 1986 and the other important one which is marketed under the name of Chaya, an oral contraceptive. Dr Nityanand has been awarded numerous awards including Padma Shri in 2012.

Dr. Nityanand has been associated with us and it was his love and affection that he dedicated the fully developed KPC from primary to 10+2 in the year 2002 and later the building of GIPER in the year 2020 to the society. He has been a guide in the development of GIPER through his message at many international conferences organised by the institute.

During my stay at CDRI, one other eminent scientist and a gem of a person who had been my close senior colleague and friend was Dr. Ved Prakash Kamboj. Dr Kamboj was born in Jalandhar on 1st April 1937 and did his B.Sc., M.Sc., and Ph.D. from the Punjab University, Chandigarh. He joined CDRI in the year 1961 as a research scientist and later also worked as director of the institute from 1995–1998. He obtained advanced training in reproductive biology from West Germany in the year 1975 and has been a pioneer in the country when it comes to the use of reproductive medicine. During his scientific tenure, he published over 300 research papers and had been very closely associated with the development of Chaya as a biologist.

He has also been closely associated with KPC and GIPER on various occasions including international conferences and annual functions.

Unfortunately, Dr Kamboj at age of 86 and Dr Nityanand at age of 99 left us on 21 November 2023 and 27 January 2024, respectively, towards immortality. We are missing both of them physically from this world, but I am sure their blessings are with us. I am sure their journey and dedication to science will be a source of inspiration to researchers and young students in years to come and their legacy will continue.

—Anil Kumar Saxena, Kashipur, India

Preface

The international symposium on Global Trends in Health, Technology and Management (GTHTM-2024) was organized by Global Health Techno Management Forum (GHTMF), Global Institute of Pharmaceutical Education and Research (GIPER), Kashipur, and Veer Madho Singh Bhandari Uttarakhand Technical University at Dehradun during March 15–17, in continuation of the previous international symposiums: Drug Design and Development Research (DDDR-2021), Current Trends in Pharmaceutical and Medical Sciences (CTPMS-2020), Computational Methods in Toxicology and Pharmacology Integrating Internet Resources (CMTPI-2017), and International Seminar on Pharmaceutical Education and Research (ISPER-2010). The symposium GTHTM-2024 (online and offline mode) started with the welcome of chief guest and participants from 14 countries by Hon'ble Vice-chancellor Prof. (Dr.) Onkar Singh followed by a tribute to eminent scientists Dr. Nityanand and Dr. V. P. Kamboj who had been deeply associated with the symposiums by me. In the inaugural session, lectures were delivered by His Excellency, the Hon'ble Governor of Uttarakhand, Lt. General Gurmit Singh, the Vice-Chancellor of KGMU, Prof. (Dr.) Soniya Nityanand, Dr. Omer Saka, a Health Economist and Business Development expert from Switzerland, and an online lecture by Prof. (Dr.) Gunda Georg, Regents Professor, University of Minnesota, USA.

There were 23 invited/plenary lectures and several oral and poster presentations covering a wide range of topics in the field of drug discovery and development including diseases like cancer, tropical and lifestyle diseases, agroecology, artificial intelligence, machine learning, environmental protection, green agriculture, global warming, digitalization in health care. Out of these 15 abstracts of invited lectures, one of the best oral and two abstracts of the best poster presentations along with the details of organizing committee and advisory board members have been included in the proceedings as the back and front materials respectively in addition to 22 chapters as the major contents in the proceeding of GTHTM-2024.

It has been a great pleasure in editing and compiling this proceeding along with my co-editor Dr. Sisir Nandi. I am very thankful to him and to all authors of 22 chapters for their contributions and valuable time in preparing the chapters. I also thank all the reviewers for their timely feedback and constructive comments. I gratefully acknowledge all the series editors of Topics in Medicinal Chemistry, particularly Prof. (Dr.) Gunda Georg, for the cooperation and support. Last but not least, thanks are also due to Dr. Sofia Costa and Ms. Neelofar Yasmeen and their team from Springer for their valuable assistance. I wish that this volume of proceeding will be useful to the researchers working in the field of health and its techno-management.

Kashipur, India — Anil Kumar Saxena

Patrons

Prof. (Dr.) Onkar Singh, **Hon'ble Vice Chancellor, VMSBUTU, Dehradun**
Prof. (Dr.) Anil Kumar Saxena, **Hon'ble Chairman GIPER, Kashipur**

Organising Committee

Chairman
Prof. (Dr.) Deepak Teotia

Co-Chairman
Prof. (Dr.) Yusra Ahmad

Organising Secretary
Prof. (Dr.) Sisir Nandi

Co-Organising Secretary
Mr. Sarfaraz Ahmed

Members

Prof. (Dr.) Vineet Kumar, Prof. (Dr.) Kuldeep K Roy, Mr. Sumit Kumar, Mrs. Shivani Lodha, Mr. Arun Kumar, Mr. Muneer Alam, Mrs. Swati Tamta, Mrs. Mahvish Jamal, Ms. Anjali Negi, Mr. Sahjad, Mr. Atul Kumar, Ms. Bhawna Bisht, Mr. Kuldeep Singh, Mr. Rahul Kumar

International Advisory Committee

Prof. Alain Carpy (CNRS, Bordeaux, France)
Prof. Athina Geronikaki (Aristotle University, Greece)
Prof. Emil Reisinger (University Medical Center, Rostock, Germany)
Dr. Imran Ahmad (Jina Pharmaceuticals Inc, USA)
Dr. James Devillers (CTIS, France)
Prof. Johann Gasteiger (Friedrich-Alexander-Universität, Germany)
Prof. Krishnan Balasubramanian (Arizona State University Tempe, AZ)
Mr. Ludy Suryantoro (Team lead, Multisectoral engagement, World Health Organization, Switzerland)
Prof. Marjan Vracko (National Institute of Chemistry, Slovenia)
Prof. Om Prakash (Kansas State University, KS)
Dr. Omer Saka (EY Parthenon Health Sciences and Wellness, Switzerland)

Prof. Subhash C. Basak (University of Minnesota, USA)
Dr. Ved Srivastava (Chemistry Intarcia Therapeutics Inc, USA)
Prof. (Dr.) Vladimir Poroikov (Institute of Biomedical Chemistry, Russia)

National Advisory Committee

Dr. Alok Dhawan (Director, Centre of BioMedical Research, Lucknow)
Dr. Alok Saxena (Advisor, GTC, Vorro Technology, United States)
Prof. B. Jayaram (IIT, New Delhi)
Dr. Dinesh Patel (MD and CEO, Themis Medicare, Haridwar)
Prof. Indira Ghosh (Ex- Professor, SCIS JNU, New Delhi)
Prof. Kunal Ganguly (Dean, IIM, Kashipur)
Prof. Kunal Roy (Professor and Ex-Head Pharmaceutical Technology, Jadavpur University, Kolkata)
Dr. Madhu Dikshit (Ex-Director, CSIR-CDRI, Lucknow)
Prof. P.V. Bharatam (Professor and Head, NIPER, Mohali)
Dr. Rajesh Upadhyay (Secretary, Uttarakhand Board of Technical Education, Roorkee)
Prof. Soniya Nityanand (Vice Chancellor, KGMU, Lucknow)
Dr. Sanjay Batra (Chief Scientist, CSIR-CDRI, Lucknow)
Prof. T.K. Chakraborty (IISC, Bangalore)
Dr. V.P. Kamboj (Ex-Director, CSIR-CDRI, Lucknow)
Dr. Yatendra Kumar (Ex-President, Wockhardt)
Prof. Y. K. Gupta (Ex- Head, Dept. of Pharmacology, AIIMS, New Delhi)

LT GEN GURMIT SINGH
PVSM, UYSM, AVSM, VSM (Retd)
GOVERNOR UTTARAKHAND

सत्यमेव जयते

RAJ BHAWAN
UTTARAKHAND
Phone : 0135-2757400
2757403

12 March, 2024

Message

It is interesting to note that the Veer Madho Singh Bhandari Uttarakhand Technical University is organizing an International Symposium on "Global Trends in Health, Technology and Management" (GTHTM-2024) in association with Global Techno Management Forum and Global Institute of Pharmaceutical Education and Research (GIPER), Kashipur between March 15-17, 2024.

Global trends in health, technology, and management are rapidly evolving, reshaping industries and societies worldwide. In health, there's a shift towards personalized medicine, leveraging genomics and data analytics to tailor treatments. Telemedicine is also on the rise, improving access to healthcare, especially in remote areas. Technology trends like artificial intelligence (AI) and machine learning are revolutionizing management practices, enabling data-driven decision-making and automation of routine tasks. Virtual and augmented reality are transforming training and simulation in various industries, including healthcare and education. These trends highlight the need for continuous adaptation and innovation to stay competitive in the global landscape.

My best wishes to the organizers for the success of the international symposium, which will help accelerate sustainable growth in the field of health and its techno-management.

Gurmit

Lt Gen Gurmit Singh
PVSM, UYSM, AVSM, VSM(Retd)

वीर माधो सिंह भण्डारी
उत्तराखण्ड प्रौद्योगिकी विश्वविद्यालय
(उत्तराखण्ड सरकार द्वारा अधिनियम 415/2005 द्वारा स्थापित पूर्ववर्ती उत्तराखण्ड तकनीकी विश्वविद्यालय)
VEER MADHO SINGH BHANDARI
UTTARAKHAND TECHNICAL UNIVERSITY
(Formerly Uttarakhand Technical University Established by Act no. 415/2005 by Uttarakhand Government)

Prof. (Dr.) Onkar Singh
Vice Chancellor
प्रो० (डॉ०) ओंकार सिंह
कुलपति

Office Address :
Chakrata Road, P.O. Chandanwari, Premnagar, Suddhowala
Dehradun - 248007, Uttarakhand (India)
Phone No. : 0135-2774068,
Email : vc@uktech.ac.in, Website : www.uktech.ac.in

Ref No. 133/VC/UTU/2024

Date 06/03/2024

It is my great pleasure to note that **Global Institute of Pharmaceutical Education and Research (GIPER), Kashipur** has formed an international forum by the name of Global Health Techno Management Forum (GHTMF) and is organizing an international symposium entitled **"International Symposium on Global Trends in Health, Technology and Management" (GTHTM-2024)** under the aegis of Veer Madho Singh Bhandari Uttarakhand Technical University from March 15-17, 2024. It makes a very good sense for excellent in the begining of 2024 because this both offline and online symposia will be a brain storming for the leading scientists, young researchers, academicians and students globally to create an open platform for discussing and resolving pertinent issues in the field of health, medicine, technology and management. I complement you all for the great efforts which may create enough well-trained human resources in the field so that these benefits are available to society all the time.

With warm wishes for the success of this symposia.

6/03/2024

Vice-Chancellor
Veer Madho Singh Bhandari Uttarakhand Technical University,
Dehradun, India

GLOBAL INSTITUTE OF PHARMACEUTICAL EDUCATION & RESEARCH (GIPER)

CONDUCTING BACHELOR IN PHARMACY (B. PHARM) & MASTER IN PHARMACY (M. PHARM)

Approved by AICTE & PCI, New Delhi. Affiliated to Uttarakhand Technical University, Dehradun

It is of immense pleasure that the Global Institute of Pharmaceutical Education and Research (GIPER), Kashipur, has formed an international forum by the name of Global Health Techno Management Forum (GHTMF) and is organizing an international symposium entitled **"International Symposium on Global Trends in Health, Technology and Management" (GTHTM-2024)** under the aegis of Veer Madho Singh Bhandari Uttarakhand Technical University from March 15-17, 2024. The conference covers several topics under the broad research heads pharmaceuticals, medical and biomedical sciences, health and technology management, environmental protection, natural resources, green agriculture, cyber security, sustainable development goals, digitalization in health care and public health. The recent COVID-19 pandemic has shown the need for multi-sectoral coordination and collaboration when it comes to solving unpredicted challenges such as natural calamities or disease outbreaks. It will be useful to provide a common platform for the innovative minds of a large number of eminent thinkers, scientists, educationists, business leaders, researchers and students to develop collaborative relationships among different specialities, industrial sectors and public-private organizations. One must think from a border perspective and there is a need for collaboration at a global level. This International Symposium provides a platform to bring stakeholders from different fields of expertise to come together, exchange their ideas and work together for the betterment of society. I wish that due emphasis on these aspects shall be helpful to translate the real benefits of the current developments to the common man for improving health and the environment in which GIPER may play an important role.

It is very graceful to combine the inauguration of this conference with the blessings of my mentor Dr. Nitya Anand who passed away at 99 years.

I welcome you all and look forward to your active participation in making this event a success. I take this opportunity to thank an elite gathering of scientists, educationists, business leaders and other professionals for taking pains to grace this occasion with their benign presence. I appreciate the sincere efforts of the members of the organizing committee and staff and students of the Global Institute of Pharmaceutical Education and Research, Kashipur and Veer Madho Singh Bhandari Uttarakhand Technical University for their active participation. I hope this event is a success.

Dr Anil Kumar Saxena

Chairman

Jaspur Road, Kashipur-244713, Udham Singh Nagar (Uttarakhand)
Ph. : 9219476108, 9839012951, 9997722274, www.giper.edu.in, E-mail : kmt.giper@gmail.com

Contents

Quantitative Structure Interaction Activity Relationship (QSIAR) as a Novel Approach to Drug Design: A Case Study of Anti-tubercular Agents

Anil K. Saxena[1] (✉)

[1] Department of Pharmaceutical Chemistry, Global Institute of Pharmaceutical Education and Research, Kashipur, Uttarakhand, India

Abstract Computer-aided drug design (CADD)—an effective tool in the process of drug design and discovery—consists of ligand-based drug design (LBDD) and structure-based drug design (SBDD). The SBDD has gained importance in recent years due to the developments in molecular biology including genomics, proteomics, and structural information of new targets. Docking scores play a key role in analysing the results in terms of interactions between the structural components of the interacting molecule with the target protein. The docking scores generally do not correlate with the observed biological activity. This may be attributed to the limitations in scoring functions used in docking algorithms that often fail to account for crucial factors like entropy change, solvation effects, and interactions, contributing to binding affinity limiting the accurate prediction of binding energy changes, which may result in the poor correlations between observed biological activity and docking scoring functions. In this context, a maiden novel approach (quantitative structure interaction activity relationship (QSIAR)) has been applied to address this issue by taking into account the specific interactions between a ligand and the amino acid residues present at the active site as independent and biological activity as dependent parameter(s) in quantitative terms to explain the observed anti-tubercular activity as mycobacterium ATP synthase inhibitory activity in diverse molecules such as 4-substituted amino sulphonyl-2-methyl-7-chloroquinolines, bisquinoline, imidazo[1,2-a]pyridine ethers, and squaramides. The developed and validated quantitative model(s) have led to the identification of novel leads through virtual screening of the focussed libraries for the development of new anti-tubercular agents.

Keywords Tuberculosis · ATP synthase · QSIAR · Virtual screening · SBDD

A. K. Saxena, S. Nandi (eds.), *Global Trends in Health, Technology and Management*, https://doi.org/10.1007/978-3-031-75457-9_1

1 Introduction

The quest for identifying novel chemical entities (NCEs) as therapeutic agents requires exhaustive research. It is intensive both in terms of time and expenses and requires the integration of chemistry and biology. As per the current updates, the time required for a new chemical to reach clinical trials is approximately 8.3 years with a median cost of $985 million dollars [1]. The estimated cost, of introducing a new drug from the starting of the process of drug discovery and development till the approval of drug for marketing, in terms of time and money is approximately 12 years and 2.6 billion dollars, respectively [2, 3]. The principles of early drug discovery during the preclinical development phase involve target identification, target validation, hit discovery, or NCE identification for optimization leading to candidate drug for development [4]. In the past, the above process depended on chance findings or conventional trial-and-error techniques, frequently requiring decades and significant resources with little assurance of success [5]. It is understandable that both time and money need saving and unsurprisingly the research and development units of pharmaceutical companies sought ways to overcome these limitations through the advancements in computational technology and molecular biology in the late twentieth century. In recent years, there has been an increase in computing capacity of the machines, development in machine learning, and accumulation of omics data [6]. More recently, the emergence of artificial intelligence (AI) has been a game changer [7–9] in the identification of hits/lead compounds and faster validation of drug targets for improved optimization in the process of drug design and discovery [10, 11]. The drug design and development through computers also known as computer-aided drug design (CADD) has been a cornerstone in the process of drug discovery. CADD combines the complex biological systems with predictive computational algorithms and helps in the development of selected databases containing chemical and biological data [12]. The CADD has resulted in the discovery and optimization of various drugs such as saquinavir, boceprevir, oseltamivir, rupintrivir, zanamivir, aliskiren, captopril, dorzolamide, nolatrexed, etc. [13]. The identification and optimization of ligands for therapeutic application has been a major goal of computer-aided drug design, which incorporates both structure-based and ligand-based techniques. When a group of bioactive chemicals share a similar target but lack protein structure information, ligand-based (indirect) drug design makes it easier to identify important chemical structural features in the group. The main aim of ligand-based drug design (LBDD) is to create predictive models capable of quantitatively grading the molecules between actives and inactives. The LBDD was the main technique used in CADD until the middle of the 1980s, and from 1990s onwards, structure-based drug design (SBDD) began to appear in the horizon, which with the advent of identification of molecular targets through molecular biology and genomics began to flourish [14]. This then further led to the integration of both LBDD and SBDD in developing more robust models than the ones developed through either of the approaches [15]. This integrated approach has been widely used for virtual screening, which is an efficient tool for drug discovery process [16]. The SBDD involves the prior knowledge of a validated protein target (associated with a particular human disease) and also gives information about the important amino acid residues involved in ligand-target interactions resulting through the binding of the ligand to the target. Molecular docking is routinely used to understand the interaction pattern of the ligand binding at the target site and to enumerate important contacts, such as van der Waals interactions, polar contacts, etc. The typical algorithm for molecular docking involves using sampling methods to generate potential ligand binding conformations, after which a scoring function is employed to rank the ligands based on their binding affinity, determined by energy terms (binding energy) [17, 18]. The ligand-receptor docking is performed through computer programs that rely on some scoring functions, which essentially describe the interactions between the ligand and the protein target (receptor). The SBDD offers several advantages over the LBDD, the main being the consideration of protein structure making it more biologically relevant, in addition to ligand structures, along with the determination of active conformation of the ligand. It also suffers from certain disadvantages such as non-consideration of protein flexibility in most of the software, accuracy in the pose prediction, and non-proportionality between the docking scores (binding energy) [16]. Pose prediction is linked with the scoring function since the most probable orientation of the ligand in the active site is ranked as per the scoring functions obtained through docking [19]. The scoring functions used in molecular docking are classically categorized as force field and knowledge-based, in addition to empirical scoring functions. These scoring functions evaluate the binding affinity of the compounds with the target protein [20–22]. The limitations of these scoring functions have been reported and may be attributed to (i) neglecting the overall entropic changes and solvation effects, (ii) oversimplification of binding affinity between the ligand and the protein only as a linear relationship between binding affinity and ligand-protein interactions with an additive approach, and (iii) inability to predict uncommon protein-ligand interactions [23–25]. Despite advancements in the classical scoring functions, unsatisfactory correlations with experimental binding affinity still exist with some of the classical scoring functions overestimating the accuracies of protein-ligand interactions in certain cases [26]. In recent years, these limitations in the classical scoring functions have been addressed by the introduction of machine learning scoring functions based on neural network approximations of the binding affinity as a function of the positions and identities of atoms of protein and ligand [27]. However, increasing concerns have emerged about the generalizability of machine learning models in terms of dataset biases instead of actual physical interactions leading to scoring errors and incorrect predictions of the biological activity [28–30]. It is also known that due to some of these problems, the binding energy often does not correlate with the observed biological activity. Moreover, the binding preference of a ligand at the receptor site is influenced not only by a favourable complementary interaction pattern but also by its competing interactions with the solvent and the flexibility at the receptor site [31]. Hence, it has become a necessity to formulate a novel methodology to address this chronic issue. In this paper, we discuss a novel methodology, known as quantitative structure interaction activity relationship (QSIAR), where the docking interactions between the ligand and the amino acid residues (AARs) present at the active site of the protein target are given weightage, which circumvents the reliance on the scoring functions. The weighed interactions are treated as independent and the observed biological activity as dependent variable for the development of predictive models through multiple regression analysis. This methodology also addresses a very peculiar issue of dealing with enantiomers where the activity of racemic compounds is only given, which makes the prediction of individual enantiomer a grave issue. We also present the case studies wherein this methodology has been used for the identification of novel chemical entities as ATPase inhibitors for the prevention of tuberculosis (TB) [32, 33].

2 Quantitative Structure Interaction Activity Relationship (QSIAR)

The novel maiden methodology to address the issues highlighted above, known as quantitative structure interaction activity relationship (QSIAR), takes into account the specific interactions between a ligand and the amino acid residues present in the active site. This methodology is based on the utilization of the results of the docking studies in terms of interactions between the ligand and amino acid residues (AARs). Thus, the interactions of the ligand molecule with the amino acid residues of the target protein are given weightage and are considered as independent parameters, while the binding energy/affinity, docking scores/biological activity, is considered as a dependent parameter. However, a particular amino acid residue may sometimes show more than one interaction with different atoms/groups constituting the ligand molecule, while in some cases, it may show just one interaction. A protocol to give weightage to such type of interactions was devised in such a way that in case of multiple interactions of one AAR with the ligands, the interactions were considered by adding all the interactions to yield the parameter describing the combined interactions, whereas, in case of single interaction protocol, just the presence or absence of interaction (instead of counting all the interactions) is considered and assigned the value of '1' and '0', respectively. Hence, an interaction matrix is designed where the number of interactions present with each AAR is meticulously noted and the resultant data is organized in two ways, namely, the single weightage (S) and combined weightage (C). So, for 'S', any number of interactions with a particular residue was considered as a single interaction parameter describing the presence of interaction(s) with a particular AAR and is assigned a value of '1' for this parameter, while in the absence of any interaction, the same was given a value of '0'. In case of the combined weightage (C), all interactions with a particular AAR were added, considering each additional interaction to contribute to its score to get the value of (C). The above solution is easily applicable for normal molecules without any asymmetric atom or achiral molecules; however, in case of chiral molecules, all enantiomers need to be considered for docking studies in spite of the available biological data for racemic compounds. Hence, for the racemic compounds, the interactions are considered for constituting enantiomers, and the combined interaction parameter for each AAR present in one or all enantiomers is calculated by adding the number of interactions present in each enantiomer for that AAR. Thus, these weighted parameters describing the number of interactions of the ligand molecule with the amino acid residues of the target protein are considered as independent parameters, while the binding energy/affinity, docking scores/biological activity, is considered as a dependent parameter. The QSIAR model(s) are then developed on the curated data using MLR or other statistical analyses used in QSAR. The QSIAR model can be presented in the following form:

$$\mathbf{BA} = \mathbf{Const.} + \mathbf{a_1 AAR_1} + \mathbf{a_2 AAR_2} + \ldots\ldots\ldots \mathbf{a_n AAR_n} \tag{1}$$

where:BA = binding energy/affinity, docking scores/biological activityAAR = interacting amino acid residue

The protocol described above is explained in Fig. 1 where the interactions of Glu65b AAR are described. In the case of compound 1, the AARs are Glu65b (AAR_1), Leu63a (AAR_2), Phe69b (AAR_3), Leu72b (AAR_4), and HOHb (AAR_5), which display interactions with the ligand molecule. The Glu65b residue shows one interaction each with the hydroxyl group, methylene group, nitrogen of the amino group, N-ethyl group, and phenylene group of the ligand molecule, thus accounting for a total of five interactions. As per the protocol discussed above, the combined weightage (C) of Glu65b will be 5; hence, the value of the independent combined interaction parameter for AAR_1 will also be 5. In the case of single weightage (S), we just see the presence and absence of interaction; hence, the value of independent single interaction parameter AAR_1 will be 1. Similarly, the independent combined and independent single interaction parameter values for each of AAR_2 (Leu63a), AAR_4 (Leu72b), and AAR_5 (HOHb) will be 1 and 1, respectively, while for AAR_3 (Phe69b), which has two interactions, these values will be 2 for (C) and 1 for (S). In the case of enantiomers, the interactions for each enantiomer were noted, and after obtaining the parameter values for both enantiomers by (C) and (S) protocols, the interactions of both enantiomers were combined separately for both (C) and (S) protocols to yield the final composite weightage of interaction parameters for the racemic molecule. This is described taking the example of compound 2 for which the docking results of both the enantiomers (R and S) were analysed. The (R) enantiomer of compound 2 showed interactions with Glu65b (AAR_1), Leu63a (AAR_2), Leu72b (AAR_3), HOHb (AAR_4), Tyr68b (AAR_5), and Ile70a (AAR_6). So, the (C) values for AAR_1, AAR_2, and AAR_6 were 7, 3, and 2, respectively, while the (C) values for the rest of AARs were 1 due to the presence of just one interaction. The (S) values for all the AARs will be 1 based on the presence of the interaction. The (S) enantiomer also

Comp. No.	R/S	Activity	Glu65b	HOHb	Tyr68b	Ala66b	Phe69b	Asp32b	Leu72b	Ile70b	Val61b	Gly62b	Phe57b	Phe58b	Phe74b	Leu63a	Ile70a	Ala66a	Phe74a	Ile59a	Gly62a	Glu65c	Phe69c	Leu72c	Docking score
01	-	0.443	5	1	0	0	2	0	1	0	0	0	0	0	0	1	0	0	0	0	0	0	0	0	-4.767
02	R	1.301	7	1	1	0	0	0	1	0	0	0	0	0	0	3	2	0	0	0	0	0	0	0	-6.778
	S	1.301	7	1	1	0	0	0	1	0	0	0	0	0	0	3	1	0	1	0	0	0	0	0	-7.332

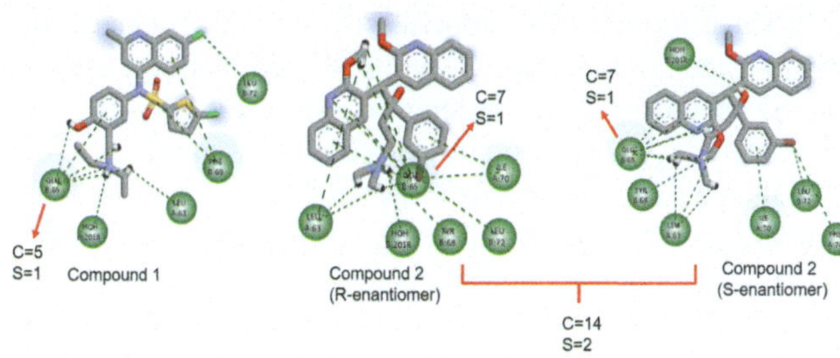

Fig. 1 Detailed discussion of providing weightage through the combined 'C' and single 'S' protocols

shows the above AARs in addition to Phe74a (AAR_7). The (C) values for AAR_1 and AAR_2 were 7 and 3, while for the rest of AARs, the (C) values were 1. The (S) value will be 1 for all the AARs. These individual values of AARs obtained for both the enantiomers are added to give the combined weightage of the racemic compound in which the (C) values for AAR_1, AAR_2, AAR_3, AAR_4, AAR_5, AAR_6, and AAR_7 will be 14, 6, 2, 2, 2, 3, and 1, respectively, while the (S) values will be 2 each for AAR_1 to AAR_6 and 1 for AAR_7.

3 Case Study on the Docking Analysis of M.Tb. ATP Synthase Inhibitors

Tuberculosis (TB) is a serious infectious disease that affected around 10.6 million and claimed 1.3 million lives in 2022 [34]. The severity of TB increased with the resistance to normally used TB drugs leading to multi-drug-resistant tuberculosis (MDR-TB) and extensively drug-resistant tuberculosis (XDR-TB). There have been numerous studies on the chemotherapeutics to contain this dreadful disease [35–38]. There has also been development towards the identification of novel drug targets against which NCEs can be developed [39, 40]. Among the different drug targets, the ATP synthase of mycobacterium is particularly significant because it has recently been identified as a target for newly introduced potential anti-TB compounds, notably bedaquiline (BDQ). ATP synthase is an important enzyme in the process of oxidative phosphorylation involved in the production of adenosine triphosphate (ATP) [41, 42]. BDQ became a drug of choice for the MDR-TB but has recently been reported for resistance through various mechanisms [43–45]. The emergence of resistance has prompted researchers to search for new molecules against ATP synthase.

In quest for finding NCEs active against TB, the QSIAR methodology was applied for the first time on the docking studies on the dataset of analogues of 4-substituted amino sulphonyl-2-methyl-7-chloroquinolines (1–16) and bisquinoline (17–25) cores [46, 47] (Table 1) as M.Tb. ATP synthase inhibitors. The docking studies were performed using the crystal structure of a mycobacterial ATP synthase rotor ring in complex with bedaquiline (pdb id: 4V1F) [48] imported from the protein data bank (www.rcsb.org).

As discussed above, the interactions were noted using both as discussed above and described in Fig. 1. Since the dataset comprised of inhibitory activity of the racemic molecules also, the average docking scores were obtained from the respective enantiomers (both R and S) of the racemic molecules. Since the docking scores and the observed biological activities do not usually correlate, the same was observed for this dataset, which also showed a poor correlation of 0.0351 and 0.0617 for compounds 1–16 and 17–25, respectively. Both the datasets (compounds 1–16 and 17–25) were later combined so that a larger number of independent parameters as compared to individual datasets may be used in the QSIAR analysis. The correlation studies between the independent interaction parameters developed as per the (S) and (C) protocols and the observed biological activity as dependent parameter were done through the Minitab

Table 1 Compounds from the dataset with ATP synthase inhibitory activity reported in the literature [44, 45, 48]

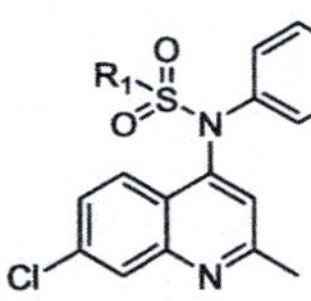

1-16

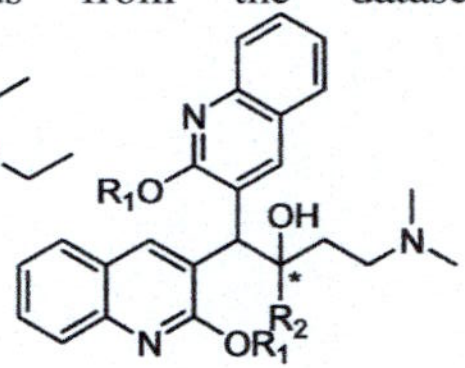

17-24

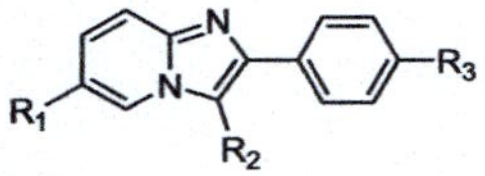

25-46

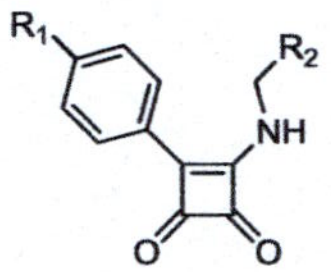

47-52

				Activity in pIC_{50} (μM)		Combined interaction parameter (C) values				
Comp no.	R_1	R_2	R_3	Observed	Estimated/predicted[a]	Glu65b	HOHB	Ala66b	Phe69b	I
1	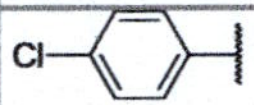	–	–	0.275	−0.263	3	1	0	0	1
2[b]	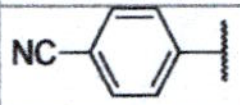	–	–	0.13	−0.268	2	0	1	2	1
3[b]	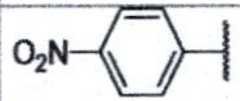	–	–	0.408	0.405	4	1	1	0	1
4[b]	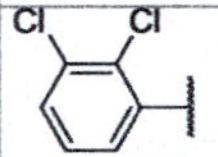	–	–	0.292	0.91	4	0	0	0	1
5[b]	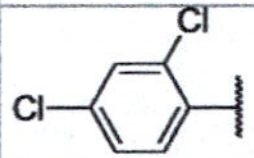	–	–	0.2	−0.648	5	2	0	1	1
6[b]	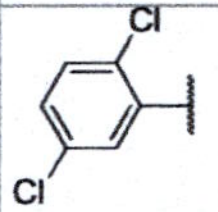	–	–	−0.127	−0.681	3	1	1	2	1
7	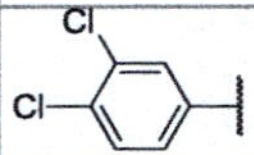	–	–	−0.1	−0.964	0	0	0	1	1

(continued)

Table 1 (continued)

Comp no.	R_1	R_2	R_3	Activity in pIC_{50} (μM)		Combined interaction parameter (C) values				
				Observed	Estimated/predicted[a]	Glu65b	HOHB	Ala66b	Phe69b	I
8	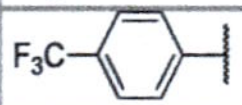	–	–	−0.133	−0.203	2	0	0	1	1
9	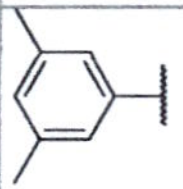	–	–	−0.262	0.177	3	0	0	1	1
10[b]	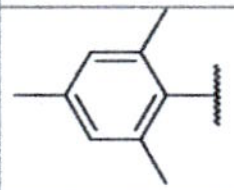	–	–	−0.017	0.139	4	0	1	3	1
11	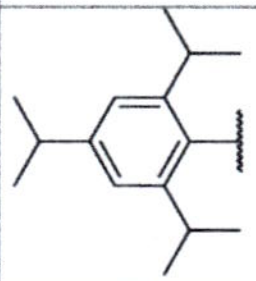	–	–	−0.517	−0.324	0	0	1	0	1
12	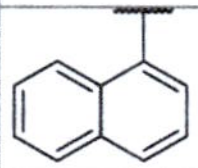	–	–	0	−0.0152	4	1	2	2	1
13[b]	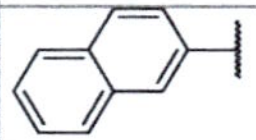	–	–	0.036	0.084	2	0	1	1	1
14[b]	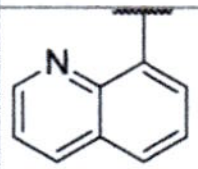	–	–	−0.267	−0.356	2	1	1	0	1
15[b]	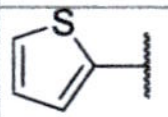	–	–	−0.195	−0.241	3	0	1	3	1
16	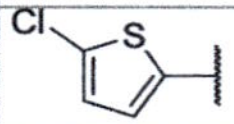	–	–	0.443	−0.209	5	1	0	2	1
17	Me	Phenyl	–	1.154	2.179	10	2	2	0	1
18	Me	3-Bromophenyl	–	1.301	3.1258	14	2	0	0	1

Comp no.	R_1	R_2	R_3	Activity in pIC_{50} (μM)		Combined interaction parameter (C) values				
				Observed	Estimated/predicted[a]	Glu65b	HOHB	Ala66b	Phe69b	I
19[b]	CH_2 - $CH = CH_2$	Phenyl	–	1.397	1.634	11	2	0	1	1
20	CH_2 - $CH = CH_2$	3-Bromophenyl	–	1.522	1.5662	11	2	1	2	1
21[b]	CH_2 - $CH = CH_2$	4-Bromophenyl	–	0.522	1.448	9	2	2	1	1
22[b]	CH_2 - CH_2 - $CH = CH_2$	4-Bromophenyl	–	1.096	1.182	9	1	2	4	1
23[b]	Me	2-Naphthyl	–	1.522	1.388	10	3	2	1	1
24	Me	4-Bromophenyl	–	2	1.252	10	2	0	1	1
25[b]	H	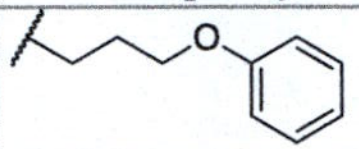	H	0.886	0.926	1	0	1	0	0
26[b]	CH_3	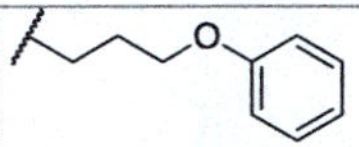	H	1.522	1.307	2	0	1	0	0
27	H	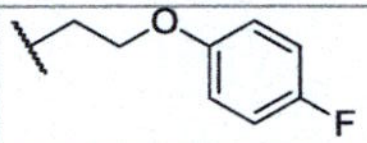	H	1.097	0.926	1	0	1	0	0
28	Cl	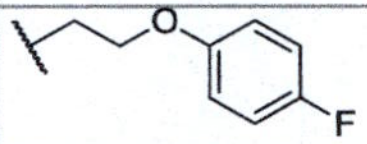	H	1.398	1.019	2	0	0	0	0
29[b]	CH_3	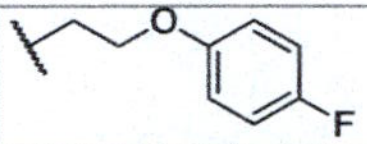	H	1.301	0.227	2	1	0	0	0
30[b]	CH_3	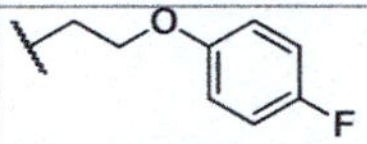	F	1.699	0.514	2	1	1	0	0
31[b]	Cl	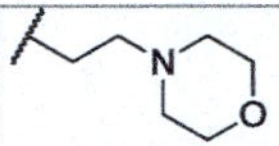	H	−1.557	−1.04	0	2	1	0	0
32	H	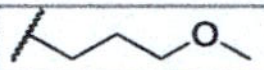	H	−1.276	−1.04	0	2	1	0	0

(continued)

Table 1 (continued)

Comp no.	R_1	R_2	R_3	Activity in pIC_{50} (μM)		Combined interaction parameter (C) values				
				Observed	Estimated/predicted[a]	Glu65b	HOHB	Ala66b	Phe69b	I
33[b]	H	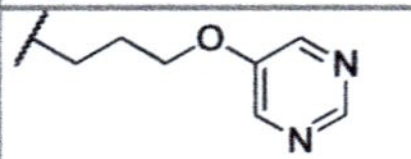	H	−0.748	−0.599	0	1	1	1	0
34	CH_3	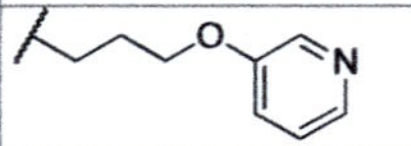	F	0.796	0.926	1	0	1	0	0
35	CH_3	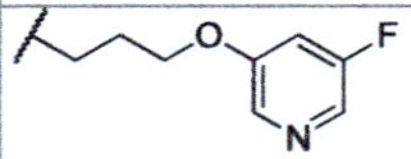	F	1.222	1.306	2	0	1	0	0
36	CH_3	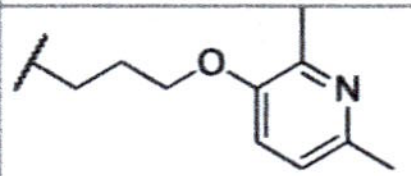	F	1.046	1.213	1	0	2	0	0
37	CH_3	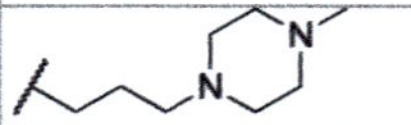	H	−0.806	−0.507	1	1	0	1	0
38[b]	CH_3	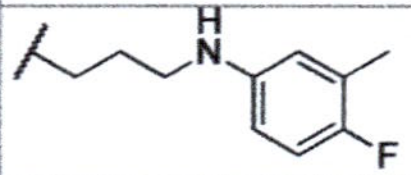	F	2.301	1.687	3	0	1	0	0
39[b]	CH_3	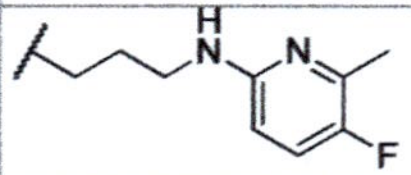	F	1.222	1.019	2	0	0	0	0
40	CH_3	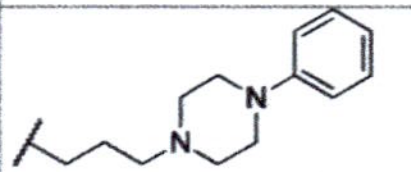	F	1.301	1.306	2	0	1	0	0
41[b]	CH_3	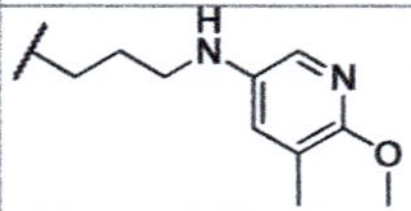	F	1.097	0.536	2	0	2	3	0

Comp no.	R_1	R_2	R_3	Activity in pIC_{50} (μM)		Combined interaction parameter (C) values				
				Observed	Estimated/predicted[a]	Glu65b	HOHB	Ala66b	Phe69b	I
42[b]	CH_3	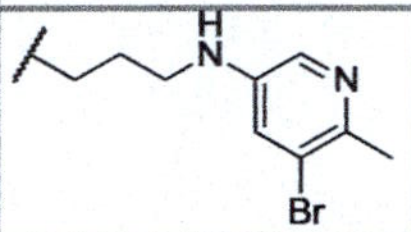	F	0.678	1.148	1	0	3	1	0
43	CH_3	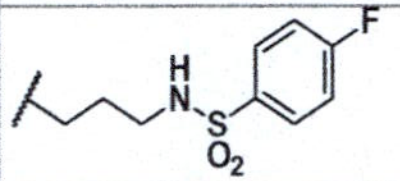	F	0.155	−0.313	0	1	2	1	0
44[b]	CH_3	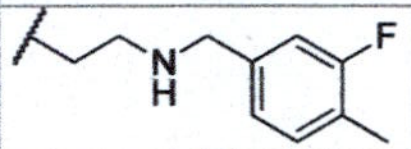	F	1.155	1.307	2	0	1	0	0
45	CH_3	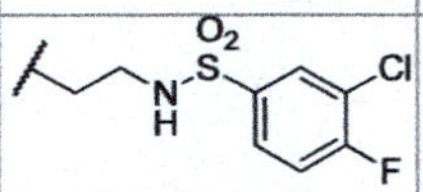	F	0.398	0.926	1	0	1	0	0
46[b]	CH_3	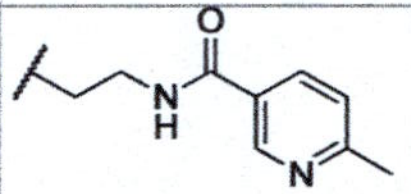	F	−0.857	0.259	0	0	0	0	0
47	CF_3	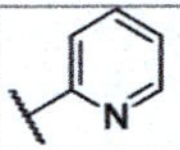	–	0.523	−0.286	1	1	2	2	0
48	CF_3	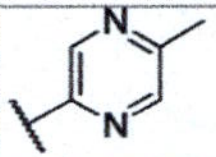	–	−0.924	−0.094	0	0	0	1	0
49[b]	CN	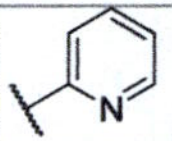	–	−1.823	−0.859	1	1	0	2	0
50	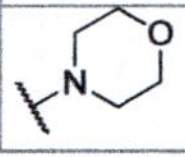	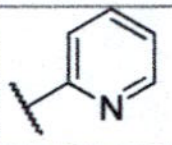	–	1.523	0.226	2	1	0	0	0

(continued)

Table 1 (continued)

Comp no.	R_1	R_2	R_3	Activity in pIC_{50} (μM)		Combined interaction parameter (C) values				
				Observed	Estimated/predicted[a]	Glu65b	HOHB	Ala66b	Phe69b	I
51	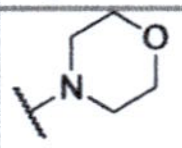	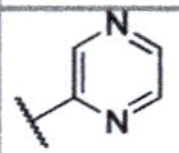	–	0.113	−0.573	1	1	1	2	0
52[b]	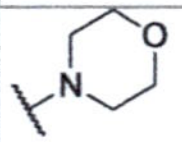	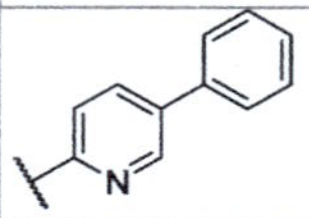	–	−0.978	−0.534	0	1	0	0	0

[a]Estimated and predicted activity of the training and test compounds, respectively, using Eq. (5)
[b]Training set compounds (remaining compounds comprised the test set)

software and are discussed in detail in the paper [32]. Firstly, the (S) protocol was used for the model development. However, the initially developed model containing all the parameters was not considered in view of very poor degrees of freedom as per the concept of five observations per parameter that is to say three degrees of freedom per parameter. Hence, the independent parameters describing the correlation were subsequently reduced in a backward stepwise regression manner, which gave a statistically significant (>99.9%) equation (Eq. (2)) having R^2 value of 82.4% and good correlation coefficient (R = 0.907).

$$\mathbf{pIC_{50}} = -0.881 + 0.659(\pm 0.1999)\mathrm{Glu65b} + 0.595(\pm 0.1738)\ \mathrm{HOHb} \\ -0.468(\pm 0.2444)\mathrm{Ala66a} + 0.569(\pm 0.2083)\mathrm{Gly62a} \tag{2}$$

$$n = 25, R^2 = 82.4\%, R - Sq(adj) = 78.9\%, F = 23.37, S = 0.3945, R - Sq(pred) = 68.09\%$$

Secondly, to check the importance of number of interactions between the AARs and the ligand molecule and to compare the results between the (S) and (C) protocols, a model based on the independent combined interaction (C) parameter was developed containing all the parameters, which was also reduced on similar grounds as described for (S) protocol to derive Eq. (2). During the reduction process, one compound acting as outlier because of its exceptional behaviour was removed [32]. It resulted in the development of single parameter equation (Eq. (3)), which was statistically significant (>99.9%) and explained 82.5% variation in activity with a good correlation coefficient value of R = 0.908.

$$\mathbf{pIC_{50}} = -0.428 + 0.161(\pm 0.01582)\mathrm{Glu65b} \tag{3}$$

$$n = 24, R^2 = 82.5\% R^2(adj) = 81.7\%, S = 0.301, F = 103.89, R - Sq(pred) = 77.71\%$$

Although both Eq. (2) and Eq. (3) explained the variance in activity, there is a slight advantage of Eq. (3) over Eq. (2). The activity variance was explained by just one parameter (Glu65b) in Eq. (3), which was being explained by four parameters in Eq. (2), and moreover, the regression coefficient for the combined interaction parameter (C) for Glu65b is 100% significant in Eq. (3), unlike the regression coefficients for the (S) parameter of the four AARs in Eq. (2). The importance of Glu65b has also been explained in previous reported literature [48]. Nonetheless, both the developed models (Eq. (2) and Eq. (3)) were used to predict an external dataset [49] where the biological activity was reported in IC_{90}. An unbiased selection of 12 compounds from external 71 compounds comprising of the most, the least, and some evenly spaced compounds was used for the purpose of prediction. The results were overwhelmingly in favour of the protocol (C) (Eq. (3)), which showed correlation coefficient (R_{ext}) of 0.851 as compared to 0.14 obtained through (S) protocol (Eq. (2)) (Fig. 2).

Further to explore the QSIAR methodology described above in the development of predictive models, a diverse set of 28 compounds belonging to imidazo[1,2-a]pyridine

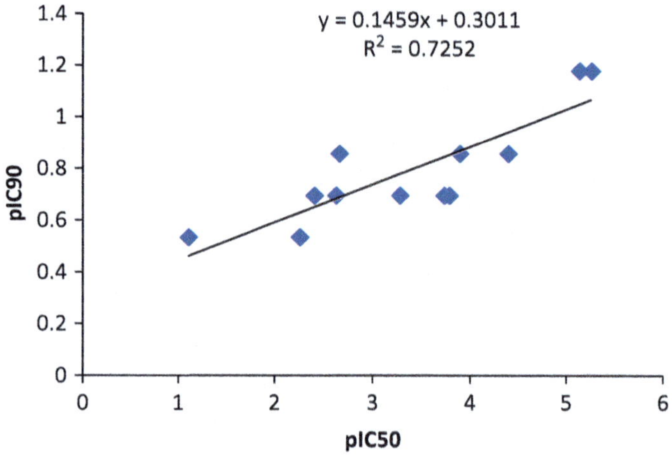

Fig. 2 Correlation between the predicted pIC_{50} vs observed pIC_{90} values of the external dataset (12 compounds)

ethers and squaramides [50] (Table 1; compound no. 25–52) was considered. In view of the better results of the protocol (C) than (S), the same procedure using the (C) protocol was applied, which resulted in the initial equation with all the parameters, which on reduction (as discussed for Eq. (2) and Eq. (3)), till no further improvement was possible, resulted in equation (Eq. (4)) with a correlation coefficient (R) value of 0.84 explaining 71% variation in activity.

$$\mathbf{pIC_{50}} = -0.845 + 1.094(\pm 0.137)\mathrm{Glu65b} \quad (4)$$

$$n = 28, R^2 = 71.00\%, R - Sq(adj) = 69.88\%, S = 0.615, F = 63.65, R - Sq(pred) = 67.75\%$$

The results obtained in Eq. (4) were similar to Eq. (3) since both the equations explained the variance in activity by just one similar parameter, i.e. Glu65b. Encouraged by these results, the datasets used in developing Eq. (3) and Eq. (4) were combined to yield a larger dataset of 52 compounds, which was divided into training and test set for further analysis. All the 52 molecules were first arranged in decreasing order of activity and divided into 2 sets [training set (27 molecules) included the most and the least active molecules along with every alternate molecule, while the test set (25 molecules) included the rest of the molecules]. To account for the difference in the structures and to also give weightage to the important quinoline substructure, an indicator variable (I) was added where the value of (I) was assigned as 1 and 0 for the presence and absence of quinoline substructure, respectively. The modelling on the training set of 27 molecules yielded Eq. (5) with five parameters, which was >99.9% statistically significant having correlation coefficient (R) value of 0.82 with explained variance of 67.7%.

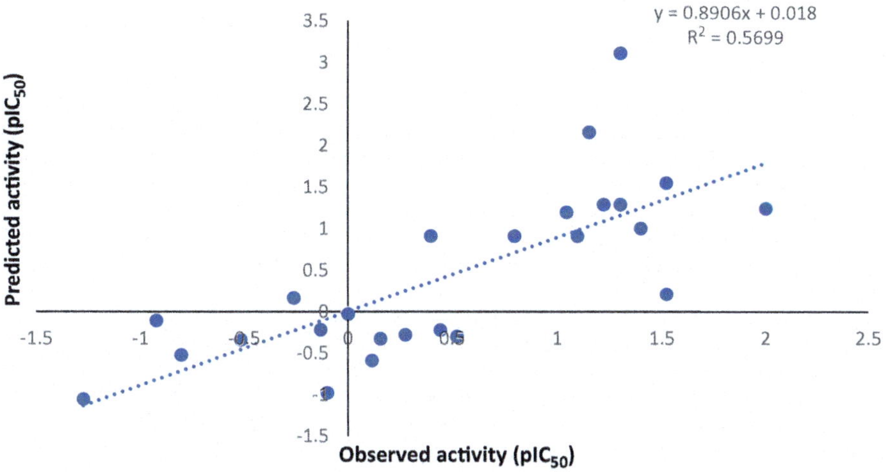

Fig. 3 Correlation between observed and predicted activities for 25 test set compounds obtained from Eq. (5)

$$\mathbf{pIC_{50}} = 0.259 + 0.3802(\pm 0.0655)\text{Glu65b} + 0.287(\pm 0.176)\text{Ala66b} - 0.793(\pm 0.191)\text{HOHb} - 0.353(\pm 0.125)\text{Phe69b} - 0.870(\pm 0.348)\text{I} \qquad (5)$$

$$n = 27, R^2 = 67.74\%, R - Sq(adj) = 60.06\%, F = 8.82, S = 0.647, R - Sq(pred) = 48.60\%$$

The developed training set model (Eq. (5)) well predicted the activity of the test set molecules, as evidenced by the correlation coefficient (R) value of 0.755 between the observed and predicted activities of the test set molecules (Fig. 3).

The model (Eq. (5)) was also validated by four molecules of an external dataset [49] comprising of the most, the least, and two more compounds with in-between activity. A correlation analysis showed a good correlation ($R_{ext} = 0.76$) between the predicted pIC_{50} and the pIC_{90} values, which further validated this model proving its robustness (Fig. 4).

4 Predictions of Molecules as Novel Anti-tubercular Agents

The developed model (Eq. (5)) also predicted three molecules from a focussed library developed by incorporating the essential features required for the inhibition of myco-bacterium ATP synthase. The predicted compounds showed the pIC_{50} values in the range of 0.0508–0.1494 μM. The docked poses of all the three compounds are given in Fig. 5.

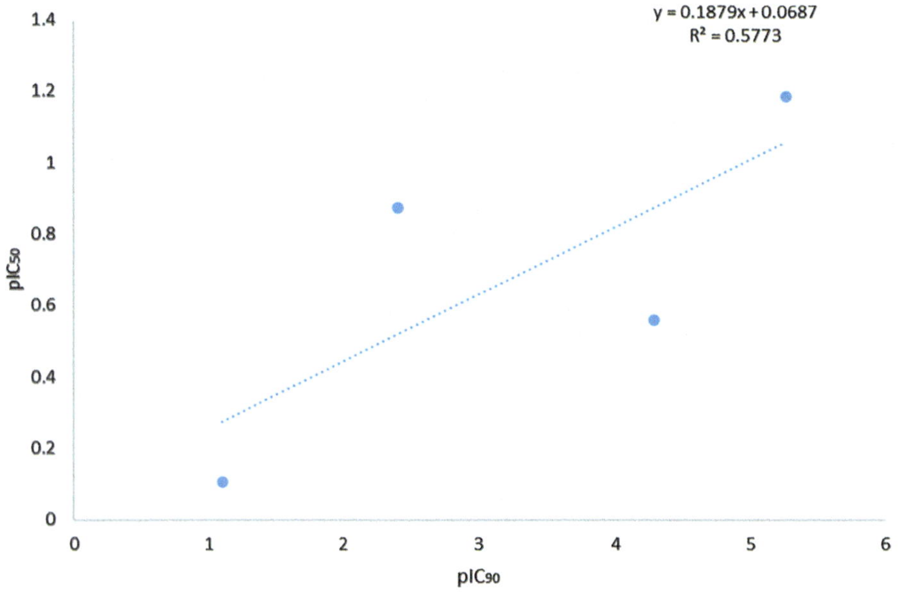

Fig. 4 Relationship between the observed (pIC_{90}) values and the predicted (pIC_{50}) values of the external dataset of four compounds

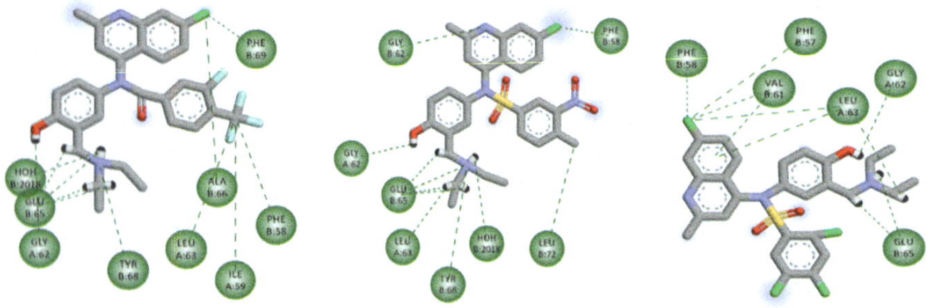

(a) Predicted pIC_{50}= 0.0508 μM

(b) Predicted pIC_{50}= 0.1168 μM

(c) Predicted pIC_{50}= 0.1494 μM

Fig. 5 The docked poses of the three predicted compounds with their predicted pIC_{50} values

5 Conclusions

The docking scores usually do not correlate with the biological activities, which leads to incorrect prediction of the biological activity. To address this issue, a novel approach, the quantitative structure interaction activity relationship (QSIAR), has been developed where the observed interactions of the ligand molecule with the amino acid residues of the target protein are used as independent parameters, and the binding energy/affinity, docking scores/biological activity, is used as a dependent parameter. This approach was applied on a dataset of 52 mycobacterium ATP synthase inhibitors comprising of diverse class of 4-substituted amino sulphonyl-2-methyl-7-chloroquinolines, bisquinoline core, imidazo[1,2-a]pyridine ethers, and squaramide molecules. The developed and validated models well explained the variation in ATP synthase inhibitory activity and also predicted the molecules of the external dataset. Three molecules from a focussed library showed good prediction indicative of their high activity. Thus, the QSIAR approach has a high potential in the development of predictive models using structure-based drug design, which was hitherto not possible. Such models may thus result in novel lead molecules for drug design and development.

References

1. Wouters, O.J., McKee, M., Luyten, J.: Estimated research and development investment needed to bring a new medicine to market, 2009–2018. JAMA. **323**(9), 844–853 (2020). https://doi.org/10.1001/jama.2020.1166
2. DiMasi, J.A., Feldman, L., Seckler, A., Wilson, A.: Trends in risks associated with new drug development: success rates for investigational drugs. Clin. Pharmacol. Ther. **87**(3), 272–277 (2010). https://doi.org/10.1038/clpt.2009.295
3. DiMasi, J.A., Grabowski, H.G., Hansen, R.W.: Innovation in the pharmaceutical industry: new estimates of R&D costs. J. Health Econ. **47**, 20–33 (2016). https://doi.org/10.1016/j.jhealeco.2016.01.012
4. Hughes, J.P., Rees, S., Kalindjian, S.B., Philpott, K.L.: Principles of early drug discovery. Br. J. Pharmacol. **162**(6), 1239–1249 (2011). https://doi.org/10.1111/j.1476-5381.2010.01127.x
5. Pina, A.S., Hussain, A., Roque, A.C.: An historical overview of drug discovery. In: Methods in Molecular Biology, vol. 572, pp. 3–12, Clifton, NJ (2009). https://doi.org/10.1007/978-1-60761-244-5_1
6. Jia, P., Pei, J., Wang, G., Pan, X., Zhu, Y., Wu, Y., Ouyang, L.: The roles of computer-aided drug synthesis in drug development. Green Synth. Catal. **3**(1), 11–24 (2022). https://doi.org/10.1016/j.gresc.2021.11.007
7. Struble, T.J., Alvarez, J.C., Brown, S.P., Chytil, M., Cisar, J., DesJarlais, R.L., Engkvist, O., Frank, S.A., Greve, D.R., Griffin, D.J., Hou, X., Johannes, J.W., Kreatsoulas, C., Lahue, B., Mathea, M., Mogk, G., Nicolaou, C.A., Palmer, A.D., Price, D.J., Robinson, R.I., Salentin, S., Xing, L., Jaakkola, T., Green, W.H., Barzilay, R., Coley, C.W., Jensen, K.F.: Current and future roles of artificial intelligence in medicinal chemistry synthesis. J. Med. Chem. **63**(16), 8667–8682 (2020). https://doi.org/10.1021/acs.jmedchem.9b02120
8. Moret, M., Friedrich, L., Grisoni, F., Merk, D., Schneider, G.: Generative molecular design in low data regimes. Nat. Mach. Intell. **2**(3), 171–180 (2020). https://doi.org/10.1038/s42256-020-0160-y
9. Paul, D., Sanap, G., Shenoy, S., Kalyane, D., Kalia, K., Tekade, R.K.: Artificial intelligence in drug discovery and development. Drug Discov. Today. **26**(1), 80–93 (2021). https://doi.org/10.1016/j.drudis.2020.10.010
10. Mak, K.K., Pichika, M.R.: Artificial intelligence in drug development: present status and future prospects. Drug Discov. Today. **24**(3), 773–780 (2019). https://doi.org/10.1016/j.drudis.2018.11.014
11. Sellwood, M.A., Ahmed, M., Segler, M.H., Brown, N.: Artificial intelligence in drug discovery. Future Med. Chem. **10**(17), 2025–2028 (2018). https://doi.org/10.4155/fmc-2018-0212

12. Niazi, S.K., Mariam, Z.: Computer-aided drug design and drug discovery: a prospective analysis. Pharmaceuticals. **17**(1), 22 (2024)
13. Talele, T.T., Khedkar, S.A., Rigby, A.C.: Successful applications of computer aided drug discovery: moving drugs from concept to the clinic. Curr. Top. Med. Chem. **10**(1), 127–141 (2010). https://doi.org/10.2174/156802610790232251
14. Saxena, A.K., Bhunia, S.S., Saxena, M.: Integration on ligand and structure based approaches in GPCRs. In: Lebon, G. (ed.) Structure and Function of GPCRs, pp. 101–161. Springer International Publishing, Cham (2019). https://doi.org/10.1007/7355_2016_24
15. Prathipati, P., Dixit, A., Saxena, A.K.: Computer-aided drug design: integration of structure-based and ligand-based approaches in drug design. Curr. Comput. Aided Drug Des. **3**(2), 133–148 (2007). https://doi.org/10.2174/157340907780809516
16. Bhunia, S.S., Saxena, M., Saxena, A.K.: Ligand- and structure-based virtual screening in drug discovery. In: Saxena, A.K. (ed.) Biophysical and Computational Tools in Drug Discovery, pp. 281–339. Springer International Publishing, Cham (2021). https://doi.org/10.1007/7355_2021_130
17. Choudhury, C., Priyakumar, U.D., Sastry, G.N.: Dynamic ligand-based pharmacophore modeling and virtual screening to identify mycobacterial cyclopropane synthase inhibitors. J. Chem. Sci. **128**(5), 719–732 (2016). https://doi.org/10.1007/s12039-016-1069-1
18. Sohn, Y.S., Park, C., Lee, Y., Kim, S., Thangapandian, S., Kim, Y., Kim, H.H., Suh, J.K., Lee, K.W.: Multi-conformation dynamic pharmacophore modeling of the peroxisome proliferator-activated receptor γ for the discovery of novel agonists. J. Mol. Graph. Model. **46**, 1–9 (2013). https://doi.org/10.1016/j.jmgm.2013.08.012
19. Chen, H., Lyne, P.D., Giordanetto, F., Lovell, T., Li, J.: On evaluating molecular-docking methods for pose prediction and enrichment factors. J. Chem. Inf. Model. **46**(1), 401–415 (2006). https://doi.org/10.1021/ci0503255
20. Kroemer, R.T.: Structure-based drug design: docking and scoring. Curr. Protein Pept. Sci. **8**(4), 312–328 (2007). https://doi.org/10.2174/138920307781369382
21. Li, J., Fu, A., Zhang, L.: An overview of scoring functions used for protein-ligand interactions in molecular docking. Interdiscip. Sci.: Comput. Life Sci. **11**(2), 320–328 (2019). https://doi.org/10.1007/s12539-019-00327-w
22. Muegge, I., Rarey, M.: Small molecule docking and scoring. Rev. Comput. Chem. **17**, 1–60 (2001). https://doi.org/10.1002/0471224413.ch1
23. Huang, S.Y., Grinter, S.Z., Zou, X.: Scoring functions and their evaluation methods for protein-ligand docking: recent advances and future directions. Phys. Chem. Chem. Phys.: PCCP. **12**(40), 12899–12908 (2010). https://doi.org/10.1039/c0cp00151a
24. Pons, C., Grosdidier, S., Solernou, A., Pérez-Cano, L., Fernández-Recio, J.: Present and future challenges and limitations in protein-protein docking. Proteins. **78**(1), 95–108 (2010). https://doi.org/10.1002/prot.22564
25. Guedes, I.A., de Magalhães, C.S., Dardenne, L.E.: Receptor-ligand molecular docking. Biophys. Rev. **6**(1), 75–87 (2014). https://doi.org/10.1007/s12551-013-0130-2
26. Guedes, I.A., Barreto, A.M.S., Marinho, D., Krempser, E., Kuenemann, M.A., Sperandio, O., Dardenne, L.E., Miteva, M.A.: New machine learning and physics-based scoring functions for drug discovery. Sci. Rep. **11**(1), 3198 (2021). https://doi.org/10.1038/s41598-021-82410-1
27. Scantlebury, J., Vost, L., Carbery, A., Hadfield, T.E., Turnbull, O.M., Brown, N., Chenthamarakshan, V., Das, P., Grosjean, H., von Delft, F., Deane, C.M.: A small step toward generalizability: training a machine learning scoring function for structure-based virtual screening. J. Chem. Inf. Model. **63**(10), 2960–2974 (2023). https://doi.org/10.1021/acs.jcim.3c00322
28. Chen, L., Cruz, A., Ramsey, S., Dickson, C.J., Duca, J.S., Hornak, V., Koes, D.R., Kurtzman, T.: Hidden bias in the DUD-E dataset leads to misleading performance of deep learning

in structure-based virtual screening. PLoS One. **14**(8), e0220113 (2019). https://doi.org/10.1371/journal.pone.0220113
29. Wallach, I., Heifets, A.: Most ligand-based classification benchmarks reward memorization rather than generalization. J. Chem. Inf. Model. **58**(5), 916–932 (2018). https://doi.org/10.1021/acs.jcim.7b00403
30. Boyles, F., Deane, C.M., Morris, G.M.: Learning from docked ligands: ligand-based features rescue structure-based scoring functions when trained on docked poses. J. Chem. Inf. Model. **62**(22), 5329–5341 (2022). https://doi.org/10.1021/acs.jcim.1c00096
31. Cross, J.B., Thompson, D.C., Rai, B.K., Baber, J.C., Fan, K.Y., Hu, Y., Humblet, C.: Comparison of several molecular docking programs: pose prediction and virtual screening accuracy. J. Chem. Inf. Model. **49**(6), 1455–1474 (2009). https://doi.org/10.1021/ci900056c
32. Ahmed, S., Prabahar, A.E., Saxena, A.K.: Molecular docking-based interactions in QSAR studies on Mycobacterium tuberculosis ATP synthase inhibitors. SAR QSAR Environ. Res. **33**(4), 289–305 (2022). https://doi.org/10.1080/1062936X.2022.2066175
33. Ahmed, S., Prabahar, A.E., Saxena, A.K.: Molecular docking-based interaction studies on imidazo[1,2-a] pyridine ethers and squaramides as anti-tubercular agents. SAR QSAR Environ. Res. **34**(6), 435–457 (2023). https://doi.org/10.1080/1062936X.2023.2225872
34. WHO World Tuberculosis Day 2024. https://www.who.int/campaigns/world-tb-day/2024 (2024). Accessed 30 Jul 2024
35. Roy, K.K., Singh, S., Sharma, S.K., Srivastava, R., Chaturvedi, V., Saxena, A.K.: Synthesis and biological evaluation of substituted 4-arylthiazol-2-amino derivatives as potent growth inhibitors of replicating Mycobacterium tuberculosis H37RV. Bioorg. Med. Chem. Lett. **21**(18), 5589–5593 (2011). https://doi.org/10.1016/j.bmcl.2011.06.076
36. Saxena, A.K., Roy, K.K., Singh, S., Vishnoi, S.P., Kumar, A., Kashyap, V.K., Kremer, L., Srivastava, R., Srivastava, B.S.: Identification and characterisation of small-molecule inhibitors of Rv3097c-encoded lipase (LipY) of Mycobacterium tuberculosis that selectively inhibit growth of bacilli in hypoxia. Int. J. Antimicrob. Agents. **42**(1), 27–35 (2013). https://doi.org/10.1016/j.ijantimicag.2013.03.007
37. Dey, R., Nandi, S., Samadder, A., Saxena, A., Saxena, A.K.: Exploring the potential inhibition of candidate drug molecules for clinical investigation based on their docking or crystallographic analyses against M. tuberculosis enzyme targets. Curr. Top. Med. Chem. **20**(29), 2662–2680 (2020). https://doi.org/10.2174/1568026620666200903163921
38. Ahmed, S., Nandi, S., Saxena, A.K.: An updated patent review on drugs for the treatment of tuberculosis (2018-present). Expert Opin. Ther. Pat. **32**(3), 243–260 (2022). https://doi.org/10.1080/13543776.2022.2012151
39. Nandi, S., Saxena, M., Saxena, A.K.: Important targets and inhibitors of *Mycobacterium tuberculosis*. In: Rezaei, N. (ed.) Tuberculosis: Integrated Studies for a Complex Disease, pp. 429–457. Springer International Publishing, Cham (2023). https://doi.org/10.1007/978-3-031-15955-8_21
40. Saxena, A.K., Singh, A.: Mycobacterial tuberculosis enzyme targets and their inhibitors. Curr. Top. Med. Chem. **19**(5), 337–355 (2019). https://doi.org/10.2174/1568026619666190219105722
41. Kelam, L.M., Wani, M.A., Dhaked, D.K.: An update on ATP synthase inhibitors: a unique target for drug development in *M. tuberculosis*. Prog. Biophys. Mol. Biol. **180–181**, 87–104 (2023). https://doi.org/10.1016/j.pbiomolbio.2023.04.009
42. Andries, K., Verhasselt, P., Guillemont, J., Göhlmann, H.W., Neefs, J.M., Winkler, H., Van Gestel, J., Timmerman, P., Zhu, M., Lee, E., Williams, P., de Chaffoy, D., Huitric, E., Hoffner, S., Cambau, E., Truffot-Pernot, C., Lounis, N., Jarlier, V.: A diarylquinoline drug active on the ATP synthase of Mycobacterium tuberculosis. Science. **307**(5707), 223–227 (2005). https://doi.org/10.1126/science.1106753

43. Satapathy, P., Itumalla, R., Neyazi, A., Nabizai Taraki, A.M., Khatib, M.N., Gaidhane, S., Zahiruddin, Q.S., Rustagi, S., Neyazi, M.: Emerging bedaquiline resistance: a threat to the global fight against drug-resistant tuberculosis. J. Biosaf. Biosecurity. **6**(1), 13–15 (2024). https://doi.org/10.1016/j.jobb.2024.01.001
44. Chesov, E., Chesov, D., Maurer, F.P., Andres, S., Utpatel, C., Barilar, I., Donica, A., Reimann, M., Niemann, S., Lange, C., Crudu, V., Heyckendorf, J., Merker, M.: Emergence of bedaquiline resistance in a high tuberculosis burden country. Eur. Respir. J. **59**(3), 2100621 (2022). https://doi.org/10.1183/13993003.00621-2021
45. Nguyen, T.V.A., Anthony, R.M., Bañuls, A.L., Nguyen, T.V.A., Vu, D.H., Alffenaar, J.C.: Bedaquiline resistance: its emergence, mechanism, and prevention. Clin. Infect. Dis. **66**(10), 1625–1630 (2018). https://doi.org/10.1093/cid/cix992
46. Singh, S., Roy, K.K., Khan, S.R., Kashyap, V.K., Sharma, A., Jaiswal, S., Sharma, S.K., Krishnan, M.Y., Chaturvedi, V., Lal, J., Sinha, S., Dasgupta, A., Srivastava, R., Saxena, A.K.: Novel, potent, orally bioavailable and selective mycobacterial ATP synthase inhibitors that demonstrated activity against both replicating and non-replicating *M. tuberculosis*. Bioorg. Med. Chem. **23**(4), 742–752 (2015). https://doi.org/10.1016/j.bmc.2014.12.060
47. Kalia, D., KS, A.K., Meena, G., Sethi, K.P., Sharma, R., Trivedi, P., Khan, S.R., Verma, A.S., Singh, S., Sharma, S., Roy, K.K., Kant, R., Krishnan, M.Y., Singh, B.N., Sinha, S., Chaturvedi, V., Saxena, A.K., Dikshit, D.K.: Synthesis and anti-tubercular activity of conformationally-constrained and bisquinoline analogs of TMC207. MedChemComm. **6**(8), 1554–1563 (2015). https://doi.org/10.1039/C5MD00131E
48. Preiss, L., Langer, J.D., Yildiz, Ö., Eckhardt-Strelau, L., Guillemont, J.E., Koul, A., Meier, T.: Structure of the mycobacterial ATP synthase Fo rotor ring in complex with the anti-TB drug bedaquiline. Sci. Adv. **1**(4), e1500106 (2015). https://doi.org/10.1126/sciadv.1500106
49. Guillemont, J., Meyer, C., Poncelet, A., Bourdrez, X., Andries, K.: Diarylquinolines, synthesis pathways and quantitative structure–activity relationship studies leading to the discovery of TMC207. Future Med. Chem. **3**(11), 1345–1360 (2011). https://doi.org/10.4155/fmc.11.79
50. Tantry, S.J., Markad, S.D., Shinde, V., Bhat, J., Balakrishnan, G., Gupta, A.K., Ambady, A., Raichurkar, A., Kedari, C., Sharma, S., Mudugal, N.V., Narayan, A., Naveen Kumar, C.N., Nanduri, R., Bharath, S., Reddy, J., Panduga, V., Prabhakar, K.R., Kandaswamy, K., Saralaya, R., Kaur, P., Dinesh, N., Guptha, S., Rich, K., Murray, D., Plant, H., Preston, M., Ashton, H., Plant, D., Walsh, J., Alcock, P., Naylor, K., Collier, M., Whiteaker, J., McLaughlin, R.E., Mallya, M., Panda, M., Rudrapatna, S., Ramachandran, V., Shandil, R., Sambandamurthy, V.K., Mdluli, K., Cooper, C.B., Rubin, H., Yano, T., Iyer, P., Narayanan, S., Kavanagh, S., Mukherjee, K., Balasubramanian, V., Hosagrahara, V.P., Solapure, S., Ravishankar, S., Hameed, P.S.: Discovery of imidazo[1,2-a]pyridine ethers and squaramides as selective and potent inhibitors of mycobacterial adenosine triphosphate (ATP) synthesis. J. Med. Chem. **60**(4), 1379–1399 (2017). https://doi.org/10.1021/acs.jmedchem.6b01358

ONCORx: A Digital App for Cancer Treatment Planning

A. Saxena[1] (✉), R. Gupta[2], C. Moss[1], C. Harper-Wynne[1], and R. Burcombe[1]

[1] Maidstone & Tunbridge Wells NHS Trust, Maidstone, UK
[2] Jaypee Institute of Information Technology, Noida, India

Abstract Treatment for breast cancer patients usually comprises multiple modalities (e.g., surgery, chemotherapy, etc.). Treatments are used at varying frequencies and durations, in combination or sequence, and over an extended period. At an initial consultation, patients are presented with new and unfamiliar information on diagnosis and prognosis. Communicating an adaptive neoadjuvant, adjuvant, or palliative treatment plan in this environment, together with potential options, is challenging for oncologists and often overwhelming for patients. Patients' lack of understanding of the proposed treatment can compromise informed choice, consent, and compliance. To simplify the process, ONCORx has been developed. The MIRO web interface was used to develop the flowcharts for web tool design. Flowcharts were based on local treatment protocols/NICE guidelines. A web application was developed in Flutter-based technology. Dart language-based programming was conducted to develop the logical framework based on the MIRO decision tree followed by web application development with the option of two user logins: one for user and another for an administrator. ONCORx App allows oncologists to select relevant treatments according to the patient's breast cancer phenotype and display a simple illustration/flowchart of their proposed treatment plan along with a treatment sequence timeline. The final output offers editing options in both Word and PDF format. Considering rapid innovation in oncology, especially with an increase in the number of treatment lines, this user-friendly app will help oncologists communicate treatment plans to patients with a timeline to avoid patients missing appointments and make things simple for patients to understand.

Keywords Communication · Breast cancer · Treatment planning · Digital health · Technology

A. K. Saxena, S. Nandi (eds.), *Global Trends in Health, Technology and Management*, https://doi.org/10.1007/978-3-031-75457-9_2

1 Introduction

1.1 Breast Cancer and Patient Communication

Breast cancer treatment is a multifaceted process requiring meticulous coordination among a diverse healthcare team. It involves various modalities such as surgery, chemotherapy, radiation therapy, and hormone therapy, often administered in a sequential manner. Effective communication of treatment plans, dates, and times between clinicians and patients is paramount to ensure adherence to the prescribed treatment regimens. However, this communication is frequently fraught with challenges that can negatively impact clinical outcomes.

One major challenge is the complexity of treatment schedules, which can be overwhelming for patients to comprehend and follow [1]. Treatment plans are typically conveyed during consultations, which might not be sufficient for patients to fully grasp the details due to the emotional and cognitive load experienced during such discussions. Additionally, any modifications to the schedule due to unforeseen circumstances such as medical emergencies, changes in treatment response, or logistical issues add to this complexity.

For clinicians, managing and conveying these intricate schedules is time-consuming and prone to errors, especially in busy oncology practices. Miscommunications can lead to missed appointments, delays in treatment, and, ultimately, suboptimal clinical outcomes. The lack of a streamlined and reliable communication system can exacerbate the inherent stress and anxiety associated with breast cancer diagnosis and treatment. Patients often experience heightened anxiety due to uncertainty and fear of the disease progression, which can be intensified by confusion over treatment schedules and a perceived lack of control over their care journey.

These communication challenges underscore the necessity for a robust digital solution. A digital app designed specifically for breast cancer treatment planning and communication can bridge this gap effectively. Such an app can provide real-time updates, reminders, and a clear overview of the treatment schedule, thereby reducing the cognitive burden on patients and minimizing the risk of errors in scheduling. Moreover, it can facilitate better engagement and empowerment for patients, allowing them to take an active role in their treatment process. This, in turn, can alleviate anxiety and stress, improving overall patient experience and potentially enhancing clinical outcomes.

In this study, we present a digital application developed to address these challenges. We will explore its design, functionality, and the impact it has on improving communication between clinicians and patients. Furthermore, we will discuss the app's role in reducing patient anxiety and stress around the treatment schedule, ultimately contributing to better adherence to treatment protocols and improved clinical outcomes.

2 Materials and Methods

2.1 Conceptual Framework and Flowchart Development

The initial phase of the digital application development involved the conceptualization and design of flowcharts using the MIRO web interface. MIRO, a versatile online collaborative platform, was employed to create detailed flowcharts that mapped out the user journey and the decision-making process within the app [2]. These flowcharts were grounded in local treatment protocols and the National Institute for Health and Care Excellence (NICE) guidelines for early breast cancer treatment to ensure that the app adhered to standardized, evidence-based practices for early breast cancer treatment [3].

2.2 Technological Foundation and Development Environment

The web application was built using Flutter [4], a UI toolkit for building natively compiled applications for mobile, web, and desktop from a single codebase. Flutter's flexibility and comprehensive set of predesigned widgets allowed for rapid development and a seamless user experience. The decision tree logic, derived from the MIRO-developed flowcharts, was meticulously translated into the Flutter framework, ensuring that all decision nodes and pathways were accurately represented.

2.3 Web Application Development

The development process involved several key stages:

a. Architecture Design

The application's architecture was designed to support only administrators (clinicians).

b. User Interface (UI) Design

The UI was designed to be intuitive and user-friendly, with a focus on accessibility and ease of navigation. Key UI components included:

- A dashboard providing options for the treatment
- Interactive modules for selecting the desired treatment based on the available treatment lines

c. Backend Development

The backend infrastructure was developed to support robust data management and secure user authentication. Key backend functionalities included a database to store user information and different types of treatment regimens practiced treating early breast cancer, APIs to facilitate real-time communication between the front end and back end, and secure login systems with role-based access controls to ensure data privacy and integrity.

d. Integration of Decision Tree Logic

The decision tree logic, as defined in the MIRO flowcharts, was integrated into the app's functionality. This logic guided users through a series of questions and decision points, ultimately providing personalized treatment schedules based on their responses.

2.4 User Login and Role Management

One distinct login portal was developed as part of the prototype:

Admin Login: Designed for clinicians and administrators, this portal allowed for the management of patient data, updating of treatment protocols, and monitoring of patient adherence and progress.

2.5 Testing and Quality Assurance

The application underwent rigorous testing to ensure functionality, reliability, and security. Testing phases included:

***Unit Testing*:** To verify the functionality of individual components.
***Integration Testing*:** To ensure seamless interaction between various components and modules.
***User Acceptance Testing (UAT)*:** To gather feedback from clinicians and define the application based on their input. This phase is still ongoing.

2.6 Deployment and Futureproofing

The final stage involved deploying the web application on a secure server, ensuring scalability to accommodate future updates and additional treatment regimens. The app's architecture was designed to be flexible, allowing for easy integration of new features and updates as medical guidelines and treatment protocols evolve.

By following this structured methodology, we developed a comprehensive digital application tailored to enhance the planning and communication of early breast cancer treatment, ultimately aiming to improve patient outcomes and streamline clinician workflow.

3 Results and Discussion

3.1 Results

To ensure the functionality of the application, three medical cases have been tested to see the generated output. The output generated through the prototype is meant to communicate the treatment plan to the patient and to ensure that the patient can easily understand the sequence of different vital components of it. The treatment for the following three cases was tested using the application.

Test Case 1: A 43-year-old (premenopausal) with high-risk, ER (estrogen receptor)-positive, HER2-negative breast cancer following surgery

The complete treatment plan for test case 1 is shown in Table 1. The dates are arbitrary and entered into the system to test the application.

The standard treatment given to treat premenopausal women with high-risk early breast cancer following surgery includes adjuvant chemotherapy, adjuvant hormone therapy, and adjuvant CDK 4/6 inhibitor. Adjuvant chemotherapy includes epirubicin and cyclophosphamide, which are given twice weekly for four cycles and are followed by accelerated paclitaxel, which is given twice weekly for four cycles [5]. The adjuvant hormone therapy includes tamoxifen, which works by competing with 17β-estradiol (E2) at the receptor site and blocking the promotional role of E2 in breast cancer [6]. After completion of the hormone therapy, adjuvant radiotherapy is given for five cycles [7]. Considering the high risk of early breast cancer, CDK 4/6 inhibitor is given for 2 years [8].

The output of the application for test case 1 is shown in Fig. 1.

Test Case 2: A 60-year-old (postmenopausal) with node-positive, ER-positive, and HER2-positive disease post-surgery

The standard treatment for a postmenopausal includes adjuvant chemotherapy with docetaxel plus carboplatin, which is given thrice weekly for 6 cycles plus Phesgo, which is given as a subcutaneous injection every 3 weeks for a total of 18 cycles (Table 2). Phesgo is a gold standard treatment for HER2 receptor-positive breast cancer [9]. Both adjuvant chemotherapy and anti-HER2 treatment are proven game changers when it comes to adjuvant treatment of node-positive and HER2+ breast cancer. Phesgo is a combination of pertuzumab and trastuzumab, which is administered intravenously [10]. Hormone therapy is a case of postmenopausal women consisting mainly of letrozole [11]. Often radiotherapy is given for 3 weeks in total with 5 days per week, which is shown to improve overall outcome [12]. Recently scientific studies have shown that treatment with bisphosphonate not only helps in minimizing bone fractures but also makes it difficult for the cancer cells to enter through the periosteum [13].

The output of the application for test case 2 is shown in Fig. 2.

Table 1 The necessary inputs fed into the digital application ONCORx to generate output for test case 1 as shown in Fig. 1

Step 1: Adjuvant chemotherapy—dose-dense (dd) EC (two drugs—epirubicin and cyclophosphamide) given twice weekly for four cycles, followed by accelerated paclitaxel given twice weekly for four cycles **(date 20.06.2024–15.10.2024)**
Step 2: On completion of chemotherapy—adjuvant radiotherapy (five daily treatments) **(25.10.2024–30.10.2024)**
Step 3: Adjuvant hormone therapy—e.g., tamoxifen 20 mg (once a day) **(25.10.2024–24.10.2029)**
Step 4: Adjuvant CDK 4/6 tablet—abemaciclib for 2 years (once a day) **(01.11.2024–30.10.2026)**

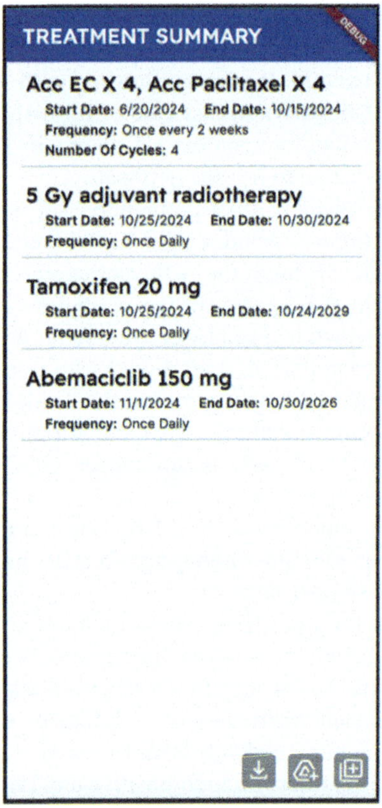

Fig. 1 The standard output of case 1 in the application

Table 2 The necessary inputs fed into the digital application ONCORx to generate output for test case 2 as shown in Fig. 2

Step 1: Adjuvant chemotherapy—two drugs, docetaxel and carboplatin given thrice weekly for six cycles (20.06.2024–05.11.2024), followed by pertuzumab + trastuzumab given thrice weekly for six cycles **(20.06.2024–05.11.2024)**
Step 2: Anti-HER2 therapy—after completing adjuvant chemotherapy cycles 7 to 18, Phesgo (every 3 weeks) **(20.06.2024–02.09.2025)**
Step 3: Adjuvant bisphosphonate therapy—intravenous injection of zoledronic acid for 3 years (every 6 months) **(10.11.2024–10.11.2027)**
Step 4: Adjuvant radiotherapy—daily treatment 5 days per week given for 3 weeks **(30.04.2024–23.05.2024)**
Step 6: Adjuvant hormone therapy—letrozole 2.5 mg (OD) for 10 years **(10.11.2024–10.11.2034)**

Test case 3: A 32-year-old (premenopausal) with node-negative and ER-positive disease and HER2+ neoadjuvant treatment?

Neoadjuvant chemotherapy treatment is often used in breast cancer to shrink the tumor before surgery [14] (Table 3). Neoadjuvant chemotherapy helps to make

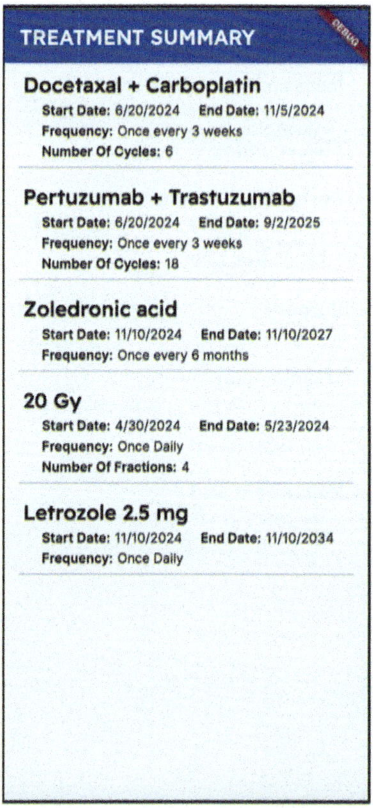

Fig. 2 The standard output of case 2 in the application

Table 3 The necessary input fed into the digital application ONCORx to generate output for test case 3 as shown in Fig. 3

Step 1: Neoadjuvant chemotherapy—docetaxel plus carboplatin given thrice weekly for six cycles (20.06.2024–05.11.2024). In addition to it, Phesgo (pertuzumab + trastuzumab) is given as a subcutaneous injection every 3 weeks up to eight cycles (20.06.2024–20.12.2024)
Step 2: Surgery happens on 27.12.2024
Step 3: Adjuvant Herceptin (only trastuzumab)—subcutaneous injection every 3 weeks up to ten cycles (10.01.2025–25.08.2025)
Step 4: Adjuvant bisphosphonate treatment—6 monthly intravenous zoledronic acid for 3 years (10.01.2025–10.01.2028)

breast-conserving surgery more feasible and reduce the risk of micrometastasis [15]. The neoadjuvant chemotherapy comprises of docetaxel and carboplatin, which is given thrice weekly for six cycles followed by Phesgo given as subcutaneous infection every 3 weeks for up to eight cycles. The neoadjuvant chemotherapy is followed by surgery and later with adjuvant Herceptin, hormone, and bisphosphonate treatment.

The output of the application for test case 3 is shown in Fig. 3.

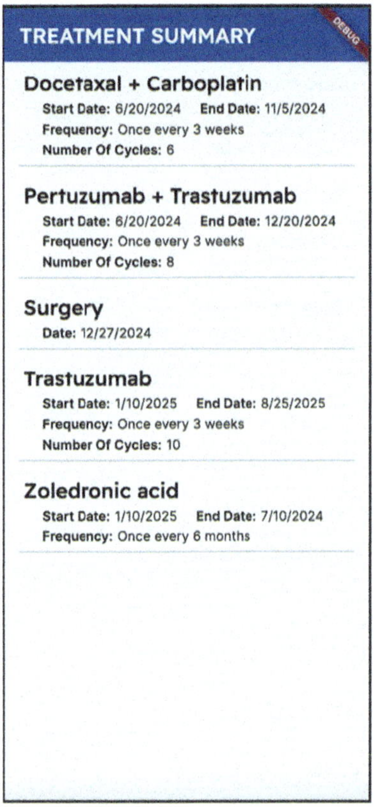

Fig. 3 The standard output of case 3 in the application

3.2 Discussion

With the advent of computer-aided drug design and development, the process of identifying novel targets and possible targeted drugs has increased exponentially. During the last decade, advancement in medical oncology has been multifold with the emergence of immunotherapy and biologics. However, this has also increased the complexity of understanding treatments, especially for the patients. Equally, it is a new challenge for medical oncologists to communicate the treatment plans to the patients. The key challenges that oncologists face while communicating treatment plans to the patient include multiple modalities, technical language, psychological and emotional impact, communicating individualized treatment approaches, balancing hope and realism to the patients, cultural and language barriers, time constraints during a consultation, and technological advances. Challenges particularly about the complexity of the treatment plan are explaining the integration and timing of different therapies (e.g., neoadjuvant chemotherapy, hormone therapy, biologics, and radiotherapy), which is often overwhelming for patients, and at the same time avoiding the use of medical jargon as far as possible [16]. Often oncologists are advised to use simple explanations, break complex information into small steps, and use visual aids for communication [17]. Nevertheless, especially in the case of cancer patients, it's important to consider the emotional distress associated with knowing that one suffers from cancer and its effect on understanding the different types of treatment at the same time, which leads to cognitive overload [18]. To overcome the emotional distress, empathy and involving mental health professionals at an early stage might help to overcome the anxiety and stress of the patients [19]. Especially in oncology, often tailoring the treatment to the needs and priority of the patient is important especially when the intent of the treatment is palliative or a patient has unique needs. This becomes a challenge for clinicians as there is a need of personalized communication and also shared decision-making. Again clear communication is essential to overcome this hurdle [20]. A patient-centered approach is required from the beginning to make individual treatment plans [21]. In the case of cancer patients, it is essential to keep a balance between hope and realism. Patients often need a realistic understanding of their prognosis and the potential side effects and outcomes of treatment. It's crucial to foster hope without providing false assurance, which requires a delicate balance. It's important to have an honest dialogue and allow a patient to ask questions or any concerns at every stage during the consultation [22]. In the era of globalization, it is common to have patients from different cultures, beliefs, and literacy. This becomes a challenge and can hinder effective communication. Patients from various cultural backgrounds may have different beliefs about illness and treatment, and language barriers can hinder effective communication. Varying levels of health literacy among patients can affect their ability to understand complex medical information. It is important to be aware of cultural differences and if required use interpreter services to avoid any confusion or misleading information [23, 24]. Another logistic challenge faced by oncologists nowadays is restricted consultation time for each patient. Oncologists often have limited time for each patient, making it challenging to thoroughly explain complex treatment plans. Adequate time is required to discuss all aspects of a multifaceted treatment plan. To overcome this challenge, oncologists often prioritize key information and the most critical aspects of the treatment plan and involve cancer specialist nurses to help patients understand complex information in a short time [25]. With the technological advances in oncology and rapidly evolving treatments, often explaining the latest advancement in cancer treatment to the patient can be challenging. Similarly, it's hard for patients to understand specialized therapies such as biologics and targeted treatments. To overcome this, it is often recommended to provide written information on different types of novel therapies to the patient during the consultation [26].

The ONCORx cancer treatment planning application addresses several challenges in communicating complex treatment plans to patients. This digital tool offers a visual approach to presenting treatment information, which can be particularly effective in overcoming common communication barriers. The ONCORx helps in the visual representation of treatment plans in the form of charts and sequencing to map out different treatment modalities (like neoadjuvant chemotherapy, hormone therapy, biologics, and radiotherapy) and their sequence. To improve patient understanding, the complex treatment is broken down into clear steps. Each treatment plan is tailored to the patient diagnosis and preference, and this application is based on patient-centered approach. At this point, access to the application is confined to the oncologist, but in the future, there could be a possibility to allow patients to interact with the application. The digital application helps to minimize anxiety by providing a clear road map of different treatments planned for the patients.

3.3 Current Challenges and Future Direction

At this stage of development, the ONCORx digital application is a prototype confined for treatment planning in case of early breast cancer. The output of different treatments needs to incorporate timelines for different treatments. It is undergoing development to improve further visualization of the treatments.

4 Conclusion

The ONCORx digital application represents a significant advancement in the realm of early breast cancer treatment planning. Developed with Flutter Flow technology, ONCORx is designed to be used exclusively by oncologists, enhancing their ability to deliver comprehensive and individualized care plans to patients.

To summarize, the key benefits of ONCORx include:

(i) **Streamlined and Intuitive Visualization**: ONCORx employs sophisticated charts and sequencing tools that allow oncologists to visually present the intricate details of treatment regimens. This visual approach simplifies the explanation of complex treatment protocols, such as the integration of neoadjuvant chemotherapy, hormone therapy, biologics, and radiotherapy, making it easier for patients to grasp and follow their treatment plans.

(ii) **Personalization and Flexibility**: The application supports the customization of treatment plans to align with each patient's unique medical needs and preferences. This adaptability ensures that patients receive a tailored approach to their care, which is critical in managing early breast cancer where treatment strategies can vary widely based on individual diagnoses and responses.

(iii) **Enhanced Patient Engagement and Empowerment**: By providing a clear and interactive road map of their treatment journey, ONCORx empowers patients to engage more actively in their care. This can help mitigate the emotional stress and cognitive overload often associated with cancer treatment, as patients can better understand what to expect and feel more in control of their health decisions.

(iv) **Improved Communication and Coordination**: ONCORx serves as a centralized platform for both oncologists and patients, facilitating consistent and transparent communication. This centralization helps ensure that all members of the care team are aligned and that patients are kept well-informed throughout their treatment process.

(v) **Technology-Driven Efficiency**: Leveraging Flutter Flow technology, ONCORx is built to provide a seamless and responsive user experience. This technology not only enhances the application's performance and usability but also allows for rapid updates and modifications to treatment plans, which is essential in the dynamic field of oncology.

4.1 Future Prospects of Development

As ONCORx continues to evolve, there is considerable potential for expanding its capabilities. Future developments could include integrating patient-facing features to allow patients' direct access to their treatment plans, expanding its use beyond oncologists to include other healthcare providers, and incorporating real-time data analytics to further personalize and optimize treatment strategies.

In conclusion, ONCORx is a pioneering tool in the digital transformation of cancer care, offering oncologists a powerful resource to improve the clarity, precision, and effectiveness of treatment planning for early breast cancer. By harnessing the capabilities of modern technology, ONCORx not only enhances the oncologist's workflow but also significantly contributes to better patient outcomes and satisfaction.

Acknowledgment We would like to express our sincere gratitude to Dr. Karim Syed, Medical Oncologist at Nottingham University Hospitals, for his invaluable inputs in the development of this application. His extensive clinical experience and insightful contributions have been instrumental in shaping the direction and focus of our work. Without his support and guidance, this project would not have reached its current level of success.

References

1. Surbone, A.: New Challenges in Communication with Cancer Patients, pp. 1–528. Springer (2013)
2. Johnson, E.K., Miro, M.: Student perceptions of a visual discussion board. In: Proceedings of the 40th ACM International Conference on Design of Communication, pp. 96–101. Association for Computing Machinery, Boston, MA, USA (2022)
3. Harnett, A., Smallwood, J., Titshall, V., Champion, A., On behalf of the Guideline Development Group: Diagnosis and treatment of early breast cancer, including locally advanced disease – summary of NICE guidance. BMJ. **338**, b438–b438 (2009). https://doi.org/10.1136/bmj.b438
4. Faust, S.: Using Google's Flutter Framework for the Development of a Large-Scale Reference Application. 85–S86 (2020)
5. Liu, D., Wu, J., Lin, C., Ding, S., Lu, S., Fang, Y., Huang, J., Hong, J., Gao, W., Zhu, S., Chen, X., Huang, O., He, J., Chen, W., Li, Y., Shen, K., Zhu, L.: The comparative safety of epirubicin and cyclophosphamide versus docetaxel and cyclophosphamide in lymph node-negative, HR-positive, HER2-negative breast cancer (ELEGANT): a randomized trial. Cancers (Basel). **14**, 3221 (2022). https://doi.org/10.3390/cancers14133221
6. Yu, F., Bender, W.: The mechanism of tamoxifen in breast cancer prevention. Breast Cancer Res. **3**, A74 (2001). https://doi.org/10.1186/bcr404
7. Kolářová, I., Melichar, B., Vaňásek, J., Sirák, I., Petera, J., Horáčková, K., Pohanková, D., Šinkorová, Z., Hošek, O., Vošmik, M.: Special techniques of adjuvant breast carcinoma radiotherapy. Cancers (Basel). **15**, 298 (2022). https://doi.org/10.3390/cancers15010298

8. Akhade, A., Van Wambeke, S., Gyawali, B.: CDK 4/6 inhibitors for adjuvant therapy in early breast cancer-do we have a clear winner? Ecancermedicalscience. **16**, ed124 (2022). https://doi.org/10.3332/ecancer.2022.ed124
9. Patel, A., Unni, N., Peng, Y.: The changing paradigm for the treatment of HER2-positive breast cancer. Cancers (Basel). **12**, 2081 (2020). https://doi.org/10.3390/cancers12082081
10. DuMond, B., Patel, V., Gross, A., Fung, A., Weber, S.: Fixed-dose combination of pertuzumab and trastuzumab for subcutaneous injection in patients with HER2-positive breast cancer: a multidisciplinary approach. J. Oncol. Pharm. Pract. **27**, 1214–1221 (2021). https://doi.org/10.1177/1078155221999712
11. Simpson, D., Curran, M.P., Perry, C.M.: Letrozole: a review of its use in postmenopausal women with breast cancer. Drugs. **64**, 1213–1230 (2004)
12. Violet, J.A., Harmer, C.: Breast cancer: improving outcome following adjuvant radiotherapy. Br. J. Radiol. **77**, 811–820 (2004). https://doi.org/10.1259/bjr/44576710
13. Yang, M., Yu, X.: Management of bone metastasis with intravenous bisphosphonates in breast cancer: a systematic review and meta-analysis of dosing frequency. Support Care Cancer. **28**, 2533–2540 (2020). https://doi.org/10.1007/s00520-020-05355-7
14. Gori, S., Fabi, A., Angiolini, C., Turazza, M., Salvini, P., Ferretti, G., Cretella, E., Gianni, L., Bighin, C., Toss, A., Zamagni, C., Vici, P., De Rossi, C., Russo, A., Bisagni, G., Frassoldati, A., Borgato, L., Cariello, A., Cappelletti, C., Bordonaro, R., Cinieri, S., Modena, A., Valerio, M., Alvisi, M.F., De Simone, I., Galli, F., Rulli, E., Santoni, A., Nicolis, F.: Neoadjuvant systemic therapy in early breast cancer: results of a prospective observational multicenter BRIDE study. Cancers (Basel). **15**, 4852 (2023). https://doi.org/10.3390/cancers15194852
15. Asselain, B., Barlow, W., Bartlett, J., Bergh, J., Bergsten-Nordström, E., Bliss, J., Boccardo, F., Boddington, C., Bogaerts, J., Bonadonna, G., Bradley, R., Brain, E., Braybrooke, J., Broet, P., Bryant, J., Burrett, J., Cameron, D., Clarke, M., Coates, A., Coleman, R., Coombes, R.C., Correa, C., Costantino, J., Cuzick, J., Danforth, D., Davidson, N., Davies, C., Davies, L., Di Leo, A., Dodwell, D., Dowsett, M., Duane, F., Evans, V., Ewertz, M., Fisher, B., Forbes, J., Ford, L., Gazet, J.-C., Gelber, R., Gettins, L., Gianni, L., Gnant, M., Godwin, J., Goldhirsch, A., Goodwin, P., Gray, R., Hayes, D., Hill, C., Ingle, J., Jagsi, R., Jakesz, R., James, S., Janni, W., Liu, H., Liu, Z., Lohrisch, C., Loibl, S., MacKinnon, L., Makris, A., Mamounas, E., Mannu, G., Martín, M., Mathoulin, S., Mauriac, L., McGale, P., McHugh, T., Morris, P., Mukai, H., Norton, L., Ohashi, Y., Olivotto, I., Paik, S., Pan, H., Peto, R., Piccart, M., Pierce, L., Poortmans, P., Powles, T., Pritchard, K., Ragaz, J., Raina, V., Ravdin, P., Read, S., Regan, M., Robertson, J., Rutgers, E., Scholl, S., Slamon, D., Sölkner, L., Sparano, J., Steinberg, S., Sutcliffe, R., Swain, S., Taylor, C., Tutt, A., Valagussa, P., van de Velde, C., van der Hage, J., Viale, G., von Minckwitz, G., Wang, Y., Wang, Z., Wang, X., Whelan, T., Wilcken, N., Winer, E., Wolmark, N., Wood, W., Zambetti, M., Zujewski, J.A.: Long-term outcomes for neoadjuvant versus adjuvant chemotherapy in early breast cancer: meta-analysis of individual patient data from ten randomised trials. Lancet Oncol. **19**, 27–39 (2018). https://doi.org/10.1016/s1470-2045(17)30777-5
16. Schofield, P.E., Butow, P.N., Thompson, J.F., Tattersall, M.H.N., Beeney, L.J., Dunn, S.M.: Psychological responses of patients receiving a diagnosis of cancer. Ann. Oncol. **14**, 48–56 (2003). https://doi.org/10.1093/annonc/mdg010
17. Baile, W.F., Aaron, J.: Patient-physician communication in oncology: past, present, and future. Curr. Opin. Oncol. **17**, 331–335 (2005). https://doi.org/10.1097/01.cco.0000167738.49325.2c
18. Kissane, D.W.: Unrecognised and untreated depression in cancer care. Lancet Psychiat. **1**, 320–321 (2014). https://doi.org/10.1016/S2215-0366(14)70345-1
19. Kesbakhi, M.S., Rohani, C.: Exploring oncology nurses' perception of the consequences of clinical empathy in patients and nurses: a qualitative study. Support Care Cancer. **28**, 2985–2993 (2020). https://doi.org/10.1007/s00520-019-05118-z

20. McAllister, M., Dunn, G., Payne, K., Davies, L., Todd, C.: Patient empowerment: the need to consider it as a measurable patient-reported outcome for chronic conditions. BMC Health Serv. Res. **12**, 157 (2012). https://doi.org/10.1186/1472-6963-12-157
21. Epstein, R.M., Street Jr., R.L.: The values and value of patient-centered care. Ann. Fam. Med. **9**, 100–103 (2011). https://doi.org/10.1370/afm.1239
22. Back, A.L., Arnold, R.M., Quill, T.E.: Hope for the best, and prepare for the worst. Ann. Intern. Med. **138**, 439–443 (2003). https://doi.org/10.7326/0003-4819-138-5-200303040-00028
23. Betancourt, J.R., Green, A.R., Carrillo, J.E., Ananeh-Firempong 2nd, O.: Defining cultural competence: a practical framework for addressing racial/ethnic disparities in health and health care. Public Health Rep. **118**, 293–302 (2003). https://doi.org/10.1093/phr/118.4.293
24. Schenker, Y., Wang, F., Selig, S.J., Ng, R., Fernandez, A.: The impact of language barriers on documentation of informed consent at a hospital with on-site interpreter services. J. Gen. Intern. Med. **22**(Suppl 2), 294–299 (2007). https://doi.org/10.1007/s11606-007-0359-1
25. Committee on Improving the Quality of Cancer Care: Addressing the Challenges of an Aging Population, Board on Health Care Services, Institute of Medicine: Delivering High-Quality Cancer Care: Charting a New Course for a System in Crisis. National Academies Press, Washington, DC (2013) Copyright 2013 by the National Academy of Sciences. All rights reserved.: Washington (DC)
26. Pazdur, R.: Endpoints for assessing drug activity in clinical trials. Oncologist. **13**(Suppl 2), 19–21 (2008). https://doi.org/10.1634/theoncologist.13-s2-19

Amprenavir-Ritonavir Combination Is More Effective in *Leishmania donovani* Compared to *Leishmania major*

Amit Roy[1] (✉)

[1] Department of Biotechnology, Savitribai Phule Pune University, Pune, India

Abstract Leishmaniasis is an infectious disease caused by unicellular protozoan parasite *Leishmania*. The disease is transmitted to humans through infected female *Phlebotomus* sand flies, which feed on blood. People belonging to low socioeconomic status are generally more prone to *Leishmania* infection, due to their weaker immune system. Ironically, drugs that are in clinical use against leishmaniasis are toxic and very expensive. There is also an increased incidence of resistance against conventional drugs. Due to these reasons, there is a need for development of new drugs for the treatment of leishmaniasis, which is affordable, non-toxic, and effective against those parasites that are resistant to conventional drugs. The antileishmanial activity of HIV-1 protease inhibitors (PIs) is reported. Therefore, the present study focused on the combinational therapy of HIV-1 protease inhibitors, amprenavir (APV) in combination with ritonavir (RTV), against *Leishmania donovani* and *Leishmania major*. This combination effectively killed the parasites, which is more potent against *L. donovani* compared to *L. major*. The combination of PIs exerted antileishmanial effect by increased ROS generation followed by elevated nitric oxide activation leading to oxidative stress. Therefore, the study might be exploited for the therapeutic development against leishmaniasis as a substitute to replace the conventional drugs.

Keywords *Leishmania donovani* · *Leishmania major* · HIV-1 protease inhibitors · Amprenavir · Ritonavir

1 Introduction

Leishmaniasis is a vector-borne infectious disease, and its causative agent is an obligate intracellular protozoan of the genus *Leishmania*. The tropical and subtropical disease is transmitted by the bite of infected phlebotomine sand flies. Cases of infection have been reported in 98 countries in Europe, Africa, Asia, and America, and according to the reports, 90% of new cases occur in just 13 countries (Afghanistan, Algeria, Bangladesh, Bolivia, Brazil, Columbia, Ethiopia, India, Iran, Peru, South Sudan, Sudan,

A. K. Saxena, S. Nandi (eds.), *Global Trends in Health, Technology and Management*, https://doi.org/10.1007/978-3-031-75457-9_3

and Syria) [1–3]. Depending upon the virulence of parasite and host's immune response, leishmaniasis can have several clinical forms and subtypes with wide-ranging symptoms. The two main forms of leishmaniasis are cutaneous leishmaniasis (CL) and visceral leishmaniasis (VL). CL is the most common form, whereas VL is the most severe form. VL, if not treated, is fatal in over 95% of cases [3]. According to the WHO report for the year 2023, more than 5000 cases were reported from 8 countries, namely, Afghanistan, Algeria, Brazil, Colombia, Iran, Iraq, Peru, and the Syrian Arab Republic, and this together accounted for the 85% of CL incidences reported worldwide. Further, seven countries such as Brazil, Ethiopia, India, Kenya, Somalia, South Sudan, and Sudan were reported to contribute for about 85% of the global VL burden. Post-kala-azar dermal leishmaniasis (PKDL) and co-infections such HIV co-infection further develop as reservoirs for leishmaniasis, which is a threat for the complete eradication of the disease [1–3]. Moreover, the conventional antileishmanial drugs such as amphotericin B, miltefosine, paromomycin, and pentavalent antimonial have demonstrated several side effects with higher toxicity and even high relapse rate. Also, the cost of treatment is a major factor that needs to be considered as majority of the affected people belong to the lower socioeconomic strata [4]. Therefore, there is no efficient treatment for leishmaniasis as yet, and hence, there is an urgent need to develop newer drugs, which are more efficacious, less toxic, and affordable.

HIV-1 protease inhibitors (PIs) are FDA-approved anti-retroviral, whose anti-parasitic activity is already reported [5–8]. However, a detailed study can elucidate the mechanism of action of PIs against the *Leishmania* parasites. Ritonavir (RTV), one of the HIV-1 protease inhibitors, is commonly prescribed for the treatment of HIV as it has higher bioavailability and is reported to inhibit the activity of cytochrome P450 (CYP3A4), which is involved in the metabolism of endogenous compounds and xenobiotics [9]. Therefore, ritonavir is being used in combination with other PIs in order to enhance the bioavailability of other PIs. So, to understand the exact mechanism of action of HIV-1 protease inhibitors against *L. donovani* and *L. major*, we have investigated the antileishmanial activity of amprenavir (APV) in combination with RTV.

2 Materials and Methods

2.1 Chemicals

Amprenavir (APV), ritonavir (RTV), and N-acetyl-L-cysteine (NAC) were purchased from Sigma-Aldrich. The drugs were dissolved in 100% dimethyl sulfoxide (DMSO) at different concentrations and stored at −20 °C. The M199 and DMEM liquid medium, fetal bovine serum (FBS), and penicillin-streptomycin were purchased from Thermo Fisher Scientific.

2.2 Parasite Culture and Maintenance

The *Leishmania donovani* and *Leishmania major* parasites were cultured and maintained at 22 °C in M199 liquid medium supplemented with 10% (*v*/v) fetal bovine serum and 1% penicillin-streptomycin [10].

2.3 Cell Viability Test by MTT Assay

To investigate the effect of PIs in combination on viability of *L. donovani* and *L. major* promastigotes, the MTT assay was performed using 3-(4,5-dimethylthiazol-2-yl)-2,5-diphenyltetrazolium bromide (MTT) as described previously [10]. Briefly, the exponential phase cells (approximately 2×10^7 cells/well) were incubated for 12, 24, and 48 h in the presence of APV (5, 10, 20, 40, and 80 μM), RTV (1, 2, 4, and 8 μM), and combination of APV-RTV. The cells were centrifuged after the treatment period, and the supernatant was aspirated. The cell pellet was given two washes with PBS (1X), and then finally the pellet was suspended in PBS (1X) and plated in 96-well plates. The MTT solution (10 μg/mL) was added in each sample of 96-well plates, and samples were incubated for 3 h. Post incubation, DMSO was added to the samples. The optical density was measured at 570 nm on multimode microplate reader.

2.4 Measurement of Cellular ROS Levels

The cellular ROS produced in drug-treated and drug-untreated *L. donovani* and *L. major* promastigotes was measured by CM-H_2DCFDA as per manufacturer's protocol (Invitrogen). In brief, *L. donovani* and *L. major* promastigotes (2×10^7 cells) were treated with APV (5 μM), RTV (4 μM), and APV in combination with RTV. Cells were also treated with anti-oxidant like NAC (20 mM) prior to APV-RTV for 12, 24, and 48 h. Leishmanial cells were also treated with H_2O_2, which was a positive control. Post the treatment duration, samples were centrifuged and washed, and the pellet was resuspended in M199 medium. Later, CM-H_2DCFDA (5 μM final concentration) was added, and the samples were incubated for approximately 1 h at RT in dark. The samples were then centrifuged at 2000 rpm for 10 min at RT, and the cell pellets were resuspended in PBS. Samples were given PBS wash by centrifugation at 5000 rpm for 5 min at RT, and the supernatant was discarded. Finally, the pellets were resuspended in lysis buffer and incubated in dark at RT for about 10 to 15 min. Further, the samples were centrifuged at 5000 rpm for 5 min at RT, and the supernatant was collected in a 96-well plate. The fluorescent product formed due to oxidization of CM-H_2DCFDA in the presence of ROS was measured by spectrofluorometer at 530 nm using excitation wavelength of 507 nm [10].

2.5 Nitric Oxide (NO) Generation

J774A.1 cells were infected with *L. donovani* and *L. major* and then were treated with APV, RTV, and APV in combination with RTV. The cells were grown in DMEM supplemented with 10% FBS and 1% penicillin-streptomycin, seeded in a 24-well plate, and incubated at 37 °C with 5% CO_2 for 48 h. The supernatants (containing lymphocytes) were gently removed and used for the measurement of nitric oxide (NO). The NO content in supernatants was measured by Griess reagent as per manufacturer's protocol (Thermo Fisher Scientific) and expressed in micromole nitrite. Standard and samples were incubated with equal amount of Griess reagent for 20 min at RT to develop color, and absorbance was read at 540 nm.

3 Results

3.1 Amprenavir in Combination with Ritonavir Causes Increased Parasite Killing

The MTT assay was carried out to examine the efficacy of APV and APV in combination with RTV on viability of *L. donovani* and *L. major*. The leishmanial promastigotes (2×10^7 cells/mL) were incubated with different concentrations of APV (5, 10, 20, 40, and 80 μM) and combination of APV-RTV for 12, 24, and 48 h, respectively. The absorbance (OD) at 570 nm for drug-treated *Leishmania* cells indicated inhibition of promastigotes' growth in a concentration-dependent manner. It was observed that individual drug dose as well as combination of PIs had more prominent effect on *L. donovani* compared to *L. major*. When the cells were treated with varying concentrations of APV individually, 50% inhibition of *L. donovani* cells was observed at 5 μM of APV after 24 h, whereas 58% cell viability was seen in *L. major* cells after APV (5 μM) treatment for 24 h. The percentage of inhibition increased with the increasing concentration of APV. About 95% inhibition was observed when *L. donovani* cells were treated with 20 μM of APV for 48 h, while in the case of *L. major* cells, only 89% of cell growth inhibition was observed after 48 h. Furthermore, the maximum inhibitory effect, i.e., 99% inhibition of cell viability, was observed when *L. donovani* cells were treated with 40 μM of APV for 24 h, while 96% inhibition of cell growth was seen in *L. major* cells after 24 h (Fig. 1a). The IC_{50} values of APV after 12 and 24 h were calculated to be 6.0 ± 0.6 μM and 4.7 ± 0.4 μM, respectively, for *L. donovani* cells, whereas the IC_{50} values of APV for *L. major* promastigotes after 12 and 24 h were calculated to be 11.2 ± 0.3 μM and 6 ± 0.4 μM, respectively.

We also checked the antileishmanial effect of RTV by performing the cell viability assay with increasing concentration of RTV at different time duration. It was observed that RTV at a concentration of 4 μM showed minimal inhibitory effect against *L. donovani* and *L. major* (data not shown). Therefore, 4 μM of RTV was considered as the minimal concentration to be used in combination with APV for the further assays.

From the above results, we performed MTT assay using the APV (5 μM), RTV (4 μM), and APV (5 μM) in combination with RTV (4 μM). It was observed that APV (5 μM) individually showed 50% cell growth inhibition after a 24-h treatment in *L. donovani* promastigotes. However, this inhibition was not evident in *L. major* cells after 24 h. When *L. donovani* and *L. major* promastigotes were treated with RTV, it was observed that the parasite growth decreased to only up to 44% and 43% in *L. donovani* and *L. major* after 48 h, respectively. Furthermore, APV (5 μM) + RTV (4 μM) inhibited 91% and 90% parasite growth in *L. donovani* and *L. major* after 12 h and 24 h, respectively (Fig. 1b). The IC_{50} values of APV (5 μM) in combination with RTV (4 μM) were calculated to be 2.8 ± 0.4 μM at 12 h and 1.9 ± 0.3 μM at 24 h, respectively, for *L. donovani*. On the other hand, IC_{50} values of APV (5 μM) in combination with RTV (4 μM) for *L. major* promastigotes were calculated as 4.6 ± 1 μM and 3.2 ± 0.4 μM.

3.2 Cellular ROS Generation After APV and RTV Treatment

The rate of ROS production is known to increase greatly upon drug treatment; therefore, to determine whether APV and combination of PIs also lead to an increase in ROS production, the cellular ROS level in *L. donovani* and *L. major* promastigotes was measured. The leishmanial parasites were treated separately with APV (5 μM), RTV (4 μM), and APV (5 μM) in combination with RTV (4 μM) and anti-oxidant NAC (20 mM) prior to the treatment with PIs. Cellular ROS generated after treatment was measured fluorometrically by conversion of CM-H_2DCFDA to highly fluorescent DCFDA. It was observed that cellular ROS levels were significantly higher when *L. donovani* parasites were treated with APV (5 μM) in combination with RTV (4 μM) compared to *L. major*. The level of peroxide radicals was significantly increased inside the parasites within 12 h of treatment, and the highest level of ROS was observed at 48 h. However, ROS generation significantly decreased when *L. donovani* and *L. major* parasites were treated with NAC (negative control) (Fig. 2).

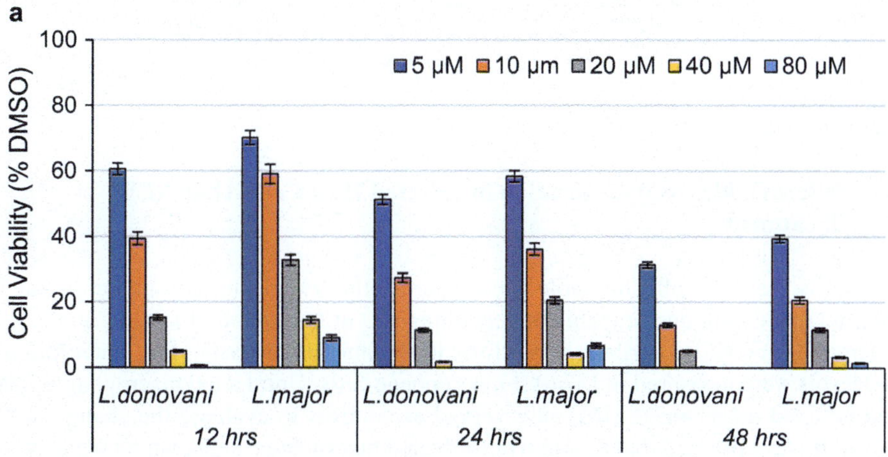

Fig. 1 (**a**) Cell viability by MTT assay. The effect of APV, RTV, and APV-RTV was evaluated by measuring the cell viability by MTT assay. It was observed that APV (5 μM) resulted in 50% and 42% inhibition of *L. donovani* and *L. major* after 24 h, respectively. APV (5 μM) + RTV (4 μM) inhibited 91% parasite growth in *L. donovani* after 12 h and 90% inhibition in *L. major* after 24 h, respectively. (**b**) Cell viability by MTT assay. The effect of APV, RTV, and APV-RTV was evaluated by measuring the cell viability by MTT assay. It was observed that APV (5 μM) + RTV (4 μM) inhibited 91% parasite growth in *L. donovani* after 12 h and 90% inhibition in *L. major* after 24 h, respectively

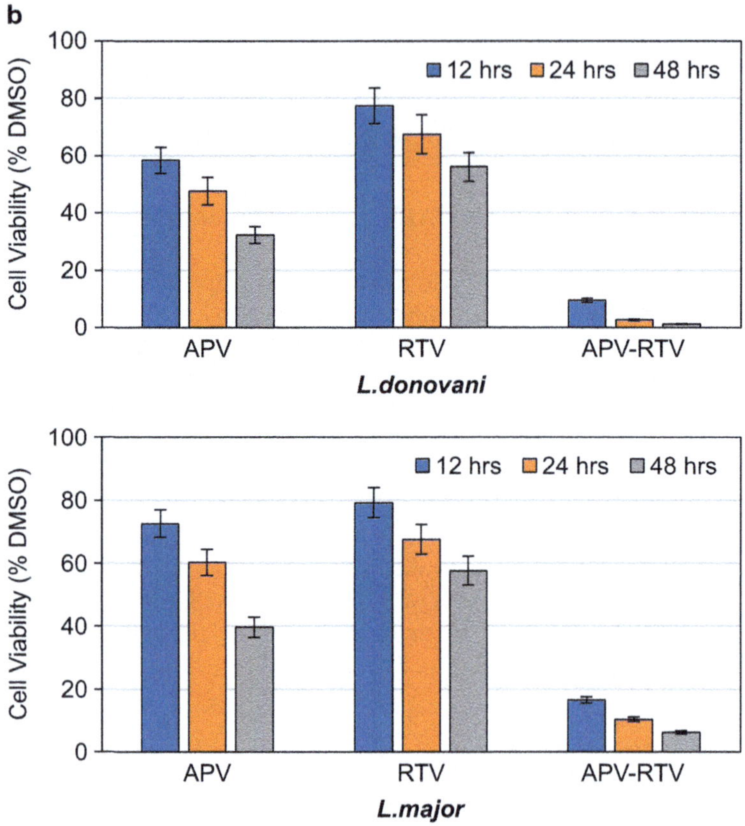

Fig. 1 (continued)

3.3 Nitric Oxide (NO) Generation in Infected Host Cells After APV Treatment

As NO is the main effector molecule for intracellular parasite elimination, it can be inferred that compounds capable of generating NO in *Leishmania* parasites and/or host macrophages could contribute toward the leishmanicidal activity [11]. Therefore, the NO levels were measured in J774A.1 macrophage cells. Uninfected macrophage (UIM) and infected macrophage (IM) cells served as controls to evaluate the change in NO levels. It was observed that *L. donovani*-infected macrophage cells, which were treated with APV in combination with RTV, caused a significant increase in NO (26.01 μM) compared to individual dose of APV and RTV to that of 22.3 μM NO generation in *L. major*-infected macrophages that were treated with APV in combination with RTV for 48 h (Fig. 3).

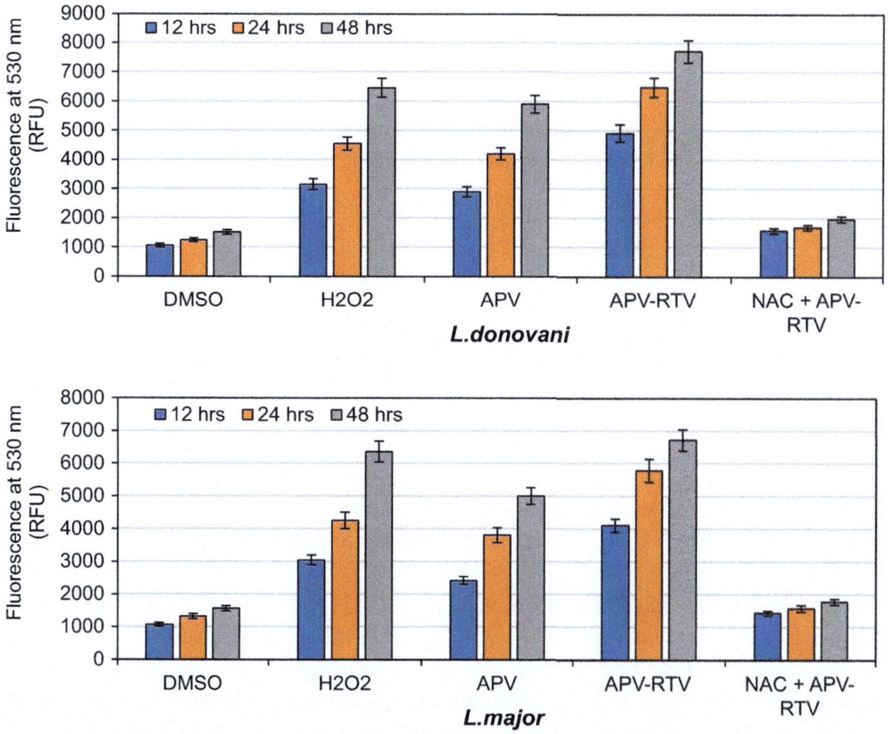

Fig. 2 Measurement of cellular ROS levels. The ROS levels within the *L. donovani* and *L. major* promastigotes were measured by CM-H_2DCFDA, an indicator of ROS. The ROS levels significantly increased after APV (5 μM) in combination with RTV (4 μM) treatment in the case of *L. donovani* compared to *L. major*. Moreover, combination of PIs had enhanced effect on the parasites rather than the individual doses of APV and RTV, respectively

4 Discussion

The prevalence of leishmaniasis is still a major health issue in the countries where the disease is endemic. The conventional therapy against CL and VL is unsatisfactory and ineffective owing to the efficacy variance, the toxicity, and the emerged resistance. Therefore, development of efficacious and affordable drug for the treatment of these two forms of leishmaniasis is crucial.

The aim of this study was to investigate the antileishmanial activity of HIV-1 protease inhibitors against *L. donovani* and *L. major*. As mentioned earlier, RTV is capable of inactivating the drug-metabolizing enzyme, CYP3A4 [9]. Thus, RTV is responsible for enhancing the bioavailability of the other PIs. Hence, our present study is focused on the effect of APV in combination with RTV in *Leishmania donovani* and *Leishmania major*. This strategy of repurposing HIV-1 protease inhibitors can be beneficial for the treatment of leishmaniasis.

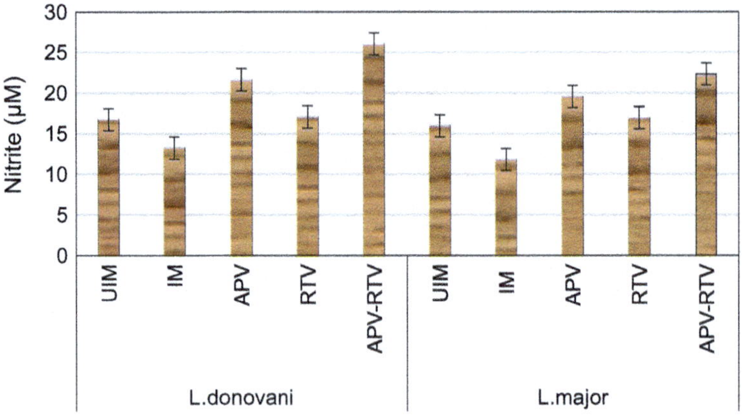

Fig. 3 Measurement of nitric oxide (NO) levels in host cells by Griess reagent. The NO levels were elevated in *L. donovani*-infected macrophage cells, treated with APV in combination with RTV, compared to *L. major*-infected macrophages that were treated with APV in combination with RTV for 48 h. The NO levels were measured in micromole nitrite

In this study, we investigated the efficacy of APV and APV in combination with RTV against *L. donovani* and *L. major* by evaluating the inhibition of the parasite growth. It was observed that the HIV-1 protease inhibitors, APV and APV in combination with RTV, had more potent leishmanicidal effect against *L. donovani* as compared to *L. major*. Moreover, the combination had more pronounced effect than the individual dose of APV or RTV. APV (5 μM) + RTV (4 μM) inhibited 91% *L. donovani* parasite growth and 90% *L. major* growth after 24 h (Fig. 1b).

Reactive oxygen species (ROS) are highly reactive and modify the proteins, lipids, and nucleic acids. So, ROS-mediated cell damage is well known in apoptosis. We speculated that APV and the combination might generate cellular ROS. Therefore, we measured the endogenous ROS in leishmanial cells after treatment with APV and APV-RTV at different time points. It was observed that APV-RTV induced the highest ROS generation at 48 h in *L. donovani* promastigotes. *L. major* promastigotes too had the increased levels of ROS after 48 h of treatment; however, it was not that high as that of *L. donovani* (Fig. 2).

It is reported previously that oxidative burst generated in macrophages is a strategy developed by macrophages for *Leishmania* eradication. O_2^- and NO are the main molecules utilized by the oxidative burst [12–14]. Thus, parasites may induce a stronger oxidative burst in monocytes from CL and VL patients. Moreover, NO is known to be produced during active leishmaniasis and is a major metabolite that has a role in controlling the parasite burden and disease prevention [14]. Therefore, we evaluated the NO levels in J774A.1 macrophage cells infected with *L. donovani* and *L. major*, respectively. Macrophage cells that were infected with *L. donovani* and treated with APV in combination with RTV caused a significant increase in NO compared to individual dose of APV and RTV and to that of NO generated in *L. major*-infected macrophage cells treated with APV in combination with RTV for 48 h (Fig. 3). This indicates that an increase in ROS levels within the parasites leads to oxidative stress, whereas the higher

NO levels within the host cells suggest its role in parasite clearance. So, oxidative stress and activation of nitric oxide might induce apoptotic cascade of unicellular kinetoplastid parasites.

Thus, induction of ROS generation and NO activation by APV in combination with RTV might be responsible for the parasite cell death. Moreover, APV has more effective leishmanicidal activity against *L. donovani* than *L. major*. Furthermore, the combination of APV-RTV induces leishmanicidal effect against *Leishmania* parasites at low concentration of APV in combination in comparison to individual higher concentration of APV. Thus, repurposing of APV as a potential therapeutic agent for the treatment of leishmaniasis can be an impactful consideration. In conclusion, this study provides evidence for superiority of drug combination compared to single drug for the effective treatment of leishmaniasis.

5 Conclusion

Based on the experimental observations, the present study demonstrated that APV-RTV induces oxidative burst that results in programmed cell death (PCD) in *Leishmania.* The effect of the combination is more pronounced in *L. donovani* compared to *L. major.* So, APV-RTV in combination can effectively kill parasites via ROS and NO generation in *Leishmania* parasites as compared to single dose of APV and RTV, respectively. APV in combination with RTV in lower dose was much more effective in killing the parasites as compared to the single dose of APV. This indicates that the combination treatment can reduce the treatment duration and thus the drug toxicity can be minimized.

References

1. Steverding, D.: The History of Leishmaniasis. Vol. 10, Parasites & Vectors. Springer Science and Business Media LLC (2017). https://doi.org/10.1186/s13071-017-2028-5
2. Mazire, P., Agarwal, V., Roy, A.: Road-Map of Pre-Clinical Treatment for Visceral Leishmaniasis, vol. 83, pp. 317–327. Drug Development Research. Wiley (2021). https://doi.org/10.1002/ddr.21907
3. The Global Health Observatory. https://www.who.int/data/gho/data/themes/topics/indicator-groups/indicator-group-details/GHO/leishmaniasis
4. Shmueli, M., Ben-Shimol, S.: Review of leishmaniasis treatment: can we see the forest through the trees? Pharmacy (Basel). **12**(1), 30 (2024). https://doi.org/10.3390/pharmacy12010030
5. Santos, L.O., Marinho, F.A., Altoe', E.F., Vito Rio, B.S., Alves, C.R., Britto, C., Motta, M.C.M., Branquinha, M.H., Santos, A.'.L.S., d'Avila-Levy, C.M.: HIV aspartyl peptidase inhibitors interfere with cellular proliferation, ultrastructure and macrophage infection of *Leishmania amazonensis*. PLoS One. **4**(3), e4918 (2009)
6. Kumar, P., Lodge, R., Trudel, N., Ouellet, M., Ouellette, M., Tremblay, M.J.: Nelfinavir, an HIV-1 protease inhibitor, induces oxidative stress-mediated, caspase-independent apoptosis in *Leishmania* amastigotes. PLoS Negl. Trop. Dis. **4**(3), e642 (2010)

7. Rebello, K.M., Valter, V., Zuma, A.-N.A.A., Motta, M.C.M., Gomes, C.R.B., de Souza, M.V.N., Atella, G.C., Branquinha, M.H., Santos, A.L.S., Torres-Santos, E.C., d'Avila-Levy, C.M.: Lopinavir, an HIV-1 peptidase inhibitor, induces alteration on the lipid metabolism of *Leishmania amazonensis* promastigotes. Parasitology. **145**, 1304–1310 (2018)
8. Rebello, K.M., Andrade-Neto, V.V., Claudia, R.B., Gomes, M.V., de Souza, N., Branquinha, M.H., Santos, A.L.S., Torres-Santos, E.C., d'Avila-Levy, C.M.: Miltefosine-Lopinavir combination therapy against *Leishmania infantum* infection: in vitro and in vivo approaches. Cell. Infect. Microbiol. **9**, 229 (2019)
9. Sevrioukova, I.F.: Poulos TL structure and mechanism of the complex between cytochrome P4503A4 and ritonavir. Proc. Natl. Acad. Sci. USA. **107**(43), 18422–18427 (2010)
10. Roy, A., Behera, S., Mazire, P.H., Kumari, B., Mandal, A., Purkait, B., Ghosh, P., Das, P., Das, P.: The HIV-1 protease inhibitor amprenavir targets *Leishmania donovani* topoisomerase I and induces oxidative stress-mediated programmed cell death. Parasitol. Int. **82**, 102287 (2021). https://doi.org/10.1016/j.parint.2021.102287
11. Bogdan, N.C., Rollinghoff, M.: The immune response to Leishmania. Mechanism of parasite control and evasion. Int. J. Parasitol. **28**, 121–134 (1998)
12. Gantt, K.R., Goldman, T.L., McCormick, M.L., Miller, M.A., Jeronimo, S.M., Nascimento, E.T., et al.: Oxidative responses of human and murine macrophages during phagocytosis of Leishmania chagasi. J. Immunol. **167**(2), 893–901 (2001)
13. Carneiro, P.P., Conceição, J., Macedo, M., Magalhães, V., Carvalho, E.M., Bacellar, O.: The role of nitric oxide and reactive oxygen species in the killing of Leishmania braziliensis by monocytes from patients with cutaneous Leishmaniasis. PLoS One. **11**(2), e0148084 (2016)
14. Formaglio, P., Alabdullah, M., Siokis, A., Handschuh, J., Sauerland, I., Fu, Y., Krone, A., Gintschel, P., Stettin, J., Heyde, S., Mohr, J., Philipsen, L., Schröder, A., Robert, P.A., Zhao, G., Khailaie, S., Dudeck, A., Bertrand, J., Späth, G.F., Kahlfuß, S., Bousso, P., Schraven, B., Huehn, J., Binder, S., Meyer-Hermann, M., Müller, A.J.: Nitric oxide controls proliferation of Leishmania major by inhibiting the recruitment of permissive host cells. Immunity. **54**(12), 2724–2739.e10 (2021). https://doi.org/10.1016/j.immuni.2021.09.021

A Hands-On Perspective on Physico-chemical Versus AI/ML Methods Along the Genome to Drug Pathway

B. Jayaram[1] (✉)

[1] Supercomputing Facility for Bioinformatics & Computational Biology (SCFBio), IIT Delhi, New Delhi, India
bjayaram@chemistry.iitd.ac.in

Abstract The availability of genomic sequences in public domain has spurred intense algorithmic developments for identifying protein-coding regions (gene finding) and for establishing their functions. This in turn accelerated protein structure prediction efforts, particularly relevant for proteins crucial to pathogens and those involved in disease conditions, to further enable structure-based drug design endeavours. This modern drug discovery pathway (genome -> gene -> protein -> drug) in our hands became a set of software suites collectively called 'Dhanvantari', which embodies *Chemgenome*, *Bhageerath* and *Sanjeevini* web suites to traverse through the genome to drug pathway with entry along any point. While developing the above science and software suites, our focus has been on energy, forcefield and molecular simulation-based methods, which are referred to here as physico-chemical methods. AI/ML methods have literally stormed in during the last few years into these research areas making it almost difficult to ignore their strengths. The result is the emergence of integrated AI/ML and physico-chemical methods for improved accuracies and greater success rates in new molecule predictions against drug targets. This brief report sketches how these methods have evolved in our hands to help accelerate drug discovery.

Keywords Genome annotation · Protein structure prediction (PSP) · Computer-aided drug design (CADD) · Energy-based physico-chemical methods (PCM) · Artificial intelligence (AI)/machine learning (ML) methods

A. K. Saxena, S. Nandi (eds.), *Global Trends in Health, Technology and Management*, https://doi.org/10.1007/978-3-031-75457-9_4

1 The Genome to Drug Pathway

Over the past score and more years, we were focused on developing physico-chemical methods from genome to drug discovery [1–3] with entry at any step along the pathway as shown in Fig. 1. While this work was finding strong roots, remarkable progresses were made in the applications of AI/ML in drug discovery in recent years. We trace the advancements in computer-aided drug discovery with these distinct approaches in our laboratory and how integration of these methods is yielding reliable predictions of drug-like candidate molecules for biomolecular targets.

2 Gene Prediction

During the first few years of the century, several methods were proposed for gene finding. GLIMMER, GENSCAN, GENESCAN, GenomeThreader, GRAIL, and Augustus are a few to name [4–9]. An exhaustive list of annotation software is available in ref. [10]. These are typically based on higher-order Markov models, hidden Markov

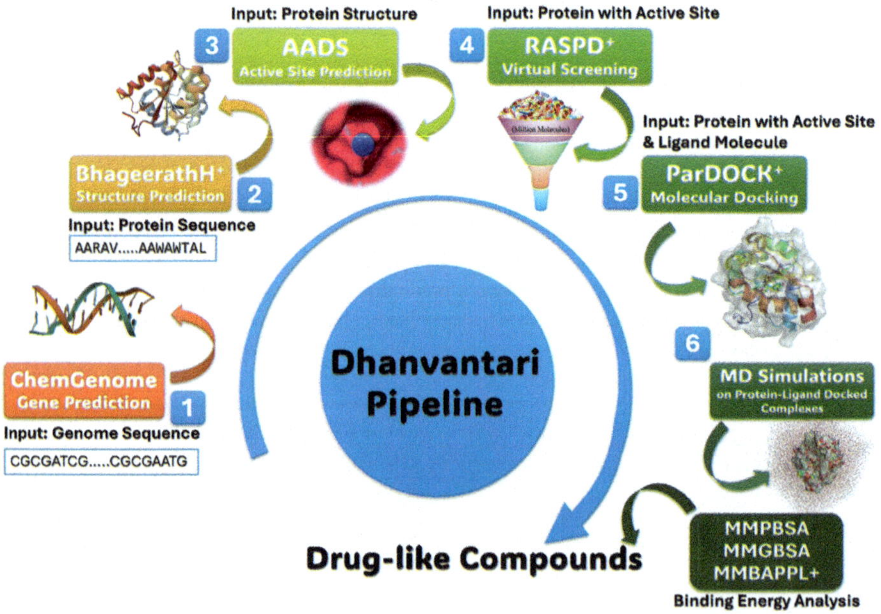

Fig. 1 In silico drug discovery assembly line developed at SCFBio. Step 1 is for genome annotation, finding protein-coding genes here, step 2 is for protein tertiary structure prediction, and steps 3, 4 and 5, which are part of *Sanjeevini* suite, are for identifying lead compounds, which are further processed through simulations and *post facto* binding free energy analyses to suggest candidate molecules for experimental testing. User can enter at any step in the above *Dhanvantari* pipeline

models, pattern recognition, discriminant analysis, neural networks, Fourier transforms and several other sophisticated statistical and mathematical techniques mostly in nucleotide sequence space. A common problem encountered with these early methods was the lack of universality and heavy dependence on sparse experimental datasets for training [11]. It was clear that we needed to go beyond sequences into property space for achieving universality. This led us to develop *Chemgenome* based on forcefields [12] and molecular dynamic simulation-derived DNA energetics, focusing on universality [13, 14]. The method was able to capture sensitivity and specificity up to 90% independent of the species. While this was the case during early part of the century, most of the above-mentioned gene finding methods, augmented with newer ML techniques and expanded experimental training sets, now deliver >95% sensitivity for any specific species, although universality is lacking. Our inspiration to continue with physico-chemical methods for deciphering the language of DNA and its apparent success hinges on two important discoveries. One was the discovery of the conjugate rule [15] and the second was the observation that structural and energetic features of DNA conveyed their functional destiny [16–19]. While physico-chemical methods (PCMs) with both energetics and structural information superposed on sequence information promise broader applicability, the accuracies are still hovering around 90%. A couple of areas where PCMs are leading are in promoter prediction and in intron-exon boundary detection wherein sequence level promiscuity is high. The ChemEXIN [20] sketched in Fig. 2 shows unprecedented accuracies in intron-exon boundary detection. We expect that PCM and ML combinations will get us to a genome reader very soon.

3 Protein Structure Prediction

We have developed an all-atom energy-based methodology named *Bhageerath* for tertiary structure prediction of small proteins and subsequently extended it to larger proteins by combining homology methods (*BhageerathH*) [21–23]. More than 100 research groups have also made significant contributions to this area as chronicled in the biennial CASP experiments [24, 25]. Some of the popular servers are Rosetta, QUARK, I-TASSER, MULTICOM, INTFOLD, Distill, HHGG, Phyre, Floudas, Raptor, Pcons and so on. A full list is available at https://predictioncenter.org/. The overall accuracy, not long ago, however, was around 30% in terms of less than 3 Å RMSD (root mean square deviation) of predictions in relation to crystal structures. This was the time when ANNs, although not fully ready for protein tertiary structure prediction (PSP), were touching near 80% accuracy in secondary structure prediction [26]. In came AI-based AlphaFold and its next version a couple of years ago developed by DeepMind, which took the PSP field by storm with near 100% accuracy for soluble proteins [27, 28]. Interestingly, most successful PSP software participating in CASP15 experiment assimilated AlphaFold codes into their algorithms. *BhageerathH*, however, fielded forcefield and homology-based models, and encouragingly enough, 49 out of 94 structures predicted by *BhageerathH* were within 3 Å of AlphaFold structures (Fig. 3). Noting that the average size of the targets released in CASP15 was 478, the 52% accuracy by PCMs was a good progress. The accuracies reported are underestimates since

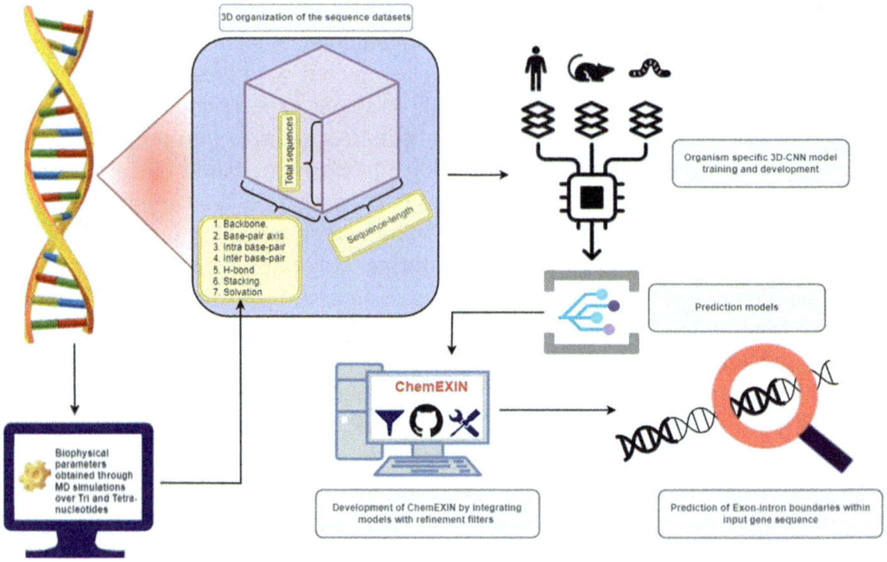

Fig. 2 An illustration of ChemEXIN, an intron-exon boundary detection scheme with physico-chemical approach supplemented with machine learning. The physico-chemical parameters obtained from MD simulations are used to train a 3D convolutional neural network (3D-CNN) model involving deep learning. This model outperforms the other state-of-the-art tools across the three organisms studied. The parameter scores are sensitivity = 93.1 and specificity = 91.9 for *H. sapiens*, sensitivity = 80.4 and specificity = 79.4 for *M. musculus* and sensitivity = 91.9 and specificity = 92.0 for *C. elegans* [20]

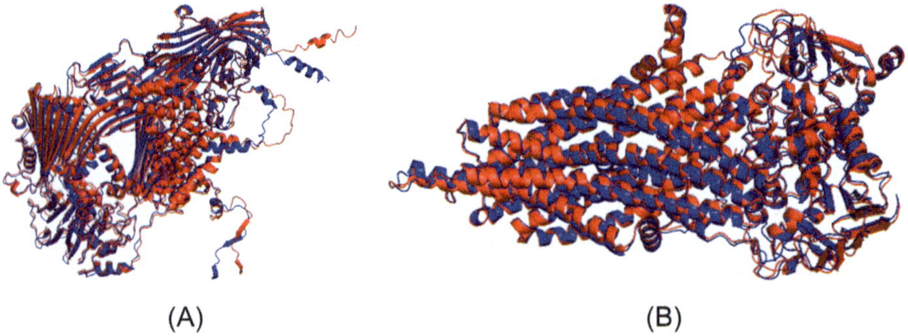

Fig. 3 A superposition of the tertiary structures predicted by BhageerathH-Pro (blue) and AlphaFold (red) for CASP15 targets: (**a**) Target T1191 with 1770 residues (RMSD: 0.97 Å) and (**b**) Target T1193 with 1297 residues (RMSD: 0.65 Å)

these are CASP targets with higher difficulty level in modelability [29]. Continuing with *Bhageerath*, we conceived of creating a computational PDB [30] and did so for Pvax in 2018 comprising over 2205 structures of soluble proteins and 3664 structures of soluble proteins of PvP01 in 2020 [31, 32]. Soon enough in 2022–2023, AlphaFold created tertiary structures of the entire uniport protein sequence database available from https://alphafold.ebi.ac.uk/. During our journey along the physico-chemical pathway, we were excited to discover that just four physico-chemical properties of amino acids not only explained the existence of the magic number 20 for naturally occurring amino acids but also set very high water marks for homology modelling and function prediction [33, 34]. We also discovered that higher-order Ramachandran maps carried sufficient information to build tertiary structures [35]. Further analyses of the configurational space of proteins suggested, when higher-order spatial correlations of amino acid residue neighbours were included, that protein folding was a convergent problem as opposed to the then conventional wisdom [36]. AlphaFold has now proved it. We were equally excited to discover that millions of proteins chronicled till date share a similar stoichiometry with very small standard deviations in their amino acid compositions, which we called as the margin of life [37, 38]. Equally interesting was the observation that there was a universality in the spatial distribution of the C-alpha atoms for any pair of amino acids defying the notion that preferential interactions among amino acid side chains dictated protein folding. This suggested to us a new view that inter-side chain interactions in proteins were offset by desolvation effects and folding thus was dictated by shape considerations of side chains along the sequence and exclusion by solvent water. Much remains to be done with PCMs in the PSP area to further understand the mechanism of folding at a molecular level. In the area of PSP, it is needless to say that AI-inspired AlphaFold and its variants, as of now, are leading the way.

4 Drug Design

As a part of our efforts to develop a freely accessible, experimentally validated, complete drug discovery web suite *Sanjeevini* [39–42] and particularly to facilitate a rapid screening of candidate molecules in the binding pocket of a target protein in search of good inhibitors, we developed an all-atom binding energy-based method for scoring named BAPPL [43] as an alternative to the conventional molecular free energy simulations [44, 45]. Given the small size of experimental datasets then, and fewer groups in the field, BAPPL performed reasonably well. A decade later, data size increased, groups have increased, and this has led us to revisit BAPPL, augment with ML methods (random forest in particular) and create BAPPL+, which seemed to outperform several other similar utilities [46]. We, however, continue to recommend the performance of molecular dynamic (MD) simulations on the protein-ligand docked complexes followed by BAPPL+ analyses of the binding energies as ensemble averages over the MD trajectories for more reliable results [47]. The atomic level modelling of protein-ligand interactions is feasible if the number of candidates is less than 100 or 1000 given today's compute power. To scan large libraries, such as a million compound library, against a target protein, in a matter of minutes, without actually docking, some novel strategies are required. We thus built a method (RASPD) based on a QSAR-type equation in the space of physico-chemical properties lining the binding pocket of the target protein and the candidate molecules [48]. More recently, i.e. almost a decade later, we have introduced RASPD+, which incorporates several ML algorithms. eRF together with PCMs is seen to perform the best [49]. Figure 4 illustrates the changing times for virtual

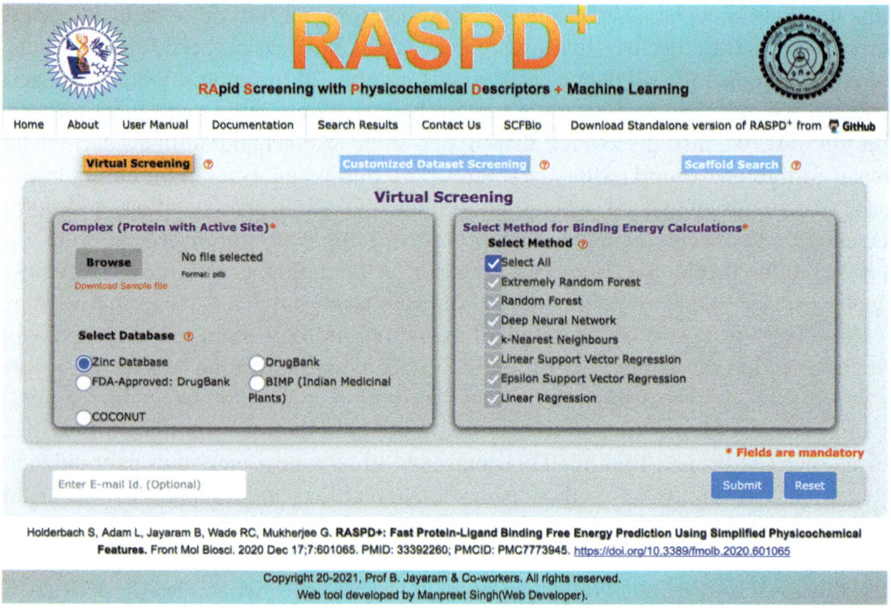

Fig. 4 A screenshot of the front end of the RASPD+ web server developed at SCFBio for a rapid virtual screening of small molecule libraries against a protein target (a million compounds in minutes!). The methodology searches for complementarity in the physico-chemical properties of the active site residues of protein targets and small molecules. User has the option to choose the scoring function trained with several models such as extremely random forest, deep neural network, linear support vector regression, etc. [48, 49]. RASPD+ is freely accessible at http://www.scfbio-iitd.res.in/raspd+/

screening. It is almost as if integration of PCMs and MLs is the way to go for improved accuracies and especially for accelerating new molecule discovery utilizing large libraries of small molecules. It may be noted that *Sanjeevini* protocols have already delivered experimentally validated low micromolar compounds against HAV [50], HBV [51, 52] and CHIKV [53] infections, breast cancer [54], Alzheimer's disease [55] and malaria [56] and fungal infections [57].

5 Conclusions

Physico-chemical methods and molecular simulations have a long history in protein structure area and in computer-aided drug discovery. They are relatively more recent in genome annotation. AI/ML methods have been the mainstay for genome annotation for more than two decades. However, these can significantly benefit with structural energetic feature inclusion as mentioned in Sect. 2 above. Despite the emergence of AlphaFold, work must go on with physico-chemical methods to arrive at unique native structures from sequences as there is no substitute to PCMs for new knowledge generation as adverted to briefly in Sect. 3 above. There are many areas that have been impacted by AI/ML in CADD [58–63]. One area is toxicity. The chemistry and biology of toxicity remain a black box hindering the growth of PCMs although a few ideas and hypotheses prevail such as off-target binding [64], metabolism of xenobiotics [65], etc. This is where AI/ML can play a tremendous role [66–68] in curtailing the failure rates of promising candidates bringing down the cost and time involved in drug discovery.

Acknowledgements The author acknowledges the CoE support received from the Department of Biotechnology, Govt. of India, to SCFBio, IIT Delhi, and support from the Ministry of Information Technology, Govt. of India, and CDAC under NSM project to SCFBio, IIT Delhi. The author sincerely places on record his gratitude to all his students (http://www.scfbio-iitd.res.in/training/training.htm) who have contributed to the various stages of development of the *Dhanvantari* pathway over the years summarized in this article and also to his mentors and collaborators, Prof. David Beveridge, Prof. Barry Honig and Prof. Aditya Mittal, in particular, for making pursuit of science even more exciting. The author is also grateful to the guest editors, Dr. A.K. Saxena and Dr. S. Nandi, for giving us the opportunity to share our research findings with the readership of the GTHTM-2024 conference proceedings sponsored by Springer.

References

1. Bhat, R., Kaushik, R., Singh, A., DasGupta, D., Jayaraj, A., Soni, A., Shandilya, A., Shekhar, V., Shekhar, S., Jayaram, B.: A comprehensive automated computer-aided discovery pipeline from genomes to hit molecules. Chem. Eng. Sci. **222**, 115711 (2020). https://doi.org/10.1016/j.ces.2020.115711
2. Soni, A., Menaria, K., Ray, P., Jayaram, B.: Genomes to hits in silico – a country path today, a highway tomorrow: a case study of chikungunya. Curr. Pharma. Des. **19**, 4687–4700 (2013) http://www.eurekaselect.com/111973/article
3. Shenoy, S., Jayaram, B., Latha, N., Narang, P., et al.: From gene to drug: a proof of concept for a plausible computational pathway. In: *Proceedings, IEEE Computer Society,* 1147–1152 (2006), 6th international conference on intelligent systems design and applications, Jinan, China, 16–18 Oct 2006. https://dl.acm.org/doi/abs/10.1109/ISDA.2006.154
4. Salzberg, S., Delcher, A., Kasif, S., White, O.: Microbial gene identification using interpolated Markov models. Nucleic Acids Res. **26**(2), 544–548 (1998)
5. Burge, C., Karlin, S.: Prediction of complete gene structures in human genomic DNA. J. Mol. Biol. **268**, 78–94 (1997)
6. Bhattacharya, A., Bhattacharya, S., Joshi, A., Ramachandran, S., Ramaswamy, R.: Identification of parasitic genes by computational methods. Parasitol. Today. **16**, 127–130 (2000)
7. Gremme, G., Brendel, V., Sparks, M.E., Kurtz, S.: Engineering a software tool for gene structure prediction in higher organisms. Inf. Softw. Technol. **47**(15), 965–978 (2005)
8. Uberbacher, E.C., Hyatt, D., Shah, M.: GrailEXP and genome analysis pipeline for genome annotation. Curr. Protoc. Hum. Genet., Wiley. **39**(1), 6 (2004). https://doi.org/10.1002/0471142905.hg0605s39
9. Stanke, M., Steinkamp, R., Waack, S., Morgenstern, B.: AUGUSTUS: a web server for gene finding in eukaryotes. Nucleic Acids Res. **32**, W309–W312 (2004)

10. Wang, R., Bao, L., Liu, S., Liu, Z.: Genome annotation: determination of the coding potential of the genome. In: Liu, Z.J. (ed.) Bioinformatics in Aquaculture, Principles and Methods, Chapter 4. Wiley Online Library (2017). https://doi.org/10.1002/9781118782392.ch4
11. Korf, I.: Gene finding in novel genomes. BMC Bioinform. **5**, 59 (2004)
12. Dutta, S., Singhal, P., Agrawal, P., Tomer, R., Kritee, K., Jayaram, B.: A physico-chemical model for analyzing DNA sequences. J. Chem. Inf. Model. **46**, 78–85 (2006) http://pubs.acs.org/doi/abs/10.1021/ci050119x
13. Singhal, P., Jayaram, B., Dixit, S.B., Beveridge, D.L.: Prokaryotic gene finding based on physicochemical characteristics of codons calculated from molecular dynamics simulations. Biophys. J. **94**, 4173–4183 (2008) http://www.cell.com/biophysj/retrieve/pii/S0006349508700754
14. Khandelwal, G., Gupta, J., Jayaram, B.: DNA energetics based analyses suggest additional genes in prokaryotes. J. Bio Sc. **37**, 433–444 (2012) http://link.springer.com/article/10.1007/s12038-012-9221-7
15. Jayaram, B.: Beyond the wobble: the rule of conjugates. J. Mol. Evol. **45**, 704–705 (1997)
16. Singh, A., Mishra, A., Khosravi, A., Khandelwal, G., Jayaram, B.: Physico-chemical fingerprinting of RNA genes. Nucleic Acids Res. **45**(7), e47 (2016). https://doi.org/10.1093/nar/gkw1236
17. Mishra, A., Siwach, P., Misra, P., Jayaram, B., Bansal, M., Olson, W.K., Thayer, K.M., Beveridge, D.L.: Towards a universal structural and energetic model for prokaryotic promoters. Biophys. J. **115**(7), 1180–1189 (2018). https://doi.org/10.1016/j.bpj.2018.08.002
18. Mishra, A., Siwach, P., Misra, P., Dhiman, S., Pandey, A.K., Srivastava, P., Jayaram, B.: Intron exon boundary junctions in human genome have in-built unique structural and energetic signals. Nucleic Acids Res. **49**(5), 2674–2683 (2021). https://doi.org/10.1093/nar/gkab098
19. Sharma, D., Sharma, K., Mishra, A., Siwach, P., Mittal, A., Jayaram, B.: Molecular dynamics simulation based trinucleotide and tetranucleotide level structural and energy characterization of functional units of genomic DNA. Phys. Chem. Chem. Phys. (PCCP). **25**, 7323–7337 (2023). https://doi.org/10.1039/D2CP04820E
20. Jayaram, B., Sharma, D., Aslam, D., Sharma, K., Mittal, A.: Exon-intron boundary detection made easy by physicochemical properties of DNA, Manuscript Submitted. https://www.researchsquare.com/article/rs-4359229/v1 (2024)
21. Narang, P., Bhushan, K., Bose, S., Jayaram, B.: A computational pathway for bracketing native-like structures for small alpha helical globular proteins. Phys. Chem. Chem. Phys. **7**, 2364–2375 (2005) http://pubs.rsc.org/en/content/articlelanding/2005/CP/b502226f
22. Jayaram, B., Bhushan, K., Shenoy, S., Narang, P., et al.: Bhageerath: an energy based web enabled computer software suite for limiting the search space of tertiary structures of small globular proteins. Nucl. Acids Res. **34**, 6195–6204 (2006) http://nar.oxfordjournals.org/content/34/21/6195.long
23. Jayaram, B., Dhingra, P., Mishra, A., Kaushik, R., Mukherjee, M., Singh, A., Shekhar, S.: Bhageerath-H: a homology ab initio hybrid server for predicting tertiary structures of monomeric soluble proteins. BMC Bioinform. **15**(Suppl 16), S7 (2014) http://www.biomedcentral.com/1471-2105/15/S16/S7
24. Moult, J., Fidelis, K., Kryshtafovych, A., Tramontano, A.: Critical assessment of methods of protein structure prediction (CASP)—round IX. Proteins Struct. Funct. Bioinform. **79**, 1–5 (2011). https://doi.org/10.1002/prot.23200
25. Simpkin, A.J., Mesdaghi, S., Rodríguez, F.S., Elliott, L., Murphy, D.L., Kryshtafovych, A., Keegan, R.M., Rigden, D.J.: Tertiary structure assessment at CASP15. Proteins. **91**(12), 1616–1635 (2023). https://doi.org/10.1002/prot.26593
26. McGuffin, L.J., Bryson, K., Jones, D.T.: The PSIPRED protein structure prediction server. Bioinformatics. **16**(4), 404–405 (2000). https://doi.org/10.1093/bioinformatics/16.4.404

27. Jumper, J., Evans, R., Pritzel, A., et al.: Highly accurate protein structure prediction with AlphaFold. Nature. **596**, 583–589 (2021). https://doi.org/10.1038/s41586-021-03819-2
28. Yang, Z., Zeng, X., Zhao, Y., et al.: AlphaFold2 and its applications in the fields of biology and medicine. Signal Transduct. Target Ther. **8**, 115 (2023). https://doi.org/10.1038/s41392-023-01381-z
29. Kaushik, R., Jayaram, B.: Structural difficulty index: a reliable measure for modelability of protein tertiary structures. Protein Eng. Des. Sel. **29**(9), 391–397 (2016). https://doi.org/10.1093/protein/gzw025
30. Jayaram, B., Dhingra, P.: Towards creating complete proteomic structural databases of whole organisms. Curr. Bioinform. **7**, 424–435 (2012). https://doi.org/10.2174/157489312803900992
31. Singh, A., Kaushik, R., Kuntal, H., Jayaram, B.: PvaxDB: a comprehensive structural repository of Plasmodium vivax proteome. *Database*. **2018**, bay021 (2018). https://doi.org/10.1093/database/bay021
32. Singh, A., Kaushik, R., Chaurasia, D.K., Singh, M., Jayaram, B.: PvP01-DB: computational structural and functional characterization of soluble proteome of PvP01 strain of Plasmodium vivax. *Database*. **2020**, baaa036 (2020). https://doi.org/10.1093/database/baaa036
33. Kaushik, R., Singh, A., Jayaram, B.: Where informatics lags chemistry leads. Biochemistry. **55**(5), 503–505 (2018). https://doi.org/10.1021/acs.biochem.7b01073
34. Amita, P., Toran, R., Avinash, E., Jayaram, B.: Mask blast with a new chemical logic of amino acids for improved protein function prediction. Proteins Struct. Funct. Bioinform. **89**(8), 922–924 (2021). https://doi.org/10.1002/prot.26069
35. DasGupta, D., Kaushik, R., Jayaram, B.: From Ramachandran maps to tertiary structures of proteins. J. Phys. Chem. B. **119**(34), 11136–11145 (2015). https://doi.org/10.1021/acs.jpcb.5b02999
36. DasGupta, D., Kaushik, R., Jayaram, B.: Protein folding is a convergent problem! Biochem. Biophys. Res. Commun. **480**(4), 741–744 (2016). https://doi.org/10.1016/j.bbrc.2016.10.119
37. Mittal, A., Jayaram, B., Shenoy, S.R., Bawa, T.S.: A stoichiometry driven universal spatial organization of backbones of folded proteins: are there Chargaff's rules for protein folding? J. Biomol. Struct. Dyn. **28**, 133–142 (2010) http://www.tandfonline.com/doi/abs/10.1080/07391102.2010.10507349
38. Mittal, A., Jayaram, B.: Backbones of folded proteins reveal novel invariant amino-acid neighborhoods. J. Biomol. Struct. Dyn. **28**, 443–454 (2011) http://www.tandfonline.com/doi/abs/10.1080/073911011010524954
39. Jayaram, B., Latha, N., Jain, T., Sharma, P., Gandhimathi, A., Pandey, V.S.: Sanjeevini: a comprehensive active-site directed lead design software. Indian J. Chem. **45A**, 1834–1837 (2006) http://www.niscair.res.in/sciencecommunication/researchjournals/rejour/ijca/ijca2k6/ijca_jan06.asp#p21
40. Jayaram, B., Singh, T., Mukherjee, G., Mathur, A., Shekhar, S., Shekhar, V.: Sanjeevini: a freely accessible web-server for target directed lead molecule discovery. BMC Bioinform. **13**, S7 (2012) http://www.biomedcentral.com/1471-2105/13/S17/S7
41. Singh, A., Shekhar, S., Jayaram, B.: CADD: some success stories from Sanjeevini and the way forward. In: Singh, S.K. (ed.) Innovations and Implementations of Computer Aided Drug Discovery Strategies in Rational Drug Design. Springer Nature, Singapore (2021). https://doi.org/10.1007/978-981-15-8936-2_1
42. Jayaram, B., Chaurasia, D. K., Anjum, R., Kesarwani, A., Shekhar, S., Mittal, A.: *Sanjeevini* 3.0: a comprehensive automated web-server for computer aided drug design, manuscript submitted (2024)

43. Jain, T., Jayaram, B.: An all atom energy based computational protocol for predicting binding affinities of protein-ligand complexes. FEBS Lett. **579**, 6659–6666 (2005) https://www.sciencedirect.com/science/article/pii/S0014579305012913
44. Kalra, P., Reddy, T.V., Jayaram, B.: Free energy component analysis for drug design: a case study of HIV-1 protease-inhibitor binding. J. Med. Chem. **44**, 4325–4338 (2001)
45. Latha, N., Jain, T., Sharma, P., Jayaram, B.: A free energy based computational pathway from chemical templates to lead compounds: a case study of COX-2 inhibitors. J. Biomol. Struct. Dyn. **21**, 791–804 (2004)
46. Soni, A., Bhat, R., Jayaram, B.: Improving the binding affinity estimations of protein-ligand complexes using machine-learning facilitated force field method. J. Comput. Aided Mol. Des. **34**, 817–830 (2020). https://doi.org/10.1007/s10822-020-00305-1
47. Bhat, R., Jayaraj, A., Soni, A., Jayaram, B.: An overview of protein-ligand docking and scoring algorithms. In: Gromiha, M. (ed.) Protein Interactions: Computational Methods, Analysis and Applications, pp. 371–389. World Scientific Publishing, Singapore (2019). https://doi.org/10.1142/11596
48. Mukherjee, G., Jayaram, B.: A rapid identification of hit molecules for target proteins via physico-chemical descriptors. Phys. Chem. Chem. Phys. **15**, 9107–9116 (2013) http://pubs.rsc.org/en/Content/ArticleLanding/2013/CP/C3CP44697B
49. Holderbach, S., Adam, L., Jayaram, B., Wade, R.C., Mukherjee, G.: RASPD+: fast protein-ligand binding free energy prediction using simplified physicochemical features. Front. Mol. Biosci. **7**, 601065 (2020). https://doi.org/10.3389/fmolb.2020.601065
50. Banerjee, K., Bhat, R., Rao, V.U.B., Nain, A., Rallapalli, K.L., Gangopadhyay, S., Singh, R.P., Banerjee, M., Jayaram, B.: Towards development of generic inhibitors against the 3C proteases of picornaviruses. FEBS J. **286**(4), 765–787 (2019)
51. Kiruthika, S., Bhat, R., Dash, R., Rathore, A.S., Perumal, V., Jayaram, B.: A novel piperazine derivative that targets hepatitis B surface antigen effectively inhibits tenofovir resistant hepatitis B virus. Sci. Rep. **11**, 11723 (2021). https://doi.org/10.1038/s41598-021-91196-1
52. Kiruthika, S., Bhat, R., Jayaram, B., Perumal, V.: A small molecule targeting hepatitis B surface antigen inhibits clinically relevant drug-resistant hepatitis B virus. J. Antimicrob. Chemother. **77**(8), 2120–2124 (2022)
53. Tripathi, P.K., Soni, A., Yadav, S.P.S., Kumar, A., Gaurav, N., Siva, R.B., Sharma, P., Sunil, S., Jayaram, B., Patel, A.K.: Evaluation of novobiocin and telmisartan for anti-CHIKV activity. Virology. **548**, 250–260 (2020). https://doi.org/10.1016/j.virol.2020.05.010
54. Bhatnagar, S., Soni, A., Kaushik, S., Rikhi, M., Kumar, T.R.S., Jayaram, B.: Non-steroidal estrogen receptor isoform selective biphenyls. Chem. Biol. Drug Des. **91**(2), 620–630 (2018)
55. Kumar, J., Gill, A., Shaikh, M., Singh, A., Shandilya, A., Jameel, E., Sharma, N., Mrinal, N., Hoda, N., Jayaram, B.: Pyrimidine-triazolopyrimidine and pyrimidine-pyridine hybrids as potential acetylcholinesterase inhibitors for Alzheimer's disease. Chem. Select: Med. Chem. Drug Dis. **3**(2), 736–747 (2018)
56. Shandilya, A., Hoda, N., Khan, S., Jameel, E., Kumar, J., Jayaram, B.: De novo lead optimization of triazine derivatives identifies potent antimalarial. J. Mol. Graph. Model. **71**, 96–103 (2017)
57. Nishikawa, J.L., et al.: Inhibiting fungal multidrug resistance by disrupting an activator-mediator interaction. Nature. **530**, 485–489 (2016)
58. Mak, K.K., Wong, Y.H., Pichika, M.R.: Artificial intelligence in drug discovery and development. In: Hock, F.J., Pugsley, M.K. (eds.) Drug Discovery and Evaluation: Safety and Pharmacokinetic Assays. Springer, Cham (2023). https://doi.org/10.1007/978-3-030-73317-9_92-1

59. Blanco-González, A., Cabezón, A., Seco-González, A., Conde-Torres, D., Antelo-Riveiro, P., Piñeiro, A., Garcia-Fandino, R.: The role of AI in drug discovery: challenges, opportunities, and strategies. Pharmaceuticals. **16**(6), 891 (2023). https://doi.org/10.3390/ph16060891
60. Walters, W.P., Barzilay, R.: Critical assessment of AI in drug discovery. Expert Opin. Drug Discov. **16**(9), 937–947 (2021). https://doi.org/10.1080/17460441.2021.1915982
61. Paul, D., Sanap, G., Shenoy, S., Kalyane, D., Kalia, K., Tekade, R.K.: Artificial intelligence in drug discovery and development. Drug Discov. Today. **26**(1), 80–93 (2021). https://doi.org/10.1016/j.drudis.2020.10.010
62. Dara, S., Dhamercherla, S., Jadav, S.S., Babu, C.M., Ahsan, M.J.: Machine learning in drug discovery: a review. Artif. Intell. Rev. **55**(3), 1947–1999 (2022). https://doi.org/10.1007/s10462-021-10058-4
63. Abbas, M., Rassam, A., Abunora, R., Karamshahi, F., Abouseada, M.: The role of AI in drug discovery. Chembiochem. **25**, e202300816 (2024). https://doi.org/10.1002/cbic.202300816
64. Shaikh, S., Jain, T., Sandhu, G., Latha, N., Jayaram, B.: A physico-chemical pathway from targets to leads. Curr. Pharm. Des. **13**, 3454–3470 (2007)
65. Mukherjee, G., Gupta, P.L., Jayaram, B.: Predicting the binding modes and sites of metabolism of xenobiotics. Mol. BioSyst. **11**, 1914–1924 (2015). https://doi.org/10.1039/C5MB00118H
66. Sharma, A.K., Srivastava, G.N., Roy, A., Sharma, V.K.: ToxiM: a toxicity prediction tool for small molecules developed using machine learning and chemoinformatics approaches. Front. Pharmacol. **8**, 880 (2017). https://doi.org/10.3389/fphar.2017.00880
67. Tran, T.T.V., Wibowo, A.S., Tayara, H., Chong, K.T.: Artificial intelligence in drug toxicity prediction: recent advances, challenges, and future perspectives. J. Chem. Inf. Model. **63**(9), 2628–2643 (2023). https://doi.org/10.1021/acs.jcim.3c00200
68. Tonoyan, L., Siraki, A.G.: Machine learning in toxicological sciences: opportunities for assessing drug toxicity. Front. Drug Discov. **4**, 1336025 (2024). https://doi.org/10.3389/fddsv.2024.1336025

Elucidation of the Binding Modes of Known Potent Inhibitors at Mycobacterial F_OF_1-ATP Synthase Subunits *c* and *ε*

Smriti Bhardwaj[1], Rishitha Gundala[2], and Kuldeep K. Roy[1] (✉)

[1] School of Health Sciences and Technology, UPES, Dehradun, Uttarakhand, India
[2] Department of Pharmacoinformatics, National Institute of Pharmaceutical Education and Research, Kolkata, West Bengal, India
[3] School of Health Sciences and Technology, UPES, Dehradun, Uttarakhand, India[3]

Abstract The F_OF_1-adenosine triphosphate (ATP) synthase is a classical anti-tubercular target that has led to Bedaquiline (BDQ), marketed as Sirturo® in 2012 for the treatment of tuberculosis (TB). A co-crystal structure of BDQ bound within the *c* subunit of the rotary ring of *Mycobacterium phlei* F_OF_1-ATP synthase was reported in 2015. Meanwhile, a hypothesis was made for its potential binding at another site in the *ε* subunit of ATP synthase that needs further exploration. In view of the rapidly emerging drug resistance cases, a molecule targeting multiple sites of ATP synthase is highly desirable to cope with drug resistance. In fact, a thorough understanding of the F_OF_1-ATP synthase structure, various binding sites, and structural insights useful for the design of new potent inhibitors are much desired. Therefore, we performed a systematic assessment of the recently resolved structures of subunits *c* and *ε* of mycobacterial ATP synthase and explored the binding characteristics of few known potent F_OF_1-ATP synthase inhibitors at the BDQ binding site (i.e., rotary ring, subunit *c*) and the hypothesized druggable subunit *ε* site. At the subunit *c* site, we observed the importance of GLU65, TYR68, LEU63, and PHE69 amino acids of protein that were in direct or water-mediated contacts with the known inhibitors, e.g., the hydroxyl and NH groups of BDQ showed hydrogen bonding and salt-bridge interactions that corroborated well with the X-ray-based information. Alongside, other inhibitors fitted well at the subunit *c* site and showed a combination of hydrophobic and hydrophilic interactions similar to BDQ but with different affinities. On the other hand, at the subunit *ε* site, we delineated the novel potential binding characteristics of these known inhibitors. Here, in the case of subunit *ε* site 1, ARG26, GLU31, and ARG115 were found to be important residues that were

A. K. Saxena, S. Nandi (eds.), *Global Trends in Health, Technology and Management*, https://doi.org/10.1007/978-3-031-75457-9_5

involved in either H-bond or salt-bridge interactions. Overall, the structural insights gained in this study may be useful in the design of new small molecule inhibitors targeting these two sites of the mycobacterial F-type ATP synthase enzyme.

Keywords ATP synthase · Docking · Tuberculosis · Inhibition · Drug target

1 Introduction

Tuberculosis (TB) is a globally leading infectious disease caused by *Mycobacterium tuberculosis* (Mtb) that mainly affects the lungs. By 2035, the World Health Organization's (WHO) End TB policy seeks to cut 90% of new cases and 95% of TB-related fatalities. The only way to hit this goal is to improve diagnosis, preventative measures, and therapeutic approaches [1]. The rapid emergence of anti-tubercular drug-resistant strains and patient non-compliance with a combination of treatments used for TB therapy are the primary reasons why TB has become a serious global health concern. This suggests that there is a pressing need to find new medications with unique modes of action, medications that can target several pathways, or drugs that may simultaneously block several underlying biological targets [2]. In order for Mtb cells to survive and grow, ATP is needed for a number of processes [3]. Targeting multiple F_OF_1-adenosine triphosphate (ATP) synthase sites may result in a more comprehensive inhibition of ATP synthase's activity, which would more severely impair Mtb's energy metabolism. Furthermore, inhibitors that target several sites may work synergistically to increase the potency of the treatment. When a drug targets multiple sites, it reduces the likelihood that bacteria will develop resistance [4, 5].

Mycobacterial electron transport chain includes a group of membrane-bound enzymes that link oxidation of electron donors from central carbon metabolism to oxygen reduction as a terminal electron acceptor. Many of these enzymes pump protons, making it possible to create a proton motive force (PMF), which is subsequently utilized in oxidative phosphorylation to produce ATP [6]. ATP synthase in Mtb has been shown to be a druggable target and is necessary for the organism's growth. Additionally, it helps the human host's potential stress tolerance [7, 8]. One crucial distinguishing trait of bacteria is their cell wall. It is necessary for maintaining structural integrity and offering defense against antibiotics or the host immune system. ATP is required for the synthesis of the cell wall, which plays an important role in Mtb's virulence, especially mycolic acids, one of its compositions [9, 10]. Furthermore, Bedaquiline (BDQ) inhibits the synthesis of ATP by targeting ATP synthase, which ultimately results in its death. The physiology, persistence, and virulence (pathogenicity) of mycobacteria depend on the production of ATP [11]. The enzyme aids in the bacteria's resistance to the host immune system, their capacity to survive in unfavorable conditions, and their ability to continue with cellular functions essential for infection and disease development.

In 2012, because of research on the clinically proven anti-tubercular target F_OF_1-adenosine triphosphate (ATP) synthase, BDQ (Mtb ATP synthase IC_{50} = 20–25 nM; MIC = 0.002–0.013 μg/ml) was developed and released under the Sirturo® brand after the approval of the US Food and Drug Administration (FDA). BDQ is the revolutionary drug discovered by Koen Andries and co-workers at the Janssen Pharmaceuticals company; it received approval exclusively for the treatment of multidrug-resistant (MDR) and extensively drug-resistant (XDR) forms of TB [12, 13]. Genetic, biochemical, and binding assay methods revealed Mtb ATP synthase as the target of BDQ [14]. ATP synthase is an enzyme involved in the generation of ATP in the last step of the respiratory chain pathway. ATP synthase is located in the mitochondria and is comprised of F_O and F_1 subunits [15]. The F_O complex is the transmembrane portion, comprising three types of subunits *a*, *b*, and *c* in the ratio of *a: b: b': 9c*. On the other hand, the F_1 complex in the cytoplasm is composed of five types of subunits, namely, α, β, γ, δ, and ε, in the ratio of 3α: 3β: γ: δ: ε in mycobacterial species. The nine *c* subunits in *Mycobacterium* form a *c*9 rotary ring complex. The *b* and *b'* subunits of F_O and the δ subunit of F_1 form the peripheral stalk, while the γ and ε subunits of the F_1 complex form the central stalk. The peripheral stalk serves as a support connecting the transmembrane F_O and cytoplasmic F_1 parts of this enzyme. The central stalk transmits the signal from the rotary ring to the catalytic site. The F_O subunit functions in analogy to an electric motor, through which H^+ ions flow across the membrane according to their concentration gradient. The flow of H^+ ions enables the rotation of the F_O rotor ring. The α and β subunits of the F_1 complex form a catalytic head where ATP is synthesized. BDQ is known to bind at the *c* subunit (*c*9 rotary ring) of the mycobacterial F-type ATP synthase and inhibit the rotation of the F_O motor; thereby, it stops the production of ATP required for mycobacterial growth and survival [13].

Nonetheless, BDQ acts as an ionophore with the potential to stimulate respiration and has the potential to cause cardiac problems through QT prolongation [6]. Moreover, several mutations in mycobacterial strains have contributed to the development of BDQ resistance; most of these mutations are located in the subunit *c* (rotor ring) of F_OF_1-ATP synthase [6]. TBAJ-876 (*M. bovis* ATP synthase $IC_{50} = 0.031$ nM), a BDQ homologue, selectively targets the inhibition of mycobacterial ATP synthase and is under exploration in the preclinical development stage. It was very recently reported that BDQ and TBAJ-876 also show interactions with the ε subunit in the F_1 complex of the mycobacterial F-type ATP synthase [16]. A co-crystallized structure of BDQ with the *c* subunit of F-type ATP synthase was reported in 2015, which highlighted the binding site of BDQ [13]. A hypothesis was made for its binding at the ε subunit as well [17], which needs further exploration. In fact, a thorough understanding of the structure of F_OF_1-ATP synthase, various binding sites, and structural insights useful in the design of new potent inhibitors are highly desirable. Therefore, we performed a systematic assessment of the recently resolved experimental structures of subunits *c* and ε of mycobacterial ATP synthase. We then explored the inhibitor binding characteristics of a few classical F_OF_1-ATP synthase inhibitors, BDQ, TBAJ-876, CDRI compound, and squaramide derivative, at the well-known BDQ binding site, i.e., rotary ring subunit *c* (BDQ binding site), and the hypothesized druggable subunit ε site. Figure 1 depicts the chemical structures of the known Mtb ATP synthase inhibitors. TBAJ-876 was discovered in the process of finding a molecule retaining the activity of BDQ and simultaneously reducing the side effect (cardiac QT prolongation) of BDQ [18]. In 2015, the Council of Scientific and Industrial Research-Central Drug Research Institute (CSIR-CDRI), Lucknow, reported a quinolone class of aryl sulfonamide derivative 1 [IUPAC name: *N*-(7-chloro-2-methylquinolin-4-yl)-

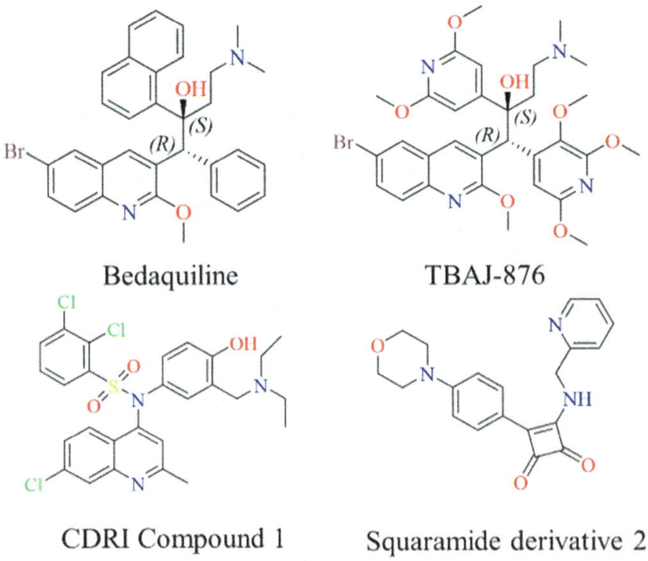

Fig. 1 Two-dimensional (2D) structures of known diverse inhibitors of mycobacterial F-type ATP synthase

N-(3-((diethylamino)methyl)-4-hydroxyphenyl)-2,3-dichlorobenzenesulfonamide] as a selective and potent inhibitor of the mycobacterial ATP synthase (*M. smegmatis* ATP synthase $IC_{50} = 0.51$ μM) [19]. Furthermore, in 2016, through high-throughput screening studies, a squaramide derivative 2 [IUPAC name: 3-(4-morpholinophenyl)-4-((pyridin-2-ylmethyl)amino)cyclobut-3-ene-1,2-dione] with Mtb ATP synthase IC_{50} of 0.03 μM was identified as a potent inhibitor of mycobacterial growth [20]. Afterward, mutation studies revealed ATP synthase as its molecular target, and authors reported that this novel class of compound may bind to a site different from BDQ (subunit *c*) or may have an additional binding site within Mtb ATP synthase, probably subunit *ε*; it needed further research. All the inhibitors described target the ATP synthase enzyme, which is crucial for the energy production and survival of *M. tuberculosis*. The compounds effectively prevent bacterial growth and viability by interfering with this process. While new compounds like TBAJ-876 demonstrated promise in clinical trials, BDQ continues to be the cornerstone of ATP synthase-targeted tuberculosis treatment. In comparison to human ATP synthase, the lead CDRI compound demonstrated potency and selectivity for mycobacterial ATP synthase, lowering the possibility of toxicity and adverse effects. The inhibitor demonstrated efficacy against both replicating (active) and non-replicating (dormant) strains of *Mycobacterium tuberculosis*. This combined action is required for treating persistent tuberculosis infections and averting recurrence. Squaramide demonstrated efficacy against *Mycobacterium tuberculosis* both in vitro and in vivo, as well as selectivity for the enzyme.

2 Materials and Methods

2.1 Protein Data

The X-ray crystal structure of BDQ-bound F_O ATP synthase (rotary ring) (PDB ID: 4V1F) (*Mycobacterium phlei*) [13] and the NMR structure of mycobacterial ATP synthase subunit ε (PDB ID: 5YIO) (*Mycobacterium tuberculosis*) [16] were downloaded from the RCSB Protein Data Bank (PDB) (www.rcsb.org) [21].

2.2 Protein and Ligand Preparation

2.2.1 Protein Preparation

The process of protein preparation entails modeling the missing residues, fixing the incomplete side chains and loops, H-bond optimization, protein minimization, and retaining or eliminating the water molecules and ions. Protein preparation was carried out using the Protein Preparation Wizard [22] in the Schrödinger software. The protein structure was preprocessed by assigning bond orders, addition of hydrogens, creating zero-order bonds to metals, creating disulfide bonds, deleting water molecules beyond 5 Å from the bound ligand, and generating ionization states using Epik at pH 7.0 ± 2.0. The protein structure was refined by optimizing the structure by sampling water orientations. PROPKA pH was set at 7.0, water with less than three H-bonds to non-waters was removed, and the root mean square deviation (RMSD) of the heavy atoms was converged to 30 Å using restrained minimization. The co-crystallized BDQ molecule along with two water molecules in the structure (PDB ID: 4VIF) was retained, while other unwanted residues were removed.

2.2.2 Ligand Preparation

Ligand preparation involves the generation of low-energy 3D conformers of the ligand. It helps explore various possible chemically/scientifically correct structures a ligand could sample, viz., ionization/tautomeric states. Stereoisomers as variations may lead to different results. LigPrep [23] was used to accomplish this task. Ligands, namely, BDQ, TBAJ-876, CDRI lead compound, and squaramide compound, were sketched in the 2D sketcher and subjected to ligand preparation. Ionization states were generated at pH 7.0 ± 2.0 using the Epik module for BDQ and TBAJ-876, while the ionizer module was used for the squaramide derivative and CDRI lead compound. It was because the default Epik module failed to generate all the possible ionization states. Stereoisomers were computed by retaining the specified chirality or varying other chiral centers. Output was saved in the maestro (.mae) file format.

2.3 Ligand Site Mapping and Analysis

When the co-crystallized protein-ligand complex is not provided and there is no experimentally demonstrable literature on the protein's binding site, the only option left is to computationally forecast the potential active site. SiteMap is one such algorithm implemented in the Schrödinger suite, which predicts the possible binding sites of the protein. The NMR structure of ATP synthase subunit ε (PDB ID: 5YIO) lacked any co-crystallized ligand. Therefore, using the SiteMap, potential ligand binding pockets were predicted, and sites with the highest score and other scores were chosen for further docking studies.

2.4 Glide Docking

2.4.1 Grid Generation

The grid defines the area of the protein (or macromolecule) where the ligand is to be docked. The protein was included in the workspace, and the co-crystallized ligand was selected to define the binding site for the generation of the receptor grid. Glide grid files of the subunit *c* were generated considering the co-crystallized ligand with default Van der Waals (vdW) radius scaling (scaling factor 1.0 and partial charge cutoff of 0.25) and grid parameters. On the other hand, in the case of ATP synthase subunit ε, the amino acids as a part of the detected binding site were used to generate the receptor grid.

2.4.2 Cross-Docking and Scoring

The glidescore (gscore) of the co-crystallized BDQ was generated by rescoring. Rescoring was performed by scoring the ligand in place through standard precision (SP) docking. In addition, it was essential to evaluate the Glide algorithm for the reproducing the binding pose close to the co-crystallized one. The sketched BDQ was subjected to flexible SP docking (Glide) and was compared with the co-crystallized conformation. The minimized protein file was split into ligands, water, and others. The generated receptor grid file was selected; for rescoring, the extracted BDQ conformation in the project table was selected, and for flexible SP, the ligand (energy-minimized conformation) to be docked was selected from the project table. The default settings (do not score ligands and rotatable bonds with 500 atoms and 100 rotatable bonds) were retained. Other default settings, *viz*., a scaling factor of 0.80 and a partial charge cutoff of 0.15, were retained.

2.4.3 Flexible Ligand Docking

Standard precision docking was performed by default settings (flexible sampling of ligands, nitrogen inversions, and ring conformations were sampled, and the torsions were bias sampled by penalizing non-polar conformations of amides). The Epik state penalties were added to the docking score, and no constraints were applied. A pose viewer file was generated, "100" poses per ligand, and post-docking minimization was performed, including "10" poses per ligand.

2.5 Induced Fit Docking (IFD)

Unlike the Glide docking, where the receptor is treated as a rigid entity, induced fit docking [24] is a ligand-flexible protein approach where both the ligand and protein are flexible. It considers all possible conformational changes that may occur in the protein's active site upon ligand binding. IFD helps to identify different sizes of the binding site in the protein, and the corresponding changes in the conformations of binding site residues can be visualized. IFD is an integrated tool for both Glide and Prime applications. In this application, first, the active ligands are docked using the Glide algorithm, and diverse ligand poses are generated by reducing the vdW radius and increasing the Coulomb-vdW cutoff. Afterward, Prime structure refinement reorients the side chains of the protein in accordance with the ligand pose. The prepared ligands (BDQ, TBAJ-876, CDRI lead compound, and squaramide derivative) were selected from the project table and docked into the subunit *c* binding site (BDQ site) and predicted sites, *viz.*, site 1 of the *ε* subunit. The standard IFD protocol was followed, which generated up to 20 poses per ligand. The grid box center for docking was the centroid of workspace ligand (BDQ) for subunit *c*, while, on the other hand, residues for *ε* subunit site 1 (22, 24–26, 31, 32, 48–51, 53, 60, 62, 112, 115, 116, 120) were defined for grid box generation. Box size was set similar to the size of the workspace ligand for subunit *c*, and a size of ≤20 Å was set for docking at subunit *ε*. Ligands' ring conformations were sampled with an energy window of 2.5 kcal/mol, and non-polar conformations of amide bonds were penalized. Glide docking was carried out with protein preparation constrained refinement, with a default value of 0.50 for receptor and ligand vdW scaling, followed by Prime refinement with the default side chain optimization algorithm (refine residues within 5 Å of ligand poses). Standard precision Glide re-docking settings were applied to re-dock ligands into structures within the 30 kcal/mol energy threshold of the best structure and within the top 20 of overall structures.

2.6 Molecular Docking

Docking is a computational technique that is used to investigate the potential orientations and conformations of two interacting partners, such as protein-ligand, protein-antigen, or protein-protein combinations. As stated differently, docking is the study of the mutual fit between two or more molecular structures, such as a drug, protein, or nucleic acid. The docking study was carried out using various tools implemented in the Schrödinger drug discovery suite. Maestro, a molecular modeling interface implemented in the Schrödinger suite, was used to visualize the structure and do docking and other related calculations [25].

3 Results

3.1 Ligand Ionization and Binding Site Identification

Bedaquiline, TBAJ-876, CDRI lead compound, and squaramide derivative were the ligands used in the study. The various ionization states of prepared ligands used in the study are depicted in Fig. 2. The reported active stereoisomers (R, S) of BDQ and TBAJ-876 were considered, while, on the other hand, the ionizer generated all the possible ionization states for the squaramide derivative. All the possible ionization states

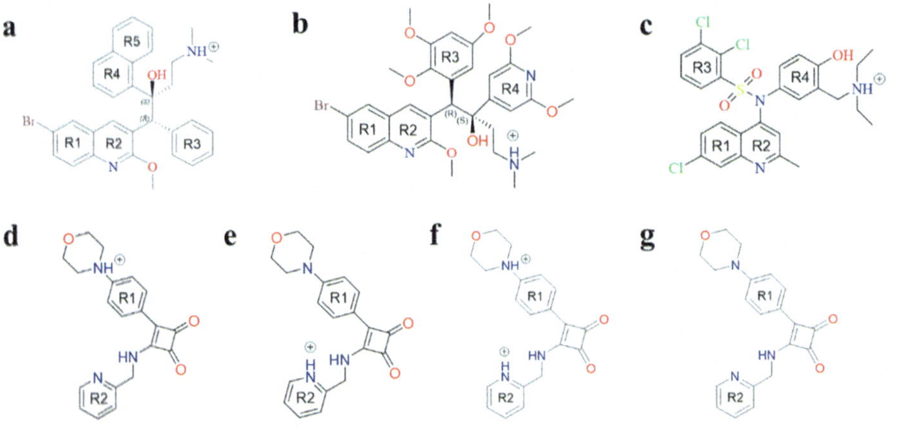

Fig. 2 Ionization states of the candidate drugs/compounds in the study: (**a**) Bedaquiline, (**b**) TBAJ-876, (**c**) CDRI lead, (**d**) ionization state 1 (IS1) ionized morpholine ring nitrogen, (**e**) ionization state 2 (IS2) ionized pyridyl ring nitrogen, (**f**) ionization state 3 (IS3) morpholine and pyridyl ring nitrogen ionized, and (**g**) ionizer state 4 (IS4) non-ionized nitrogen

for the inhibitors (Fig. 1) were generated using the ionizer (Schrödinger suite). For the *c* subunit, the co-crystallized BDQ conformation was used for grid generation; for the *ε* subunit, the five potential binding sites were mapped using the SiteMap (Schrödinger suite). The amino acids composing these five sites are summarized in Table 1. Among the five sites, the most favorable binding site identified was site 1 as it showed the highest SiteScore and Dscore. This site, located between the α-helix and β-sheath of the *ε* subunit, was used for grid generation. Therefore, site 1 of the *ε* subunit was selected for further study.

3.2 Docking at Mtb ATP Synthase Subunit *c*

The amino acid residues within 4 Å of BDQ in subunit *c* included hydrophobic and negatively charged amino acids and two water residues mediating hydrogen bonding. The BDQ binding conformations generated by Glide SP docking were quite similar to the co-crystallized conformation with negligible RMSD; it thus confirmed the reliability of Glide SP docking for other ligand molecules in the study. Figure 3 depicts the 3D interaction diagram representing the binding poses of ligands at the *c* subunit binding site, in which Fig. 3a depicts the comparison of the docked and co-crystallized conformations of BDQ bound with the ATP synthase subunit *c*. The docking score of BDQ co-crystallized conformation up on SP rescoring was found to be −5.560 kcal/mol, while the induced fit docking showed a docking score of −5.613 kcal/mol and IFD score of −519.11 kcal/mol. In an interaction-based study, it was observed that TBAJ-876 interacted in a similar way to BDQ. For instance, the chiral hydroxyl (-OH) group is engaged in water-mediated H-bond formation with GLU65; the dimethyl amino (DMA) group forms H-bond and salt-bridge interactions with GLU65. Furthermore, π-cationic interaction of pyridine with PHE69 was observed with TBAJ-876 but was

Table 1 Representation of the SiteMap-generated sites for the ε subunit of ATP synthase

Sites	List of amino acids	SiteScore	Size	Dscore
Site 1	PHE22, PHE24, THR25, ARG26, GLU31, LEU45, VAL46, ASP47, ASP48, ALA49, MET50, VAL51, VAL53, ARG55, GLY57, GLU58, LYS59, ASP60, ARG62, ILE63, VAL65, ALA112, ARG115, ALA116, ILE120	0.939	195	1.045
Site 2	VAL9, ALA10, VAL11, ASP12, ARG13, ASN14, ILE15, ARG26, THR28, GLY30, GLU31, GLN44, LEU45, VAL46, ASP47, ASP66, GLY67, GLY68, PHE69, LEU80, ALA81, GLU82, SER83, ALA84, GLU85, PHE86, GLU89, ALA93, ALA94, GLN97, ASP98, GLU100, ASP102, ASP103, PRO104, ARG105, ILE106, ARG109, ARG113	0.869	320	0.991
Site 3	MET1, ALA2, GLU3, LEU4, ASN5, THR19, ALA20, LYS21, ARG26, LYS35, PRO36, ARG37, HIS38, ASP48, GLU54, ARG55, GLU56, GLU58, LYS59, ASP60, LEU61, ARG62, GLU74, PHE86, GLU87, ILE90, ASP91, GLU92, ALA95, LYS96, SER99, SER101, ALA107, ALA108, GLY110, ARG111, ALA112, LEU114, ARG115, VAL117, ALA119, ILE120	0.856	457	0.975
Site 4	THR28, VAL29, ILE32, GLY33, ILE34, HIS38, LEU41, VAL42, ALA43, GLN44, PHE69, VAL72	0.757	66	0.871
Site 5	ASN5, GLU7, THR73, GLU75, GLY76, SER78	0.507	27	0.534

absent in BDQ because of the lack of a cationic center (phenyl group in BDQ). This led TBAJ-876 to exhibit a better IFD score than BDQ (IFD score of TBAJ-876, −526.39 kcal/mol; BDQ, −519.11 kcal/mol). In the case of the CDRI compound, the hydroxyl group also showed a similar kind of water-mediated H-bond formation with the backbone keto of LEU63 and π-π interaction between the R3 ring (2,3-dichlorophenyl) and the PHE69 residue. It showed an IFD score slightly better than BDQ (IFD score, −520.75 kcal/mol), but a smaller number of direct interactions than BDQ or TBAJ-876. The better score may be attributed to a stronger hydrophobic interaction. In the case of the squaramide ionization state, IS (both morpholine and pyridyl ionized), there was H-bond formation between GLU65 and squaramide –NH, while there was a water-mediated H-bond between GLU65 and squaramide –NH in the case of the IS4 structure. The π-π and π-cationic interactions were formed between TYR68 and ionized pyridyl (R2) nitrogen in IS3 and IS2 states, respectively. Despite different conformations and binding poses observed with three ionization states of squaramide, the ionized nitrogen formed H-bond and salt-bridge interactions with GLU65; ionized nitrogen and –NH groups in the squaramide, to an extent, mimicked the roles of -OH and -DMA in BDQ and TBAJ-876. They were forming a similar kind of interaction with GLU65 and TYR68 at the subunit *c* binding site. Although the ionized and non-ionized states were showing a similar IFD score range, IS1 (ionized morpholine nitrogen) had a better IFD score (−518.86 kcal/mol) compared to other states. The observed order of IFD scores was IS (nitrogen ionized) (−516.75 kcal/mol) < IS2 (pyridyl ring nitrogen ionized) (−515.29 kcal/mol) < IS4 (non-ionized) (−514.59 kcal/mol). Overall, the

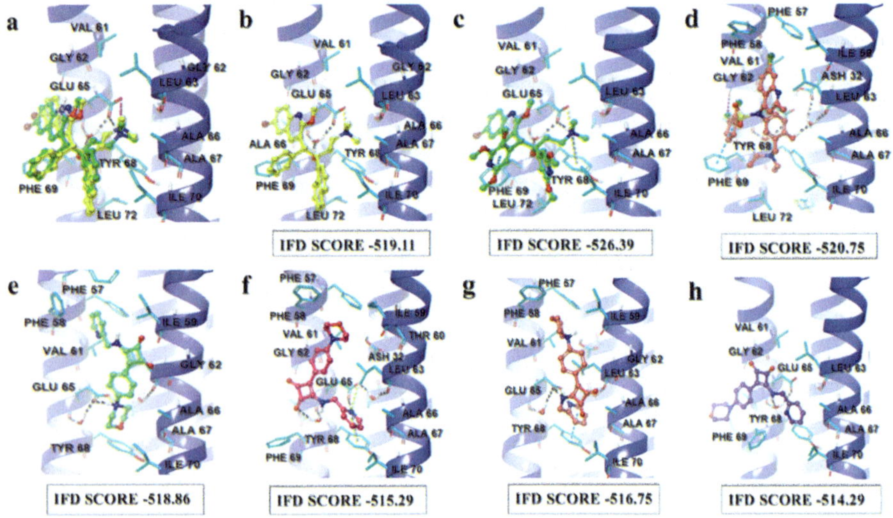

Fig. 3 3D interaction diagram representing binding poses of ligands at *c* subunit binding site, backbone helices shown as ribbon and residues in stick representation. (**a**) Superimposed image of Bedaquiline co-crystallized structure and conformation generated by Glide, (**b**) binding site of BDQ, (**c**) TBAJ-876, (**d**) CDRI lead compound, (**e**) squaramide derivative IS1 (morpholine nitrogen ionized), (**f**) squaramide derivative IS2 (pyridyl nitrogen ionized), (**g**) squaramide derivative IS3 (morpholine and pyridyl nitrogen ionized), and (**h**) squaramide derivative IS4 (non-ionized)

squaramide derivatives except IS4 (non-ionized form) and the CDRI compound showed a different binding site with reference to BDQ and TBAJ-876 (Fig. 3). Docking results at the ATP synthase subunit *c* emphasized the importance of the amino acid GLU65 and the water molecule in establishing protein-ligand interactions in all cases.

In addition, the diethylamino (-DEA) group in the CDRI compound formed H-bond with ASP48 and the sulfonyl oxygen H-bonded with ARG115. In the squaramide derivative IS1 (morpholine ionized), the amine group (-NH) formed H-bond with ASP48, while, on the other hand, in the case of IS2 (ionized pyridyl nitrogen), R2 nitrogen established H-bond and a salt bridge with ASP60. In IS3 (both ionized), the ionized nitrogen of the morpholine ring and GLU31 established a salt-bridge interaction, and the pyridyl nitrogen formed a salt bridge with ASP60 and H-bond with LYS59. In IS4 (non-ionized), the squaramide morpholine ring oxygen (2-one of squaramide) formed H-bond with ARG62, and NH formed H-bond with ASP48 (Fig. 3).

3.3 Docking at Mtb ATP Synthase Subunit ε Site 1

On the other hand, subunit ε site 1 was lined by amino acids of different characteristics, such as hydrophobic, polar, positively charged, and negatively charged, and a glycine residue (Fig. 4). In the case of BDQ, the hydroxyl group formed H-bond with ARG115, the DMA group formed H-bond with ALA116 (backbone keto) and a salt bridge with ASP60, and the R1 ring of quinolone formed a π-cationic interaction with ARG115.

TBAJ-876 forms interactions with multiple residues. For example, both the R1 and R2 rings formed π-π interactions with PHE24; the -OH group formed H-bond interaction with GLU31; the DMA group formed a salt-bridge interaction. The 2-methoxy group of the R3 (pyridine) ring formed H-bond with ARG26, and the nitrogen of the R4 ring (pyridine) formed H-bond with ARG115. At site 1 of the *ε* subunit, unlike at the *c* subunit, BDQ and TBAJ-876 interaction modes were noticeably different. Hence, it may show different activity targeting this site 1 of ATP synthase.

4 Discussion

ATP synthase in *Mycobacterium* species has proven to be a druggable target for BDQ. The ATP synthase or F_1F_O ATPase utilizes the electrochemical gradient of protons. These protons are the result of the respiratory chain, which further synthesizes ATP. The highly conserved enzymes comprising ATP synthases are present in the bacterial cytoplasmic membranes and inner membranes of the mitochondria themselves. The ATP synthase is a rotating motor that has a membrane-bound hydrophobic subunit as F_O that protrudes out with the hydrophilic subunit as F_1 in the mitochondrial matrix. Due to the concentration gradient of H^+ ions, the F_O subunit rotates, and the protons flow inside the mitochondrial matrix, and ATP generates. BDQ binds to the *c* unit, a part of F_O. This inhibits the rotation of the subunit, inhibiting the rotation of the hydrophobic part, leading to no ATP production inside the cell that ultimately causes the death of the

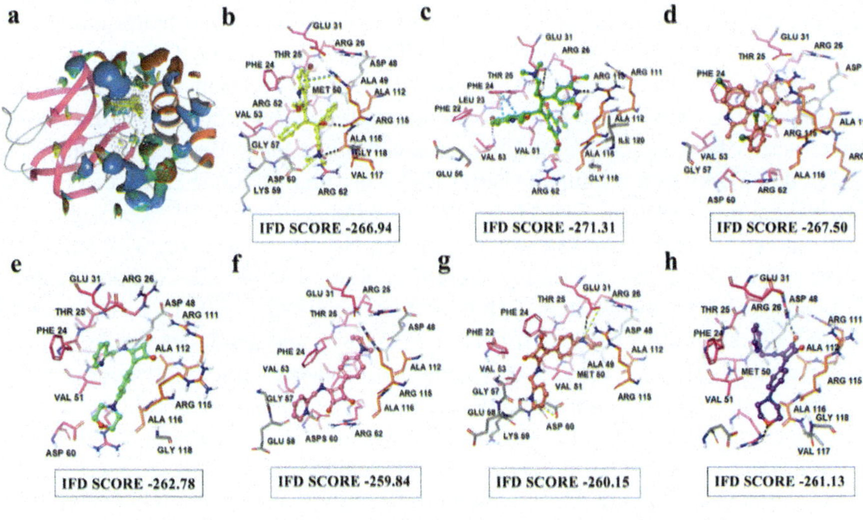

Fig. 4 3D interaction diagram representing binding poses of ligands at SiteMap (site 1) of *ε* subunit. (**a**) *ε* subunit with the SiteMap-generated site, (**b**) BDQ, (**c**) TBAJ-876, (**d**) CDRI lead compound, (**e**) squaramide derivative IS1 (morpholine nitrogen ionized), (**f**) squaramide derivative IS2 (pyridyl nitrogen ionized), (**g**) squaramide derivative IS3 (morpholine and pyridyl nitrogen ionized), and (**h**) squaramide derivative IS4 (non-ionized)

organism. It was hypothesized that BDQ also binds to the *ε* subunit, a part of F_1, which needs further evaluation. Resistant cases against BDQ have been reported. Multiple binding sites of the drugs are highly required to combat the resistant forms of Mtb.

We performed a systematic assessment of the recently resolved structures of subunits *c* and *ε* of mycobacterial ATP synthase. In addition, we explored the inhibitor binding characteristics of a few known small molecules as F_OF_1-ATP synthase inhibitors at the BDQ binding site, which is the subunit *c* of F_O, and the hypothesized druggable subunit *ε* site. We focused on three molecules for this study. Firstly, TBAJ-876 is a pre-clinical candidate as an anti-mycobacterial agent, which is a structural homologue of BDQ. It was discovered to retain the BDQ activity but reduce the side effects of QT prolongation. The second molecule is a CDRI lead compound, which is a quinolone class of arylsulfonamides 1, and the third is a squaramide derivative. All three compounds were reported to target ATP synthase in *Mycobacterium* species. We utilized several methods to study the resolved structures and understand binding sites within the enzyme, including the *c* and *ε* subunits. Proteins and ligands were prepared and docked. Site mapping and analysis were performed to predict the possible binding sites in the protein. Induced fit docking (IFD) was performed, which provided flexibility to the proteins and ligands while showing different conformational changes during docking. In the case of *c* subunit, the BDQ binding site was used for grid generation, whereas SiteMap was used to predict the binding site in *ε* subunit, for which site 1 was selected. It showed a SiteScore of 0.939 and a Dscore of 1.045. Site 1 included PHE22, PHE24, THR25, ARG26, GLU31, LEU45, VAL46, ASP47, ASP48, ALA49, MET50, VAL51, VAL53, ARG55, GLY57, GLU58, LYS59, ASP60, ARG62, ILE63, VAL65, VAL112, VAL115, VAL116, and VAL120 as the amino acid residues.

The amino acid residues within 4 Å of BDQ in subunit *c* included hydrophobic and negatively charged amino acids and two water residues mediating hydrogen bonding. When docked at *c* subunit binding site of ATP synthase, TBAJ-876 showed similar interactions as of BDQ. The –OH group of TBAJ-876 formed the water-mediated H-bonding with GLU65 and the π-cationic interaction with PHE69. BDQ did not show this interaction because of the absence of a cationic center. Because of this, it was observed that TBAJ-876 exhibited a better IFD score than BDQ (IFD score: TBAJ-876, −526.39 kcal/mol; BDQ, −519.11 kcal/mol), whereas the CDRI compound, when docked, showed a similar kind of water-mediated H-bond interaction with LEU63 and π-π interaction between the R3 ring (2,3-dichlorophenyl) and PHE69 residue. It showed an IFD score slightly better than BDQ (IFD score: BDQ, −519.11 kcal/mol; CDRI lead, −520.75 kcal/mol), but less IFD than TBAJ-876 and a smaller number of direct interactions than BDQ or TBAJ-876. Strong hydrophobic interaction might lead to a better IFD. In the case of the squaramide ionization state, IS (both morpholine and pyridyl ionized), there was H-bond formation between –NH of squaramide and GLU65, whereas a water-mediated H-bond between squaramide –NH and GLU65 was observed in the IS4 structure. The π-π and π-cationic interactions were formed between TYR68 and ionized pyridyl (R2) nitrogen of IS3 and IS2 states, respectively. Instead of different conformations and binding poses of the squaramide derivatives, similar kinds of interactions with GLU65 and TYR68 were observed. The observed order of IFD scores of squaramide derivatives was IS (nitrogen ionized) (−516.75 kcal/mol) < IS2 (pyridyl ring nitrogen ionized) (−515.29 kcal/mol) < IS4 (non-ionized) (−514.59 kcal/mol). The IFD scores of all the ligands reflected that TBAJ-876 (−526.39 kcal/mol) and CDRI lead (−520.75 kcal/mol) showed the better binding affinity. Also, docking followed by an interaction-based study at *c* subunit suggested that GLU65 was observed to be an important residue as it was involved in interaction in all the protein-ligand complexes.

On the other hand, subunit ε site 1 had amino acid residues with different characteristics, such as hydrophobic, polar, positively charged, and negatively charged. Here, ARG115 and ALA116 were found to form hydrogen bonding with BDQ and salt bridge with ASP60, whereas TBAJ-876 was found to form hydrogen bond with GLU31, ARG26, and ARG115. We observed that BDQ and TBAJ-876 at subunit ε site of ATP synthase have shown different interactions, unlike at the *c* subunit site. In addition, the IFD of TBAJ-876 showed the best docking score at −271.34 kcal/mol compared to any other compounds, including BDQ. We summarized that in the case of the subunit ε site 1, both BDQ and TBAJ-876 formed H-bond with ARG115, which might be the key residue for interaction with TBAJ-876 with a better docking score. Therefore, through our study, we found that in the case of subunit *c* site, residues such as GLU65, TYR68, LEU63, and PHE69 amino acids of protein were in either direct or water-mediated contacts with the known inhibitors, whereas, in the case of subunit ε site 1 study, ARG26, GLU31, and ARG115 were found to be important residues that were involved in either H-bond or salt-bridge interactions. The structural insights gained in this study may be useful in the design of new small molecules as inhibitors targeting these two sites of the mycobacterial F-type ATP synthase enzyme combating resistant forms of TB.

5 Conclusions

Tuberculosis is a globally leading age-old infectious disease caused by *Mycobacterium* species. Two major bottlenecks for new TB drug discovery are the permeability of the membrane and rapid emergence of drug resistance. Mycobacterial F_OF_1-ATP synthase is a novel target, leading to the drug Bedaquiline. In this study, upon MSA studies on ε subunit of ATP synthase, we have observed that there is a major sequence divergence between the human and the mycobacterial species. Mtb showed only 10.20% identity in sequences when compared with the human ε subunit. Therefore, it can be a potential target for the TB drug discovery. We have also considered the X-ray crystal structure of the rotary ring part and the NMR structure of ε subunit of mycobacterial ATP synthase. We have studied the interaction patterns of BDQ, TBAJ-876, CDRI compound, and squaramide derivative at the *c* subunit (rotary ring) binding site and at the ε subunit (hypothesized to be another binding site for BDQ and TBAJ-876). The ε subunit exists in two states: compact and extended states. The extended state helps in the transformation of the signal from the F_O to F_1 region, where the formation of ATP takes place (i.e., the alpha and beta subunits). So, it can be hypothesized that binding of the drug at site 1 would inhibit the change of conformation of the ε subunit to the extended state (or enable the ε subunit to remain in the compact state). We found that residues such as ARG26, GLU31, and ARG115 were important in ε subunit as they were found in hydrogen bonding or the salt-bridge interaction with the ATP synthase. The structural insights obtained in the study might be useful in the invention of novel small molecule inhibitors that specifically target these two locations, which are the c and ε subunits in the mycobacterial F-type ATP synthase enzyme.

Acknowledgments The author S. B. is thankful to UPES, Dehradun, for providing doctoral fellowship. R.G. is thankful to NIPER Kolkata for providing fellowship and infrastructural support. The authors S. B. and K. K. R. would like to acknowledge UPES, Dehradun, for providing other necessary infrastructural support.

References

1. Chauhan, A., Parmar, M., Dash, G.C., Solanki, H., Chauhan, S., Sharma, J., Sahoo, K.C., Mahapatra, P., Rao, R., Kumar, R., Rade, K., Pati, S.: The prevalence of tuberculosis infection in India: a systematic review and meta-analysis. Indian J. Med. Res. **157**, 135 (2023)
2. Zhang, J., Lair, C., Roubert, C., Amaning, K., Barrio, M.B., Benedetti, Y., Cui, Z., Xing, Z., Li, X., Franzblau, S.G., Baurin, N., Bordon-Pallier, F., Cantalloube, C., Sans, S., Silve, S., Blanc, I., Fraisse, L., Rak, A., Jenner, L.B., Yusupova, G., Yusupov, M., Zhang, J., Kaneko, T., Yang, T.J., Fotouhi, N., Nuermberger, E., Tyagi, S., Betoudji, F., Upton, A., Sacchettini, J.C., Lagrange, S.: Discovery of natural-product-derived sequanamycins as potent oral anti-tuberculosis agents. Cell. **186**, 1013 (2023)
3. Althaher, A.R., Alwahsh, M.: An overview of ATP synthase, inhibitors, and their toxicity. Heliyon. **9**, e22459 (2023)
4. Stelitano, G., Sammartino, J.C., Chiarelli, L.R.: Multitargeting compounds: a promising strategy to overcome multi-drug resistant tuberculosis. Molecules. **25**, 1239 (2020)
5. Suvvari, T.K.: Exploring the potential of multi-target-directed ligands for tuberculosis treatment. Ther. Adv. Infect. Dis. **10**, 20499361231188039 (2023)
6. McNeil, M.B., Cheung, C.Y., Waller, N.J.E., Adolph, C., Chapman, C.L., Seeto, N.E.J., Jowsey, W., Li, Z., Hameed, H.M.A., Zhang, T., Cook, G.M.: Uncovering interactions between mycobacterial respiratory complexes to target drug-resistant Mycobacterium tuberculosis. Front. Cell. Infect. Microbiol. **12**, 980844 (2022)
7. Vestergaard, M., Bald, D., Ingmer, H.: Targeting the ATP synthase in bacterial and fungal pathogens—beyond Mycobacterium tuberculosis. J. Glob. Antimicrob. Resist. **29**, 29–41 (2022)
8. Lu, P., Lill, H., Bald, D.: ATP synthase in mycobacteria: special features and implications for a function as drug target. Biochim. Biophys. Acta Bioenerg. **1837**, 1208 (2014)
9. Maitra, A., Munshi, T., Healy, J., Martin, L.T., Vollmer, W., Keep, N.H., Bhakta, S.: Cell wall peptidoglycan in Mycobacterium tuberculosis: an Achilles' heel for the TB-causing pathogen. FEMS Microbiol. Rev. **43**, 548 (2019)
10. Ghazaei, C.: Mycobacterium tuberculosis and lipids: insights into molecular mechanisms from persistence to virulence. J. Res. Med. Sci. **23**, 63 (2018)
11. Krah, A., Grüber, G., Bond, P.J.: Binding properties of the anti-TB drugs bedaquiline and TBAJ-876 to a mycobacterial F-ATP synthase. Curr. Res. Struct. Biol. **4**, 278 (2022)
12. Sarathy, J.P., Gruber, G., Dick, T.: Re-understanding the mechanisms of action of the anti-mycobacterial drug bedaquiline. Antibiotics (Basel). **8**, 261 (2019)

13. Preiss, L., Langer, J.D., Yildiz, O., Eckhardt-Strelau, L., Guillemont, J.E., Koul, A., Meier, T.: Structure of the mycobacterial ATP synthase Fo rotor ring in complex with the anti-TB drug bedaquiline. Sci. Adv. **1**, e1500106 (2015)
14. Koul, A., Dendouga, N., Vergauwen, K., Molenberghs, B., Vranckx, L., Willebrords, R., Ristic, Z., Lill, H., Dorange, I., Guillemont, J., Bald, D., Andries, K.: Diarylquinolines target subunit c of mycobacterial ATP synthase. Nat. Chem. Biol. **3**, 323 (2007)
15. Roy, K.K., Wani, M.A.: Emerging opportunities of exploiting mycobacterial electron transport chain pathway for drug-resistant tuberculosis drug discovery. Expert Opin. Drug Discov. **15**, 231 (2020)
16. Joon, S., Ragunathan, P., Sundararaman, L., Nartey, W., Kundu, S., Manimekalai, M.S.S., Bogdanovic, N., Dick, T., Gruber, G.: The NMR solution structure of Mycobacterium tuberculosis F-ATP synthase subunit epsilon provides new insight into energy coupling inside the rotary engine. FEBS J. **285**, 1111 (2018)
17. Sarathy, J.P., Ragunathan, P., Shin, J., Cooper, C.B., Upton, A.M., Gruber, G., Dick, T.: TBAJ-876 retains bedaquiline's activity against subunits c and epsilon of Mycobacterium tuberculosis F-ATP synthase. Antimicrob. Agents Chemother. **63**, 10–1128 (2019)
18. Sutherland, H.S., Tong, A.S.T., Choi, P.J., Blaser, A., Conole, D., Franzblau, S.G., Lotlikar, M.U., Cooper, C.B., Upton, A.M., Denny, W.A., Palmer, B.D.: 3,5-Dialkoxypyridine analogues of bedaquiline are potent antituberculosis agents with minimal inhibition of the hERG channel. Bioorg. Med. Chem. **27**, 1292 (2019)
19. Singh, S., Roy, K.K., Khan, S.R., Kashyap, V.K., Sharma, A., Jaiswal, S., Sharma, S.K., Krishnan, M.Y., Chaturvedi, V., Lal, J., Sinha, S., Dasgupta, A., Srivastava, R., Saxena, A.K.: Novel, potent, orally bioavailable and selective mycobacterial ATP synthase inhibitors that demonstrated activity against both replicating and non-replicating M. tuberculosis. Bioorg. Med. Chem. **23**, 742 (2015)
20. Tantry, S.J., Markad, S.D., Shinde, V., Bhat, J., Balakrishnan, G., Gupta, A.K., Ambady, A., Raichurkar, A., Kedari, C., Sharma, S., Mudugal, N.V., Narayan, A., Naveen Kumar, C.N., Nanduri, R., Bharath, S., Reddy, J., Panduga, V., Prabhakar, K.R., Kandaswamy, K., Saralaya, R., Kaur, P., Dinesh, N., Guptha, S., Rich, K., Murray, D., Plant, H., Preston, M., Ashton, H., Plant, D., Walsh, J., Alcock, P., Naylor, K., Collier, M., Whiteaker, J., McLaughlin, R.E., Mallya, M., Panda, M., Rudrapatna, S., Ramachandran, V., Shandil, R., Sambandamurthy, V.K., Mdluli, K., Cooper, C.B., Rubin, H., Yano, T., Iyer, P., Narayanan, S., Kavanagh, S., Mukherjee, K., Balasubramanian, V., Hosagrahara, V.P., Solapure, S., Ravishankar, S., Hameed, P.S.: Discovery of imidazo[1,2-a]pyridine ethers and squaramides as selective and potent inhibitors of mycobacterial adenosine triphosphate (ATP) synthesis. J. Med. Chem. **60**, 1379 (2017)
21. Berman, H.M., Westbrook, J., Feng, Z., Gilliland, G., Bhat, T.N., Weissig, H., Shindyalov, I.N., Bourne, P.E.: The protein data bank. Nucleic Acids Res. **28**, 235 (2000)
22. Sastry, G.M., Adzhigirey, M., Day, T., Annabhimoju, R., Sherman, W.: Protein and ligand preparation: parameters, protocols, and influence on virtual screening enrichments. J. Comput. Aided Mol. Des. **27**, 221 (2013)
23. LigPrep. Schrödinger, LLC, New York, NY (2024)
24. Induced fit docking protocol. Schrödinger, LLC, New York, NY (2024)
25. Maestro (Version v11.8). Schrödinger, LLC, New York, NY (2023)

Docking-Based Experimental Approach Examining the Effectiveness of Nano-formulations as Diabetes Therapeutics

Rishita Dey[1], Debojyoti Tarafdar[3], Sisir Nandi[2] (✉), and Asmita Samadder[1] (✉)

[1] Cytogenetics and Molecular Biology Lab., Department of Zoology, University of Kalyani, Kalyani, India
[2] Department of Chemistry, Chanchal College (Affiliated to University of Gour Banga), Malda, India
[3] Department of Pharmaceutical Chemistry, Global Institute of Pharmaceutical Education and Research (GIPER) (Affiliated to Veer Madho Singh Bhandari Uttarakhand Technical University), Kashipur, India
[4] Cytogenetics and Molecular Biology Lab., Department of Zoology, University of Kalyani, Kalyani, India
asmitazoo19@klyuniv.ac.in

Abstract Diabetes mellitus is a chronic, metabolic disorder that needs to be addressed globally. Diabetes aids in the onset of several associated disorders, including neuropathy, nephropathy, cardiomyopathy, pulmonary arterial disease, etc. Patients develop a reliance on regular medication or insulin dosage, which triggers these related issues. A theoretical validation using computational docking study-based prediction utilizing phyto-based nutraceuticals offers an edge over conventional treatment methods in preventing such situations. The present study explores the potential benefits of plant-based compounds as replacements or alternatives to synthetic pharmaceuticals to improve therapeutic effectiveness while minimizing side effects. Further, to minimize the problems of solubility and dose of phytocompound during internalization, nanotechnology-based drug delivery was explored. Pelargonidin, a phyto-based nutraceutical belonging to the anthocyanidin group known for its therapeutic properties when encapsulated in a biodegradable poly-lactide-co-glycolic acid (PLGA) polymer, demonstrated increased activity, exceeding that of its non-encapsulated equivalents. Studies using AFM, TEM, and DLS were conducted to determine its physicochemical properties. Therefore, nano-mediated delivery of drugs could improve the solubility, dissolution, retention, and stability of drugs, thereby extending their pharmacological effect. Hence,

A. K. Saxena, S. Nandi (eds.), *Global Trends in Health, Technology and Management*, https://doi.org/10.1007/978-3-031-75457-9_6

computational molecular docking studies of drug-protein interactions emerge as a new field aimed at selecting molecules that act to modulate biotargets, thereby establishing the concept of drug delivery with "target specificity" for combating a variety of diseases including diabetes and its complications.

Keywords Diabetes · Molecular docking · Nanotechnology · Pelargonidin · PLGA

1 Introduction

Type 2 diabetes is a chronic and complicated illness marked by varying degrees of insulin resistance, decreased insulin secretion, and elevated glucose production. Type 2 diabetes management necessitates rigorous commitment, which frequently leads to mental exhaustion and stress. India is the nation holding the second greatest burden of diabetes worldwide, after China. Following the International Diabetes Federation's data on the prevalence of diabetes, there were an estimated 537 million cases of diabetes worldwide among adults aged 20 to 79 years. Based on this summary, 643 million are expected to develop diabetes by 2030, and by 2045, estimates predicted the number of patients at an astounding 783 million [1]. An estimated 25 million Indians are prediabetic, meaning they have a higher chance of getting diabetes in the near future, while an estimated 77 million Indians have diabetes. Thus, nearly 1 out of every 11 Indians above the age of 18 years is believed to have diabetes [2]. Managing this disorder effectively involves a combination of pharmaceutical and multimodal behavioral interventions for maintaining a better quality of lifestyle. An elevated and unmanaged blood glucose level may adversely influence the functionality of several body organs, consequently raising the risk of nephropathy, retinopathy, neuropathy, infarction, hypertension, arteriosclerosis, and stroke [3]. Well-known risk factors, including the implementation of a modern routine, a lifestyle that is sedentary, an absence of vigorous exercise, and energy-rich eating habits, have been implicated in the spike in the occurrence of the disorder [4].

Many synthetic antidiabetic medications, namely, metformin, the first-line oral medication, and empagliflozin, a drug belonging to sodium-dependent glucose transporter 2 (SGLT-2) inhibitor, and several other medications from different classes are recommended as treatments for blood sugar regulation [5]. Despite the wide range of antidiabetic medications available, the majority of patients fail to achieve ideal control of their glucose levels. Furthermore, maintaining glucose control alone is not always enough to mitigate long-lasting consequences of diabetes. However, these artificial antidiabetic medications come with a host of negative side effects and health issues, and they might trigger or worsen other medical conditions. Consequently, these facts are leading to an increasing demand for alternatives to oral antidiabetic medications in order to reduce diabetes and associated side effects [6].

Natural sources are believed to possess "substantial" collections of bioactive ingredients. Therefore, research on plant-based components to ascertain if the bioactive components have antidiabetic characteristics is mostly unexplored [7]. The multifarious potential and structural versatility of bioactive compounds are the main benefits of utilizing phytochemicals for preventing disease progression. It is important to note that though synthetic drugs have a strong affinity for the target protein, the target specificity often enables to achieve a high probability of effectiveness toward a molecular target, as opposed to phytochemicals, whose diversity or selectivity for a multitude of different targets may prevent diabetic individuals from evolving potential resistance mechanisms [8]. However, due to inadequate lipid solubility or inappropriate molecular size, many medicinal products face minimal in vivo effects notwithstanding their great potential in *in vitro* observations. This eventually contributes to poor absorption and poor bioavailability in *in vivo* systems [9]. To address this issue, this study covers nano-based drug delivery as a way to improve the therapeutic effectiveness of drugs when administered *in vivo* by ensuring proper distribution near the area of interest and appropriate utilization by the diseased tissues. Therefore, by strategically targeting the drug at the location of action by utilizing nano-based delivery mechanisms, the drug is more effectively delivered without enduring structural changes owing to pH variations or biochemical breakdown, and unfavorable side effects are reduced. This study investigated the anti-hyperglycemic properties of nanopelargonidin, the active ingredient found in berries. Pelargonidin, a member of the class anthocyanidins, has been the focus of this study owing to its multifaceted qualities, including antioxidant, anti-inflammatory, and anti-genotoxic effects [10]. As shreds of evidence presented the anti-inflammatory, anti-genotoxic, and antioxidative properties of pelargonidin, consequently, repurposing pelargonidin to explore its potential as a hypoglycemic agent will help identify potential leads and open up new avenues for drug development. Additionally, nano-based ways of delivery prohibit uncontrolled drugs from being released at the incorrect location, which would reduce the concentration of medicine that should be present in the target tissue and in turn eliminate enhancing any undesirable adverse response [11].

Choosing bioactive molecules with specific target recognition is one of the key tasks in drug development. To predict the biochemical mechanisms involved, the molecular docking technique is essential to comprehend how the protein and the ligand attach. Therefore, before the experimental investigation, an *in silico*-based docking study was undertaken to check the possible interaction between encapsulated ligand molecule and protein. Based on the dock score metrics for ligand-protein interactions, p53 and PARP-1 proteins were selected as the study's target proteins. p53 plays a significant role in facilitating DNA repair by permitting its repair apparatus to recuperate genomic integrity. Another ubiquitous nuclear protein called PARP-1 detects and aids in damage repair through poly-ADP-ribosylation (PARylation) [12].

Thus, the therapeutic efficacy of this nano-encapsulated pelargonidin having anthocyanidin scaffold was evaluated in this work by examining whether this nanodrug can regulate normal glycemia. Thus, the objectives of this study are as follows: (i) to implement molecular docking to predict the drug-protein interaction based on the dock score value; (ii) to formulate nanoparticles of the docking-based screened compound using PLGA; (iii) to determine whether the nanodrug is capable of reversing glucose imbalance; (iv) to assess whether the nanodrug can modulate antioxidative status; and (v) to identify whether the nanodrug can prevent diabetes-induced damage by modulating repair proteins.

2 Materials and Methods

2.1 Requisite Chemicals and Software for the Study

All the chemicals used during the study were purchased from reputable companies. Sigma-Aldrich (St. Louis, MO, USA) provided pelargonidin (CAS no. 134-04-3), PLGA (CAS no. 26780-50-7), and alloxan (CAS no. 2244-11-3) utilized in this work. The antibodies were purchased from Santa Cruz Biotechnologies, USA. All compounds are of analytical grade and have ≥97% purity. The *in silico* investigation was conducted using a variety of software programs, including ChemDraw Ultra 8.0, Chem3D Ultra 8.0, ArgusLab 4.0.1, and Discovery Studio 3.5 (Dassault Systèmes BIOVIA, Discovery Studio 3.5, San Diego: Dassault Systèmes, 2016). The AFM analysis was conducted in WSxM 5.0 Develop 7.0 software.

2.2 Ligand and Protein Preparation for *In Silico* Study

Molecular docking is a theoretical approach for examining the interaction and recognition of proteins and ligands. This technique predicts the binding manner and binding strength, by visualizing the important residues of amino acids linking the ligands and receptor biomacromolecules at the active site. ChemDraw Ultra 8.0 was utilized for drawing the ligand pelargonidin, an anthocyanidin (PubChem CID: 440832) (Fig. 1a). The basis of anthocyanidins is the flavylium cation, an oxonium ion with different groups in place of its hydrogen atoms. Flavylium cation is the chromenylium cation with a phenyl group substituted at position 2. Positions 3, 5, 7, and 4′ of the anthocyanidin cation, pelargonidin, are flavylium replaced by hydroxy groups [13]. Chem3D Ultra 8.0 was used for converting this 2D image into its 3D form. Using an MM2 force field with a dielectric constant of 0.01, the energy minimization to achieve the correct geometrical alignment of the ligand was carried out. Therefore, the 3D energy minimized structure of pelargonidin (Fig. 1b) thus obtained was applied for the docking program. The preparation of the 3D configuration of the p53 and PARP-1 was done by extracting the proteins from the Protein Data Bank having PDB ID 1YC5 and 3GJW, respectively [14, 15] (Fig. 1c, d). The addition of hydrogen atoms and the simultaneous deletion of water molecules in the downloaded protein are necessary factors before effective docking. The active binding cleft of the protein was determined by identifying the location of nicotinamide and 7-(pyrrolidin-1-ylmethyl)pyrrolo[1,2-a]quinoxalin-4(5H)-one (A968427), the reference molecules within the proteins—p53 and PARP-1. Flexible docking simulation was used in ArgusLab 4.0.1 to produce a variety of ligand-receptor complex conformers based on different orientations of binding. Among the different scores obtained from different poses of the ligand-receptor complex, the one with the lowest binding score was identified as the best docked value after an effective docking attempt. The relevant amino acid residues and related connections linking the proteins to different moieties of ligands were visualized via the Discovery Studio program.

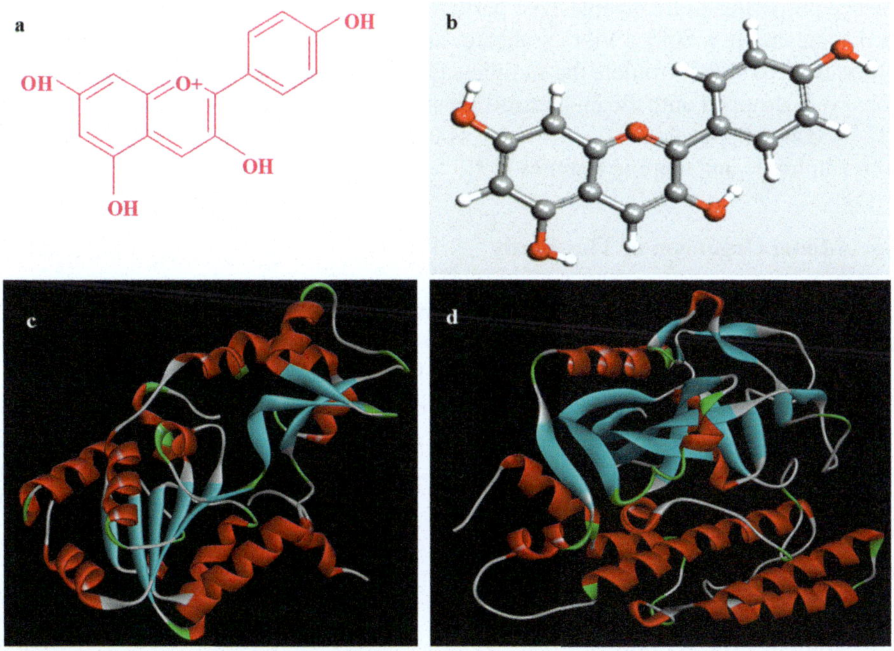

Fig. 1 Structure of ligand and protein: (**a**) 2D structure of pelargonidin, (**b**) 3D structure of pelargonidin, (**c**) 3D structure of protein p53, and (**d**) 3D structure of protein PARP-1

2.3 Preparation and Synthesis of Nanopelargonidin (NPG)

For the preparation and synthesis of monodisperse polymeric NPG, the solvent displacement approach, developed by Fessi et al. (1989) [16], was implemented. Polylactide-co-glycolic acid (PLGA), an environmentally friendly, non-hazardous, and biodegradable polymeric material, was used to encapsulate pelargonidin [17]. In general, the organic solvent was prepared in a homeopathic vial by dissolving 20 mg of the drug (pelargonidin) and 50 mg of PLGA in 2 mL of acetone. The organic solvent was poured dropwise into a beaker containing 20 mL of an aqueous solution with 0.2 g of the stabilizer F68 (1% polyoxyethylene-polyoxypropylene). The beaker was constantly stirred at ambient temperature using a magnetic stirrer until the organic solvent in it evaporated entirely. After completion of the reaction process, NPG was kept in a desiccator for drying and then stored at 4 °C for further study.

2.4 Atomic Force Microscopic (AFM) Study of Synthesized Nanopelargonidin (NPG)

For the synthesis of a nano-based delivery system, an effective size of the nanodrug is required for proper delivery and action in the target region. Therefore, before treatment with the drug, there is a need to determine the 2D and 3D structure, shape, and size of the synthesized nanopelargonidin (NPG). AFM imaging was utilized to determine the surface properties of the synthesized particle. NPG was dissolved in Milli-Q water and then sonicated in a Sonics Vibra sonicator at an amplitude of 36 Hz, pulsed for 10 s at intervals of 10 s, to formulate the samples. During sample preparation, the diluted solutions were dropped onto clean, grease-free slides and dried in a desiccator for 24 h [17, 18]. WSxM 5.0 Develop 7.0 software was used to gather and develop AFM signals in both amplitude and tapping regimes [19].

2.5 Model Organism of This Study

The model organism used for the current study comprised Swiss albino mice (*Mus musculus*) (4–6 weeks old, weighing 20–25 g) that were acquired from the State Centre for Laboratory Animal Breeding (SCLAB), Kalyani. The animals were maintained in proper clean cages and kept at a temperature of 24–26 °C, humidity at 55±5% and a 12-h light/dark cycle, and with provisions of food pellets and drinking water. The Institutional Animal Ethical Committee of the University of Kalyani (892/GO/Re/S/01/CPCSEA) granted ethical approval for the tests, which were carried out under the Animal Ethics Committee and the Animal Welfare Committee at the University of Kalyani.

2.6 Diabetes Induction

The mice model was subjected to diabetes using alloxan monohydrate (ALX). Mice that had been starved overnight were given an intraperitoneal injection of 0.04% alloxan monohydrate dissolved in 0.87% normal saline. The resultant solution was delivered intraperitoneally (i.p.) into mice at a dose of 100 mg/kg body weight [20]. The mice were administered a 5% glucose solution throughout the night to recuperate from an acute hypoglycemia shock that may have occurred soon after ALX induction [10]. On the fifth day of ALX administration, blood glucose levels were measured. Animals with blood glucose levels more than 200 mg/dL were classified as diabetics and were taken into the experiment.

2.7 Experimental Design and Treatment Groups

The 7–14 days' acclimatized mice were divided at random into five groups listed below.

Group A: Normal control mice group (untreated group) designated as **CONT**

Group B: Diabetic mice group (ALX induced) designated as **ALX**

Group C: Pre-drug-treated diabetic mice (nanopelargonidin at a dose of 15 mg/kg body weight followed by ALX induction) designated as **PG + ALX**

Group D: Pre-nanodrug-treated diabetic mice (nanopelargonidin at a dose of 15 mg/kg body weight followed by ALX induction) designated as **NPG I + ALX**

Group E: Pre-nanodrug-treated diabetic mice (nanopelargonidin at a dose of 7.5 mg/kg body weight followed by ALX induction) designated as **NPG II + ALX**

2.8 Administration of Pelargonidin and Nanopelargonidin in the Experimental Model

Pelargonidin and nanopelargonidin feeding was made orally in mice once a day for 7 days [10], and blood glucose level was measured following alloxan administration.

2.9 Measurement of Serum Glucose Level Using the GOD-POD Method

Blood samples were collected from the mice groups during the experimental period after anesthetizing the mice, and serum was collected from the blood by centrifugation at 2000 rpm for an 8-min duration. The serum glucose levels of all the experimental sets were monitored following the protocol as per the standardized GOD-POD glucose kit (Autospan) from Cogent Diagnostics Ltd., India (code: 93DP100–74) [21].

2.10 Pancreatic β Cell Isolation

The collagenase digestion procedures were followed to isolate β cells from the pancreas [17, 22]. In brief, the pancreatic tissues were perfused in DMEM (Dulbecco's Modified Eagle Medium) and then subjected to Krebs-Ringer HEPES (KRH) 10% BSA solution. Subsequently, three consecutive washes of the cells were performed using a KRH 10% HEPES buffer solution. After that, the cells underwent treatment for 10 min at 37 °C with 1 mg/mL collagenase in KRH 10% BSA and then for an additional 10 min at 4 °C. Collagenase and proteases were eliminated from the digested pancreatic cells by thoroughly washing them and centrifuging them at least twice or three times. Afterward, DMEM was used to wash the separated β cells, and they were centrifuged at 6000 rpm for 2 min. The purified islets were seeded in a culture flask containing DMEM culture medium supplemented with 1% BSA and 10% FBS and cultured for an additional night at 37 °C and 5% CO_2. The purified islets were utilized to measure biological variables like protein expression study.

2.11 Estimation of ROS Generation by DCFDA Dye

2′,7′-Dichlorodihydrofluorescein diacetate (H_2DCFDA), a cell-permeable, non-fluorescent dye, was used to detect and quantify the reactive oxygen species generation in different experimental groups [23]. An equal number of isolated pancreatic β cells were incubated at room temperature with 1 mg/mL H_2DCFDA dye (Sigma-Aldrich) in complete darkness for half an hour. The accumulation of ROS leads to the conversion of H_2DCFDA to DCF having fluorescent properties that were measured within the range of 485 nm–535 nm in a fluorescence microscope (Nikon Eclipse TiE), and the cells were scored by analysis software (NIS Elements AR, version 4.00).

2.12 Analysis of DNA Damage Using Hoechst Staining

Hoechst staining (1 μg/mL) allows analysis of the extent of DNA damage in the pancreatic tissue of experimental groups. An identical number of isolated pancreatic β cells of all the experimental sets were incubated with Hoechst dye for 5 min at 4 °C. The fluorescence microscope was used to quantify the PI-stained cells with damaged DNA [24].

2.13 Protein Expression Study

The methodology of Currier et al. (2005) [25] was followed in employing the indirect ELISA technique to investigate the protein expression of p53 and PARP-1 in the isolated pancreatic β cells. In brief, after lysis, homogenization, and centrifugation of isolated pancreatic β cells, the quantification of protein was done with the supernatant by Bradford's reagent technique. The same quantity of protein was collected for each sample. A 50 μL of the sample was taken on 96-well plates, where 100 μL of 0.05 M carbonate buffer was added and kept at 4 °C overnight. After flipping the plates to remove the unbound antigen solution, the blocking solution containing 1% BSA in 0.1 M TBS was added to each well and placed for 1 h at 37 °C. Following blocking, 200 μL TBS wash with 0.05% Tween 20, the targeted primary antibody—either anti-p53 for one set of plates or anti-PARP-1 for another—was added. The plates were incubated at 15–20 °C for 2–4 h. Following the incubation period, secondary antibodies conjugated with alkaline phosphatase (Sigma-Aldrich, USA) were added and incubated for 2 h at 37 °C. Then the plates were incubated for 20–30 min at 37 °C with a substrate solution containing 10 mM Tris, 0.5 mM $MgCl_2$, and 0.5 M para-nitrophenylphosphate to produce color. The final absorbance was determined using an ELISA reader (Thermo Electron Corporation Multi Scan EX) at a wavelength of 405 nm.

2.14 Biochemical Parameters

The biochemical parameters were examined by extraction of pancreatic tissues from autopsied mice, homogenized in 1 mL cold PBS (phosphate buffer saline, pH 7.4), and centrifuged at 12000 rpm for 45 min at 4 °C. The supernatants were taken for biochemical analysis. The SOD and CAT were studied as per the protocol of Kakkar et al. (1984) [26] and Maehly and Chance (1954) [27], respectively.

2.15 Statistical Analysis

The information provided in this study is based on the means of three separate experiments, which were statistically examined using the Student's t-test and analysis of variance (ANOVA) to assess the significance of the variations in means.

3 Results

3.1 Molecular Docking Analysis of Ligand-Protein Complex

The 3D coordinates of the proteins p53 and PARP-1 were downloaded from the Protein Data Bank, and then the docking study was elucidated in Discovery Studio (Fig. 2). Docking studies of the target protein with the ligand molecule pelargonidin showed favorable interactions as deduced from a scoring value of −5.7 Kcal/mol for p53 and −8.19 Kcal/mol for PARP-1. The favorable interactions of pelargonidin with the amino acid residues of the proteins were represented in Table 1.

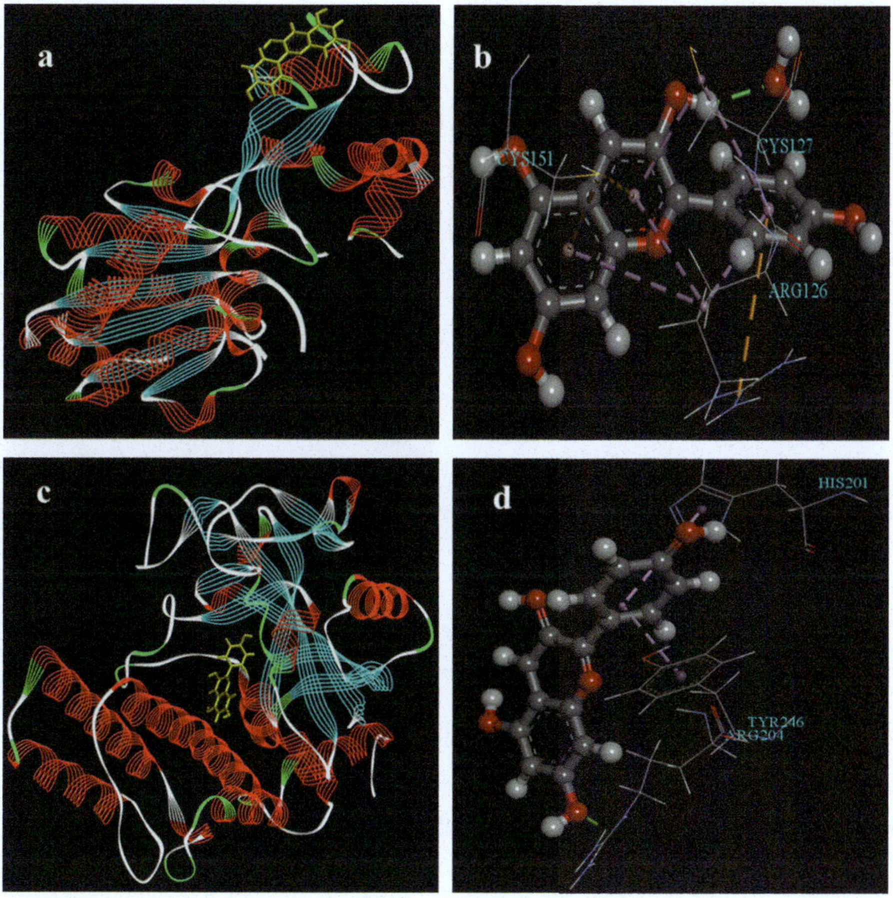

Fig. 2 (**a**) Best docked pose of p53 and PG complex, (**b**) p53-PG interaction showing amino acid residues, (**c**) best docked pose of PARP-1 and PG complex, and (**d**) PARP-1-PG interaction showing amino acid residues

Table 1 Interactions of pelargonidin with the proteins

Target protein	Amino acid residues	Interactions
p53	CYS151	Disulfide bonding (brown color)
	ARG126	Electrostatic interaction (orange color)
	ARG126 and CYS127	Hydrophobic interactions (purple color)
PARP-1	ARG204	Hydrogen bond (green color)
	HIS201 and TYR246	Aromatic hydrophobic interactions (purple color)

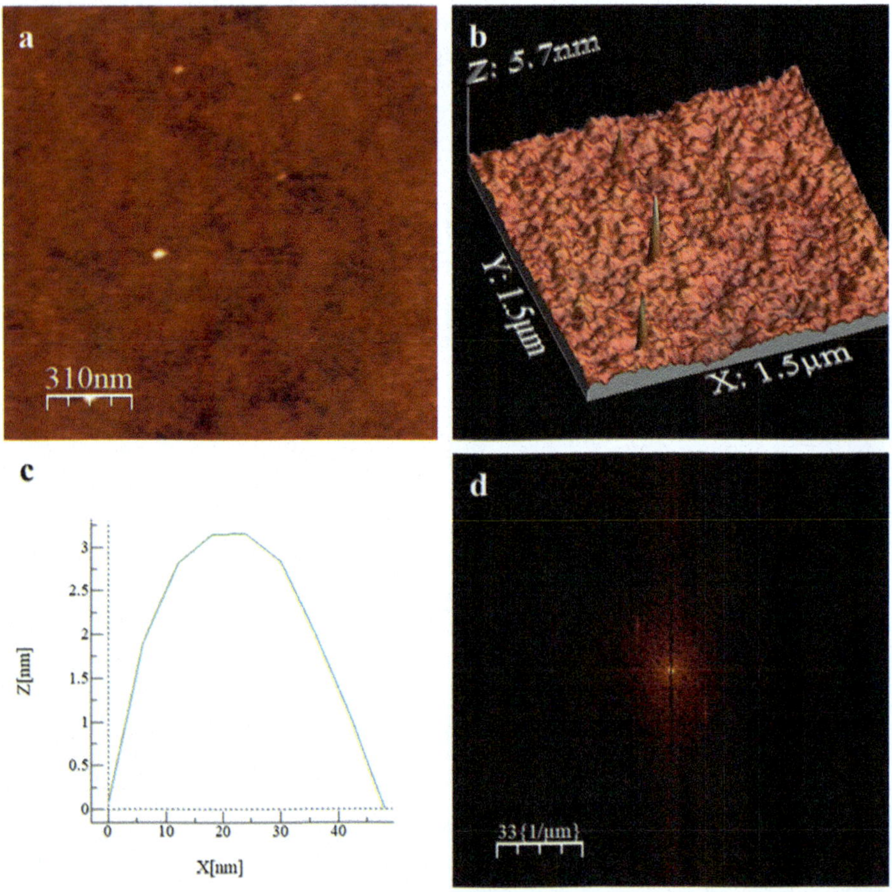

Fig. 3 (**a**) AFM 2D view of NPG, (**b**) AFM 3D view of NPG, (**c**) size profile of NPG, and (**d**) FFT study of NPG

3.2 Physicochemical Nature of Nanopelargonidin (NPG) from AFM Study

The results of the AFM analysis presented in Fig. 3 showed that nanopelargonidin (NPG) had a rounded morphology (Fig. 3a, b). The size of the nanoparticles as confirmed from the size profile of 2D and 3D scans was found to be −47 nm (Fig. 3c). The smooth surface pattern of NPG was also visible in the 3D perspective. The fast Fourier

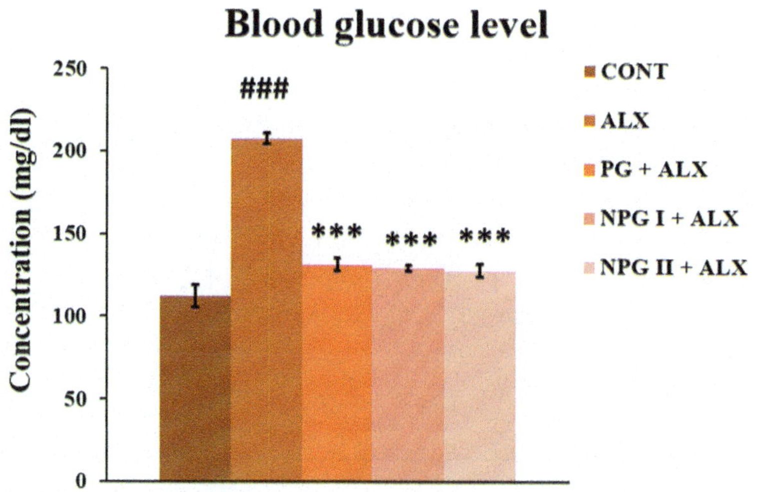

Fig. 4 Graphical representation of blood glucose level in experimental sets ($^{###}p < 0.001$ vs CONT and $^{***}p < 0.001$ vs ALX were considered significant for Student's t-test)

transformation (FFT) investigation substantiates the incidence of the formation of identical-sized nanoparticles of NPG, which confirmed that the encapsulated NPG are equal sized maintaining the homogeneity (Fig. 3d). Using the Nano-ZS device from Malvern Instruments, Southborough, UK, the zeta potential of the NPG was calculated to be −17 mV [28].

3.3 Estimation of Blood Glucose Level

ALX-induced mice exhibited a significant increase in the levels of blood glucose as compared to the control group. After administration of NPG to hyperglycemic mice, their blood glucose levels were significantly reduced. The overall results are presented in Fig. 4.

3.4 Determination of ROS Generation by H_2DCFDA Dye

The quantification in levels of intracellular reactive species (ROS) in pancreatic β cells was measured using H_2DCFDA dye by using a fluorescence microscope. The quantity of ROS produced is correlated with an upsurge in DCF level that is noticed among mice administered with only ALX, as illustrated in Fig. 5. This finding showed that the ALX-treated group generated more intracellular ROS than the control group that received no treatment. However, as shown by the decreased number of cells showing fluorescence in the NPG-pretreated mice group, pretreatment with NPG substantially inhibited ALX-induced ROS production.

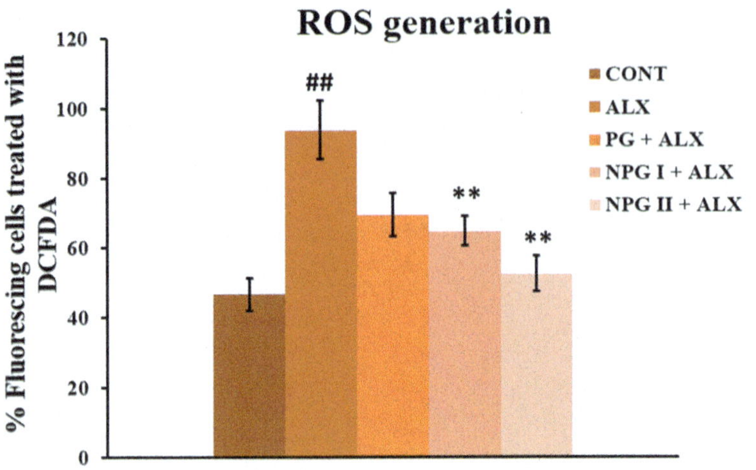

Fig. 5 Graphical representation of ROS generation in experimental sets ($^{##}p < 0.01$ vs CONT and $^{**}p < 0.01$ vs ALX were considered significant for Student's t-test)

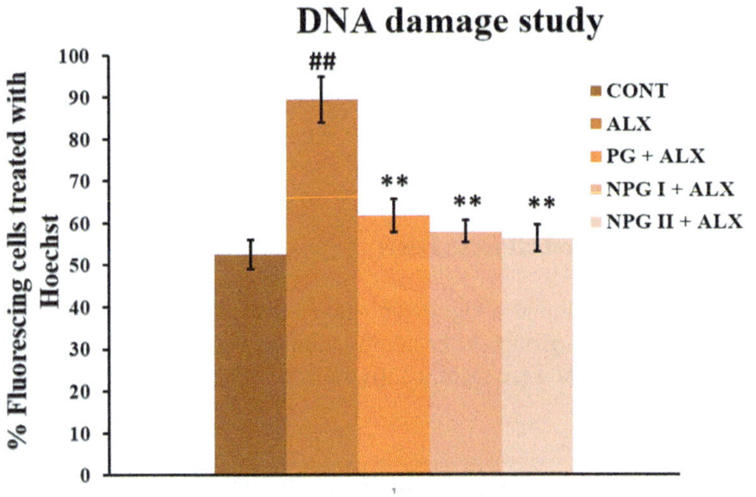

Fig. 6 Graphical representation showing DNA damage study in experimental sets ($^{##}p < 0.01$ vs CONT and $^{**}p < 0.01$ vs ALX were considered significant for Student's t-test)

3.5 Determination of DNA Damage by Hoechst Staining

To determine the effect of NPG in the reduction of DNA damage, Hoechst staining was done in pancreatic β cells. As summarized in Fig. 6, upon pretreatment with NPG, the population of cells containing intact nuclei increased as evidenced by reduced PI fluorescence intensity. Therefore, fluorescence levels in ALX-treated sets showed increased intensity determining damaged DNA. The untreated set of mice showed minimum PI intensity estimating negligible/low damage.

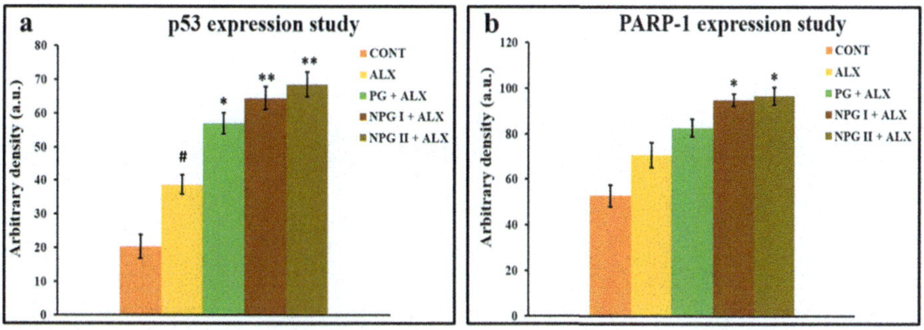

Fig. 7 (**a**) p53 protein expression study and (**b**) PARP-1 protein expression study in experimental groups of mice ($^{\#}p < 0.05$ vs CONT, $^{**}p < 0.01$ vs ALX, and $^{*}p < 0.05$ vs ALX were considered significant for Student's t-test)

3.6 Immunofluorescence Detection of p53 and PARP-1 Protein

The analysis to measure p53 and PARP-1 protein expression in the experimental groups was done by indirect ELISA. Comparing the ELISA assay findings to those of the control and ALX-treated sets, it was found that both the proteins p53 and PARP-1 were overexpressed in NPG/PG-pretreated β cells after ALX administration (Fig. 7a, b). The greater extent of expression of these proteins in the NPG group compared to the mice treated with PG indicates that pelargonidin in its encapsulated form has better activity. Therefore, by triggering specific repair mechanisms, the expression of p53 and PARP-1 proteins in NPG-treated groups expedites the elimination of damage generated by ALX. The findings of the *in silico* docking investigation thus corroborated the *in vivo* results highlighting crucial amino acid residues that prevent aberrant transmission of signal cascades and shield DNA from damage caused by free radicals.

3.7 Effect of NPG on Antioxidant Status

The levels of antioxidant enzymes—superoxide dismutase (SOD) (Fig. 8a) and catalase (CAT) (Fig. 8b)—in various experimental sets were depicted in the figure, demonstrating a substantial rise in enzyme levels indicative of enhanced function in the pretreated NPG + ALX group (Fig. 8).

4 Discussion

Type 2 diabetes is a highly diverse condition associated with hyperglycemia, insulin resistance, and a propensity for complications. An effective diabetes treatment also includes the varying costs of medicines. The implementation of plant-based resources in the execution of an inexpensive strategy was the main focus of the present research. Considering the numerous benefits of pelargonidin as an anti-genotoxic agent, with

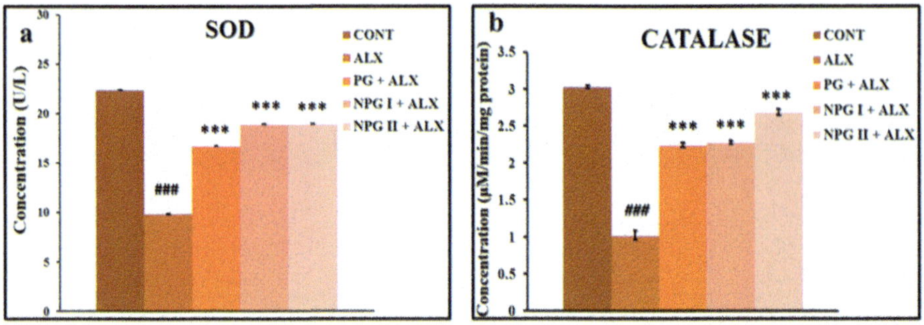

Fig. 8 Biochemical analysis in different experimental sets: (**a**) SOD and (**b**) catalase ($^{###}p < 0.001$ vs CONT and $^{***}p < 0.001$ vs ALX were considered significant for Student's t-test)

anti-inflammatory and antioxidative properties, this phyto-drug was repurposed to uncover the drug's new therapeutic possibilities by employing computational and experimental methods. Numerous reports have addressed the implementation of nanotechnologies for multiple facets of disease management, such as drug delivery, therapy, etc. The benefits of drug distribution by nanocarriers for improving antidiabetic action are the primary emphasis of this article. The advanced field of nanotechnology has rendered it feasible to formulate drug compounds with a wide range of physicochemical properties aimed at enhancing stability, bioavailability, and pharmacokinetics. PLGA, the biopolymer, was used as an encapsulating agent of a drug that exhibits no toxicity owing to its biodegradable nature. Therefore, this PLGA-encapsulated delivery technique provides a viable therapy option for hyperglycemia as the greatest concentration of drug accumulated due to its nano-size (~47 nm) that transports to the site with the lowest loss. These medication formulations and regimens are intended to improve the drug's tissue localization, leading to an enhanced activity level. This work attempted to develop a cost-effective approach for using PLGA-encapsulated drug (NPG) as an alternative for the management of diabetic complications. Therefore, this study enabled the encapsulation of pelargonidin within PLGA polymer to investigate antidiabetic effects, thereby reducing treatment costs and boosting medicinal effectiveness at lower dosages.

A primary indicator of diabetes mellitus involves elevated blood sugar levels. In this study, pre-administration of NPG to diabetic mice drastically lowered glucose imbalance, suggesting a potential hypoglycemic impact. The pancreas is a vital organ that tightly regulates blood glucose levels to maintain the optimal functioning of the body. This glucose homeostatic balance was sustained by an intricate network of insulin activity, the pancreatic hormone that lowers blood sugar. However, the degeneration of pancreatic islets by alloxan-generated reactive radicals led to anomalies in the hormone's functioning. Serious health effects may result from the impairment of this organ caused by oxidative stress. The degree of pancreatic damage is indicated by the significant drop in antioxidative enzyme parameters in diabetic mice. This implies that damage to pancreatic islets in an ALX-induced state may be a mediating factor in the occurrence of hyperglycemic conditions. The outcome of the present research showed reduced levels of SOD and CAT in the pancreatic tissue of the ALX group, which were significantly correlated with imbalanced blood sugar. Therefore, this marked decline in the antioxidant defense mechanism in the pancreas was mostly attributed to alloxan toxicity. However, PG/NPG pretreatment for alloxan diabetic groups might restore pancreatic damage by improving antioxidant enzymatic activity to normal levels. Therefore, NPG might function as a protective factor by preventing oxidative stress from disrupting pancreatic damage by elevating antioxidative enzyme levels, thereby balancing the pancreatic function of glucose regulation.

The mechanism of action of NPG was also addressed by the molecular docking technique, which is regarded as one of the fast, low-cost, and accurate structure-based computational methods to identify interactions between particular pharmacological molecules and biological targets. The binding activity of pelargonidin within the protein cavity was demonstrated by the *in silico* molecular dock value of −5.7 Kcal/mol and −8.19 Kcal/mol. Strong hydrophobic (purple color line) and hydrogen bonding (green color line) patterns were found between ligands and proteins [29–31]. Further, the immunofluorescence assay also confirmed an upregulation of p53 and PARP-1 proteins in the NPG pre-treatment set, which corroborated well with the dock score values, indicating reduction of pancreatic damage owing to ALX-induced diabetes. The expression of p53 and PARP-1 proteins in NPG-treated groups thus accelerates ALX-induced damage removal through the activation of specialized repair machinery. Therefore, the immunofluorescence investigation supports the idea that NPG may have a function in reducing free radical-induced damage of pancreatic tissue by overexpression of damage repair proteins, which further maintains the release of insulin from secretory vesicles in a diabetic state, thereby symbolizing the maintenance of blood glucose levels. Consequently, NPG may play a part in maintaining the interconnected balance between PARP-1 and p53 to initiate the repair process that was disrupted in the pathological condition. As a result, it could be deduced that the pancreatic cells of NPG-pretreated mice continued to function normally, preventing the development of hyperglycemia by modulating the antioxidative enzyme system and by elevating the expression of the PARP-1 and p53 proteins. Additionally, the decreased fluorescence of PI depicting lower DNA damage was evidenced in pancreatic β cells of NPG-treated group demonstrating that decreased ROS production and increased SOD and CAT activity suppressed the ROS-induced damage to pancreatic cells, hence reducing DNA damage.

The oxidative tension generated by ALX triggers damage to pancreatic β cells. In our study, a comparison with a normal control condition showed downregulation of the first-line defense enzyme system in mice that received ALX induction. The present study found that the antihyperglycemic therapeutic efficacy of NPG enabled protection against pancreatic damage balancing antioxidant enzyme levels and upregulating protein expression involved in reparative mechanism. As a result, NPG might be regarded as a hypoglycemic agent and could be considered a viable treatment alternative for the management of diabetes in the near future.

5 Conclusion

Diabetes is a serious condition that can lead to lifelong implications later if left untreated. This work draws forward to an alternative nano-based approach for preventing diabetes rather than relying on synthetic medications, which might avert additional issues that diabetics face. Based on the overall study, it can be inferred that at a lower dose, nanoformulated pelargonidin (NPG) may provide superior protection against diabetic complications caused by ALX compared to its non-nano-substitutes. By overexpressing p53 and PARP-1 proteins and controlling oxidative and antioxidative balance, NPG reduces ROS formation, thereby preventing islet destruction and maintenance of blood glucose levels. Therefore, NPG might prove to be a phyto-based drug lead opening up new possibilities for the treatment of diabetes and the reduction of hyperglycemic risks in the near future.

Acknowledgments The authors would extend acknowledgement to UGC, New Delhi, for the UGC-BSR (grant no. F.30-488/2019 (BSR)) and DST for SERB (DST) (grant no. ECR/2017/000355/LS) extramural research grant sanctioned to Dr. Asmita Samadder for this research work. The authors are grateful to the University of Kalyani for providing Personal Research Grant (PRG) to A.S., which was used partly for the research work. R.D. is gratified to UGC for her CSIR-NET SRF fellowship. The authors express sincere appreciation to the respective institutes for providing essential research facilities. The authors also deeply express gratitude to UGC-DAE Consortium for Scientific Research, Kolkata Centre, for providing necessary research facilities and essential infrastructural support.

Author Contribution The authors contribution in the paper are as follows: conceptualization, A.S. and S.N.; data curation, R.D.; funding acquisition, A.S.; investigation, A.S., S.N., and D.T.; software, R.D. and S.N. for *in silico* study; supervision, A.S.; validation, A.S. and S.N.; roles/writing—original draft, R.D.; and writing—review and editing, A.S. and S.N.

Declaration of Competing Interest None to declare

References

1. Jayachandran, P., Gayathri, R., Jayaraman, S., Vishnu Priya, V., Kavitha, S.: Antioxidative stress potential of piperine in the gastrocnemius muscle of high fat diet and sucrose induced type 2 diabetic rats. HIV Nurs. **23**, 81–89 (2023)
2. https://www.who.int/india/healthtopics/mobile-technology-for-preventing-ncds
3. Nitzan, O., Elias, M., Chazan, B., Saliba, W.: Urinary tract infections in patients with type 2 diabetes mellitus: review of prevalence, diagnosis, and management. Diabetes Metab. Syndr. Obes. **8**, 129–136 (2015)
4. Tsalamandris, S., Antonopoulos, A.S., Oikonomou, E., Papamikroulis, G.A., Vogiatzi, G., Papaioannou, S., Deftereos, S., Tousoulis, D.: The role of inflammation in diabetes: current concepts and future perspectives. Eur. Cardiol. **14**(1), 50–59 (2019)
5. Babiker, A., Al Dubayee, M.: Anti-diabetic medications: how to make a choice? Sudan J. Paediatr. **17**(2), 11–20 (2017)

6. Rohm, T.V., Meier, D.T., Olefsky, J.M., Donath, M.Y.: Inflammation in obesity, diabetes, and related disorders. Immunity. **55**(1), 31–55 (2022)
7. Van Vuuren, S., Viljoen, A.: Plant-based antimicrobial studies – methods and approaches to study the interaction between natural products. Planta Med. **77**, 1168–1182 (2011)
8. Álvarez-Martínez, J.F., Barrajón-Catalán, E., Encinar, A.J., Rodríguez-Díaz, C.J., Micol, V.: Antimicrobial capacity of plant polyphenols against Gram-positive bacteria: a comprehensive review. Curr. Med. Chem. **27**(15), 2576–2606 (2020)
9. Kesarwani, K., Gupta, R., Mukerjee, A.: Bioavailability enhancers of herbal origin: an overview. Asian Pac. J. Trop. Biomed. **3**(4), 253–266 (2013)
10. Dey, R., Nandi, S., Samadder, A.: Pelargonidin mediated selective activation of p53 and parp proteins in preventing food additive induced genotoxicity: an in vivo coupled in silico molecular docking study. Eur. J. Pharm. Sci. **156**, 105586 (2021)
11. Kwon, I.K., Lee, S.C., Han, B., Park, K.: Analysis on the current status of targeted drug delivery to tumors. J. Control. Release. **164**(2), 108–114 (2012)
12. Süsse, S., Scholz, C.J., Bürkle, A., Wiesmüller, L.: Poly(ADP-ribose) polymerase (PARP-1) and p53 independently function in regulating double-strand break repair in primate cells. Nucleic Acids Res. **32**(2), 669–680 (2004)
13. Praveena, R., Balasankar, A., Aruchamy, K., Oh, T., Polisetti, V., Ramasundaram, S., Anbazhakan, K.: Structural activity and HAD inhibition efficiency of pelargonidin and its glucoside—a theoretical approach. Molecules. **27**, 8016 (2022)
14. Avalos, J.L., Bever, K.M., Wolberger, C.: Mechanism of sirtuin inhibition by nicotinamide: altering the NAD(+) cosubstrate specificity of a Sir2 enzyme. J. Mol. Cell. **17**, 855–868 (2005)
15. Miyashiro, J., Woods, K.W., Park, C.H., Liu, X., Shi, Y., Johnson, E.F., Bouska, J.J., Olson, A.M., Luo, Y., Fry, E.H., Giranda, V.L., Penning, T.D.: Synthesis and SAR of novel tricyclic quinoxalinone inhibitors of poly(ADP-ribose)polymerase-1 (PARP-1). Bioorg. Med. Chem. Lett. **19**, 4050–4054 (2009)
16. Fessi, H., Puisieux, F., Devissaguet, J.P., Ammoury, N., Benita, S.: Nanocapsule formation by interfacial polymer deposition following solvent displacement. Int. J. Pharm. **55**(1), R1–R4 (1989)
17. Samadder, A., Das, S., Das, J., Khuda-Bukhsh, A.R.: Relative efficacies of insulin and poly (lactic-co-glycolic) acid encapsulated nano-insulin in modulating certain significant biomarkers in arsenic intoxicated L6 cells. Colloids Surf. B: Biointerfaces. **109**, 10–19 (2013a)
18. Samadder, A., Das, J., Das, S., De, A., Saha, S.K., Bhattacharyya, S.S., Khuda-Bukhsh, A.R.: Poly(lactic-co-glycolic) acid loaded nano-insulin has greater potentials of combating arsenic induced hyperglycemia in mice: some novel findings. Toxicol. Appl. Pharmacol. **267**(1), 57–73 (2013b)
19. Horcas, I., Fern'andez, R., G'omez-Rodríguez, J.M., Colchero, J., G'omez-Herrero, J., Baro, A.M.: WSXM: a software for scanning probe microscopy and a tool for nanotechnology. Rev. Sci. Instrum. **78**, 013705 (2017)
20. Solomon, J., Bulkley, R.J., Mayer, J.: Effect of low dose of alloxan on blood glucose, islet beta cell granulation, body weight and insulin resistance of ob/ob mice. Diabetologia. **10**(6), 709–715 (1974)
21. Kaplan, L.A.: Carbohydrates and metabolite. In: Clinical Chemistry: Theory, Analysis and Co-relation, pp. 1032–1040 (1984)
22. Smelt, M.J., Faas, M.M., de Haan, B.J., de Vos, P.: Pancreatic beta-cell purification by altering FAD and NAD (P)H metabolism. Exp. Diabetes Res. **2008**, 165360 (2008)
23. Samadder, A., Tarafdar, D., Das, R., Khuda-Bukhsh, A.R., Abraham, S.K.: Efficacy of nanoencapsulated pelargonidin in ameliorating pesticide toxicity in fish and L6 cells: modulation of oxidative stress and signalling cascade. Sci. Total Environ. **671**, 466–473 (2019)

24. Sow, P., Dey, S., Dey, R., Samadder, A., Nandi, S., Tarafdar, D., Khuda-Bukhsh, A.R.: Nano-pelargonidin modulates p53/PARP/HSP90/XRCC1 signaling axis to combat cypermethrin-induced genotoxicity and metabolic dysfunction in fish: a molecular docking predicted in vivo approach. Environ. Sci. Nano. **11**(6) (2024)
25. Currier, N., Solomon, S.E., Demicco, E.G., Chang, D.L.F., Farago, M., Ying, H., Dominguez, I., Sonenshein, G.E., Cardiff, R.D., Xiao, Z.X.J., Sherr, D.H., Seldin, D.C.: Oncogenic signaling pathways activated in DMBA-induced mouse mammary tumors. Toxicol. Pathol. **33**, 726–737 (2005)
26. Kakkar, P., Das, B., Viswanathan, P.N.: A modified spectrophotometric assay of superoxide dismutase. Indian J. Biochem. Biophys. **21**(2), 130–132 (1984)
27. Maehly, A., Chance, B.: The assay of catalases and peroxidases. Methods Biochem. Anal. **1**, 357–424 (1954)
28. Dey, R., Saha, S., Nandi, S., Molla, S.H., Samadder, A.: Exploring a new phyto-derived nanoparticle for targeting bacterial protein EF-Tu: an integrated approach to develop antimicrobial drug. Smart Sci. **12**(2), 357–372 (2024)
29. Paul, D., Basu, D., Ghosh Dastidar, S.: Multi-conformation representation of Mpro identifies promising candidates for drug repurposing against COVID-19. J. Mol. Model. **27**, 128 (2021)
30. Atasever, A.: Molecular docking and inhibition profiles of some antidepressants on human carbonic anhydrase enzymes. Results Chem. **6**, 101174 (2023)
31. Rajalakshmi, R., Lalitha, P., Sharma, S.C., Rajiv, A., Chithambharan, A., Ponnusamy, A.: In silico studies: physicochemical properties, drug score, toxicity predictions and molecular docking of organosulphur compounds against diabetes mellitus. J. Mol. Recognit. **34**, e2925 (2021)

Nano-phytocompound Combats Food Additive-Induced Genotoxicity and Mitochondrial Dysfunction in Diabetic Model: An In Silico Predicted Experimental Study

Sudatta Dey[1], Pradeepta Guptaroy[2] (✉), and Asmita Samadder[1] (✉)

[1] Cytogenetics And Molecular Biology Laboratory, Department of Zoology, University of Kalyani, Kalyani, West Bengal, India
[2] Dum Dum Motijheel College, Kolkata, West Bengal, India
[3] Cytogenetics And Molecular Biology Laboratory, Department of Zoology, University of Kalyani, Kalyani, West Bengal, India
asmitazoo19@klyuniv.ac.in

Abstract The study focuses on the therapeutic management of diabetes and associated complications induced by the food additive alloxan (ALX), which is detrimental to pancreatic β cell functioning. To address the problem, a flavonoid compound curcumin (CUR) with anti-diabetic, anti-genotoxic, and anti-inflammatory properties was considered to test its effectiveness in preventing ALX-induced diabetic complications based on evidences from molecular docking study. In silico study of CUR revealed its interactions with proteins PARP-1, p53, Bcl-2, and Cyt c suggesting a possible function of CUR in modulating cellular cascades. Further, to expedite targeted and faster drug delivery, drug bioavailability, and efficient tissue penetration, nanoencapsulation of CUR was opted as a novel drug design tool. AFM, DLS, and FESEM studies confirmed physico-chemical characteristics of newly formed nanocurcumin (NCUR). As validated by in silico docking study, pre-treatment of NCUR in experimental models showed its effective role in restricting genotoxicity and mitochondrial dysfunction by effectively binding with DNA damage repair proteins PARP-1 and p53 and mitochondrial signaling proteins Bcl-2 and Cyt c, respectively. Furthermore, a number of parametric studies have demonstrated NCUR to exert greater efficacy by preventing tissue damage and delaying incidence of diabetes, DNA damage, and chromosomal aberrations. Thus, it can be concluded that pretreatment of nano-based phyto-drugs offers promising strategy as therapeutic tool

A. K. Saxena, S. Nandi (eds.), *Global Trends in Health, Technology and Management*, https://doi.org/10.1007/978-3-031-75457-9_7

for effective regulation of food additive-induced damages and diabetes via balancing the expression of proteins involved therein.

Keywords Diabetes · Alloxan · In silico molecular docking · Nanocurcumin · Genotoxicity · Mitochondrial dysfunction

1 Introduction

Diabetes mellitus type 2 has become a lifestyle disease representing an expensive condition that affects people, economies, health-care systems, and entire society on a global scale [1]. Diabetes mellitus is marked by an accumulation of metabolic disorders characterized by high blood sugar levels caused by inadequacies in either insulin production or action or both. Uncontrolled incidence of diabetes manifests its crippling attack in human by causing major organ damage and subsequent complications like macrovascular complications (cardiovascular (CV), cerebrovascular, and peripheral artery disease) and microvascular complications (diabetic retinopathy, nephropathy, and neuropathy) [2]. End-stage renal failure, adult-onset blindness, and non-traumatic amputations are primarily caused by the long-term complications of this lifestyle disease [3]. For instance, peripheral neuropathy is the result of long-term effects of diabetes that raises the risk of Charcot joints, amputations, and foot ulcers; retinopathy results in vision loss; autonomic neuropathy causes symptoms related to the gastrointestinal tract, genitourinary system, cardiovascular system, and sexual function. Moreover, diabetes is causing health damages worldwide by raising the burden of the disease within the family, the community, and the country as a whole [4]. Therefore, diabetes is a silent killer disease that can attack any organ of the body, as the primary stage of this disease is silent but acute and chronic complications arise due to lack of prudence in patients, leading them to suffer from reduced quality of life, disability, early morbidity, and mortality [5]. This highlights both the health and financial risks that diabetes presents in several nations and the significance of established diabetes management guidelines in order to avert complications and guarantee patients a normal quality of life [6]. The management of diabetes includes a combination of pharmaceutical intervention as well as lifestyle changes for maintaining a healthy living. A variety of lifestyle factors have been recognized, such as obesity, absence of physical activity, unhealthy diet, stress, and urbanization, which have a substantial impact on the development of type 2 diabetes [4, 7].

According to recent researches related to food health and safety, the synthetic chemicals used as food additives in food industry have become one of the rising concerns due to their endocrine disruptive effects in human [8]. Food additive-induced toxicity leading to oxidative stress and genotoxicity is one of the prime causes for the onset of several complicated disorders owing to development of serious health issues as studied so far [9]. Sometimes the inability of the pancreas to produce and secrete insulin may not be solely responsible for diabetes incidence. Rather some food additives like aspartame and monosodium glutamate (MSG) are found to be associated with obesity (increasing weight gain) that ultimately contributes to insulin receptor insensitivity in fat cells and peripheral tissues by elevated levels of tumor necrosis factor α (TNFα), followed by type 2 diabetes mellitus (T2DM) condition in the subjects. Different food additives that are ingested through our daily diets may disrupt several signaling cascades resulting in conditions of prediabetes. For instance, a by-product of food processing industry, alloxan (ALX) has been reported to destroy pancreatic beta cells and hamper insulin levels in the body and is therefore chosen as a diabetes-inducing agent in model animals based on previous investigations [10]. To avoid such health-related issues, more dependency on natural foods and fruits and avoidance of synthetic colors and flavors containing foods that are very harmful and slowly causing the lifestyle diseases like cancer and diabetes should be made. Due to the presence of both metabolic and physiologic effects on different body systems, naturally obtained foods should become a more accepted option. Fruits, vegetables, and dietary spices that are naturally occurring show high potential in reducing the negative effects of several diseases along with their dietary advantages. Thus, the natural ingredients' multifarious effects function both as potential therapeutic agents for diseases and health-promoting agents [10, 11].

To pharmacologically treat the manifestations of diabetes mellitus characterized by high blood glucose, insulin resistance, inefficient insulin production and secretion, etc., several synthetic drugs belonging to different groups were recommended. The oral drugs belonging to the biguanide, thiazolidinedione, sulfonylurea, and meglitinide groups have been in use for a long time [12]. Despite the anti-diabetic functions, the majority of them have been found to be associated with long-term consequences. These synthetic drugs have been reported to leave negative side effects in the patients that further worsen the medical condition [13, 14]. This concern emphasizes the immediate need of alternative medicines that are naturally obtained and are non-toxic to biological system. Scientists are therefore keen toward screening phyto-compounds for target- or disease-specific activity to bring out the best beneficial effects of the phyto-products to combat several ailments of human kind [15]. Due to their nutritional benefits, they are also found to be beneficial in a number of chronic diseases and have been instrumental in the management of diabetes mellitus.

Products derived from natural sources can be used to treat type 2 diabetes and its complications, which might serve as viable alternatives that have no negative side effects. Many potent medicinal plants and their naturally occurring bioactive compounds have already been shown to have anti-diabetic properties [16]. Among them, the natural bioactive component curcumin (CUR) obtained from turmeric plant (*Curcuma longa*) has been used as traditional medicinal herb in Asian countries for treating various diseases and conditions caused due to inflammatory factors [17]. Moreover, several reports are present that suggest that curcumin could act as an anti-inflammatory agent in biological systems by regulating the inflammatory signaling pathways and inhibiting the production of inflammatory mediators. Apart from the anti-inflammatory properties, CUR has been studied to possess other pharmacological properties, such as antioxidant, anti-cancer, neuroprotective, as well as anti-diabetic activities [18]. With these findings, curcumin was selected as the chosen phyto-compound to prevent the onset of diabetes incidence as well as mitigation of complications associated with it. Despite having reported pharmacological roles, the major problem with using CUR as a drug is its poor water solubility, poor absorption, and bioavailability that follows its rapid metabolism in the liver and rapid elimination. Currently, there is a trend in research toward the selective formulation and design of medications that are less expensive and target-specific. Moreover, the drug safety is enhanced by the minute-sized nanoparticles of approximately 1–100 nm, which increases drug dissolving power, stability, and sustainable release [19]. Pharmaceutical substances have been improved by the use of nano-drug systems in terms of their pharmacokinetic/pharmacodynamic profile, safety, and physico-chemical characteristics. Specifically, targeted delivery to particular tissues, extended half-lives of injected drugs (by lowering immunogenicity), and improved bioavailability of oral drugs are all possible with functionalized nano-drug systems [20]. Additionally, drug activity is improved by preventing enzymatic degradation, enhancing bioavailability, and improving biocompatibility [21].

Thus, millions of people are now inclining toward nanotechnology in an effort to fulfill their dream of creating novel target-specific drug delivery tools in biological systems that are effective against a wide range of diseases, including diabetes. Therefore, to overcome the challenges faced when using CUR as a drug, nano-based drug delivery might be used to enhance the drug utilization at target tissue without undergoing structural breakdown due to biological pH changes [22, 23]. Thus, the nanotechnological approach ensures to bring out the pharmacological properties of drugs and improve their therapeutic efficacy in in vivo systems. This present study involves the nanoencapsulation of the drug CUR inside a biodegradable, biopolymer PLGA (poly-lactide-co-glycolide) that protects the core drug from degradation and ensures sustained drug release and targeted delivery at the diseased site. Therefore, this approach of drug design using nanotechnology increases the drug effectivity even at a minimum dose in managing diabetes.

Moreover, before choosing a phyto-compound to investigate its efficacy against progression of a disease, the advantage of the bioactive compound is checked on the basis of its adaptable structure and diverse potential. Based on the structural versatility, the wide diversity of target molecules in the case of the phytochemicals opens a huge probability to resist a wide range of pathological conditions [24]. Exploring the in silico molecular docking study allow us to understand the protein-ligand binding to predict the relevant proteins or signaling factors involved. Thus, molecular docking study with CUR was performed prior to the experimental studies. Different proteins were chosen based on their function in governing the proper cell function at molecular levels. Based on the previous evidences with other phyto-compounds as well as literature survey concerning CUR, it was evident that it could influence the expression of DNA repair proteins in other models, thereby regulating proteins associated directly or indirectly with DNA damage [25]. Among the fundamental proteins, p53, PARP-1, Bcl-2, and Cyt c were selected. p53 is the transcription factor responsible for its vital role in maintaining the integrity of the genome and thus governs the signaling cascade involved in DNA repair. PARP-1 is involved in transcription, chromatin structure modification, DNA repair, and numerous other biological processes. As an enzyme, PARP-1 facilitates the repair of DNA damage brought on by toxic agents or other chemotherapeutic drugs. The complex process of cell division, or mitosis, involves the proper functioning of p53 and PARP-1 proteins that maintain the DNA structure and synthesis of undamaged complementary DNA strands during DNA replication and chromosome segregation at the spindle complexes. The food additive ALX has been reported to cause lesions and bring mutations in DNA structure that hinder the proper replication of the DNA and maintenance of stable chromosomal structures with abnormalities [26]. The mitochondrial proteins like Bcl-2 and cytochrome c (Cyt c) play major roles in maintaining mitochondrial functioning and cellular homeostasis. Bcl-2 governs the cell survival by controlling permeabilization of the outer mitochondrial membrane, leading to the release of certain apoptotic activators into the cytosol that ultimately leads to cell death. The Cyt c protein present in the inter-membrane space of mitochondria plays two distinct roles in the regulation of cellular energetic metabolism and apoptosis. The induction of the apoptotic program requires cytosolic cytochrome c, indicating a potential relationship between Cyt c and Bcl-2. The overexpression of Bcl-2 thus prevents the release of Cyt c from the mitochondria and prevents initiation of apoptosis. Thus, the docking study of all these above mentioned proteins with the ligand presented the dock score values that revealed efficient interaction of the proteins with CUR.

Therefore, the main objectives that were intended to be addressed in our present study are as follows: (i) to formulate PLGA-encapsulated nanocurcumin (NCUR) drug and to characterize them for assessing their physico-chemical properties; (ii) to evaluate the interaction between CUR and target biological molecules that are related to onset/progression of genotoxicity, mitochondrial signaling dysfunction, and stress-induced hyperglycemia in mice model in in silico-based molecular docking study; (iii) to determine the relative efficacy of NCUR over CUR in maintaining the levels of hepatic biomarkers and anti-oxidative enzyme level inmice model; and (iv) to examine the modulating impact of NCUR on signaling factors involved in DNA repair and mitochondrial functioning in in vitro model.

2 Materials and Methods

2.1 Chemicals and Reagents

The majority of the solvents and chemicals of analytical grade utilized in the investigation were acquired from Sigma-Aldrich, St. Louis, MO, USA, and other well-known companies, and all of them were ≥97% pure. Additional chemicals and reagents from other companies that were used for different experiments are also mentioned in the texts accordingly.

2.2 In Silico-Based Molecular Docking Analysis of CUR with Relevant Protein

The ligand in this study is curcumin. The two-dimensional (2D) and three-dimensional (3D) chemical framework of curcumin was drawn using the ChemDraw Ultra 8.0 program (Fig. 1). The Protein Data Bank (PDB) was used to download p53 protein having PDB ID of 1YC5 [27], PARP-1 protein having PDB ID of 3GJW [28], Bcl-2 protein with a PDB ID of 6GL8 [29], and Cyt c protein with a PDB ID of 7LJX [30] having co-crystallized ligand nicotinamide, 7-(pyrrolidin-1-ylmethyl)pyrrolo[1,2-a]quinoxalin-4(5H)-one), (~{N}-(4-hydroxyphenyl)-3-[6-[[(3~{S})-3-(morpholin-4-ylmethyl)-

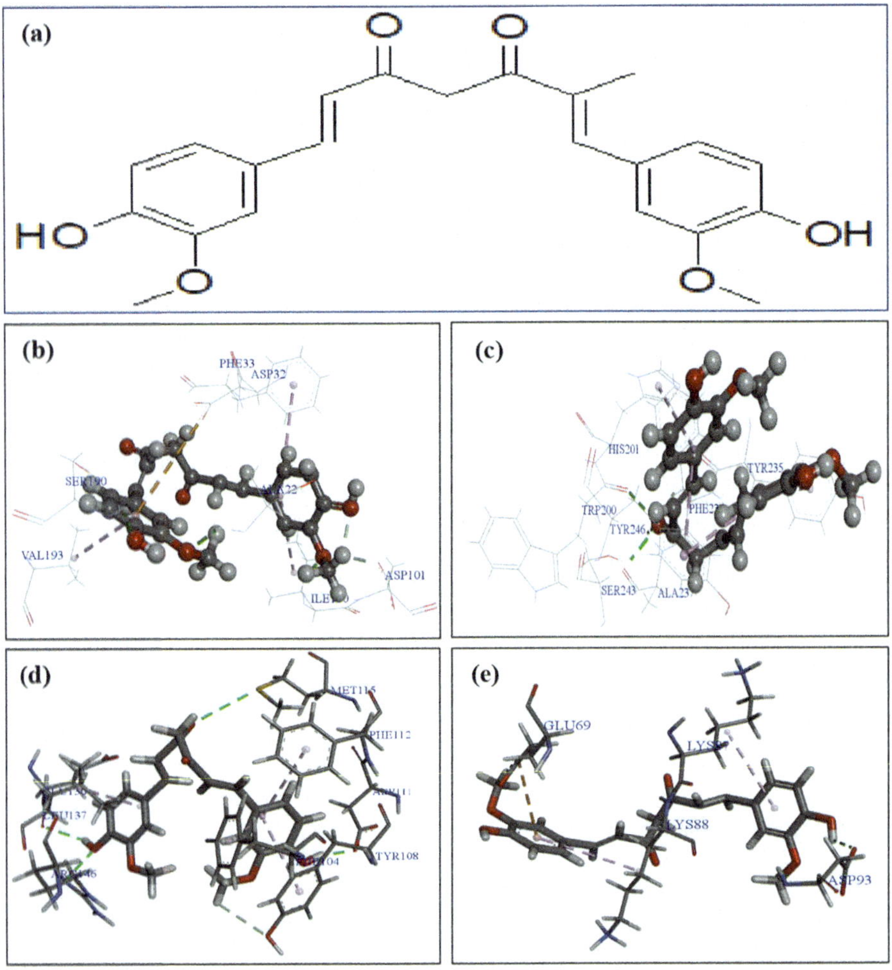

Fig. 1 Analysis of in silico docking of curcumin with p53. (**a**) 2D structure of CUR, (**b**) image showing the best docked conformation of p53-CUR complex, (**c**) image showing the best docked conformation of PARP-1-CUR complex, (**d**) image showing the best docked conformation of Bcl-2-CUR complex, and (**e**) image showing the best docked conformation of Cyt c-CUR complex

3,4-dihydro-1~{H}-isoquinolin-2-yl]carbonyl]-1,3-benzodioxol-5-yl]-~{N}-phenyl-5,6,7,8-tetrahydroindolizine-1-carboxamide, and HEME C, respectively. The required protein was built in order to dock by eliminating water molecules and incorporating hydrogen atoms. Since it allows for flexible docking, ArgusLab 4.0.1 was the preferred choice to carry out docking operations. The conformation of the ligand-receptor complex with the lowest binding energy was determined after flexible docking. Discovery Studio 3.5 (Dassault Systèmes BIOVIA, Discovery Studio 3.5, San Diego: Dassault Systèmes, 2016) was then used to analyze the interaction patterns and identify the crucial amino acid residues and associated linkages that bound the protein to the different ligand groups.

2.3 Formulation of PLGA-Encapsulated Nanocurcumin (NCUR)

The standard method of solvent displacement technique [31] was followed with slight modification to prepare NCUR encapsulated in poly-lactide-co-glycolide (PLGA) [32, 33] under ideal conditions. A net tenfold increase in the amount of drug was obtained after nano-formulation of CUR.

2.4 Characterization Using Atomic Force Microscopic (AFM) Analysis

NCUR samples were prepared for AFM imaging following the standard process [34–36] to estimate the shape, the size, as well as the uniformity in the frequency of the newly formed nanoparticles. Placed on a spotless, grease-free mica sheet, the NCUR drops were suspended in autoclaved Milli-Q water and left to dry. The WSxM 5.0 Develop 7.0 software was used to view and assess AFM signals recorded from NCUR, which were obtained as images in the amplitude and tapping modes.

2.5 Dynamic Light Scattering Spectroscopic Study and Zeta Potential Measurement

The average particle size, stability, and homogeneity of NCUR samples were examined using DLS analysis. The surface charge of NCUR particles, their intracellular stability, and interactions with other molecules are all described by their zeta potential characteristics, which were determined using Nano-ZS instrument (Malvern, UK) following standard protocol [34, 37].

2.6 Characterization by Field Emission Scanning Electron Microscope (FESEM) with Energy-Dispersive X-Ray Spectroscopy (EDX)

Analysis of the morphology and size of the newly synthesized NCUR was performed with dried powder sample taken on a carbon conductive tape by FESEM (GeminiSEM 450, Zeiss) instrument. Moreover, the EDX spectra gave additional information about the elemental composition of the PLGA-encapsulated NCUR [34].

2.7 CD Spectroscopy

Changes in the DNA structure on gradual interaction of drug with CT-DNA maintained at 37 °C temperature were monitored using circular dichroism (CD) spectra utilizing a rectangular quartz cuvette with a path length of 1 cm, in the wavelength range of 200–650 nm (Jasco spectropolarimeter; J-720 model, PC controlled; Jasco International Co. Ltd).

2.8 In Vitro Study

2.8.1 Cell Culture of Mammalian Cell Line L6

Mammalian (rat) skeletal muscle cells L6 obtained from the National Centre for Cell Science (NCCS, Pune, India) were grown and maintained in an atmosphere with 5% carbon dioxide and 37 °C temperature in Dulbecco's Modified Eagle's Medium (DMEM). The nutrient medium was supplemented with 10% fetal bovine serum and 1% antibiotic (Gibco 100X Antibiotic). For all experimental studies, cells were plated separately at a desired density and were allowed to grow to 80–90% confluency in fresh DMEM (pH 7.4). Before individual treatments in each and every culture plate, cells were re-equilibrated for 24 h [31, 38].

2.8.2 Selection of Dose of Alloxan in L6 Muscle Cells

ALX concentration of 1 mM was prepared in sterilized Milli-Q and kept for future use at 4 °C. During the treatment, the cells were incubated for 2 h with ALX at a concentration of 1 mM, freshly prepared in DMEM media [39]. Thus, this concentration and incubation time of ALX were further used in the study depending upon the results from previous range-finding trials [29].

2.8.3 Effective Dose of CUR and NCUR to Protect Against ALX-Induced Glucose Uptake in L6 Cells

The cells were administered with different CUR treatment dosages as a pre-treatment before being subjected to a normal dose of ALX for 2 h. The best effective dose of CUR was selected depending upon the extent of intracellular glucose absorption. A 1.6 mM dose of CUR was found to give the best result at 2 h pre-incubation time and therefore was further used as the standard dose of the study. The dose of NCUR was chosen to be 1.6 mM (higher dose) and 0.8 mM (lower dose), respectively, based on the previous data.

2.8.4 Evaluation of Glucose Uptake in L6 Cells

Glucose uptake by control and experimental L6 cells was estimated from the media glucose concentration following the standard GOD-POD kit technique [40]. This method was followed to ascertain the effective role of NCUR in reducing the hyperglycemic conditions in L6 cells when examined for both the control and experimental cell sets.

2.8.5 Assessment of In Vitro DNA Damage by Fragmentation Assay

The standardized phenol-chloroform method was used to extract and isolate genomic DNA from L6 cells belonging to various experimental groups. After isolation, the purity and the concentration of the extracted DNA were measured using a NanoDrop spectrophotometer. Then the DNA was made to run in 1% agarose gel electrophoresis technique to ascertain the DNA fragmentation from each experimental group [39, 41]. Then, the densitometric intensity of the bands was viewed under a UV transilluminator (Genei).

2.8.6 Analysis of Nuclear Condensation in L6 Cells by PI Staining

Several experimental cell sets were subjected to a staining process using propidium iodide (PI) to observe the nuclear morphology. The cells were treated with PI and incubated for 15 min at 4 °C (Samadder et al. 2013b) [35]. The cells were then examined under confocal microscope (Carl Zeiss LSM 800) at 10X magnification.

2.8.7 Expression Pattern of p53, PARP-1, Bcl-2, and Cyt c Proteins by Immunofluorescence Study in L6 Cell Line

The expression of proteins in the experimental and control groups was evaluated using anti-p53, anti-PARP-1, anti-Bcl-2, and anti-Cyt c primary antibodies and FITC (fluorescein-5-isothiocyanate)-labeled secondary antibodies [35]. The immunofluorescence study revealed the expression pattern of p53, PARP-1, Bcl-2, and Cyt c proteins in L6 cell line. The images thus obtained were recorded at 40X magnification using a confocal microscope (Carl Zeiss LSM 800), and the intensity graph representation was obtained using ImageJ software.

2.9 In Vivo Study

2.9.1 Mice Model

The experimental animals were acquired from Kalyani branch of West Bengal Livestock Development Corporation Ltd., Kolkata. Experiments were carried out with a 12–14-week-old, healthy inbred strain of Swiss albino mice (*Mus musculus*) weighing 22–30 g, considering both sexes, following the guidelines of CPCSEA, India. The animals were maintained at 25 ± 5°C with 12h light and dark cycle of day and night. The experimental procedure was authorized by the Institutional Animal Ethics Committee of Kalyani University (IAEC-KU) under reference 892/GO/Re/S/01/CPCSEA; Pr: 20.4.14–28.4.2019. The animals were fed a standard diet and had unlimited access to water.

2.9.2 Diabetes Induction in Mice

A single intraperitoneal (i.p.) injection of alloxan monohydrate (ALX), obtained from Sigma-Aldrich (CAS no. A7413), administered to each animal post-starvation for 12 h at a dose of 100 mg/kg b.w., was used to induce diabetes in the mice [42]. The animals were then given 5% glucose solution to drink overnight in order to prevent the initial hypoglycemic effect that occurred shortly after the administration of ALX. The development of hyperglycemic condition was observed just after 5 days of alloxan administration, which was verified after measuring the daily blood glucose level [43].

2.9.3 Dose Selection and Administration of CUR and NCUR in Mice

Based on the outcomes of a range-finding trial experiment using mice selected at random from the mice stock, the dose of curcumin (CUR) to be administered to mice was determined. After conducting a range-finding trial, the pre-treatment dose of CUR, which was 50 mg/kg b.w., was determined by comparing the percentage of viable cells in the control and experimental sets. Before administering 100 mg/kg b.w. of ALX intraperitoneally (i.p.), each mouse in their group received a single daily oral gavage feeding of CUR for 7 consecutive days. On the fifth day following the ALX injection, the blood glucose level was assessed for every treatment group. The mice specimens that were fed 50 and 60 mg/kg b.w. of CUR were discovered to be the most effective lowest and second lowest doses at limiting the significant rise in blood sugar levels following exposure to ALX. The ultimate dosage of 50 mg/kg b.w. was chosen for additional research.

2.9.4 Experimental Design

The animals were selected at random, irrespective of any sex biases, and grouped into the following categories:

Control group: A group of mice were fed with a standard diet and water ad libitum (untreated).

ALX-treated group: A group of mice were administered with a single intraperitoneal (i.p.) injection of ALX at a dose of 100 mg/kg b.w. post-starvation for 12 h.

Pre-treated CUR + ALX-treated group: CUR was pre-treated at a dose of 50 mg/kg b.w. for 7 consecutive days, followed by intraperitoneal administration of ALX at a dose of 100 mg/kg b.w. Feeding was stopped for 12 h prior to the exposure to ALX.

Pre-treated NCURI + ALX-treated group: This group was fed with NCUR at a dose of 25 mg/kg b.w. for 7 consecutive days, followed by intraperitoneal administration of ALX at a dose of 100 mg/kg b.w. Feeding was stopped for 12 h prior to the exposure to ALX.

Pre-treated NCURII + ALX-treated group: This group was fed with NCUR at a dose of 12.5 mg/kg b.w. for 7 consecutive days, followed by intraperitoneal administration of ALX at a dose of 100 mg/kg b.w. Feeding was stopped for 12 h prior to the exposure to ALX.

2.9.5 Estimation of Blood Glucose Level

Blood glucose levels from control and experimental sets of mice were determined following the standardized GOD-POD glucose kit (Autospan) from Arkray Healthcare Pvt. Ltd. (ref code: 93DP100-74) [40, 44, 45].

2.9.6 Estimation of Hepatic Biomarker Enzymes

The levels of acid phosphatase and alkaline phosphatase from hepatic tissue suspension were analyzed following the standard protocol of [46, 47], respectively. These liver biomarkers were measured as indicators of hepatotoxicity.

2.9.7 Assessment of Chromosomal Aberrations

Mice were given 0.03% colchicine intraperitoneally (i.p.) at a rate of 1 mL per 100 g body weight, and they were kept for 1 h and 10 min prior to sacrifice. Standard cytogenetical procedures were followed to analyze chromosome spreads in mouse bone marrow cells from various experimental groups [48–50]. To enable a comprehensive analysis of chromosomal abnormalities across different experimental sets, chromosome metaphase spreads were prepared, appropriately stained with Giemsa, and microscopically examined for cytogenetic abnormalities [46].

2.9.8 Assessment of Mitochondrial Membrane Potential (MMP) Using Spectrophotometer

Rhodamine 123 (Rh 123) dye was used to measure any modulation or changes in the mitochondrial membrane potential (MMP) ($\Delta\psi$ m) of cells isolated from pancreatic tissues from control and experimental sets of mice. From each experimental group, equal number of pancreatic tissue-isolated cells was incubated with 1 mM of the fluorescent probe and left in the dark for 10–15 min at room temperature. Ultimately, using the standard protocol, the result was examined spectrofluorometrically using excitation and emission at 488 and 530 nm wavelengths [47].

2.9.9 Determination of Anti-oxidative Enzyme Profiles

The levels of anti-oxidative enzymes were examined from the pancreatic tissues isolated from different experimental groups of mice. The tissues after isolation were homogenized in cold PBS and centrifuged at 12000 rpm for about 45–50 min at 4 °C. The supernatants thus obtained were taken for the enzymatic analysis. The anti-oxidative enzymes like GSH and CAT levels were determined from all experimental groups as per the standard protocol of Sedlak and Lindsey (1968), Kakkar et al. (1954), and Maehly and Chance (1954) [48–50].

2.9.10 Statistical Analysis

The experiments were conducted three times independently. After calculating the standard mean values from each experimental trial, the data were statistically evaluated using Student's t-test and one-way ANOVA (analysis of variance) between control vs. ALX treatment and ALX vs. drug-pre-treated sets. The data represented in the figure and table are as follows:

***$p < 0.001$ vs ALX, **$p < 0.01$ vs ALX, and *$p < 0.05$ vs ALX were considered significant as per Student's t-test. Further, the experiments with L6 cells involved preparing identical control and experimental culture dishes, repeating them three times independently before statistically analyzing the data.

3 Results

3.1 In Silico Prediction of the Drug-Protein Interaction

The successful docking of curcumin (CUR) into the functional binding pocket of the targeted proteins showed essential amino acid residues required for the interaction as shown in Fig. 1. The figure (Fig. 1a) represents the 2D structure of curcumin. The findings from molecular docking imply that the 3D structure of CUR interacts with p53 protein with a dock score value of −11.06 Kcal/mol (Fig. 1b) and the important amino acid residues responsible for best docking pose were SER190, VAL193, ALA22, ASP32, PHE33, ASP101, and PRO31. The best binding pose of the protein PARP-1 with the 3D structure of CUR suggests a dock score value of −9.63 Kcal/mol (Fig. 1c), and the important amino acid residues were TYR246, TYR235, SER243, ALA237, PHE236, HIS201, and TRP200. The binding score of the Bcl-2-CUR complex was found to be −10.94 Kcal/mol through docking studies (Fig. 1d), where the hydrophobic interactions were made between the ligand and the residues GLU136, LEU137, PHE104, ARG146, TYR108, ASP111, and PHE112, while the carbonyl moieties form hydrogen bonds with residues MET115 and PHE153, leading to the drug-protein complex's stability. The ligand curcumin (CUR) binds within the active cleft of Cyt c protein with a dock score value of −6.51 Kcal/mol (Fig. 1e). The amino acid residues responsible for the drug-protein stability are GLU69, LYS87, LYS88, and ASP93. These findings were further investigated in in vivo and in vitro experiments.

3.2 Characterization of PLGA-Encapsulated Nanocurcumin (NCUR)

PLGA-encapsulated NCUR was fully characterized by AFM, FESEM, and DLS techniques to validate the size of the nanoparticle (NP) that ranged between 0 and 100 nm. Atomic force microscopy exhibited the morphological characteristics of NCUR as recorded from the AFM images suggesting a spherical shape of the newly synthesized NPs (Fig. 2a, b, c, d). The mean diameter of NCUR was found to be 30 ± 0.05 nm in the subsequent 2D and 3D views of AFM images suggesting the NPs to have no crevices with smooth surface. From AFM study, NCUR also revealed a uniform spatial frequency of the topographic planar signal, as observed from the 2D view of the fast Fourier transformation (FFT). The spherical shape of the NCUR was further validated from the FESEM images (Fig. 2e). Moreover, the carbon and oxygen peaks detected in the EDX spectrum analysis attested to the purity of the recently produced NCUR. The

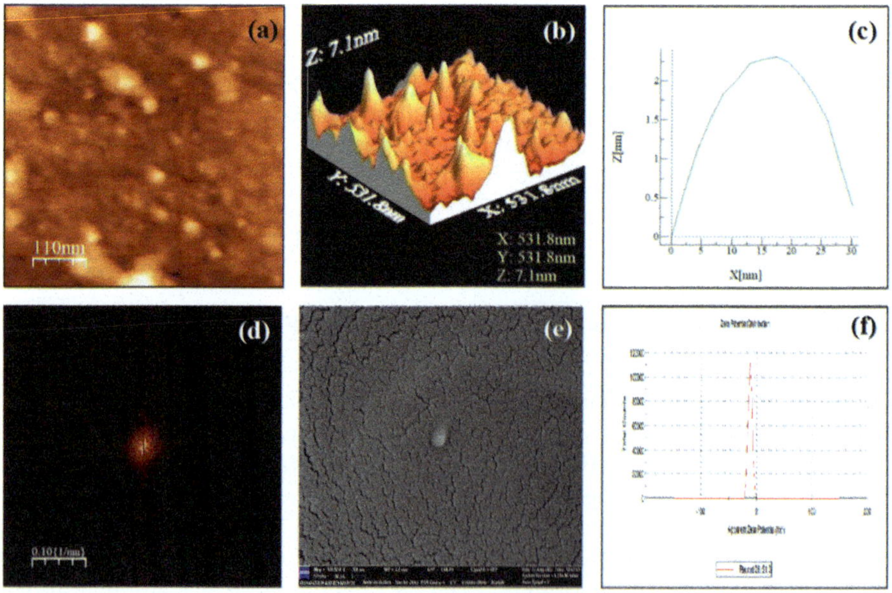

Fig. 2 (**a–d**) AFM study of NCUR: (**a**) 2D view of AFM image of NCUR, (**b**) 3D view of AFM image of NCUR, (**c**) size profile of NCUR, (**d**) FFT analysis, (**e**) FESEM image of NCUR, and (**f**) plot for zeta potential of PLGA-encapsulated NCUR

newly synthesized NPs were present below 100 nm, and the zeta potential value was determined to be −17 mV (Fig. 2f). These findings were verified by the DLS study, which also examined the size distribution of PLGA-encapsulated NCUR. Since the NCUR thus formed contained a negative charge (negative zeta value), it could enter cells readily by altering the potential of the cell membrane without compromising the integrity of the cell.

3.3 Estimation of NCUR-DNA Interaction by CD Spectroscopy

A stronger interaction between NCUR and CT-DNA than between CUR and CT-DNA alone is confirmed by the CD spectroscopic study. A strong affinity for NCUR (ten times lower dose of CUR) over CUR itself can be inferred from the differences in the upsurge in CD molar ellipticity of CT-DNA in the region of 265 nm upon addition of both CUR and NCUR to the solution.

3.4 In Vitro Study

3.4.1 Estimation of Glucose Uptake in Experimental L6 Cells (Rat Skeletal Muscle Cell Line)

The table (Table 1) demonstrated that pre-incubating cells with NCUR prior to the administration of ALX resulted in greater glucose uptake than CUR, suggesting that it may play a crucial role in mitigating hyperglycemic conditions in L6 cells. Thus, NCUR may be effective in modifying the decrease in glucose absorption in L6 cells caused by ALX (Table 1.)

3.4.2 Assessment of DNA Damage in L6 Cells

The extent of DNA damage was determined by measuring the intensity of DNA fragmentation following gel electrophoresis of genomic DNA isolated from the L6 cells from control and different treatment groups. The result revealed that CUR + ALX group could not bring about any restriction to the ALX-induced DNA damage in comparison to NCURI- and NCURII-pre-treated groups (Fig. 3). Thus, it can be said that the NCUR doses work more efficiently by binding to the grooves of the DNA isolated from L6 cells, which might be the reason for restriction in the extent of DNA damage induced

Table 1 Determination of the concentration of media glucose level in control and experimental L6 cells

Group	Media glucose level (mg/dL)
CONT	105 ± 3.53
ALX	$127 \pm 1.92^{\#\#}$
CUR + ALX	$118 \pm 2.59^{*}$
NCUR I + ALX	$112 \pm 1.24^{**}$
NCUR II + ALX	$109 \pm 3.24^{**}$

$^{\#\#}p < 0.01$ vs CONT, $^{**}p < 0.01$ vs ALX, and $^{*}p < 0.001$ vs ALX were considered significant for Student's t-test

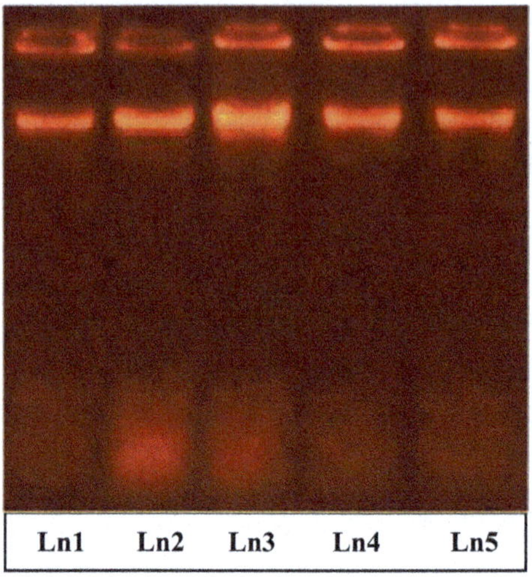

Fig. 3 DNA fragmentation assay in L6 cells of different experimental groups showing effects of NCUR on DNA damage. Ln 1, control; Ln 2, ALX; Ln 3, CUR + ALX; Ln 4, NCURI + ALX; Ln 5, NCURII + ALX

by ALX treatment. These results are in agreement with the findings of nuclear condensation study of hepatic cells.

3.4.3 Detection of Nuclear Condensation in L6 Cell

As per the images obtained from the confocal microscope, the fluorescence intensity of PI dye was evaluated for determining the effectiveness of CUR and NCUR in protection against ALX-induced DNA damage. The decreased fluorescence intensity of PI (propidium iodide) in pre-treated CUR and NCUR-pre-treated L6 cells indicates the protective efficacy of CUR and NCUR against ROS-induced nuclear condensation more effectively than the ALX-treated group of L6 cells (Fig. 4a).

3.4.4 Analysis of Expression Level of p53, PARP-1, Bcl-2, and Cyt c Proteins by Immunofluorescence Study in L6 Cells

According to the immunofluorescence investigation observed under confocal microscope, the pre-treated NCURI and NCURII groups of L6 cells showed an elevated expression of p53 protein than that of the cells exposed to ALX. Moreover, greater nuclear condensation-based damage and lower levels of protein expression were observed in the ALX-treated group (Fig. 4b). Similar results were obtained in the case of expression study with DNA repair protein PARP-1 using immunofluorescence method. Moreover, mitochondrial membrane proteins like Bcl-2 and Cyt c protein were also considered for immunofluorescence study. Results suggested that Bcl-2 protein was overexpressed and Cyt c protein expression was suppressed in the NCUR-pre-treated groups as compared to that of CUR-treated groups. The expression patterns of all the proteins PARP-1, Bcl-2, and Cyt c are represented as intensity graphs for better understanding of the interaction between these proteins (Fig. 5).

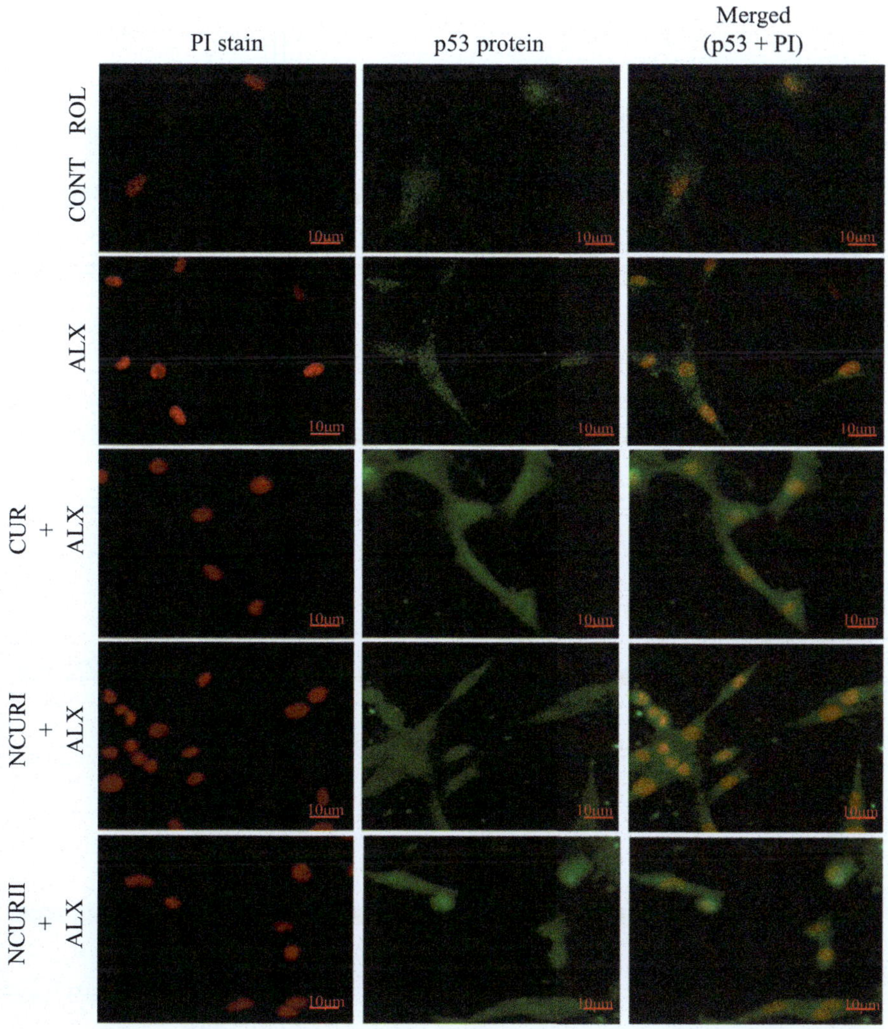

Fig. 4 Analysis of nuclear condensation and expression pattern of p53 protein in different experimental groups of L6 cells under confocal microscope. (**a**) Assessment of nuclear condensation using PI dye. (**b**) Immunofluorescence study showing expression level of p53 in in vitro L6 cell

3.5 In Vivo Study

3.5.1 Glucose Lowering Effect of CUR and NCUR

The mice were treated with CUR, NCURI, and NCURII before being exposed to ALX. As per the GOD-POD analysis, when compared to the control group, alloxan-treated mice group had considerably higher blood glucose levels, while pre-treated groups (CUR + ALX, NCURI + ALX, NCURII + ALX) showed dramatically lower or

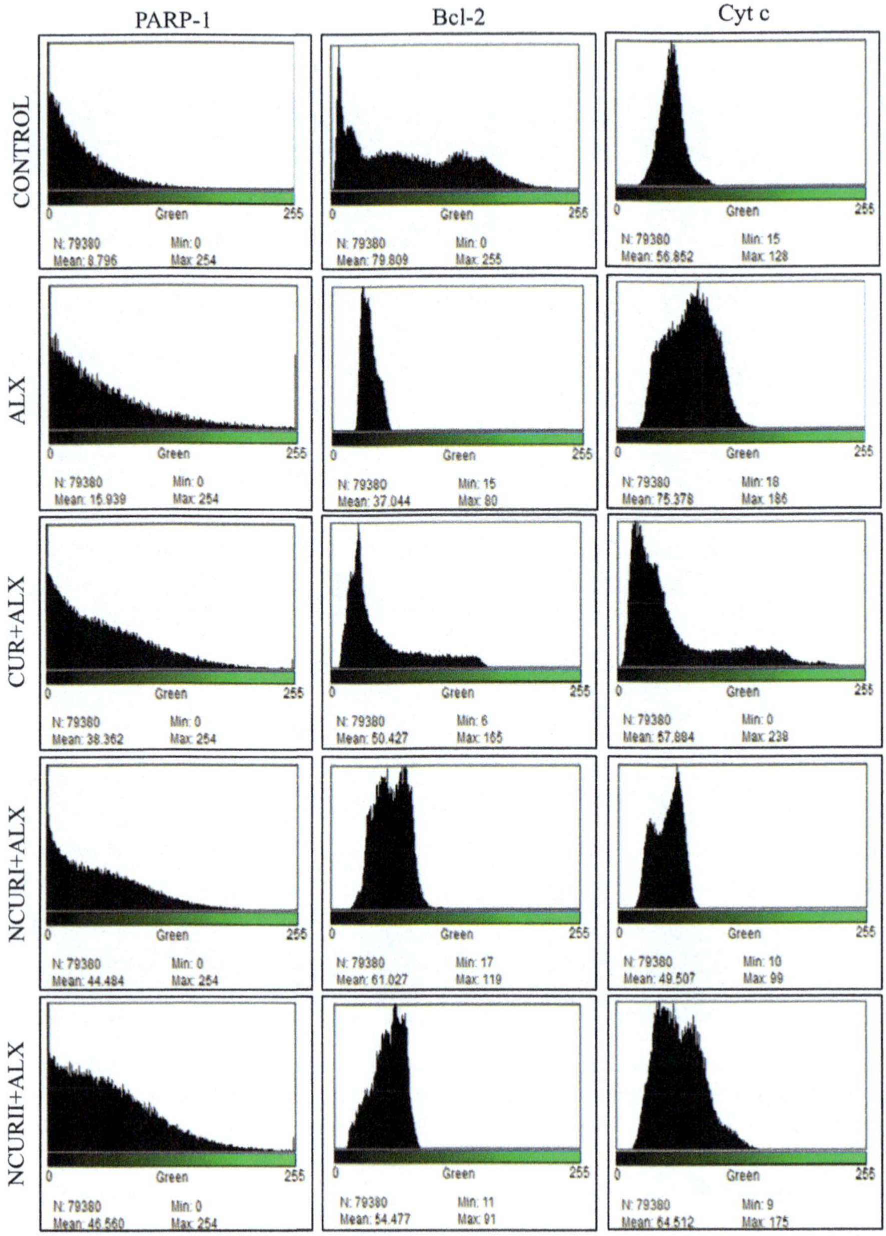

Fig. 5 Intensity plot representation of PARP-1, Bcl-2, and Cyt c protein expression pattern in L6 cells of different experimental groups. The color bar representation ranging from 0 to 255 in the image signifies the intensity of protein expression where green color represents higher value, greenish-black color represents a moderate value, and black color represents lower value

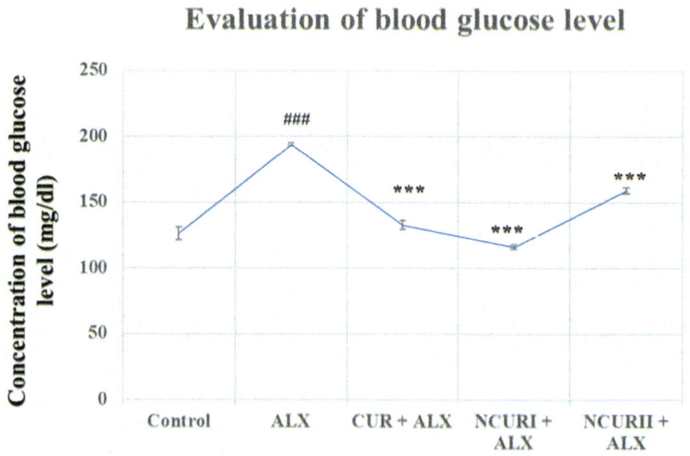

Fig. 6 Graphical representation of modulation of blood glucose level in the experimental groups. $^{###}p < 0.001$ vs control and $^{***}p < 0.001$ vs ALX were considered significant for Student's t-test

near-normal blood glucose levels, particularly pre-treated nano-groups showing the greatest reduction in blood glucose levels (Fig. 6).

3.5.2 Analysis of Hepatic Enzyme Level

Following exposure to ALX, the active biomarker enzymes were stimulated at the onset of hepatotoxicity. The decreased levels of both acid phosphatase and alkaline phosphatase in the ALX-treated groups were found to be constrained in pre-treated CUR and NCUR groups. It was discovered that the acid and alkaline phosphatase levels as measured from the homogenized liver tissues of mice were brought close to normal in pre-treated NCUR groups as compared to CUR-pre-treated group (Fig. 7).

3.5.3 Assessment of Chromosomal Aberrations in Experimental Groups

Alloxan (ALX) administration in mice model resulted in several types of chromosomal aberrations (CAs) like ring chromosome, chromatid breakage, terminal deletion, stretching, etc. as compared to the control group. Further, the aberrations were measured in all the treatment groups as number of observations per 1000 cell count. From the observations, the NCUR + ALX group revealed better results in inhibiting the ALX-induced genotoxicity by restricting the number of CAs than the CUR + ALX group as shown by the graphical representation (Fig. 8). This study suggests that alloxan also has a role in DNA damage or lesions that ultimately hinder the proper replication of the DNA and maintenance of stable chromosomal structures without mutations or abnormalities. These positive results in the case of the NCUR + ALX group imply an active DNA damage response (DDR), favoring protection of the DNA against ALX-induced DNA lesions. Thus, in the in vivo model, NCUR doses could work concomitantly against ALX-induced DNA damage and chromosomal aberrations that ultimately leads to impaired function of the entire cell.

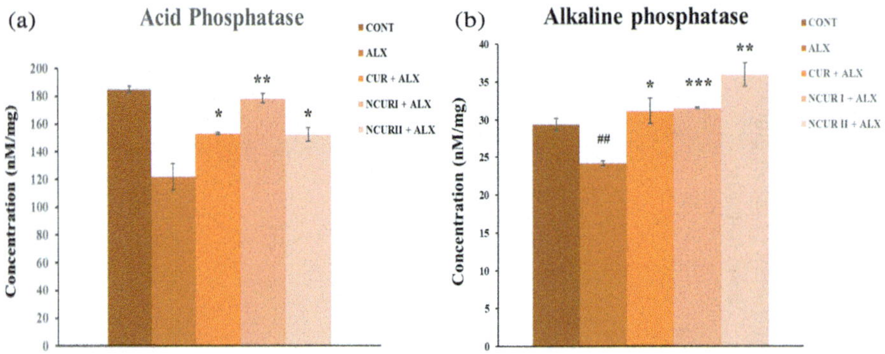

Fig. 7 Analysis of hepatic enzyme profiles of all experimental groups of mice. (**a**) Graphical representation of acid phosphatase levels in hepatic tissues of experimental groups of mice. (**b**) Graphical representation of alkaline phosphatase levels in hepatic tissues of experimental groups of mice. $^{\#\#}p < 0.01$ vs control, $^{***}p < 0.001$ vs ALX, $^{**}p < 0.01$ vs ALX, $^{*}p < 0.05$ vs ALX

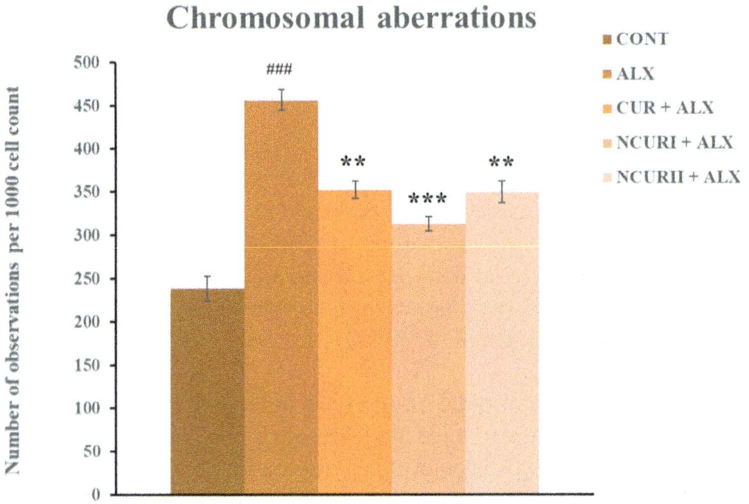

Fig. 8 Graphical representation showing the occurrence of chromosomal aberrations (CAs). $^{\#\#\#}$p < 0.001 vs control, $^{**}p < 0.01$, and $^{***}p < 0.001$ vs ALX were considered significant for Student's t-test

3.5.4 Estimation of Mitochondrial Membrane Potential (MMP)

As compared to control group of mice, the ALX-treated group showed a decrease in the MMP level by reduced fluorescence spectra indicating a depolarization of mitochondrial membrane potential. However, pre-treated CUR and NCUR groups showed an upsurge of the fluorescence intensity suggesting a restriction in the lowering of MMP value by NCUR, thereby reverting back to the near-normal hyperpolarized state of the mitochondrial membrane potential. In this instance, NCURI was a more effective candidate than NCURII for preserving the MMP level (Fig. 9).

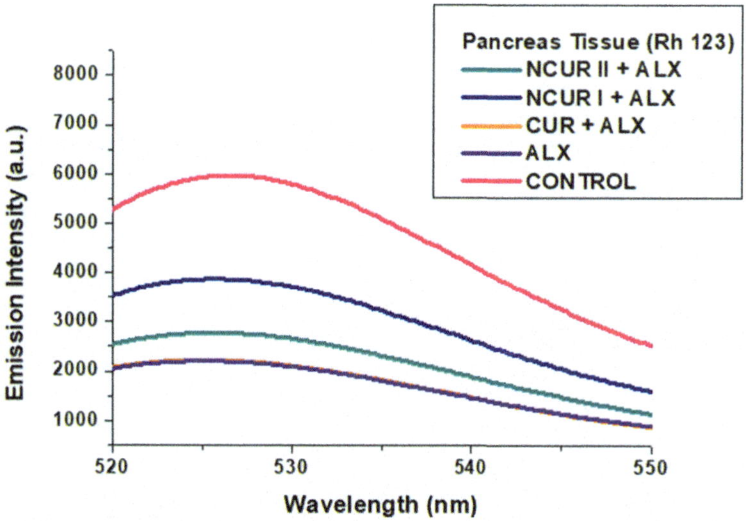

Fig. 9 Analysis of mitochondrial membrane potential in pancreatic cells by Rh 123 staining (λexi = 510 nm; λemi = 525 nm)

3.5.5 Determination of Anti-oxidative Enzyme Profile in Pancreatic Tissue

The figure shows the levels of antioxidant enzymes, specifically glutathione (GSH), in different experimental sets. It shows that the pre-treated CUR and NCUR groups had significantly higher enzyme levels than the ALX-treated group, which is indicative of enhanced function. Total thiol enzyme activity was higher in the NCURI-pre-treated group than in the NCURII-treated group (Fig. 10).

4 Discussion

The incidence of diabetes has been increasing worldwide and has become a chronic condition characterized by elevated blood glucose levels that finally cause serious harm to the vital organs of the body. When blood glucose levels are above normal but not high enough to be diagnosed as diabetes, the condition is known as prediabetes. The most common causes of prediabetes are ineffective insulin utilization by the body or insufficient insulin production by the pancreas to maintain normal blood glucose levels. Type 2 diabetes is more likely to strike people who have prediabetic condition. Therefore, to manage the symptoms of diabetes, prevention of the complications developing from diabetes has become necessary. Reports have confirmed the use of a wide range of synthetic drugs for management of diabetes. But due to the health-related shortcomings from synthetic drugs, inclination to the phyto-based drug was opted. Recently, nanotechnology has become a significant tool for molecularly altering and modifying various objects to achieve desired features. This modern technique has shown promise in the diagnosis and treatment of several diseases through improved

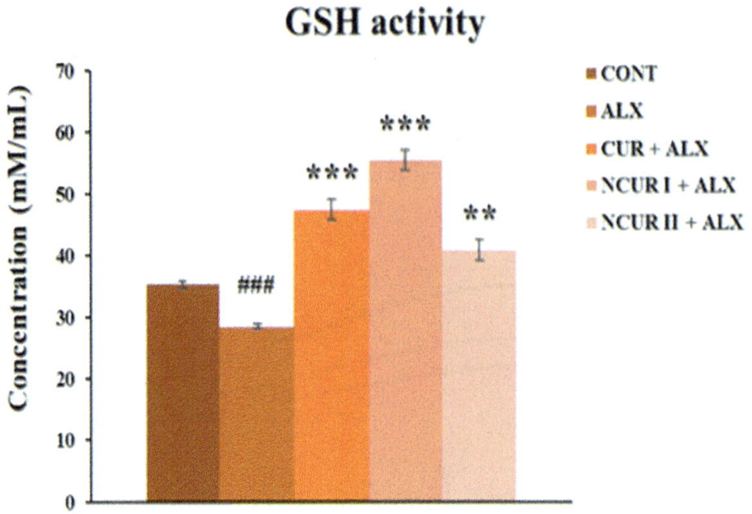

Fig. 10 Analysis of activity of anti-oxidative enzyme GSH in all experimental groups of mice. ###$p < 0.001$ vs control, ***$p < 0.001$ vs ALX

drug delivery and efficacy. Therefore, nanotechnological approach was undertaken to enhance the physical properties of the phyto-drug to ensure drug stability, targeted delivery, bioavailability, and tissue-specific drug distribution even at lower amounts. The nanocarrier used during the nano-drug formulation offered the advance benefits to the drug in terms of toxicity, drug release, and drug stability. This study presents the use of PLGA for nano-encapsulation of the phyto-drug. The biopolymer PLGA exhibits lower toxicity owing to its natural origin as well as controlled release of drug due to its biodegradable property. Moreover, the solvent displacement technique used for nano-encapsulation of the drug CUR within the nanocarrier PLGA promotes the size of the nanoparticles (NPs) to be within nanorange (0–100 nm) with a negative charge on the surface. Owing to the nano-size as well as negative charge, the functional efficacy of the nano-drug increased at lowest doses. This novel approach of nano-formulating phyto-based drugs makes this therapeutic strategy a cost-effective one without long-term consequences with high probability of therapeutic success.

Nanoparticles have attracted intense interest due to their wide range of applications. To confirm the physical features of the formed nanoparticles, namely, the average size, surface charge, elemental composition, as well as other physical characteristics, several techniques have been followed. The high surface-to-volume ratio causes an exponential increase in molecular reactivity, and nanoscale materials frequently exhibit properties that differ from those of their bulk counterparts. The attributes include chemical, mechanical, optical, and electronic ones. The properties of the nanoparticles (NPs) can also vary greatly. All of these methods, taking into account the aspects like accessibility, expense, accuracy, simplicity, non-destructiveness, and compatibility with particular nanoparticle compositions, and these characterization techniques were chosen and thoroughly examined. The size, shape, morphology, their uniform distribution, and electronic charges were determined using different techniques. Using atomic force microscopy, the size of the NP and its shape in 3D mode were recorded. Moreover, the AFM studies determined the dispersion pattern of the NPs in the surface. The height profile recorded from AFM study determined the average size of the NCUR to be about 30 nm, and FFT data confirmed the uniformity in the frequency of the formed NPs. FESEM validated the spherical shape of the formed NCUR and the average diameter of the NPs to be within nanorange (<100 nm). The findings from DLS study detected the presence of NP agglomeration and thus determined the negative zeta potential of −17 mV suggesting a good internalization and stability inside the biological cells. Thus, these physico-chemical characterizations confirmed the formed nanoparticle was of desired physical characteristics that increases the propensity of using NPs in biological system by many folds even at lower doses of drug. The circular dichroic spectra further determined the biophysical characteristics of the NCUR, which confirmed the successful binding of the NPs to the calf thymus (CT-DNA) by bringing positive helicity changes in the DNA structure as compared to the non-nano-form.

After successful nanoencapsulation of the drug with desired properties, the hypoglycemic effect of both CUR and NCUR was investigated in the rat skeletal muscle L6 cell line. The skeletal muscle L6 was chosen in this study to ascertain the effect of glucose uptake in the peripheral tissues. Moreover, due to the expression of GLUT4 transporter in the L6 cells, this study aims to determine the pharmacokinetics of the skeletal muscle cell in the presence of the drug. From the in vitro analysis of media glucose level, it was revealed that NCUR pre-exposure prior to ALX exposure could enhance the ability of the L6 myotubes to take up glucose from the culture media (DMEM). This study thus reflects an effective utilization of glucose by peripheral tissue cells in the presence of NCUR drug as compared to its non-nano-form. Moreover, the DNA damage assay performed with the DNA isolated from L6 cells suggests effective protection of NCUR against DNA fragmentation induced by ALX. These results were further confirmed by a lower intensity of PI dye depicting a reduced level of DNA damage in the L6 cell when exposed to NCUR prior to ALX. This effective protection of DNA structure by the CUR was ascertained prior to these experimental studies by addressing a virtual molecular docking technique. In this study, a prime protein responsible for DNA damage response, that is, p53, was chosen to foresee effective interactions between the ligand (CUR) and the protein (p53). The effective binding of p53 with CUR with a dock score of −11.06 Kcal/mol suggests effective interaction of CUR with the active sites of the p53 protein. The immunofluorescence study with the p53 protein further validated the in silico molecular docking study. Since p53 protein functions as DNA damage response protein, the higher expression of p53 in NCUR-pre-treated groups is indicative of protection of DNA structure and integrity from ALX-induced ROS production and DNA damage. This overexpression of p53 protein in the case of the NCUR group as compared to the ALX group suggests that NCUR could act as an anti-genotoxic drug by binding effectively with p53 protein and thus increasing the protein activity in preventing the ALX-induced DNA damage. The expression pattern of a DNA repair protein PARP-1 was considered, which showed similar observations in L6 cells. Thus, from the expression study, it becomes very clear that NCUR doses function by enhancing the p53 expression, which thereby increases the possibility of DNA repair mechanisms, which is indicative of protection of DNA structure and integrity from ALX-induced genotoxic stress and cellular dysfunctioning. On the other hand, Bcl-2 protein acts as mitochondrial membrane protein that takes part in promoting cellular survival by inhibiting the actions of important proteins like Cyt c. The differential expression patterns of Bcl-2 and Cyt c proteins occur in a synchronized fashion in the cell. Thus, protein expression study confirms that NCUR plays a crucial role in maintaining cellular homeostasis by maintaining the levels of mitochondrial proteins as well as the DNA damage response proteins prior to ALX-induced genotoxicity and mitochondrial dysfunctioning.

Both the nano- and non-nano-forms of the drug were found to dictate the elevated blood glucose levels in the ALX-induced diabetic mice models. The pre-administration of NCUR to mice with diabetes mellitus significantly reduced the glucose imbalance, indicating a possible hypoglycemic effect of NCUR as compared to CUR alone. Reports have confirmed a direct link of diabetes to liver damage and biochemical abnormalities in addition to glucose imbalance. So, any form of abnormalities at the level of hepatic biomarker enzymes was studied. The liver is known to be an important organ that the body needs to regulate many different physiological processes like gluconeogenesis and glycogenesis. So, any form of impairment in the proper liver functioning would be reflected in the circulating blood glucose level of the body. So, determining the levels of hepatic biomarker enzymes like acid phosphatase and alkaline phosphatase gave an idea of the extent of liver damages in the diabetic mice induced due to ALX-mediated oxidative stress in the liver tissue. A marked drop in the acid phosphatase and alkaline phosphatase levels obtained from the homogenized liver of diabetic mice suggests a correlation of hepatocellular damage as well as decreased liver function with the imbalanced blood sugar level in ALX-mediated diabetic mice. From the studies, it was found that NCUR pre-treatment could significantly control the ALX-mediated organ damage and thus modulate the activity levels of the enzymes near to normal.

Pancreatic tissue is directly related with the maintenance of glucose levels by balancing the production and secretion of insulin hormones. Normal functioning of the pancreatic tissues offers normal secretion of insulin and proper utilization of circulating blood glucose by the peripheral cells. So, in this study, abnormalities in the pancreatic tissues were studied by determining the mitochondrial membrane potential, which further determines the extent of functioning of pancreatic cells by maintaining optimum balance of ADP/ATP levels. Results showed that the MMP levels were decreased in the ALX-treated mice tissue. This lowering of the mitochondrial membrane potential might be the result of ALX-induced oxidative stress in the pancreas mediated by reactive oxygen species (ROS) generation. The ROS generation in the pancreas could result from the decreased activity of the anti-oxidative enzymes, thereby agreeing well with the obtained results. Since inflammatory processes have a close relationship with oxidative stress, which is brought on by the buildup of reactive oxygen species (ROS), it intensifies inflammation by activating transcription factors linked to inflammation. Thus, curcumin effectively functions by inhibiting nicotinamide adenine dinucleotide phosphate (NADPH) oxidase and increasing the activity of antioxidant enzymes, and reducing the inflammation by decreasing the generation of reactive oxygen species (ROS), which is indirectly associated with the Nrf2-Keap1 pathway [51–53]. Moreover, curcumin functions by attaching itself to toll-like receptors (TLRs) and modulates downstream signaling pathways, such as mitogen-activated protein kinases (MAPK), activator protein 1 (AP-1), and nuclear factor kappa-B (NF-κB), thus controlling inflammatory mediators and inflammatory diseases [54]. Pro-inflammatory mediators including interleukin-1 (IL-1), IL-1β, IL-6, IL-8, IL-17, IL-27, tumor necrosis factor α (TNF-α), inducible nitric oxide synthase (iNOS), etc. are also reported to be reduced in levels by curcumin [55]. Thus, it can be implied that the developed nanoparticle NCUR could more effectively protect the pancreatic tissue by its enhanced absorption and bioavailability owing to its nano-size and negative zeta potential. Moreover, the present study with NCUR corroborated with the positive reports of curcumin against inflammation by preventing the alteration of the anti-oxidative enzyme profile, namely, GSH. This depicts the decrease in MMP levels in pancreatic cells in NCUR-pre-treated mice group as compared to the CUR-pre-treated groups. Thus, NCUR functions as an efficient protective agent against ALX-induced tissue damages and thereby control the imbalance in circulating blood glucose levels.

The current investigation suggests that the NCUR could function as anti-hyperglycemic therapeutics against ALX-induced mitochondrial dysfunction by maintaining the MMP levels and levels of anti-oxidative enzymes in the pancreas. This reflects the strong hypoglycemic effect of NCUR as compared to CUR as a phyto-based drug. Thus, from the overall study, it can be concluded that nanoencapsulation of phyto-based drug might be chosen as a successful therapeutic approach to manage diabetes induced due to genotoxicity and mitochondrial dysfunction in both in vitro and in vivo models.

5 Conclusion

To summarize, this study demonstrates the overall efficacy of a PLGA-encapsulated nano-curcumin drug delivery system that uses curcumin, a bioactive phytocomponent, to prevent a variety of potentially fatal biological problems brought on by alloxan. This nanotechnological approach ensues an excellent sustainable release and enhanced uptake of the drug in the tissues. To our knowledge, the current study confirmed for the first time the usefulness of nano-curcumin (NCUR) in preventing genotoxicity caused by ALX by modifying p53, PARP-1, and mitochondrial dysfunction through modifying the expression levels of Bcl-2 and Cyt c proteins in an in vitro system. The results of the in silico drug and protein interaction prediction were well supported by the experimental data. By examining a number of signaling proteins linked to DNA repair pathways, anti-inflammatory enzymatic levels, and MMP levels, the mechanisms by which NCUR prevented ALX toxicity in the L6 cell line and mice model were also examined. Overall, the results clearly specify that NCUR could efficiently lower the genotoxicity by upregulating p53 and PARP-1 proteins and modulate Bcl-2 and Cyt c proteins that govern the normal mitochondrial functioning. Therefore, the in silico molecular docking-based therapeutic strategies using nano-formulated curcumin drug would restrict the harmful effects of the food additive in mice models, which might be replicated in human population. Thus, phyto-based nano-drug proves to serve more efficient in restricting the ill-effects of food additive-induced diabetes in mankind in a cost-effective and eco-friendly manner.

Acknowledgments The authors would extend acknowledgment to UGC, New Delhi, for the UGC-BSR (grant no. F.30-488/2019(BSR) and SERB (DST) (grant no. ECR/2017/000355/LS) sanctioned to Dr. Asmita Samadder for this research work. The authors are grateful to the University of Kalyani for providing Personal Research Grant to A.S., which was partly used for the research work. S.D. is grateful to UGC for the CSIR-NET SRF fellowship. The authors express sincere gratitude and appreciation to UGC-DAE Consortium for Scientific Research, Kolkata Centre, and Visva-Bharati University for their infrastructural support and providing necessary research facilities.

Author Contribution All the authors contributed equally to the paper as follows: conceptualization, A.S., S.N., and P.G.; data curation, S.D.; funding acquisition, A.S.; investigation, A.S. and S.N.; software, S.N. for in silico study and A.S. for in vivo study; supervision, A.S.; validation, A.S., S.N., and P.G.; roles/writing—original draft, S.D.; and writing—review and editing, A.S., S.N., and P.G.

Declaration of Competing Interest None to declare

References

1. Guariguata, L., Whiting, D.R., Hambleton, I., Beagley, J., Linnenkamp, U., Shaw, J.E.: Global estimates of diabetes prevalence for 2013 and projections for 2035. Diabetes Res. Clin. Pract. **103**, 137–149 (2014)
2. van Dieren, S., Beulens, J.W., van der Schouw, Y.T., Grobbee, D.E., Neal, B.: The global burden of diabetes and its complications: an emerging pandemic. Eur. J. Cardiovasc. Prev. Rehabil. **17**, S3–S8 (2010)
3. American Diabetes Association: Cardiovascular Disease and Risk Management. Sec. 9. Diabetes Care. **40**, S75–S87 (2017)
4. Melmed, S., Polonsky, K.S., Larsen, P.R., Kronenberg, H.M.: Williams Textbook of Endocrinology, 12th edn. Elsevier/Saunders, Philadelphia (2011)
5. Joseph, T.C., Vadasseril, J.: Diabetes—a silent killer: a threat for cardiorespiratory fitness. In: Cardiorespiratory Fitness - New Topics. IntechOpen (2023)
6. Zhang, P., Engelgau, M.M., Norris, S.L., Gregg, E.W., Narayan, K.M.: Application of economic analysis to diabetes and diabetes care. Ann. Intern. Med. **140**, 972–977 (2004)
7. World Health Organization: Definition, Diagnosis and Classification of Diabetes Mellitus and its Complications. Report of WHO consultation. Part 1. WHO, Geneva (1999)
8. Trasande, L., Shaffer, R.M., Sathyanarayana, S.: Council on environmental health. Food additives and child health. Pediatrics. **142**, e20181410 (2018)
9. Dey, S., Samadder, A., Nandi, S.: Exploring current role of nanotechnology used in food processing industry to control food additives and their biochemical mechanisms. Curr. Drug Targets. **23**, 513–539 (2022)
10. Ighodaro, O.M., Adeosun, A.M., Akinloye, O.A.: Alloxan-induced diabetes, a common model for evaluating the glycemic-control potential of therapeutic compounds and plants extracts in experimental studies. Medicina. **53**, 365–374 (2017)

11. Mohammad, A.: The role of fruits, vegetables, and spices in diabetes. Int. J. Nutr. Pharmacol. Neurol. Dis. **1**, 27–35 (2011)
12. Sagandira, C.R., Khasipo, A.Z., Sagandira, M.B., Watts, P.: An overview of the synthetic routes to essential oral anti-diabetes drugs. Tetrahedron. **96**, 132378 (2021)
13. Salehi, B., Ata, A., Anil Kumar, V.N., Sharopov, F., Ramírez-Alarcón, K., Ruiz-Ortega, A., Abdulmajid Ayatollahi, S., Tsouh Fokou, P.V., Kobarfard, F., Amiruddin Zakaria, Z., Iriti, M., Taheri, Y., Martorell, M., Sureda, A., Setzer, W.N., Durazzo, A., Lucarini, M., Santini, A., Capasso, R., Ostrander, E.A., ur-Rahman, A., Choudhary, M.I., Cho, W.C., Sharifi-Rad, J.: Antidiabetic potential of medicinal plants and their active components. Biomolecules. **9**, 551 (2019)
14. Wadkar, K., Magdum, C., Patil, S., Naikwade, N.: Antidiabetic potential and Indian medicinal plants. J. Herb. Med. Toxicol. **2**, 45–50 (2008)
15. Rao, M.U., Sreenivasulu, M., Chengaiah, B., Reddy, K.J., Chetty, C.M.: Herbal medicines for diabetes mellitus: a review. Int. J. PharmTech Res. **2**, 1883–1892 (2010)
16. Patel, D., Prasad, S., Kumar, R., Hemalatha, S.: An overview on antidiabetic medicinal plants having insulin mimetic property. Asian Pac. J. Trop. Biomed. **2**, 320–330 (2012)
17. Nosrati-Oskouie, M., Aghili-Moghaddam, N.S., Tavakoli-Rouzbehani, O.M., Jamialahmadi, T., Johnston, T.P., Sahebkar, A.: Curcumin: a dietary phytochemical for boosting exercise performance and recovery. Food Sci. Nutr. **10**, 3531–3543 (2022)
18. Nabavi, S.F., Thiagarajan, R., Rastrelli, L., Daglia, M., Sobarzo-Sánchez, E., Alinezhad, H., Nabavi, S.M.: Curcumin: a natural product for diabetes and its complications. Curr. Top. Med. Chem. **15**, 2445–2455 (2015)
19. Matoba, T., Koga, J.I., Nakano, K., Egashira, K., Tsutsui, H.: Nanoparticle-mediated drug delivery system for atherosclerotic cardiovascular disease. J. Cardiol. **70**, 206–211 (2017)
20. Onoue, S., Yamada, S., Chan, H.K.: Nanodrugs: pharmacokinetics and safety. Int. J. Nanomedicine. **9**, 1025–1037 (2014)
21. Pala, R., Anju, V.T., Dyavaiah, M., Busi, S., Nauli, S.M.: Nanoparticle-mediated drug delivery for the treatment of cardiovascular diseases. Int. J. Nanomedicine. **27**, 3741–3769 (2020)
22. Sow, P., Dey, S., Dey, R., Majumder, A., Nandi, S., Bera, M., Samadder, A.: Poly lactide-co-glycolide encapsulated nano-curcumin promoting antagonistic interactions between HSP 90 and XRCC1 proteins to prevent cypermethrin-induced toxicity: an in silico predicted in vitro and in vivo approach. Colloids Surf. B Biointerfaces. **220**, 112905 (2022)
23. Dey, R., Dey, S., Sow, P., Chakrovorty, A., Bhattacharjee, B., Nandi, S., Samadder, A.: Novel PLGA-encapsulated-nanopiperine promotes synergistic interaction of p53/PARP-1/Hsp90 axis to combat ALX-induced-hyperglycemia. Sci. Rep. **14**, 9483 (2024)
24. Álvarez-Martínez, J.F., Barrajón-Catalán, E., Encinar, A.J., Rodríguez-Díaz, C.J., Micol, V.: Antimicrobial capacity of plant polyphenols against Gram-positive bacteria: a comprehensive review. Curr. Top. Med. Chem. **27**, 2576–2606 (2020)
25. Ting, C.Y., Wang, H.E., Yu, C.C., Liu, H.C., Liu, Y.C., Chiang, I.T.: Curcumin triggers DNA damage and inhibits expression of DNA repair proteins in human lung cancer cells. Anticancer Res. **35**, 3867–3873 (2015)
26. Hoeijmakers, J.H.J.: DNA damage, aging, and cancer. N. Engl. J. Med. **361**, 1475–1485 (2009)
27. Avalos, J.L., Bever, K.M., Wolberger, C.: Mechanism of sirtuin inhibition by nicotinamide: altering the NAD(+) cosubstrate specificity of a Sir2 enzyme. J. Mol. Cell. **17**, 855–868 (2005)
28. Miyashiro, J., Woods, K.W., Park, C.H., Liu, X., Shi, Y., Johnson, E.F., Bouska, J.J., Olson, A.M., Luo, Y., Fry, E.H., Giranda, V.L., Penning, T.D.: Synthesis and SAR of novel tricyclic quinoxalinone inhibitors of poly (ADP-ribose)polymerase-1 (PARP-1). Bioorg. Med. Chem. Lett. **19**, 4050–4054 (2009)
29. Fatokun, A.A., Dawson, V.L., Dawson, T.M.: Parthanatos: mitochondrial-linked mechanisms and therapeutic opportunities. Br. J. Pharmacol. **171**, 2000–2016 (2014)

30. Rohilla, A., Ali, S.: Alloxan induced diabetes: mechanisms and effects. Int. J. Res. Pharm. Biomed. Sci. **3**, 819–823 (2012)
31. Fessi, H., Puisieux, F., Devissaquet, J.P., Ammoury, N., Benita, S.: Nanocapsule formation by interfacial polymer deposition following solvent displacement. Int. J. Pharm. **55**, R1–R4 (1989)
32. Samadder, A., Abraham, S.K., Khuda-Bukhsh, A.R.: Nanopharmaceutical approach using pelargonidin towards enhancement of efficacy for prevention of alloxan-induced DNA damage in L6 cells via activation of PARP and p53. Environ. Toxicol. Pharmacol. **43**, 27–37 (2016)
33. Das, J., Samadder, A., Das, S., Paul, A., Khuda-Bukhsh, A.R.: Nanopharmaceutical approach for enhanced anti-cancer activity of betulinic acid in lung-cancer treatment via activation of PARP: interaction with DNA as a target anti-cancer potential of nano-betulinic acid in lung cancer. J. Pharmacopunct. **19**, 37–44 (2016a)
34. Samadder, A., Das, S., Das, J., Khuda-Bukhsh, A.R.: Relative efficacies of insulin and poly (lactic-co-glycolic) acid encapsulated nano-insulin in modulating certain significant biomarkers in arsenic intoxicated L6 cells. Colloids Surf. B: Biointerfaces. **109**, 10–19 (2013a)
35. Samadder, A., Das, J., Das, S., De, A., Saha, S.K., Bhattacharyya, S.S., Khuda-Bukhsh, A.R.: Poly (lactic-co-glycolic) acid loaded nano-insulin has greater potentials of combating arsenic induced hyperglycemia in mice: some novel findings. Toxicol. Appl. Pharmacol. **267**, 57–73 (2013b)
36. Horcas, I., Fern'andez, R., G'omez-Rodrı́guez, J.M., Colchero, J., G'omez-Herrero, J., Baro, A.M.: WSXM: a software for scanning probe microscopy and a tool for nanotechnology. Rev. Sci. Instrum. **78**, 013705 (2017)
37. Lamey, C.A., Moussa, N., Helmy, M.W., Haroun, M., Sabra, S.A.: Simultaneous encapsulation of dasatinib and celecoxib into caseinate micelles towards improved in vivo anti-breast cancer efficacy with reduced drug toxicity. J. Drug Deliv. Sci. Technol. **87**, 104807 (2023)
38. Bhattacharyya, S.S., Paul, S., De, A., Das, D., Samadder, A., Boujedaini, N., Khuda-Bukhsh, A.R.: Poly (lactide-co-glycolide) acid nanoencapsulation of a synthetic coumarin: cytotoxicity and bio-distribution in mice, in cancer cell line and interaction with calf thymus DNA as target. Toxicol. Appl. Pharmacol. **253**, 270–281 (2011)
39. Chowdhury, M.M., Rahman, A.S.M., Nahar, L., Rahman, M., Reza, H.A., Ahmed, M.S.: Efficiency of different DNA extraction methods for fish tissues: a comparative analysis. IOSR J. Pharm. Biol. Sci. **11**, 11–15 (2016)
40. Kaplan, L.A.: Carbohydrates and metabolite. In: Kaplan, L.A., Pesce, A.J., Kazmierczak, S.C. (eds.) Clinical Chemistry: Theory, Analysis and Co-Relation 1984, 4th edn, pp. 1032–1040. Mosby, St. Louis (1984)
41. Das, S., Das, J., Samadder, A., Paul, A., Khuda-Bukhsh, A.R.: Strategic formulation of apigenin-loaded PLGA nanoparticles for intracellular trafficking, DNA targeting and improved therapeutic effects in skin melanoma in vitro. Toxicol. Lett. **223**, 124–138 (2013)
42. Solomon, J., Bulkley, R.J., Mayer, J.: Effect of a low dose of alloxan on blood glucose, islet beta cell granulation, body weight, and insulin resistance of Ob/Ob mice. Diabetologia. **10**, 709–715 (1974)
43. Parmar, H.S., Kar, A.: Antidiabetic potential of Citrus sinensis and Punica granatum peel extracts in alloxan treated male mice. Biofactors. **31**, 17–24 (2007)
44. Samadder, A., Chakraborty, D., De, A., Bhattacharyya, S.S., Bhadra, K., Khuda-Bukhsh, A.R.: Possible signaling cascades involved in attenuation of alloxan-induced oxidative stress and hyperglycemia in mice by ethanolic extract of Syzygium jambolanum: drug–DNA interaction with calf thymus DNA as target. Eur. J. Pharm. Sci. **44**, 207–217 (2011)
45. Samadder, A., Das, J., Das, S., Paul, A., Khuda-Bukhsh, A.R.: Ameliorative effects of Syzygium jambolanum extract and its poly (lactic-co-glycolic) acid nano-encapsulated

form on arsenic-induced hyperglycemic stress: a multi-parametric approach. J. Acupunct. Meridian Stud. **5**, 310–318 (2012a)
46. Kind, P.R., King, E.J.: Estimation of plasma phosphatase by determination of hydrolysed phenol with amino-antipyrine. J. Clin. Pathol. **7**, 322–326 (1954)
47. Weisshaar, D., Gossrau, E., Faderl, B.: Normal ranges of alpha–HBDH, LDH, AP and LAP as measured substrate-optimated test charges. Med. Welt. **26**, 387–392 (1975)
48. Das, J., Das, S., Samadder, A., Bhadra, K., Khuda-Bukhsh, A.R.: Poly (lactide-co-glycolide) encapsulated extract of Phytolacca decandra demonstrates better intervention against induced lung adenocarcinoma in mice and on A549 cells. Eur. J. Pharm. Sci. **47**, 313–324 (2012)
49. Das, J., Samadder, A., Mondal, J., Abraham, S.K., Khuda-Bukhsh, A.R.: Nano-encapsulated chlorophyllin significantly delays progression of lung cancer both in in vitro and in vivo models through activation of mitochondrial signaling cascades and drug-DNA interaction. Environ. Toxicol. Pharmacol. **46**, 147–157 (2016b)
50. Samadder, A., Dey, S., Sow, P., Das, R., Nandi, S., Das, J., Bhattacharjee, B., Chakrovorty, A., Biswas, M., Guptaroy, P.: Phyto-chlorophyllin prevents food additive induced genotoxicity and mitochondrial dysfunction via cytochrome c mediated pathway in mice model. Comb. Chem. High Throughput Screen. **24**, 1618–1627 (2021)
51. Derochette, S., Franck, T., Mouithys-Mickalad, A., Ceusters, J., Deby-Dupont, G., Lejeune, J.P., Neven, P., Serteyn, D.: Curcumin and resveratrol act by different ways on NADPH oxidase activity and reactive oxygen species produced by equine neutrophils. Chem. Biol. Interact. **206**, 186–193 (2013)
52. Lin, X., Bai, D., Wei, Z., Zhang, Y., Huang, Y., Deng, H., Huang, X.: Curcumin attenuates oxidative stress in RAW264.7 cells by increasing the activity of antioxidant enzymes and activating the Nrf2-Keap1 pathway. PLoS One. **14**, e0216711 (2019)
53. Yousefian, M., Shakour, N., Hosseinzadeh, H., Hayes, A.W., Hadizadeh, F., Karimi, G.: The natural phenolic compounds as modulators of NADPH oxidases in hypertension. Phytomedicine. **55**, 200–213 (2019)
54. Gao, Y., Zhuang, Z., Lu, Y., Tao, T., Zhou, Y., Liu, G., Wang, H., Zhang, D., Wu, L., Dai, H., Li, W., Hang, C.: Curcumin mitigates neuro-inflammation by modulating microglia polarization through inhibiting TLR4 axis signaling pathway following experimental subarachnoid hemorrhage. Front. Neurosci. **13**, 1223 (2019)
55. Chowdhury, I., Banerjee, S., Driss, A., Xu, W., Mehrabi, S., Nezhat, C., Sidell, N., Taylor, R.N., Thompson, W.E.: Curcumin attenuates proangiogenic and proinflammatory factors in human eutopic endometrial stromal cells through the NF-κB signaling pathway. J. Cell. Physiol. **234**, 6298–6312 (2019)

Optimizing Solubility Through Co-crystal Technology: A Rigorous Investigation into Tablets Loaded with Azelnidipine Co-crystals

Aparna Joshi[1] (✉) and G. Gnana Rajan[1]

[1] Department of Pharmaceutics, School of Pharmaceutical Sciences, SGRR University, Dehradun, Uttarakhand, India

Abstract In the pursuit of enhancing the therapeutic efficacy of azelnidipine, a widely prescribed antihypertensive agent with limited aqueous solubility, this research delves into the innovative realm of co-crystal technology. The formulation phase incorporates co-crystal strategies designed to optimize the azelnidipine solubility profile. Three co-formers were employed for co-crystal formulation by wet grinding and solvent evaporation methods. Best co-crystals were found with co-former saccharin by solvent evaporation methods with the help of different analytical characterization. Azelnidipine-loaded co-crystals using saccharin as a co-former were compressed in tablet dosage form using the wet granulation method. The evaluation process encompasses a spectrum of pharmaceutical attributes, including dissolution behavior and stability, ensuring a holistic understanding of the developed dosage form. In vitro bioequivalent studies form a pivotal component of this investigation, elucidating the comparative pharmacokinetics and bioavailability of the optimized azelnidipine co-crystal tablet formulation against the conventional dosage form. The findings of this research promise to contribute significantly to the field of drug delivery, offering a comprehensive exploration of co-crystal technology's role in optimizing solubility and bioavailability.

Keywords Azelnidipine · Co-crystals · Bioequivalence · Solubility enhancement · Co-former

1 Introduction

1.1 Solubility Enhancement Techniques

BCS class II drugs are characterized by high permeability but low solubility, posing significant hurdles in achieving optimal drug delivery. One of the primary challenges lies in enhancing the drug's solubility to improve its bioavailability, a critical factor in ensuring therapeutic efficacy [1, 2]. Formulation scientists must grapple with various

A. K. Saxena, S. Nandi (eds.), *Global Trends in Health, Technology and Management*, https://doi.org/10.1007/978-3-031-75457-9_8

techniques, such as the incorporation of solubilizing agents, nanotechnology, and lipid-based formulations, to overcome these hurdles. Co-crystallization is recognized as one of the most effective techniques for solubility enhancement in the field of pharmaceuticals [3, 4]. This innovative approach involves the formation of crystalline structures through the combination of an active pharmaceutical ingredient (API) with another molecule, known as a co-former [5–7].

Azelnidipine exhibits high permeability but low aqueous solubility, making it prone to poor and variable oral bioavailability. One significant challenge lies in enhancing its solubility to ensure consistent and efficient absorption, a crucial factor in achieving therapeutic efficacy [8–10].

1.2 Drug Selection and Co-former Selection

The formation of co-crystals for azelnidipine is an innovative approach aimed at enhancing its solubility, thereby improving its bioavailability [11, 12]. Co-crystals involve the association of the active pharmaceutical ingredient (API) with another molecule, known as a co-former, through non-covalent interactions [13–14].

The selection of an appropriate co-former and the optimization of the co-crystal formation process are crucial steps in achieving the desired improvement in solubility [15–16].

The important one is the selection of a suitable co-former, ensuring its safety, and the amount used in the co-crystal formulation. Since there are a variety of co-formers available under "Generally Regarded as Safe (GRAS)" and "Everything Added to Food in the United States (EAFUS)," the co-crystal formulation is quicker and involves less risk as compared to the design of new chemical entities [17]. Ascorbic acid, succinic acid, and Saccharin were selected as co-formers for co-crystal formulation. All three co-formers were selected based on the literature review and the compatibility studies between all three co-formers and the drug azelnidipine by IR spectroscopy.

Saccharin's potential carcinogenicity has been a topic of extensive research and debate. As per mechanistic insights and epidemiological studies, numerous studies performed on humans did not show a consistent link between saccharin consumption and cancer risk. Today, saccharin is considered safe for consumption by major health authorities, including the US Food and Drug Administration (FDA) and the European Food Safety Authority (EFSA) when consumed within the established acceptable daily intake (ADI) levels [17–18]. Saccharin has not been reported to show any harmful effects on humans when consumed at 5 g daily over 5 months [17].

Azelnidipine is selected as an ideal drug for the study because the chemical structure of azelnidipine allows it to form a hydrogen bond with many co-formers, which are in the list of GRAS, to form co-crystals.

The key functional groups in azelnidipine allow it to interact with various co-formers in co-crystal formation [Fig. 1, 14]:

Ester Groups: Azelnidipine has two ester functional groups, which participate in hydrogen bond formation with co-formers.

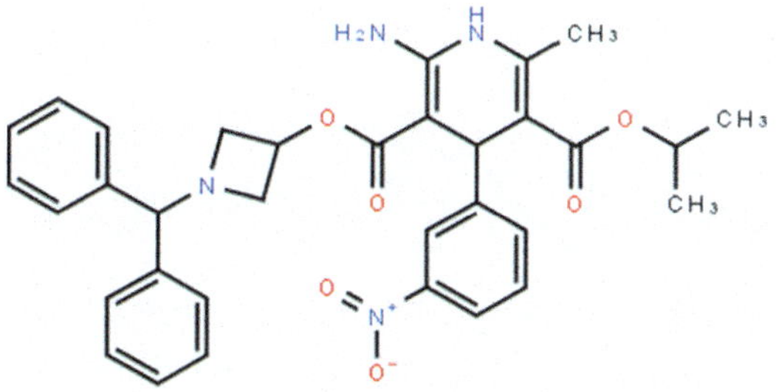

Fig. 1 Structure of azelnidipine

Carbonyl Group (C=O): Can act as a hydrogen bond acceptor.
Oxygen (O): The oxygen in the ester can participate in hydrogen bonding.
Nitrile Group (–CN): Nitrile group can interact through dipole-dipole interactions and may participate in hydrogen bonding.
Aromatic Rings: Aromatic rings can engage in π-π stacking interactions with co-formers having aromatic structures.
Amino Group (–NH2): The amino group can act as both a hydrogen bond donor and acceptor in azelnidipine.
Methyl Groups (–CH3): Methyl substituents can affect the hydrophobic characteristics and overall solubility of an API.

Azelnidipine acts by selectively inhibiting the influx of calcium ions into vascular smooth muscle cells by binding to L-type calcium channels by blocking calcium channels, and it reduces the entry of calcium into the cells, causing relaxation of the smooth muscles in blood vessels. Vasodilation results in decreased peripheral resistance, leading to a reduction in blood pressure [16].

Solubility of Pure Azelnidipine The solubility profile of azelnidipine was performed on different media at different pH to check solubility. Azelnidipine showed the best results with phosphate buffer pH 7.4 [Fig. 2].

Drug Dissolution of Pure Azelnidipine The drug dissolution profile of azelnidipine at phosphate buffer pH 7.4 has been measured over 60 min [Fig. 3]. Pure azelnidipine shows 39.17% drug release over 60 min.

The important one is the selection of a suitable co-former, ensuring its safety, and the amount used in the co-crystal formulation. Since there are a variety of co-formers available under "Generally Regarded as Safe (GRAS)" and "Everything Added to Food in the United States (EAFUS)," the co-crystal formulation is quicker and involves less risk as compared to the design of new chemical entities [17]. Ascorbic acid [19], succinic acid [19], and saccharin [16] and were selected as co-formers for co-crystal formulation. All three co-formers were selected based on the literature review and the compatibility studies between all three co-formers and the drug azelnidipine by IR spectroscopy.

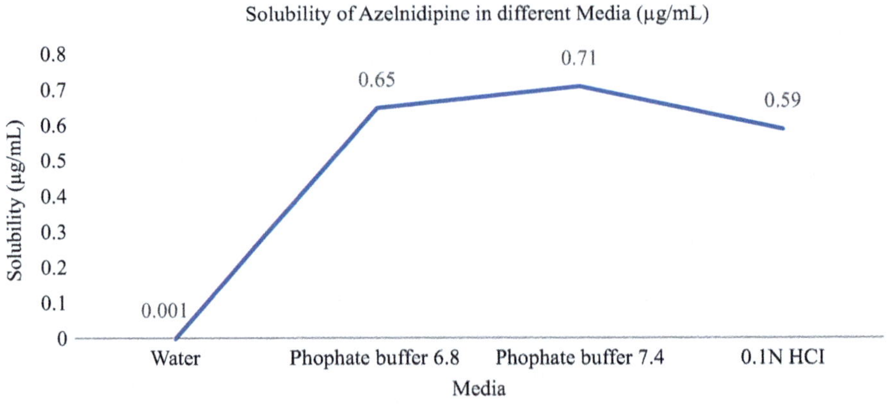

Fig. 2 Solubility of pure azelnidipine

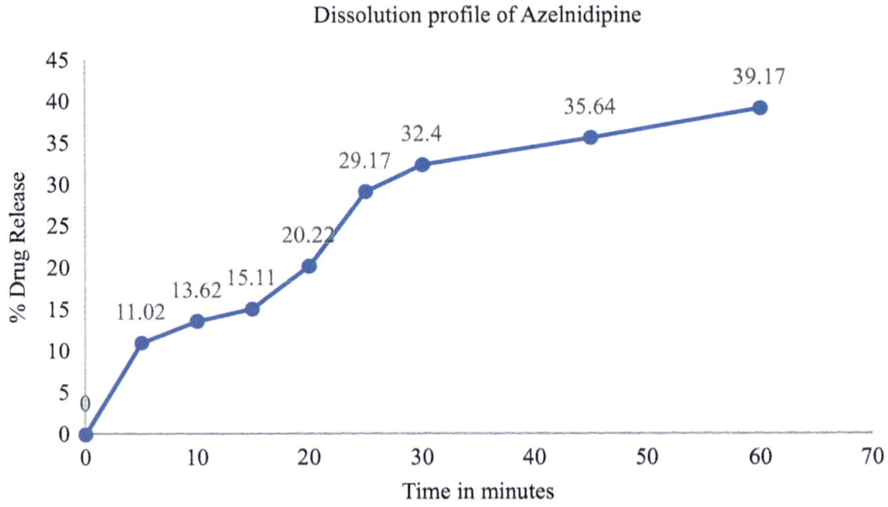

Fig. 3 Drug dissolution of pure azelnidipine

1.3 Co-crystallization and Tablet Formation

Solvent evaporation and wet grinding techniques were employed for the formation of azelnidipine co-crystals. The solvent evaporation method involves the following key steps in the preparation of co-crystals:

Solvent selection: The selection of solvent is a crucial part of the co-crystal formulation as it may affect the evaporation rate and morphology of final co-crystals.

Preparation of solution: The API and co-former are dissolved in the selected solvent to form a homogeneous solution with continuous stirring at an adequate temperature.

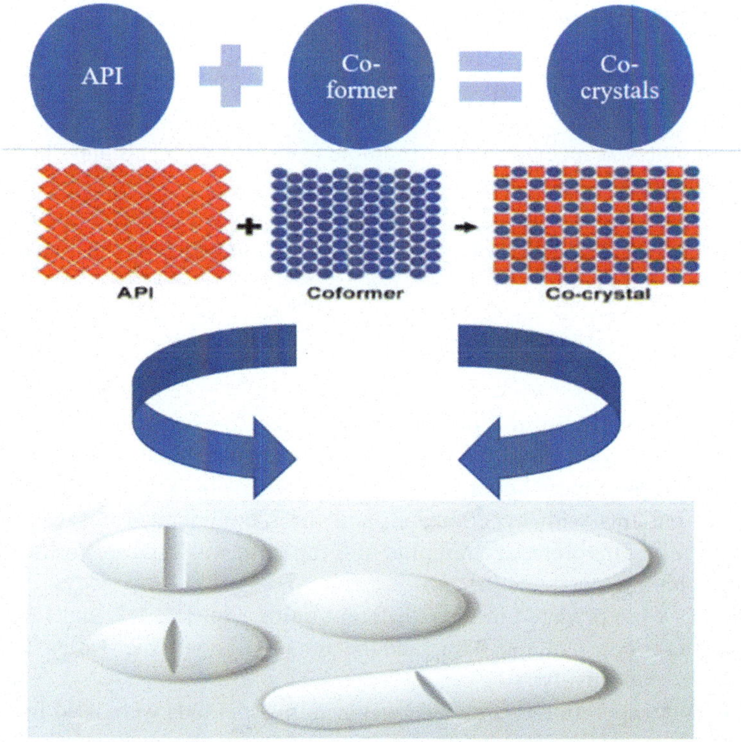

Fig. 4 Azelnidipine-co-former-loaded tablets

Evaporation of the solvent: The prepared solution is then left to evaporate under controlled conditions. Temperature is a crucial part here that can affect the formation of high-quality co-crystals.

Crystallization: As the solvent evaporates, the solution becomes supersaturated, which leads to nucleation and growth of co-crystals [19].

Isolation of co-crystals: Once crystallization is complete, the co-crystals are isolated from the remaining solvent by filtration or decantation. The crystals are then dried properly for the removal of any residual solvent.

Once azelnidipine co-crystals are successfully formed, the next step involves their incorporation into tablet dosage forms [Fig. 4]. This process requires careful consideration of factors such as co-crystal stability, compatibility with excipients, and tablet manufacturability [18].

The incorporation of azelnidipine co-crystals into tablet dosage forms aims to provide a consistent and controlled release of the drug, thereby enhancing its therapeutic efficacy. The tablets should be designed to disintegrate and release the co-crystals efficiently, ensuring optimal dissolution in the gastrointestinal tract for improved absorption [19].

2 Materials and Methods

2.1 Materials

Pure azelnidipine was received as a gift sample from JB Chemicals and Pharmaceutical Limited Mumbai.

Ascorbic acid, succinic acid, saccharin, and methanol were used as laboratory grade. Distilled water was used wherever required and was sourced from the distillation apparatus at the laboratory.

2.2 Methods

Preparation of Azelnidipine-Loaded Co-crystals Co-crystals were prepared using three co-formers ascorbic acid, succinic acid, and saccharin in different stoichiometric ratios of azelnidipine and all three co-formers, respectively. Azelnidipine is used in the treatment of patients with hypertension, and the recommended dosage is 8–16 mg orally once daily. Therefore, azelnidipine of 8 mg dose was used for formulating azelnidipine-saccharin co-crystals [6].

Co-crystals were prepared in three different ratios, i.e., 1:1, 1:2, and 1:3, where the amount of drug is constant, i.e., 8 mg, and the amount of co-former varies, i.e., 500 mg, 1 g, and 1.5 g, respectively.

The solvent evaporation method and wet grinding methods were used for the preparation of co-crystals [Fig. 5]. Methanol was used as a standard solvent [7].

Methanol was used as a solvent for the preparation of azelnidipine-saccharin co-crystals. Methanol is poisonous in nature. To detect the presence of methanol in the final azelnidipine and saccharin co-crystals, the co-crystals were analyzed by IR spectroscopy.

The x-axis represents the wavenumber in cm^{-1}, and the y-axis represents the absorbance. This spectrum includes characteristic peaks for azelnidipine and saccharin, such as:

C=O stretching around 1694 cm^{-1}.

N-H bending around 1500 cm^{-1}.

S=O stretching around 1310 cm^{-1}.

C-N stretching around 1176 cm^{-1}.

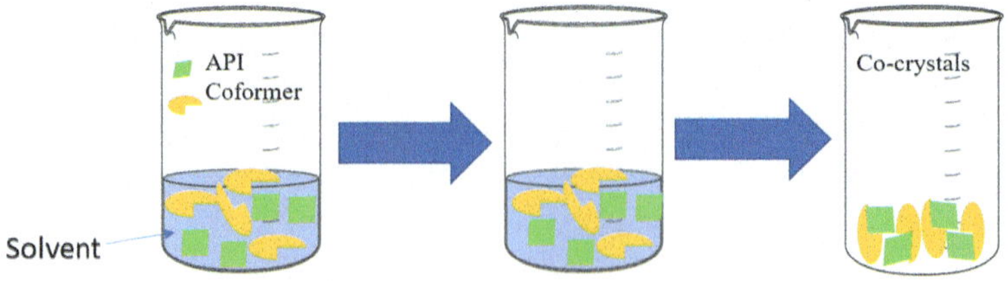

Fig. 5 Solvent evaporation method

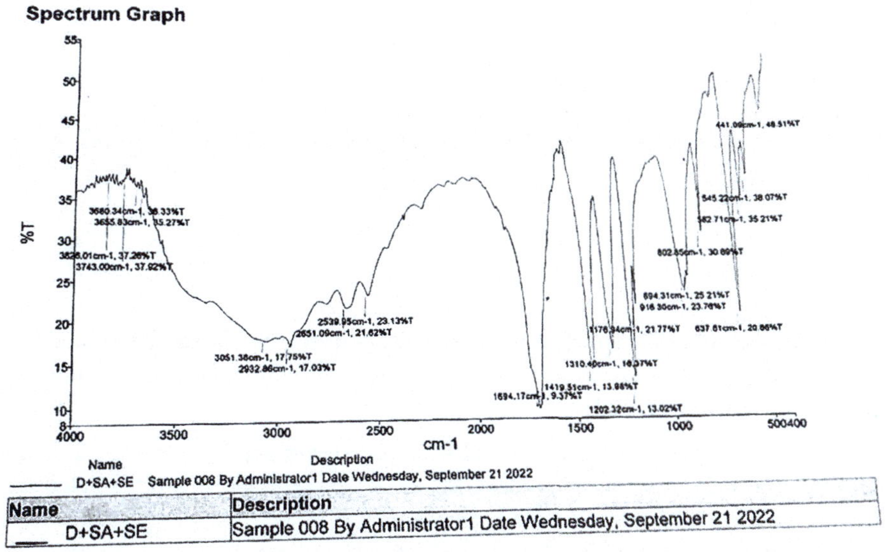

Fig. 6 IR spectrum of azelnidipine-saccharin co-crystals

No significant absorption is seen around 3200–3600 cm^{-1}, indicating the absence of methanol [Fig. 6].

Prepared co-crystals were characterized by different analytical techniques—IR, XRD, SEM, and DSC.

All the prepared azelnidipine-loaded co-crystals were also evaluated for melting point analysis, solubility profile, and dissolution profile. Based on evaluation and characterization, azelnidipine-saccharin co-crystals were found the best co-crystals prepared by solvent evaporation [7].

Preparation of Tablet Dosage Form Using Azelnidipine-Saccharin Co-crystals The azelnidipine-saccharin co-crystal-loaded tablets were prepared by wet granulation method as per the given formulation in the tablet below [Table 1]. All the excipients including drug co-crystals were mixed properly. Granulation was done using distilled water. Prepared granules were dried at 45 °C in a hot air oven for 3 h. Dried granules were again passed through different sieve sizes for uniformity. 100 mg tablets were obtained using a multi-punch tablet punching machine [18–21].

2.3 Evaluation

Pre-compression Evaluation of Azelnidipine-Saccharin Granules The mixture of azelnidipine-saccharin co-crystals and the excipient blend were evaluated for different pre-compression parameters, e.g., flow properties and compressibility, tapped density, bulk density, angle of repose, and Carr's index.

Table 1 Azelnidipine-saccharin co-crystal tablet formulation design

S. no	Ingredients	Mg/tablet		
1	Azelnidipine-saccharin	8 mg [F1] (1:1)	8 mg [F2] (1:2)	8 mg [F3] (1:3)
2	Microcrystalline cellulose	30 mg	30 mg	30 mg
3	Lactose	7 mg	7 mg	7 mg
4	Magnesium stearate	21 mg	21 mg	21 mg
5	Talc	22 mg	22 mg	22 mg
6	Crospovidone	12 mg	12 mg	12 mg

Post-compression Evaluation of Azelnidipine-Saccharin Tablet Drug-loaded co-crystal tablets were evaluated for hardness, friability, weight variation, dissolution, and disintegration studies [21–24].

Conduction of In Vitro Pharmaceutical Bioequivalence Studies In vitro dissolution studies will be conducted between drug-loaded co-crystal tablets and marketed tablet formulations having the same API [22, 23, 24, 25].

Accelerated Stability Studies Accelerated stability studies will be performed for optimized tablet formulation as per ICH guidance. The stability profile of optimized co-crystal tablets includes the change in the stored co-crystals for accelerated stability study testing as per ICH guidelines. Co-crystal tablets were packed in HDPE bottles. These bottles were placed in stable conditions for 6 months. Samples were withdrawn after 1 M/3 M/6 M and evaluated for stability parameters such as physical changes and the amount of drug in the formulation.

3 Results and Discussion

Azelnidipine-saccharin co-crystal formation was confirmed by different analytical evaluation parameters. The thermal behavior of pure azelnidipine, saccharin, and co-crystals was analyzed with DSC analysis. Pure azelnidipine had an endothermic peak at 190° and saccharin at 220° that indicated its melting point, whereas the endothermic peak for azel-sac co-crystals was found at 178°. Decreased endothermic peak indicates the formation of a new component, a new crystalline component.

The sharpest peaks from XRD for azelnidipine-saccharin co-crystals were identified at (2θ values) 12.12, 16.19, 21.02, 23.43, 30.23, and 44.03. These sharp peaks were different from both individual azelnidipine and saccharin. The above sharp peaks indicate the formation of a new crystalline complex.

Azelnidipine and saccharin and azelnidipine: saccharin co-crystals were also investigated by scanning electron microscopy for surface morphology. Prepared co-crystals were identified as needle-shaped and irregular in size and shape, which also confirms the formation of co-crystals.

All pre-compression parameters of azelnidipine-saccharin co-crystal granules in different ratios were analyzed. Tapped density, bulk density, angle of repose, Hausner's

Table 2 Pre-compression parameters for azelnidipine-saccharin co-crystal granules in different ratios

S. no	Pre-compression parameters	Azel-sac F1	Azel-sac F2	Azel-sac F3
1	Tapped density(g/mL)	0.602	0.611	0.616
2	Bulk density(g/mL)	0.521	0.501	0.508
3	Angle of repose (degree)	28.020	30.120	29.220
4	Hauser's ratio	1.15	1.21	1.21
5	Compressibility index	8.1	11	10.8

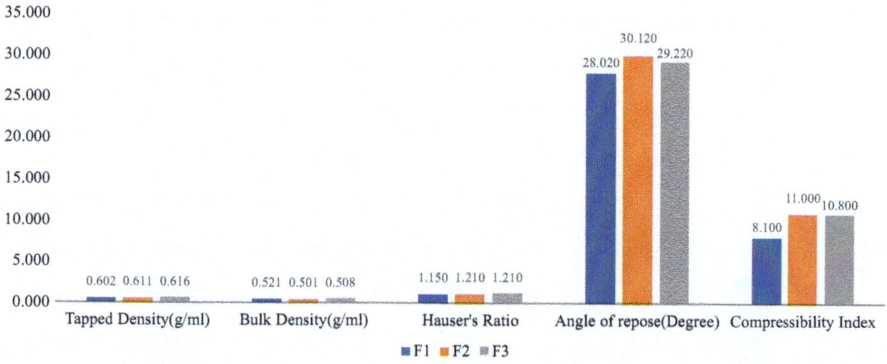

Fig. 7 Pre-compression parameters

ratio, and compressibility index were calculated, and all observation was found within the advised range of IP [Table 2 and Fig. 7].

The angle of repose found between 28.02° and 30.12°.

As per IP, angle of repose <30°➔ very free-flowing, 30° < angle of repose <38°➔ free-flowing,

The compressibility index found between 8 and 11 m^2/N.

As per IP, compressibility index <10 ➔ excellent flow characteristics, 11 < compressibility index <10 ➔ good flow characteristics.

Tapped density and bulk density were observed between 0.0602 g/mL–0.616 g/mL and 0.501 g/mL–0.521 g/mL, respectively.

Hausner's ratio observation lies within 1.1 to 1.2. As per IP, Hausner's ratio <1.25 ➔ fair flow.

After formulation of tablets of azelnidipine-saccharin-loaded co-crystal, all tablets were evaluated for different post-compression parameters such as weight variation, disintegration time, hardness, friability, and thickness of tablets. All observations were found within the specified range of IP [Tables 3 and 4 and Figs. 8 and 9].

Friability ➔a maximum mean weight loss from the three samples of not more than 1.0% is considered acceptable. All prepared tablets range from 0.45% to 0.049%, which is in the acceptable range.

Thickness → in the range of 2.11 mm to 2.81 mm ± 0.015.

Disintegration time ➔ within limits of 15 min, which complies with IP.

Table 3 Post-compression parameters for azelnidipine-saccharin co-crystal tablets in different ratios

S. no	Formulation	Weight variation (mg)	Disintegration time (min)
1	F1	100.12	14.89
2	F2	100.02	11.12
3	F3	100.06	13.89

Table 4 Post-compression parameters for azelnidipine-saccharin co-crystal tablets in different ratios

S. no	Formulation	Hardness (kg/mm)	Friability (%)	Thickness (mm)
1	F1	4.5	0.45	2.11
2	F2	5	0.421	2.12
3	F3	5.5	0.49	2.81

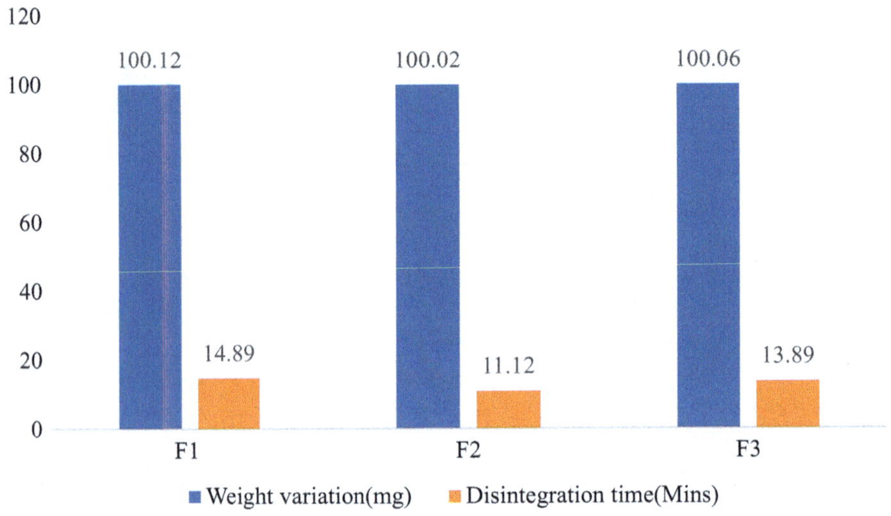

Fig. 8 Pre-compression parameters

Hardness → found in a range of 4.5–5.5 kg/cm^2 within IP limits.

Weight variation also complies with IP limits.

All the formations demonstrated good characteristics (as per IP standards) during pre- and post-compression analysis.

All the prepared co-crystal tablets were analyzed for drug dissolution profile for 60 min. A graph was plotted between time on the x-axis and cumulative percent drug release on the y-axis. All tablets exhibit drug release more than 80%. Formulation with azelnidipine-saccharin in 1:2 showed the highest drug release of more than 98% [Figs. 10 and 11].

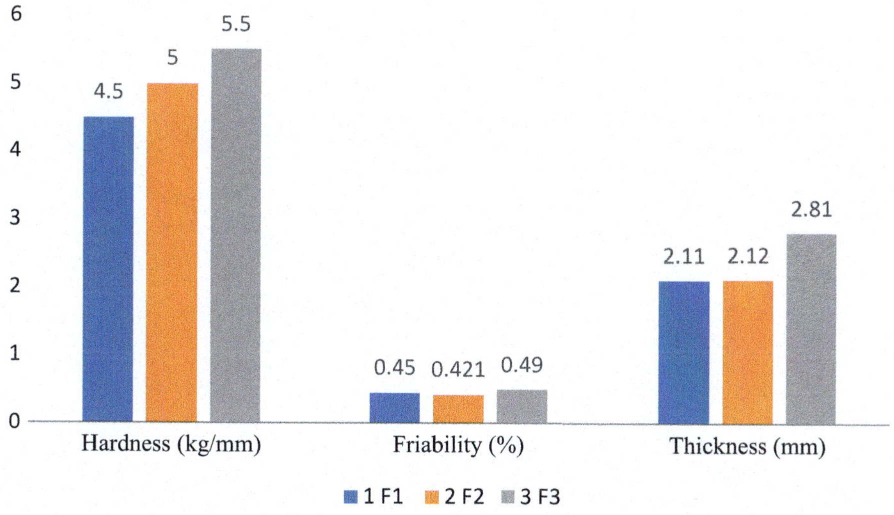

Fig. 9 Post-compression parameters

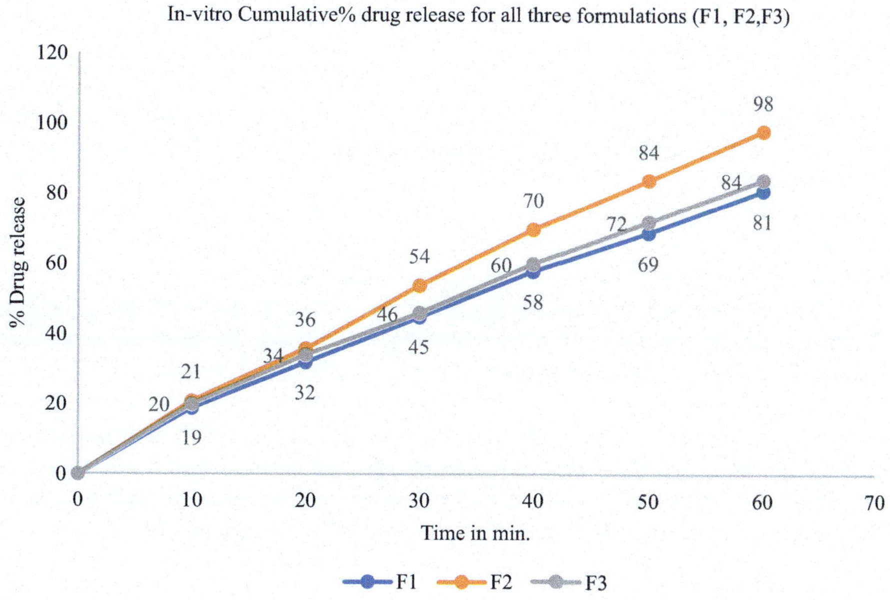

Fig. 10 In vitro cumulative percent drug release for all three formulations

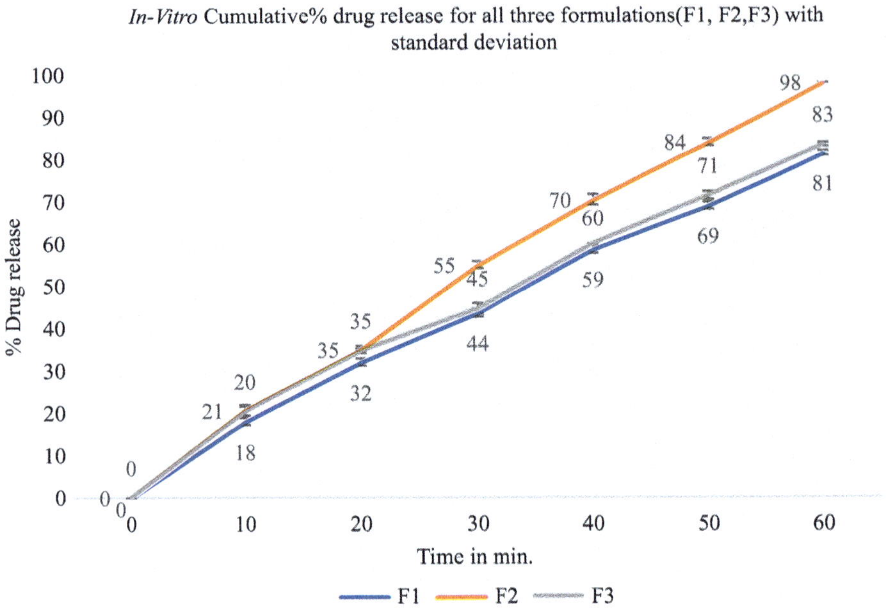

Fig. 11 In vitro cumulative percent drug release for all three formulations with a standard deviation

In vitro pharmaceutical bioequivalence was performed with an optimized formulation, i.e., F2, with a marketed formulation of azelnidipine having the same pharmaceutical dose [Fig. 12].

Optimized tablet formulations were kept for accelerated stability studies at accelerated conditions (40 ± 2 °C/75 ± 5%RH). At different time intervals, samples were withdrawn at 15 days, 1 month, 3 months, and 6 months [Table 5].

No physical change was found during sampling at 1 month and 3 months. There was a slight discoloration was found in the 6 months of sampling. The drug dissolution rate and content uniformity were found constant, i.e., 100% after 1 month.

There was a slight decrease to 99.10% at 3 months and 98.30% at 6 months. Content uniformity slightly decreased to 99.60% at 3 months and 99.10% at 6 months. The minimal variation in percent drug dissolution and content uniformity indicates excellent stability in the dissolution rate, ensuring consistent drug release over time and reliable uniformity in the distribution of the active ingredient, respectively.

4 Conclusion

Azelnidipine-saccharin co-crystal-loaded tablets were formulated with the wet grinding method. Prepared tablets were analyzed for different pre- and post-compression parameters including in vitro drug release studies. Based on all observations found by all, parameter formulation F2 was found the best formulation with 98% drug release.

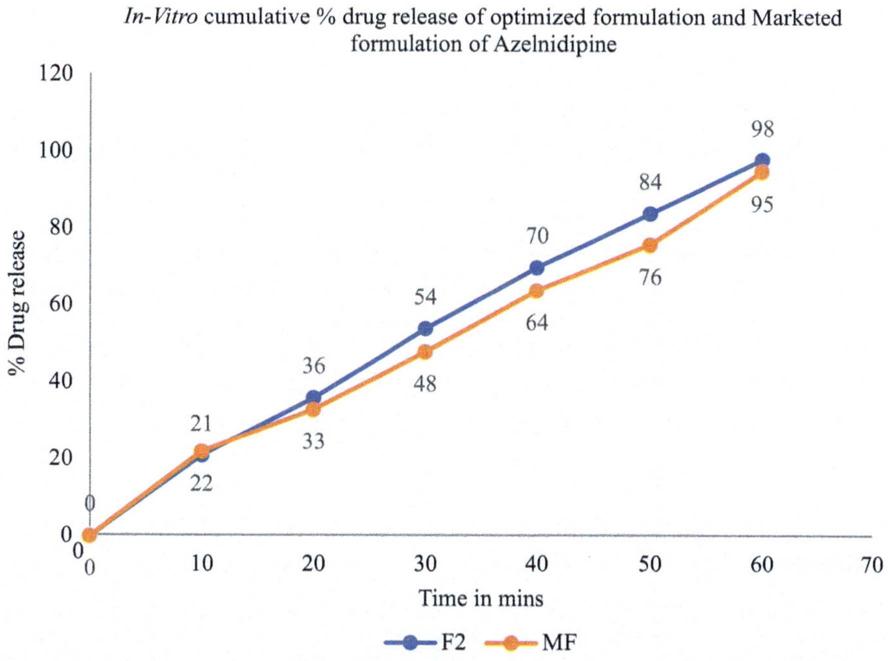

Fig. 12 In vitro cumulative percent drug release of optimized formulation and marketed formulation of azelnidipine

Table 5 Accelerated stability study data of optimized azelnidipine-saccharin-loaded co-crystal tablets

S. no	Temp/humidity condition	Sample taken time (months)	Physical appearance	Rate of dissolution (%)	Content uniformity
1	40 °C ± 2 °C/75% RH ± 5% RH	0	No change	100%	100%
2	40 °C ± 2 °C/75% RH ± 5% RH	1	No change	100%	100%
3	40 °C ± 2 °C/75% RH ± 5% RH	3	No change	99.10%	99.60%
4	40 °C ± 2 °C/75% RH ± 5% RH	6	Slight discoloration	98.30%	99.10%

Optimized formulation F2 is also compared with the marketed formulation of azelnidipine with the same dose, which shows an improved dissolution rate than the marketed formulation, indicating higher bioavailability. Between the 10th and 60th minute of drug release, the optimized formulation exhibits a significant difference of 10% against the marketed formulation, which relates to a faster pharmacological action of the drug. Azelnidipine-loaded co-crystal tablets can improve the overall therapeutic efficacy and patient compliance.

The stability data for azelnidipine-saccharin co-crystal tablets shows minimal changes in both the rate of drug dissolution and content uniformity over 6 months. This indicates that the formulation is stable and maintains its efficacy and safety with time at accelerated temperature and humidity conditions.

This study suggests that co-crystallization is a novel approach toward solubility enhancement of any poorly soluble drug molecules using an appropriate co-former.

Areas of Improvements and Prospects Co-crystallization is an evolving technique in the area of drug manufacturing. Continuous improvements through interdisciplinary collaborations involving various research areas would be instrumental in fully realizing its potential. Utilization of AI-based data modeling techniques such as CSD for determining co-former and co-crystal compatibility may lead to minimizing the duration and cost of the study. Co-crystallization technology is also an enabler for multi-drug regimens, simplifying patient compliance and minimizing medication errors and adverse drug effects.

References

1. Budiman, A., Husni, P., Shafira, A.T.Q.: The development of glibenclamide-saccharin cocrystal tablet formulations to increase the dissolution rate of the drug. Int. J. Appl. Pharm. **11**, 2–3 (2019)
2. Gaikwad, S.S., Avari, J.G.: Improved bioavailability of azelnidipine gastro retentive tablets-optimization and in-vivo assessment. Mater. Sci. Eng. C Mater. Biol. Appl. **103**, 4–8 (2019)
3. Galam, A.G., Shahi, S.R., Zadbuke, N.: Enhancement of aqueous solubility of azelnidipine using solvent evaporation technique. World J. Pharm. Res. **9**, 5–8 (2020)
4. Kawabata, K., Sakaue, M., Akimoto, S., Miyara, M., Kotake, Y., Nishi, H.: Evaluation of photostability of azelnidipine tablets and structure determination of its photoproducts. J. Pharm. Biomed. Anal. **233**, 2–3 (2023)
5. Rajbhar, P., et al.: Formulation and evaluation of clarithromycin co-crystals tablets dosage forms to enhance the bioavailability. Pharm. Innov. **7**, 3–7 (2016)
6. Chen, B.L., Zhang, Y.Z., Luo, J.Q., Zhang, W.: Clinical use of azelnidipine in the treatment of hypertension in Chinese patients. Ther. Clin. Risk Manag. **11**, 2–3 (2015)
7. Shete, A.S., Yadav, A.V., Murthy, M.S.: Evaluation of performance of co crystals of mefloquine hydrochloride in tablet dosage form. Drug Dev. Ind. Pharm. **39**, 3–5 (2013)
8. Ullah, M., Hussain, I., Sun, C.: The development of carbamazepine-succinic acid cocrystal tablet formulations with improved in vitro and in vivo performance. Drug Dev. Ind. Pharm. **42**, 6 (2016)
9. Gaikwad, S.S., Kshirsagar, S.J.: Review on tablet in tablet techniques. Beni-Suef Univ. J. Basic Appl. Sci. **9**, 1–5 (2020)
10. Bhakare, S., Pise, A.G.: Deep insights into the developmental progress of orally disintegrating tablet formulation, techniques, and products: a comprehensive review. Int. J. Med. Pharm. Sci. **12**, 2–3 (2022)
11. Shi, G., et al.: Pharmaceutical application of multivariate modeling techniques: a review on the manufacturing of tablets. RSC Adv. **14**, 4 (2021)
12. Inigo, P., Ramprasad, R., Senthilkumar, S.R., Venkatesan, N.: A review of tablet-in-table vitamin D3 formulation techniques. Am. J. Pharm. Health Res. **6**, 8–11 (2018)

13. Bera, A., Shah, K.: Advanced forced degradation technique for simultaneous estimation of azelnidipine and telmisartan implementing AQbD approach in its tablet dosage form. Int. J. Health Sci. (IJHS). **6**, 6–8 (2022)
14. Ratmalana, J., Kotadiya, R.: Separation and quantification of azelnidipine and chlorthalidone in a synthetic mixture using optimized HPTLC method. Ann. Pharm. Fr. **83**, 4–7 (2023)
15. Arora, D., Taneja, Y., Dhingra, A.K., Guarve, K., Chauhan, M., Nagpal, K.: Solubility enhancement and antioxidant potential of silymarin: a poorly water-soluble drug. Curr. Drug. **31**, 4 (2024)
16. Shonman, M., Shi, X., Kang, M., Wang, Z., Li, X., Dahbura, A.: Using a computational cognitive model to understand phishing classification decisions. In: Electronic Workshops in Computing, vol. 6, p. 4 (2022)
17. Wang, I.C., Lee, M.J., Sim, S.J., Kim, W.S., Chun, N.H., Choi, G.J.: Anti-solvent co-crystallization of carbamazepine and saccharin. Int. J. Pharm. **450**, 25–28 (2013)
18. Touyz, L.Z.G.: Saccharin deemed "not hazardous" in the United States and abroad. Curr. Oncol. **18**, 3–5 (2011)
19. EFSA Panel on Dietetic Products, Nutrition and Allergies (NDA): Scientific opinion on the substantiation of health claims related to fruits and/or vegetables "Mediterranean diet". EFSA J. **9**, 19 (2011)
20. Zhang, H., Zeng, H., Li, M., Song, Y., Tian, S., Xiong, J., He, L., Liu, Y., Wu, X.: Novel ascorbic acid co-crystal formulations for improved stability. Molecules. **27**, 7998 (2022)
21. Kumar, S., Bhargava, D., Thakkar, A., Arora, S.: Drug carrier systems for solubility enhancement of BCS class II drugs: a critical review. Crit. Rev. Ther. Drug Carrier Syst. **30**, 220 (2013)
22. Maclean, N., Khadra, I., Mann, J., Williams, H., Abbott, A., Mead, H., Markl, D.: Investigating the role of excipients on the physical stability of directly compressed tablets. Int. J. Pharm. **X**, 5–8 (2021)
23. Han, J., Xiao, B., Le, P.K., Mangwandi, C.: Enhancement of the solubility of BS class II drugs with MOF and MOF/GO composite materials: case studies of felodipine, ketoprofen, and ibuprofen. Materials (Basel). **16**, 8–10 (2023)
24. Chen, W., et al.: Enhancement of the solubility and oral bioavailability of altrenogest through complexation with hydroxypropyl-β-cyclodextrin. Eur. J. Pharm. Sci. **194**, 4–5 (2024)
25. Rajbhar, K., et al.: Comparative assessment of solubility enhancement of itraconazole by solid dispersion and co-crystallization technique: investigation of simultaneous effect of media composition on drug dissolution. Ann. Pharm. Fr. **81**, 840 (2023)

Natural Products and Their Derivatives: Promising Caspase-3 Activators as Anticancer Agents

Nitin Srivastava[1] and Anil K. Saxena[2] (✉)

[1] Amity University, Lucknow, India
[2] Department of Chemistry, GIPER, Kashipur, India

Abstract Cancer is one of the major causes of global morbidity and mortality. Among various targets for cancer treatment, caspase-3 is an important target whose activation may induce apoptosis and ultimately cell death in cancerous cells. Apart from many synthetic small molecules, several natural products have also been reported as caspase-3 activators. In an extension of our work on caspase-3 activation by small synthetic molecules, we have now focused to review the work on the development of natural products as caspase-3 activators with anticancer activity. Hence, the natural compounds reported in the last 8 years have been broadly classified in different chemical classes to provide some insights into the structural features responsible for caspase-3 activation and their anticancer activities. The study may provide lead molecules for designing novel molecules with enhanced caspase-3 activation with improved cancer therapeutics.

Keywords Cancer · Caspase-3 · Apoptosis · Natural product

1 Introduction

Cancer is among the top diseases causing worldwide deaths. According to the latest report, the worldwide total new cancer cases in 2022 were almost 20 million with 9.7 million cancer-related deaths. Further the estimated new cancer cases per year by 2040 are likely to be 29.9 million, causing 15.3 million deaths [1]. Interference in hemostasis gives way to unbridled growth in cells causing cancers [2, 3]. The first and the second line of cancer chemotherapy targets DNA replication, which suffers from the development of resistance by cancer cells and the toxicity of prolonged treatment [2].

A. K. Saxena, S. Nandi (eds.), *Global Trends in Health, Technology and Management*, https://doi.org/10.1007/978-3-031-75457-9_9

There are several promising targets for designing anticancer agents like mTOR [4], proteasome [5], BCR-ABL kinase [6], and EFGR kinase [7], but they suffer from poor blood-brain barrier penetration and are unfit for neurological cancers [8]. In view of the importance of the apoptosis that involves DNA fragmentation and prevents uncontrolled growth of cancer cells and metastasis [9], the cysteine-aspartyl proteases (caspases) [10] have been considered for the development of their activators as important targets for designing cancer therapeutics. The caspases are of three types, namely, initiator (2, 8, 9, 10), executionary (3, 6, 7), and inflammatory (1, 4, 5, 11, 12) [11–13]. The onset of apoptosis begins with initiator caspases, which activate executionary caspases to carry out apoptosis. The executionary caspases hit at the aspartate residue to rupture the substrate (proteins) and the cell structure [14]. Executionary caspase-3 plays a key role in apoptosis in mammalian cancer cells [15, 16]. Hence, caspase-3 has been considered as an important target, and several reviews have been published describing its activators [17–20]. Most of them, particularly the recent one, describes synthetic small molecules. Hence, it appeared of interest to review the developments in natural products as caspase-3 activators during the last 8 years, and the results are summarized in this review.

2 Materials and Method

A thorough literature search has been carried out using different search engines like RSC, ScienceDirect, Elsevier, ACS, PubMed, etc. for the selection of research papers describing the natural products and their derivatives exhibiting the caspase-3 activation for their anticancer activity.

3 Natural Products as Caspase-3 Activators

3.1 α,β-Unsaturated Ketone Compounds

Several α,β-unsaturated ketones including curcumin, gingerol, ginger, sesquiterpenes, etc. have been reported as caspase-3 activators with anticancer activities.

Curcumin **1**, a diferuloylmethane found in *Curcuma longa* (Linn.) plant, possesses anticancer activity in various cancer cells including breast, colon, liver, and skin. It inhibited phosphorylation of AKT and mTOR, decreased Bcl-2 and increased Bax, and cleaved caspase-3 in the MCF-7 and MDA-MB-231 cancer cells and thus induced apoptosis in cancer cells [21]. Its treatment to the B-pre-ALL cell lines (leukemia) raised Bax/Bcl-2 ratio to release cytochrome c from the mitochondria to the cytoplasm, activated caspase-3, and cleaved PARP [22]. It caused dephosphorylation of the phosphorylated AKT/PKB and downregulated cIAP1 and XIAP expressions. It increased reactive oxygen species. Its combination with cisplatin enhanced the anticancer potential of cisplatin. The curcumin also inhibited the cell growth of immortalized human KCs (HaCaT). It arrested the cells at the G2/M phase and exhibited apoptosis of 33.95% at a dose of 24 μM [23]. Treatment with curcumin resulted in mitochondrial membrane depolarization and release of cytochrome c into the cytoplasm, along with the increase in the ratio of Bax to Bcl-2 in HaCaT cells and subsequent caspase-3 activation. Curcumin-induced apoptosis in HaCaT cells was reported to occur via both caspase-dependent and caspase-independent pathways. This research suggested further exploration with curcumin for proliferative skin diseases, including skin cancer.

It has also been found to induce apoptosis in the colon cancer cell, and at 10 μM, curcumin inhibited the migration of colon cancer cells [24]. It upregulated the expression for Bax and pro-apoptotic proteins caspase-3, caspase-9, and p53 and downregulated the expression for Bcl-2. It has been highly selective for cancer cells and nontoxic to normal colon mucosa epithelial cells. Cytotoxicity of curcumin has also been reported in the human epidermal growth factor receptor-positive breast cancer cells (BT474) with CC_{50} value of 49.96 μM through extrinsic pathway (caspase-8 and caspase-3 positive and caspase-9 negative) [25].

Among pyrazoline-substituted curcumin analogues screened against human cervical cancer HeLa cell line, the chloro-bromo analogue **2** exhibited fivefold improvement in the potency with IC_{50} value of 8.7 μg/mL as compared to curcumin (IC_{50} = 42.4 μg/mL). However, it was less potent than the standard drug, paclitaxel (IC_{50} = 0.008 μg/mL). It induced 70.5% apoptosis through cleavage of caspase-3 enzyme [26].

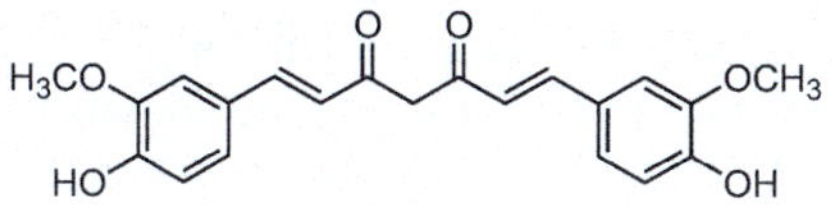

1

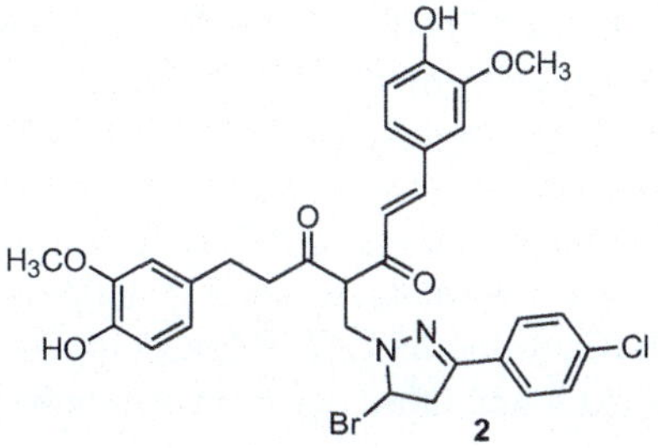

2

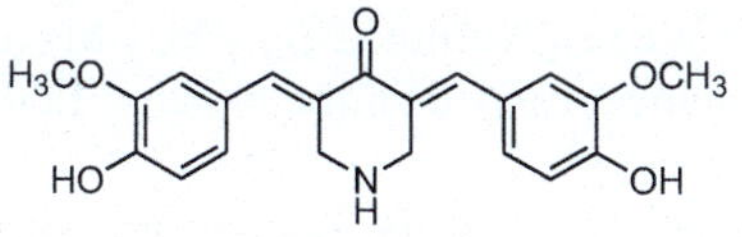

3 (FLD-5)

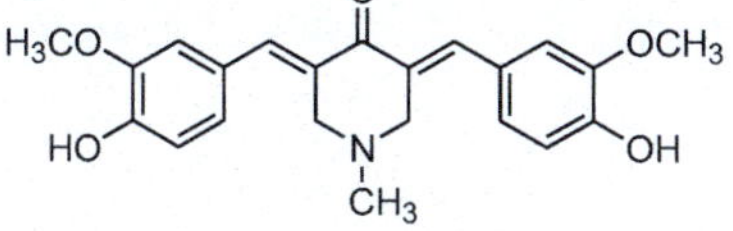

4 (FLD-8)

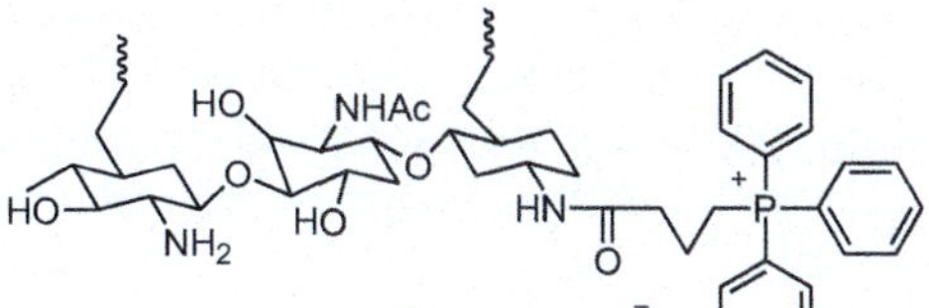

5

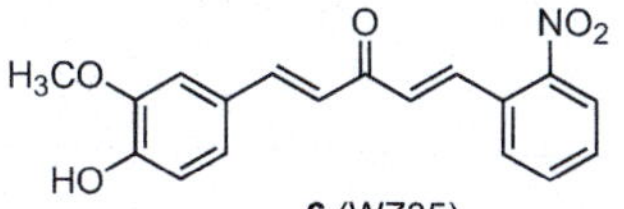

6 (WZ35)

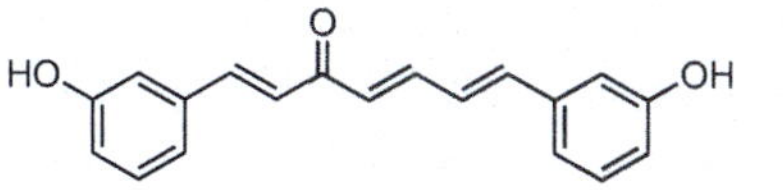

7

Curcumin piperidone derivatives **3** (FLDP-5) and **4** (FLDP-8) have been reported as more potent in inhibiting the proliferation and migration of the glioblastoma LN-18 cells than the curcumin itself [27]. The apoptosis induced here was due to the activation of caspase-3. The downregulation of miRNA was also responsible for the cancer cell death. FLDP-5 and FLDP-8, and curcumin, induced apoptosis in LN-18 cells in a dose-dependent manner from the dose as low as 0.625 μM for curcuminoid analogues than 6.25 μM for curcumin. The induction of apoptosis was 47.37%, 54.37%, and 53.57% at 2.5 μM, 5 μM, and 25 μM for FLDP-5, FLDP-8, and curcumin, respectively.

Mitochondrion-targeting delivery system based on triphenylphosphine bromide (TPP)-chitosan-g-poly-(N-3-carbobenzyloxy-l-lysine) (CZL; **5**) having TPP functional on the surface and loaded with curcumin induced apoptosis in the liver cancer cells [28]. Curcumin/TPP-CZL nano-micelles reduced the membrane potential of mitochondria, enhanced pro-apoptotic protein Bcl-2, and reduced anti-apoptotic Bax protein more effectively than curcumin/CZL nano-micelles and curcumin. This drug delivery system was highly efficient in targeting mitochondria of liver cancer cells. It was found that this treatment significantly lowered the anti-apoptotic protein Bcl-2, while the expressions of pro-apoptotic protein Bax and caspase-3 were increased in curcumin/TPP-CZL nano-micelle group.

Compound **6** (WZ35), a monocarbonyl analogue of curcumin (MACs), is reported to have advantage over curcumin in chemical stability and antitumor activity [29]. It enhanced the expression of eIF2α-ATF4-CHOP signaling and JNK in SGC-7901 cells. Moreover, mitochondria-mediated apoptotic factor including Bax and cleavage of caspase-3 was observed. These results suggested that ROS generation may interplay with its ER stress-mediated apoptosis. In another report, compound **6** (WZ35) not only inhibited cell growth but also induced apoptosis in H1975 cancer cells [30]. A dose-dependent apoptosis was observed in WZ35-treated cancer cells, and the apoptosis in the cell was further verified by the cleavage of PARP (poly (ADP-ribose) polymerase) and caspase-3 along with a decrease in Bcl-2 and procaspase-3. Analogue **6** induced ROS generation and ultimately apoptosis in gastric, colon, and pancreatic cancer cells with IC_{50} value of 1.17 μM as compared to erlotinib and gefitinib (IC_{50} > 20 μM) and thus exhibited more potential to inhibit growth of cancer cells than the existing EGFR TKIs.

Trienone analogue **7** was explored for the anticancer activity against multidrug-resistant (CLS-354/DX) oral cancer cells [31] and was found three times more active in inducing apoptosis in the cancer cells as compared to curcumin. It increased activity of the caspase-3/caspase-7, caspase-8, and caspase-9 by 5.2, 2.0, and 1.9 times, respectively.

3.2 Flavonoids

Dihydromyricetin or ampelopsin **8**, a flavonoid obtained from the leaves and stems of *Ampelopsis grossedentata*, exhibited diverse pharmacological properties like anticancer, anti-fatigue, anti-inflammatory, and antioxidant [32]. Recently, it is reported to prevent prostate and breast cancer by suppressing the growth and proliferation of affected cancerous cells [33, 34].

Ampelopsin dramatically activated caspase-8, caspase-9, and caspase-3, contributing to PARP cleavage in a dose- and time-dependent manner in both U251 and A172 cells [35]. At low concentration (5 μM), it displayed minimal influences on the survival of U251, A172, and SHG44, especially for 24 h. Obviously, ampelopsin in the range of 10 to 100 μM suppressed the cell proliferation of these cancer cell lines with IC_{50} of 39.6, 35.8, and 33.6 μM for U251, A172, and SHG44 for 24 h and with IC_{50} value of 15.8, 18.6, and 17.9 uM for 48 h, respectively, in a dose-dependent manner. The apoptosis here involved both intrinsic and extrinsic pathways involving caspase-3.

Gambogenic acid **9**, a flavonoid from gamboge plants, increased the expression for the cleaved caspase-3, caspase-8, and caspase-9 and Bax, while it decreased the expression of anti-apoptotic protein Bcl-2 in small-cell lung cancer cell line. These results were also confirmed in the mouse tumor xenograft with low toxicity to the normal cells [36].

Bioactive flavonol molecule fisetin (3,3′,4′,7-tetrahydroxyflavone), found in many vegetables and fruits like onion, cucumber, grape, apple, strawberry, etc., reduced the phosphorylation of PI3K, AKT, mTOR, and p70S6K and upregulation of apoptotic bodies like Bax and caspase-3/caspase-8/caspase-9 in breast cancer cells [37].

Anticancer activities of fisetin **10** have also been explored in the HeLa cells, in which it arrested cell cycle at G_2/M phase. It enhanced pro-apoptotic genes such as APAF1, Bad, Bax, Bid, and BIK at both transcript and protein levels and downregulated anti-apoptotic genes Bcl-2, BIRC8, MCL-1, XIAP/BIRC4, Livin/BIRC7, clap-2/BIRC3, etc. to avert cell proliferation. It induced apoptosis in cancer cells and activated caspase-3, caspase-8, and caspase-9 [38].

Chrysin **11** is a naturally occurring flavonoid found in plants. N′-alkylidene/arylidene acetohydrazide derivatives of **11** were synthesized and evaluated for their anticancer activity in MCF-7 and MDA-MB-231 cell lines [39] where it showed anticancer activity in MCF-7 and MDA-MB-231 with IC_{50} value of 4.2 and 3.3 μM, respectively. It was found that the apoptosis in the cancer cells was induced by the rise in the caspase-3 and caspase-7 activities.

Hyperoside **12** induced apoptosis in MCF-7 and MCF-4 T1 cells through the ROS-mediated NF-κB signaling pathway, which was attributed to reduced levels of Bcl-2 and XIAP and increased expression of Bax and cleaved caspase-3 [40].

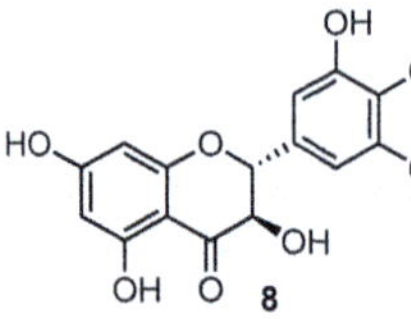
8

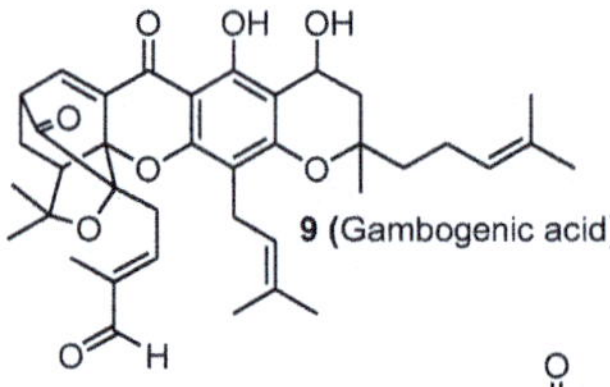
9 (Gambogenic acid)

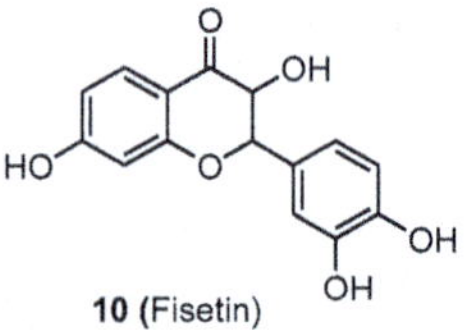
10 (Fisetin)

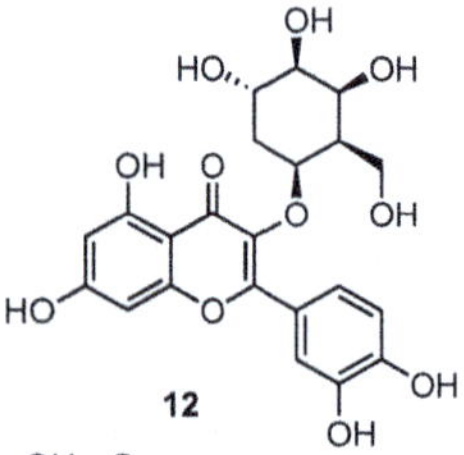
12

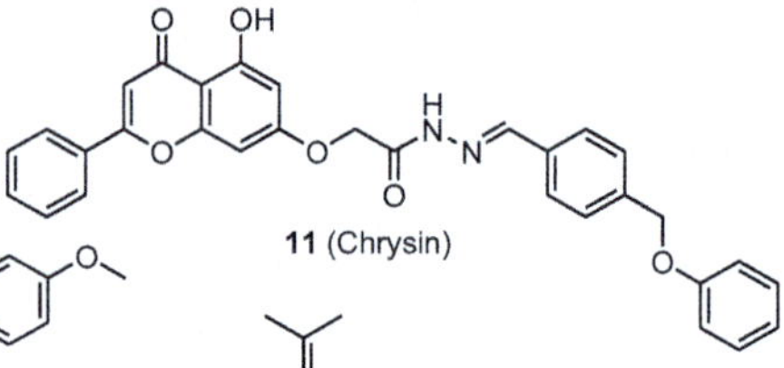
11 (Chrysin)

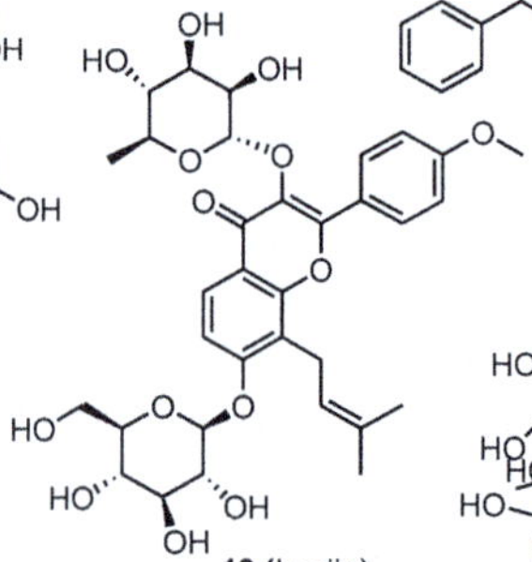
13 (Icariin)

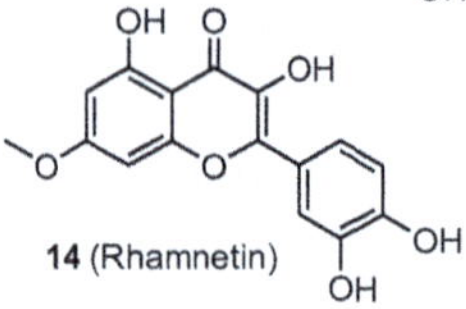
14 (Rhamnetin)

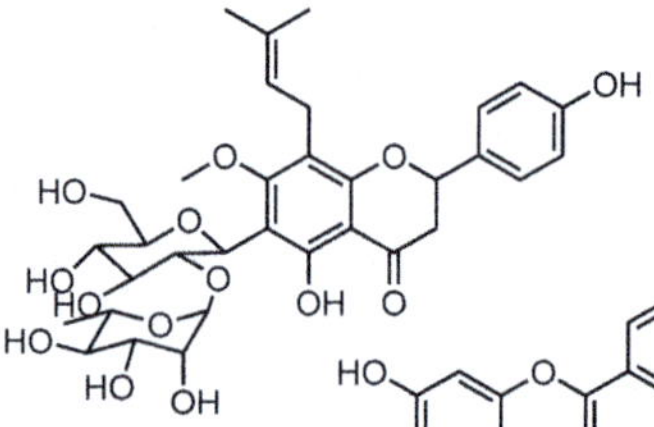
16

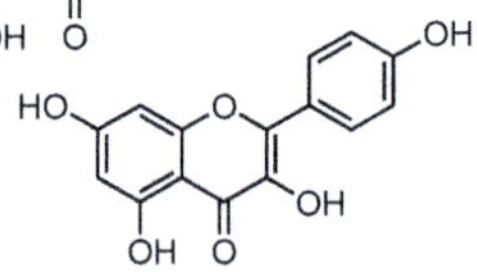
19

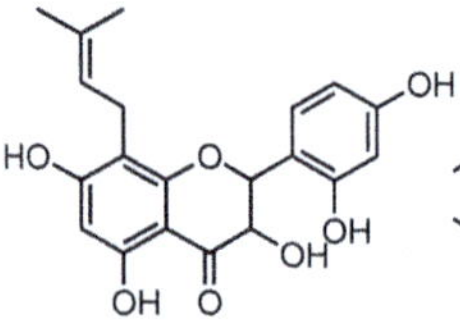
15

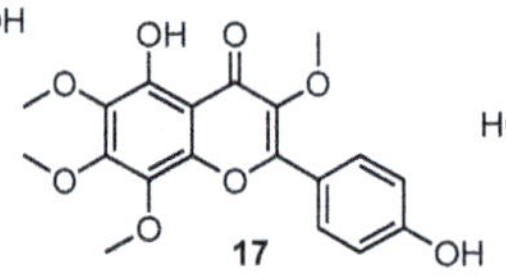
17

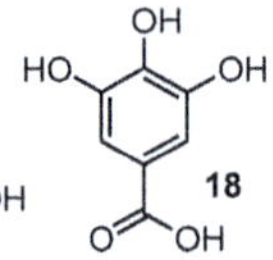
18

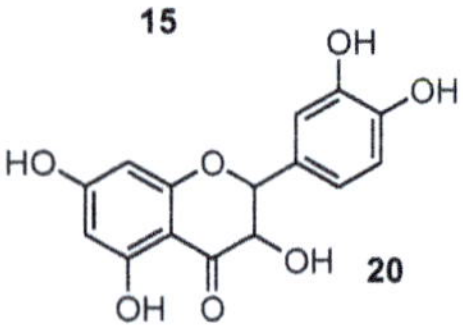
20

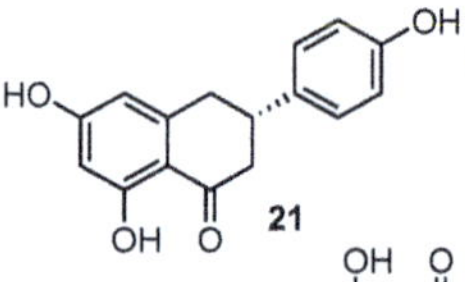
21

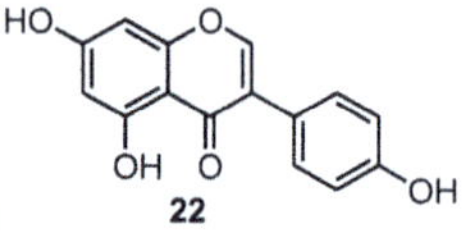
22

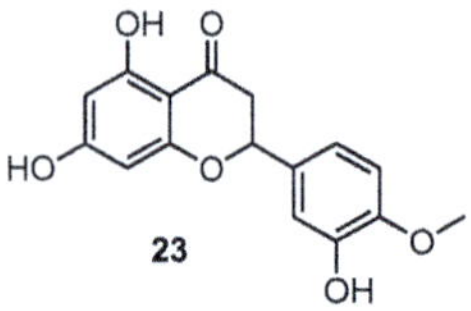
23

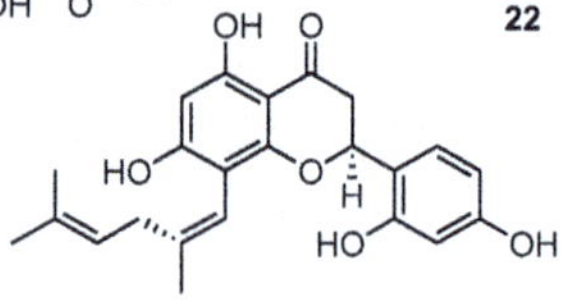
24

Icariin **13** can significantly downregulate CDK2, CDK4, cyclin D1, and Bcl-2 and upregulate cleaved caspase-3 and PARP, thereby inducing apoptosis of the tamoxifen-resistant MCF-7/TAM BC cell line [41].

Rhamnetin **14** increased caspase-3/caspase-9 activity, upregulated p53 protein and miR-34a expression, and downregulated Notch-1 expression, suggesting that rhamnetin induced apoptosis of human BC cells through miR-34a/Notch-1 signaling pathway [42].

Descaudatine A is a phenolic glycoside, and its analogue and ten flavonoids were isolated from the whole plant of *Desmodium caudatum*. Among them, compounds **15** and **16** inhibited the proliferation of HeLa cells with IC_{50} values of 56.14 μM and 69.04 μM, respectively [43]. The mechanism of apoptosis revealed that **16** and **17** induce PARP cleavage and trigger caspase-3, caspase-8, and caspase-9 activation in a dose-dependent manner.

Flavonoid calycopterin **17** induced MDA-MB-231 cell apoptosis by increasing the caspase-3/caspase-8 expression. In addition, it also increased the ratio of Bax/Bcl-2 [44].

Gallic acid **18** decreased the expression of cyclin D1/CDK4 and cyclin E/CDK2, increased the expression of p21 and p27, induced the activation of caspase-3/caspase-9 in MDA-MB-231 cells, and induced apoptosis. The results indicated that this apoptosis was related to the p38/p21/p27 axis [45].

Kaempferol **19** exhibited anti-proliferative effects on human ovarian cancer cell line OVACAR-3. It significantly affected the viability of the cells with cleavage of caspase-9 and caspase-3. It also inhibited PI3K/AKT/mTOR and STAT3 signaling [46].

Quercetin **20** increased the expression of GAS5, Bax, and caspase-3 and decreased the expressions of Notch-1, Jagged1, Hes1, and Bcl-2 in MCF-7 cells, suggesting that quercetin can induce BC cell apoptosis by the GAS5/Notch-1 signaling pathway [47].

Naringenin **21**, a natural flavonoid found in the citrus fruits, has been reported as an anti-breast cancer drug. It promotes the activation of caspase-9 and caspase-3, which further induces apoptosis in the MDA-MB-231 breast cancer cells [48]. Its administration with cyclophosphamide enhanced antitumor properties with upregulation of BAX and decrease in Bcl-2. It activated caspase-3 and caspase-9 and attenuated STAT3 phosphorylation in the cancer cells. It also suppressed the function of IL-6 to modulate apoptosis-associated gene expression [49].

Genistein **22** ($C_{15}H_{10}O_5$), a naturally occurring isoflavone, modulated the p38 MAPK and caspase-3 signaling pathway to promote apoptosis in HT-29 colon cancer cells [50]. It also decreased the concentration of lung cancer cells A549. Its increasing concentration enhanced the expressions of cleaved caspase-9 and caspase-3 and promoted apoptosis in LC cells [51].

Diosmetin **23**, a monomethoxy flavonoid obtained from citrus fruits, exhibited diversified activities. It upregulated the expression of p53, Bax, and caspase-3, while it downregulated the expression for Bcl-2 and MMP and promoted ROS accumulation in MCF-7 cells. It promoted apoptosis in MCF-7 cells through ROS- and p53-mediated mitochondrial pathways [52].

Sophoraflavanone G **24** suppressed the MAPK-related pathway to increase cleaved caspase-3/caspase-8/caspase-9 and Bax expression, decreased Bcl-2 and BclXL expression, and prompted release of cyto C from mitochondria to the cytoplasm in MDA-MB-231 cells [53].

3.3 Isoflavones

Isoflavones, found in plants as glycosides in inactive form, are non-steroid phytoestrogens possessing estrogenic and anti-estrogenic bioactivities and get hydrolyzed by intestinal bacteria and get transformed to bioactive form aglycones like daidzein and genistein. They get easily absorbed in blood from the intestine [54]. Isoflavones possess anticancer activities against various hormone-dependent and hormone-independent cancers like endometrium, ovary, breast, prostate, and lung and melanoma, respectively.

Among isoflavones, calycosin **25** promoted apoptosis in tumor cells MCF-7, possibly in part by reducing SIRT1 levels, thereby increasing p53 and cleaving caspase-3 expression to promote apoptosis [55].

Formononetin **26** induced mitochondrial damage with decrease of MMP in BC cells, interrupted with antioxidant enzyme activity by increasing ROS levels, and induced apoptosis by activating caspase-3 in the cancer cells [56].

Novel ERb-selective daidzein analogues (NCE) were synthesized and screened for their anticancer activity along with their apoptosis mechanism. Among them, NCE1 **27** and NCE2 **28** showed anti-prostate cancer activity in DU145, LNCaP, and PC3 cell lines and were 50% more active than genistein. The NCEs increased caspase-3 and caspase-9 and downregulated cMYC and cyclin D1 genes [57].

Sorafenib (SOR), the first-line treatment for hepatocellular carcinoma (HCC), suffers the constraint of high toxicity. The toxicity of SOR has been decreased by using it in combination with other less toxic agents like biochanin A (Bio-A) **29**, a promising isoflavone. This combination arrested cell cycle in pre-G and G0/G1 phase and decreased cyclin D1 proteins and Bcl-2/Bax ratio. It enhanced apoptotic markers caspase-3 and caspase-9 and decreased proliferative and anti-apoptotic marker Ki-67 and survivin. Enhanced activity of caspase-3 that was found in HepG2, SNU-449, and Huh-7 confirmed the synergistic approach [58].

3.4 Polyphenolic Compounds

Ellagic acid **30** is an herbal polyphenolic compound found to be effective against carbon tetrachloride (CCl_4)-induced testicular tissue damage in rats. It enhanced caspase-3 and Nrf-2 protein synthesis levels in EA + CCl_4 group but decreased VEGF, Bcl-2, NF-κB, TNF-α, and Akt protein synthesis. The studies suggested that EA can be used as a drug in the future for curing testicular damage [59].

Podophyllotoxin (podo) **31** is a natural product that exhibits good in vitro cytotoxicity profile and is an apoptosis inducer against different types of cancer cells. However, its use is limited due to toxicity and side effects [60]. Podo and etoposide are two structurally related antitumor agents with different anti-proliferative mechanisms of action [61]. Podo exhibited a strong β-tubulin polymerization inhibition, responsible for antimitotic effect, while etoposide showed topo enzyme inhibition, which prevents DNA synthesis [62]. Structure investigation of podo and etoposide showed similar pharmacophoric fragments as 3,4,5-trimethoxyphenyl (TMP) moiety and orthogonal tetrahydrofuronaphthodioxolone ring system in their skeletons. Both features describe key pharmacophoric points responsible for interaction with colchicine binding site, leading to β-tubulin polymerization inhibition [63].

Polyphenolic compounds like lignans (sesamin and arctigenin) possess potential anti-proliferative activities against several cancers [64]. Sesamin **32**, a natural substance lignin obtained from sesame (*Sesamum indicum L.*) seeds, causes apoptosis in MCF-7 cells by upregulating the expression of the apoptosis markers Bax, caspase-3, and p53 [65]. It also induced apoptosis in human breast cancer MCF-7 cells by enhancing the expression levels of Bax and caspase-3 [65].

Arctigenin **33** has been found to promote apoptosis in HT-29 colon cancer cell line via ROS production, loss of mitochondrial membrane potential (MMP), and activation of p38 MAPK, caspase-3, and caspase-9 [66]. The study demonstrated that like sesamin, it also triggers apoptosis in breast cancer cells by activating caspase-9 and caspase-3 [67]. It also enhanced caspase-3 activity in lung cancer A549 cells (50 μg/mL) and human ovarian cancer cells (10 μM) and suppressed their viability [68]. It also induced apoptosis in both HepG2 and SMMC-7721 cells through mitochondrial pathway, followed by caspase-3 and caspase-9 activation succeeded by PARP cleavage.

It also prevented the growth of cancers in different organs like the colon, liver, lung, and stomach as well as leukocytes [69] and induced apoptosis by the activation of caspase-3 in cancer cells.

MDA-MB-231 and MDA-MB-468 cells with varying concentrations of **33** (0.4, 0.8, or 1.2 μM) cleaved PARP and procaspase-3 and procaspase-9 in a dose-dependent manner [67].

6-Shogaol **34**, the bioactive component of ginger, induced apoptosis of SW872 cells at 20 μM by initiating nuclear DNA fragmentation followed by accumulation of sub-G1 phase cells, leading to the cleavage of PARP and activation of caspase-3/caspase-9. The apoptotic mechanism followed both extrinsic (DR-dependent activation of caspase-8) and intrinsic pathways (mitochondria-mediated activation of caspase-9) [70, 71].

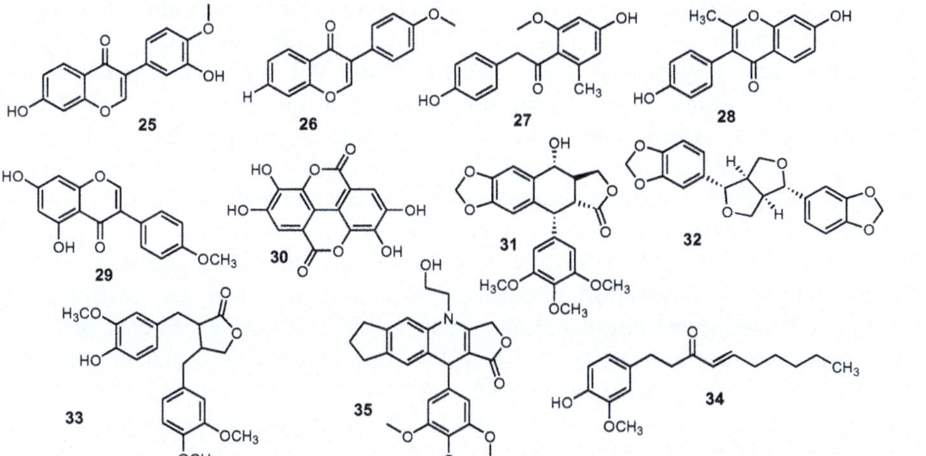

Treatment of head and neck squamous cell carcinomas (HNSCCs), SCC4 and SCC25 cells, with 10 mM of 6-shogaol **34** initiated phosphorylation in SCC4 and SCC25 cells within 2 h in a dose- and time-dependent manner. It arrested the cell cycle in G2/M phase in SCC4 cells. Cells co-treated with ERK1/2 or p38 signaling inhibitors and 6-shogaol also decreased the distribution of the G0/G1 phase but increased the distribution of both S and G2/M phases in SCC4 cells. The cells pretreated with ERK1/2 or p38 signaling inhibitors also block the 6-shogaol effect on inducing apoptosis and suppressed cleaved caspase-3 level in SCC4 cells [72]. HeLa cancer cells when treated with 6-shogaol **34** significantly raised the caspase-3 activity in the cell and induced apoptosis via activation of the intrinsic apoptotic pathway. Inhibition of TrxR activity in HeLa cells was seen in the 6-shogaol (0, 5, 10, 20, and 50 μM concentration)-treated cell for 24 h, and the activity of TrxR was reduced to almost 12% at 50 μM concentration [73].

Novel aza-podophyllotoxin derivative **35** (SU212) induced oxidative phosphorylation and cell death via AMPK activation in triple-negative breast cancer. Its treatment increased the expression of p-AMPKα, Bax, and c-caspase-3 and inhibited the expression of Ki-67 and LDHA but did not affect blood glucose levels indicating that its effect is independent of energy stress. It showed IC_{50} values of 0.25 and 0.58 μM for MDA-MB-468 and MDA-MB-231 cells, respectively, with minimal toxicity to normal breast cell lines and nerve cells [74].

3.5 Carbohydrates

Carbohydrate-based antitumor therapeutics have gained significance for improving immune responses and to target the antigens. Tumor cells consume more energy than normal cells for proliferation through high sugar binding or via transporting receptors located in the cellular membrane. The Warburg effect has inspired the glycoconjugation strategies for development of anticancer drugs [75].

Amygdalin **36**, a natural compound found in seeds of apricot, bitter almond, peaches, and apples, is a cyanogenic glycoside. It has been investigated for its anticancer activities where it induced apoptosis of Hs578T TNBC cells by downregulating Bcl-2, upregulating Bax, activating caspase-3, and cleaving PARP [76]. It also activated caspase-3 and BAX and inhibited Bcl-2 and poly (ADP-ribose) polymerase-1 (PARP-1). Molecular dynamic investigation revealed that it acts as a caspase-3 activator [77].

Polyphyllin D **37**, a saponin steroid, is the main constituent of traditional Chinese medicinal herb *Paris polyphylla* reported to be potential anticancer agent against several tumors, including bladder cancer, pancreatic cancer, breast cancer, liver cancer, and lung cancer [78, 79]. It decreased the levels of Bcl-xL and Bcl-2 and increased the levels of Bax in CT26 cells [80]. It reduced cell viability in human neuroblastoma cultured cells in a concentration-dependent manner. There was enhanced caspase-3/caspase-7 and caspase-8 activation in NB-69 cells. Thus, polyphyllin D may be employed for the treatment of not only bladder, pancreatic, breast, liver, and lung cancers, but as it also crosses the blood-brain barrier, it may be used in treatment of neurogenic cancers like neuroblastoma [81].

Ginsenosides are steroid glycosides obtained from ginseng plant of genus *Panax*. Many ginsenosides, viz., Rh2 **38**, Rk1 **39**, Rb1 **40**, Rg5 **41**, Rg3 **42**, Rd **43**, and Rk1 **44**, have shown promising anticancer activities through the activation of caspase-3. Rh2 **38** inhibited the proliferation of ovarian cancer cell line SKOV3 by activation of caspase-3 and PARP and inhibition of Bcl-2 [82]. It also inhibited the proliferation of osteosarcoma cells (U20S) in a dose-dependent manner and prevented metastasis by decreasing Bcl-2, caspase-3, and caspase-9 expressions while increasing Bax levels in osteosarcoma cells. It induced apoptosis in U20S cells through MAPK pathway and suppressing the activities of PI3K/AKT/mTOR and NF-κB pathways [83]. Rk1 **39** enhanced the expressions for Bax and cyto C, cleaved caspase-3/caspase-8/caspase-9, and reduced Bcl-2 by blocking the PI3K/AKT pathway [84]. On screening ginsenosides **40** (Rb1),

41 (Rg5), and **42** (Rg3) against five human cancer cell lines, the **41** (Rg5) was highly cytotoxic to breast cancer cell (MCF-7). It inhibited proliferation of cancer cells through mitochondria-mediated apoptosis and autophagy of cancer cells. It minimized the phosphorylation levels of PI3K, Akt, mTOR, and Bad with suppression of PI3K/Akt signaling. It upregulated the cleavage of caspase-3, caspase-8, caspase-9, PARP, Bax, and cytochrome C, while it downregulated Bcl-2 protein in a dose-dependent manner [85]. The (Rg3) **42** inhibited differentiation and adipogenesis and induced ER stress-mediated apoptosis in mouse primary brown adipocytes (MPBAs) by increasing ER stress and pro-apoptotic markers in MPBAs. It showed selective cytotoxicity in differentiated MPBAs keeping undifferentiated MPBAs as such. It enhanced the cleavage of procaspase-3 in MPBAs, leading to apoptosis in the cancer cells [86]. Rg3 **42** inhibited the proliferation of gallbladder cell lines (NOZ and GBC-SD) in a dose-dependent manner both in vitro and in vivo and arrested cell cycle at G0 /G1 phase with enhanced p53 and p21. It enhanced apoptosis through mitochondrial-mediated intrinsic caspase pathway by decreasing PCNA expression and enhancing caspase-3 activity in cancer cells. Thus, compound **42** may be a potential chemotherapeutic agent for GBC therapy [87].

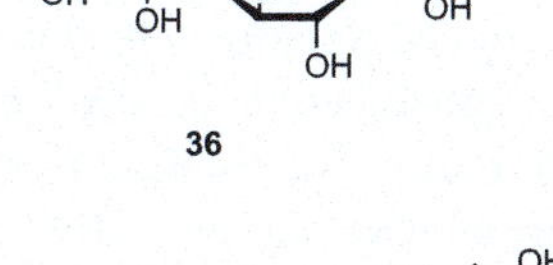

36

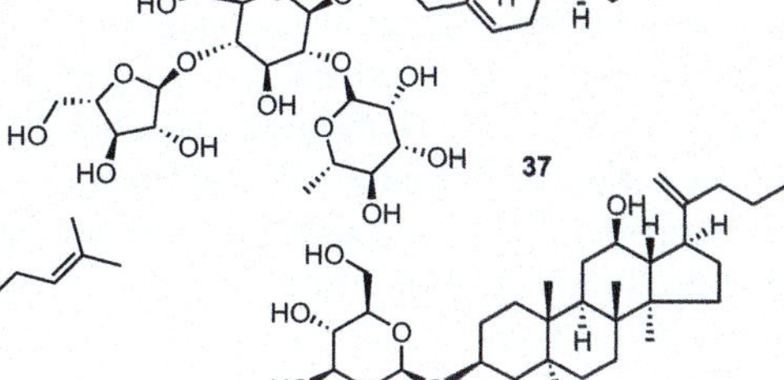

37

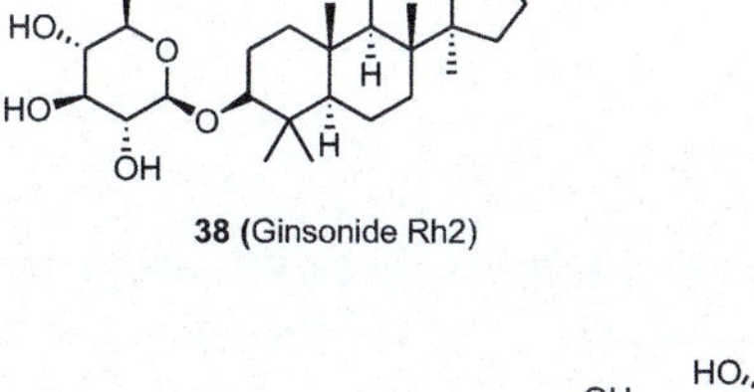

38 (Ginsonide Rh2)

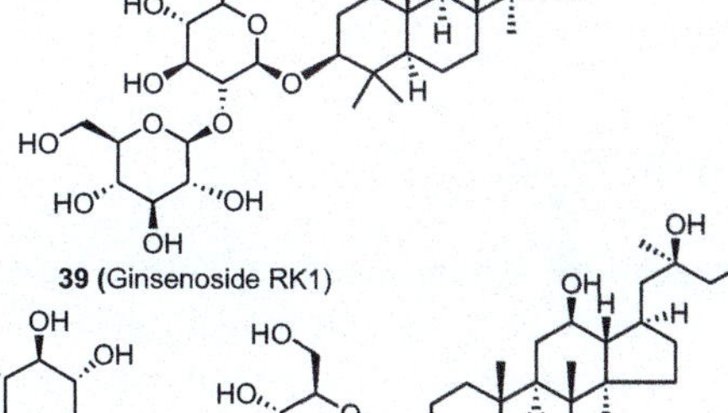

39 (Ginsenoside RK1)

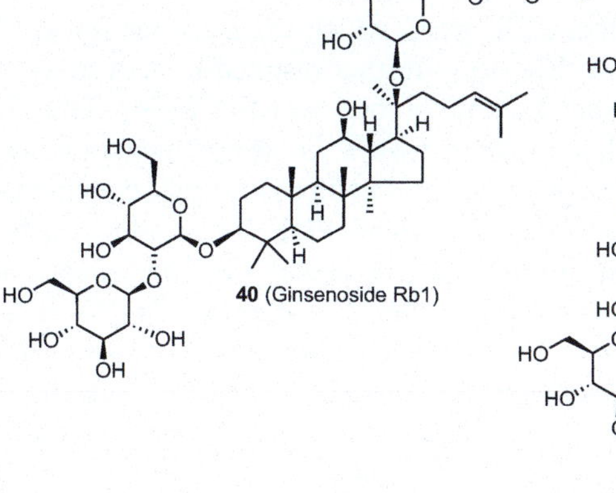

40 (Ginsenoside Rb1)

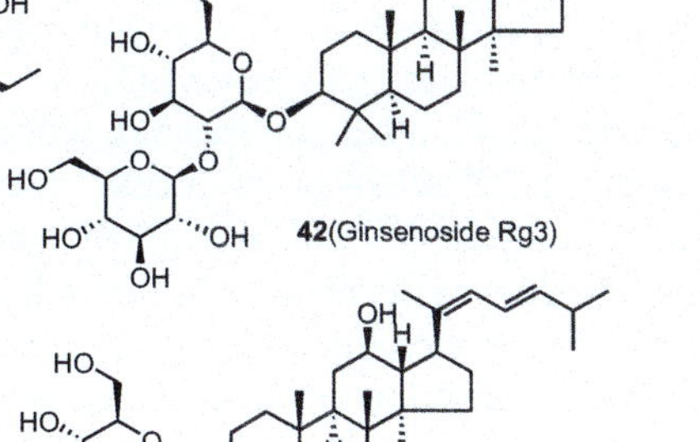

42(Ginsenoside Rg3)

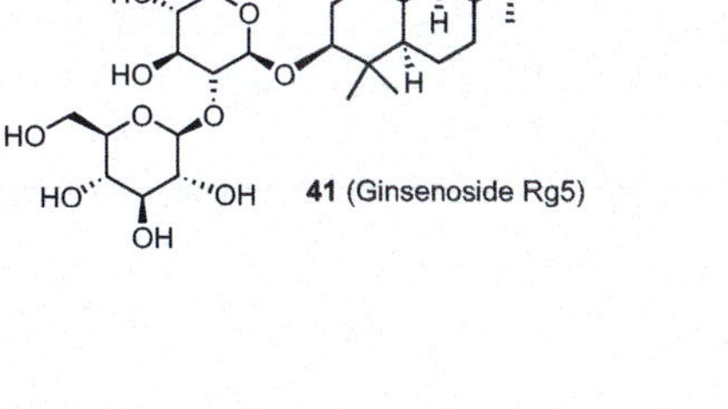

41 (Ginsenoside Rg5)

Rd **43** inhibited gastric cancer (SGC-7901 and MKN-45 cells). It also prevented cancer cell proliferation, arrested cell cycle at G0/G1 phase, and induced apoptosis in them. It increased Bax/Bcl-2 ratio and caspase-3 and caspase-9 expressions, while it decreased cyclin D1 expressions. It may be developed as a potential therapeutic agent against gastric cancer [88].

Rk1 **44** induced apoptosis in neuroblastoma cancer cells. It reduced the mitochondrial membrane potential, arrested the cancer cell cycle at G0/G1 phase, and prevented metastasis. Its injections (30 mg/kg) remarkably inhibited xenograft tumor growth. It induced apoptosis via both intrinsic and extrinsic pathways by upregulating caspase-3, caspase-8, Bak, tBID, and PARP and downregulating Bcl-xL and Bcl-2 [89].

Betanin **45**, an important betacyanin, inhibited 50% of the cell growth of U87MG human glioma cells after 12 h of its administration at 7 μg/mL. It induced apoptosis via activation of caspase-3 through the mitochondrial pathway [90].

3.6 Terpenoids

Terpenoids exhibit anticancer activities by arresting the cell cycle at various stages and inducing apoptosis. They also modulate the signaling pathways of cells and inhibit metastasis and angiogenesis [91].

The triterpene arnidiol **46** obtained from the bloom of the plants *Arnica montana* and *Taraxacum officinale* has shown anticancer activities. It induced apoptosis in human breast cancer cells MDA-MB-231 through mitochondrial translocation of Drp1 and cofilin, responsible for downregulation of PARP expression. It also increased caspase-3 cleavage and synergistic effect with cepharanthine and epirubicin by significantly promoting mitochondrial translocation of cofilin, thus inducing apoptosis in TNBC cells via the mitochondrial pathway [92].

3.7 Sesquiterpene

Sesquiterpenes are synthesized by plants in leaves, stems, fruits, roots, etc. and defend plants from bacterial, viral, and fungal infections. These possess significant anticancer activities to almost all cancers by inhibiting lipid peroxidation and generation of reactive oxygen and nitrogen species [93].

Zerumbone **47**, a sesquiterpenoid, induced ER stress, which enhanced the levels of CHOP and GRP78/BiP in PC-3 cells. The ER stress further resulted in activation of caspase-3, caspase-7, caspase-9, PARP, and Bid and inhibition of Bcl-2 [94]. Thus, it inhibits the growth of cancer cells through multiple pathways. In the treatment of melanoma, it upregulated the Bax and cytochrome c gene-related proteins and downregulated Bcl-2 and induced apoptosis by activating caspase-3 [95]. It also blocked the activity of cyclophilin A and thus promoted mitochondrial pathway-mediated apoptosis in gastric cancer cells through activation of caspase-3 [96]. It also induced DNA damage and activated the cleavage of procaspase-9, procaspase-3, and PARP by increasing the intracellular free Ca^{2+} and soluble histone H2AX in K562 chronic myelogenous leukemia cells, induced apoptosis in cancer cells, and prevented formation of their colonies [97].

Renal cell carcinoma (RCC) is a malignant cancer, which does not respond to the treatments in use and is the most common form of cancer in the kidney [98]. The **47** induced apoptosis in the cancer cells by caspase-3/caspase-9 activation, cleavage of PARP, and downregulation of Gli-1 and Bcl-2. It is a potential anticancer agent for inhibiting proliferation and inducing apoptosis on pediatric rhabdomyosarcoma cell lines by enhancing caspase-3/caspase-7 activity, ROS production, and modulation of NF-κB pathway. Further, it was found that caspase-3/caspase-7 activity also increased significantly in RH30 and RD cells.

The natural compound parthenolide **48**, a sesquiterpene lactone, known for the treatment of migraines and arthritis, and a potent NF-κB signaling inhibitor, prevented the growth of SiHa and MCF-7 cell lines in a concentration-dependent manner with IC_{50} values of 8.42 and 9.54 μM, respectively. It induced apoptosis in the cancer cells by upregulating p53, Bax, caspase-3, caspase-6, and caspase-3 genes and downregulating Bcl-2 [99] and also modulated the PI3K/AKT signaling pathway [100]. The compound induced ROS-mediated apoptosis in lymphoid malignancies and decreased metabolic activities in cell lines in a time-, dose-, and cell line-dependent manner and induced apoptosis with GSH decrease, reduction in Ψ_{mit}, and caspase-3 activation [101]. It also inhibited GBC cell growth in vitro and induced apoptosis in the GBC-SD and NOZ cell lines in a dose-dependent manner by upregulating BAX, cleaved PARP-1, cleaved caspase-3 and caspase-9, and p53 and downregulating the expression for Bcl-2, phosphorylated ERK, and phosphorylated MEK [102].

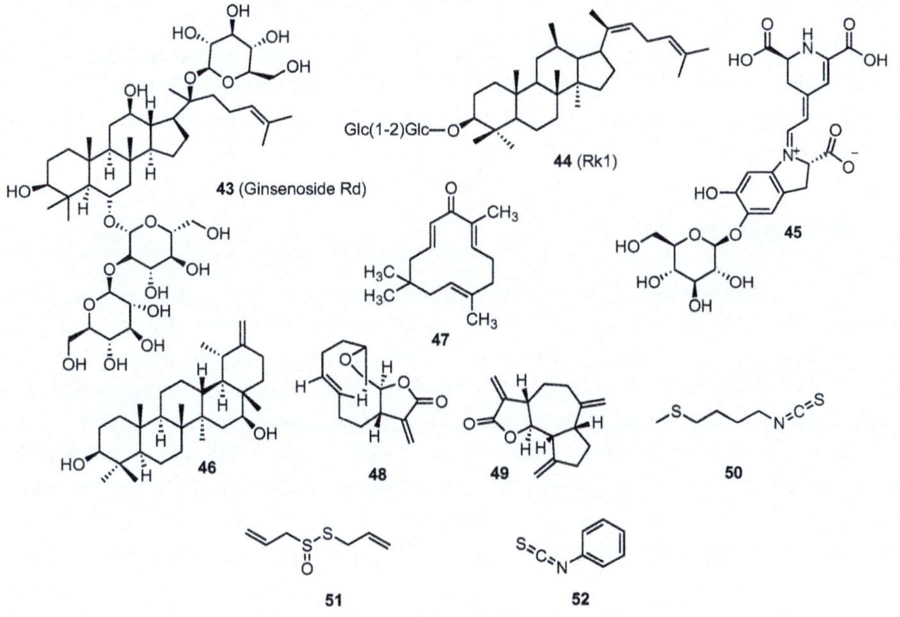

Dehydrocostus lactone (DHE) **49**, a natural sesquiterpene lactone derived from Chinese plants like *Inula helenium* L. and *Saussurea lappa*, has shown significant anticancer activity by suppressing the growth and promoting apoptosis in SK-BR-3 BC cells, associated with the inhibition of the anti-apoptotic ability of BC cells by upregulation of Bax/Bcl-2 and caspase-3 expression [103].

3.8 Organosulfur Compounds

Organosulfur compounds have also been found to participate in cellular functions and are also useful in anticancer therapeutics [104]. 4-Methylthiobutyl isothiocyanate (erucin) **50**, a bioactive compound found in cruciferous vegetables, induced mitochondrial division and apoptosis in human BC cells through mitochondrial translocation of cofilin, which was correlated with downregulation of PARP and upregulation of caspase-3 and caspase-9 cleavage [105]. Allicin **51**, a bioactive compound present in garlic, reduced cell viability and induced apoptosis by activating caspase-3, caspase-8, and caspase-9 in the breast cancer cells by upregulation of NOXA, p21, and Bak and downregulation of Bcl-xL expression in BC cells [106].

Phenylethyl isothiocyanate **52**, obtained from cruciferous plants, increased intercellular ROS and was effective against cervical cancer cells. It also exhibited synergistic effects in combination with cisplatin. Its administration increased caspase-3 activities in cervical cancer cells by 62.53, 81.89, and 112.06% at 20, 25, and 30 μM, respectively, compared to the untreated control [107].

3.9 Vitamin E

Vitamin E δ-tocotrienol (VEDT) **53**, separated from annatto beans and palm fruit, showed anticancer activities against MiaPaCa-2, BxPC-3, and SW1990 cell lines. It cleaved procaspase-3 and procaspase-8 and induced apoptosis via caspase-3 and caspase-8 activation. It may be investigated further for treating pancreatic cancer [108].

3.10 Benzoquinone

Natural products containing benzoquinone moiety have also shown anticancer activity through caspase-3 activation. Emodin (1,3,8-trihydroxy-6-methylanthraquinone) **54**, an anthraquinone derivative, reduced Bcl-2 levels and increased cleaved caspase-3, PARP, p53, and Bax levels in BCAP-37 and ZR-75-30 BC cells. The study suggested that it could induce apoptosis through the p53-mediated mitochondrial pathway [109].

Chrysophanol **55** upregulated caspase-3 and PARP cleavage and downregulated the apoptosis regulators Bcl-2, p-NF-κBp65, and p-IκB, indicating that the pro-apoptotic function of chrysophanol was achieved by regulating the NF-κB/Bcl-2 signaling cascade for the anticancer effect in breast cancer cell lines [110].

3.11 Macrocyclic Compounds

Tetrandrine **56**, a bis-benzylisoquinoline separated from *Cyclea peltata*, induced cytotoxicity and apoptosis by increasing ROS and caspase-8/caspase-9/caspase-3 in MDA-MB-231 and PANC-1 cells through intrinsic pathway [111]. It showed dose- and time-dependent cytotoxicity on both cancer cells with IC_{50} values in the range of 51–54 μM and 22–27 μM for 24 h and 48 h, respectively.

3.12 Alkaloids

The alkaloids are secondary plant metabolites that have justified themselves as potential drug candidates based on their efficacy in cancer treatment [112].

Neferine **57** is an alkaloid derived from the lotus and is bis-benzylisoquinoline alkaloid with diverse therapeutic activities in addition to anticancer like anti-HIV, anti-inflammatory, anti-thrombotic, anti-arrhythmic, antimicrobial, anti-aging, and antidiabetic activities [113]. It arrested cell cycle in hepatocellular carcinoma (Hep3B) but did not affect the normal liver cells. The apoptosis here is caused by ER stress and activation of caspase-3, caspase-6, caspase-7, caspase-8, Bak, Bax, Bid, Bim, and PARP [114]. Its combination with dehydroepiandrosterone (DHEA) synergistically enhanced the potency of the DHEA and significantly increased apoptosis by 36.7% in MCF-7 cells. Neferine-treated MCF-7 cells exhibited a significant increase in expression of caspase-3, caspase-8, caspase-9, E2F1 mRNAs, FasL, Fas, p53, p21, and Bax and notable decrease in HIAP-2, HIAP-1, Bcl-2, and Bcl-xL expressions [115].

It has also synergistically enhanced the anticancer activities of drugs like Taxol, Adriamycin, cisplatin, etc. It is also found to overcome chemo-resistance of cancer cells toward the established drugs and sensitizes the cancer cells toward anticancer drugs. This co-treatment enhances cytochrome c release and activates the cleavage of caspase-3, caspase-9, and PARP [116].

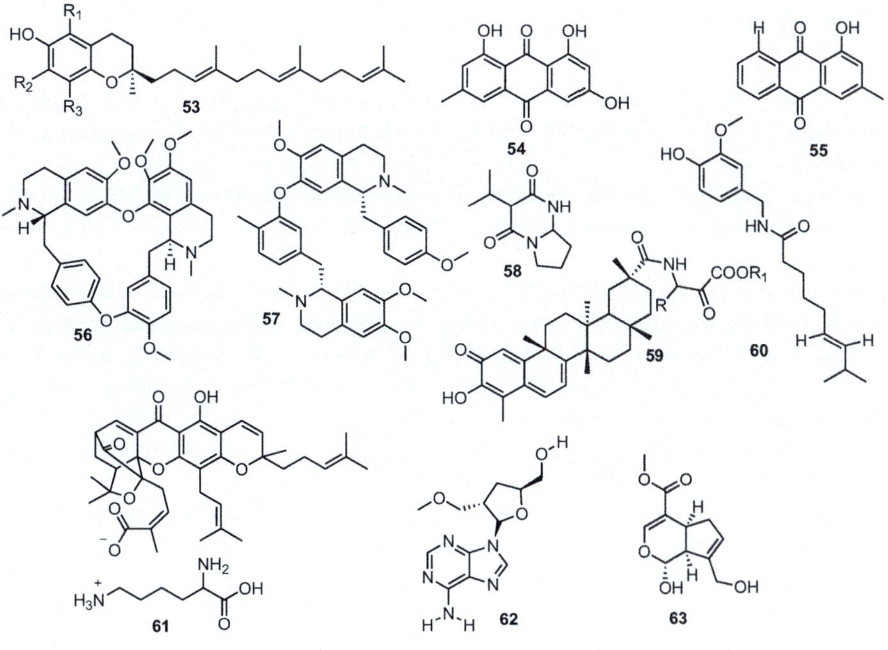

3.13 Amides

Naturally occurring amides have also enhanced caspase-3 activation and thus behave as anticancer agents. The naturally occurring 3-isopropyl-tetrahydropyrrolo[1, 2-a] pyrimidine-2,4(1H,3H)-dione (ITPD) **58** has been totally synthesized and has shown promising in vitro anticancer activity in SMMC-7721 cell lines through mitochondrial pathway by activating caspase-3/caspase-9 and increasing the ratio of Bax/Bcl-2 to induce cell apoptosis through DNA damage. It showed anticancer activity against SMMC-7721 cells with IC_{50} at 37.85 μM as compared to 5-FU with IC_{50} at 19.73 μM. It caused a remarkable increase in accumulation of cells in S phase and reduced cells in G1 and G2 phase at 20 μM [117].

The synthesized novel celastrol derivatives were screened for their anticancer activities against AGS, MGC-803, SGC-7901, HCT-116, A549, HeLa, BEL-7402, and HepG-2 cell lines. Among them, the compound 3-hydroxy-9b,13a-dimethyl-2-oxo-24,25,26-trinoroleana-1(10),3,5,7-tetraen-29-oic amide N-(R)-methyl-3-(1H-indol-2-yl)propanoate **59** showed significant anticancer activity against AGS, HCT-116, BEL-7402, and HepG-2 cell lines with IC_{50} values of 0.44, 0.78, 0.63, and 0.76 μM, respectively. The mechanism of apoptosis induced by this compound was through activation of caspase-3 [118].

Capsaicin **60** upregulated the expressions for Bax protein and activated caspase-3, while it downregulated the expression of FBI-1, Ki-67, Bcl-2, and survivin in BC cells. It also deactivated target gene NF-κB of FBI-1 [119].

4 Miscellaneous

Apart from the natural products discussed above, the following are some other natural product molecules that have shown significant anticancer activities through caspase-3 activation:

Gambogic acid lysinate **61** induced dose-dependent apoptosis in MCF-7 cells through upregulation of cleaved caspase-3 and inhibition of expressions of SIRT1 protein [120].

Cordycepin **62**, a major compound isolated from *Cordyceps sinensis*, induced apoptosis in BC cells by the activation of pro-apoptotic proteins, such as caspase-8, caspase-9, caspase-3, and Bax, and inhibition of Bcl-2 expression [121]. Recently, it has been reported to induce apoptosis in breast cancer cells by inhibiting the hedgehog pathway, and this effect was associated with elevation of PUMA, cyto C, Fas, and DR4/5 levels; decrease in Bcl-2, XIAP, and PDGFR-α; and cleavage of caspase-3 in cancer cells [122].

Genipin **63** present in *Genipa americana* fruit extract is a bioactive natural cross-linker for proteins, collagen, gelatin, and chitosan. It promoted the apoptosis in SCC-25 and SCC-9 cells through cleavage of caspase-3 and PARP [123].

5 Conclusions

Drugs of plant origin being less toxic are promising for the cancer chemotherapy. It is evident that caspase-3 is a validated target for development of cancer therapeutics and several molecules belonging to diverse classes of natural products have shown very promising anticancer activities through caspase-3 activation and induction of apoptosis in cancer cells. Since natural products have established themselves as successful drugs, the described potential lead candidates based on their in vitro and in vivo anticancer activities through caspase-3 activation may be useful molecules for optimization of caspase-3 activation and anticancer activities by crucial structural modifications for enhancing the activity and reducing the toxicity through induction of selectivity toward cancer cells and improving the pharmacokinetics of the lead molecules. Insight into the essential structural features for the activity as well as toxicity may be useful in designing novel chemical entities as caspase-3 activators for anticancer drug development.Conflict of InterestThe authors declare no potential conflict of interest.

Funding The authors declare that there was no funding for this work.

References

1. http://www.Cancer.gov/about-cancer/understanding/statistics
2. Dynlacht, B.: Regulation of transcription by proteins that control the cell cycle. Nature. **389**, 149–152 (1997). https://doi.org/10.1038/38225
3. Elledge, S.J.: Cell cycle checkpoints: preventing an identity crisis. Science. **274**, 1664–1671 (1996). https://doi.org/10.1126/science.274.5293.1664
4. Rini, B.I.: Temsirolimus, an inhibitor of mammalian target of rapamycin. Clin. Cancer Res. **14**, 1286–1290 (2008). https://doi.org/10.1158/1078-0432.CCR-07-4719
5. Roccaro, A.M., Vacca, A., Ribatti, D.: Bortezomib in the treatment of cancer. Recent Pat. Anticancer Drug Discov. **1**, 397–403 (2006). https://doi.org/10.2174/157489206778776925
6. Cross, S.A., Lyseng-Williamson, K.A.: Imatinib: in relapsed or refractory Philadelphia chromosome-positive acute lymphoblastic leukemia. Drugs. **67**, 2645–2654 (2007). https://doi.org/10.2165/00003495-200767170-00013
7. Baselga, J.: Targeting tyrosine kinases in cancer: the second wave. Science. **312**, 1175–1178 (2006). https://doi.org/10.1126/science.1125951
8. Deeken, J.F., Löscher, W.: The blood-brain barrier and cancer: transporters, treatment, and Trojan horses. Clin. Cancer Res. **13**, 1663–1674 (2007)
9. Jiang, W.G., Sanders, A.J., Katoh, M., Ungefroren, H., et al.: Tissue invasion and metastasis: molecular, biological and clinical perspectives. Semin. Cancer Biol. **35**(Supplement), S244–S275 (2015). https://doi.org/10.1016/j.semcancer.2015.03.008
10. Earnshaw, W.C., Martins, L.M., Kaufmann, S.H.: Mammalian caspases: structure, activation, substrates, and functions during apoptosis. Annu. Rev. Biochem. **68**, 383–424 (1999). https://doi.org/10.1146/annurev.biochem.68.1.383
11. Jiménez Fernández, D., Lamkanfi, M.: Inflammatory caspases: key regulators of inflammation and cell death. Biol. Chem. **396**, 193–203 (2015). https://doi.org/10.1515/hsz-2014-0253

12. Rendl, M., Ban, J., Mrass, P., Mayer, C., et al.: Caspase-14 expression by epidermal keratinocytes is regulated by retinoids in a differentiation-associated manner. J. Invest. Dermatol. **119**, 1150–1155 (2002). https://doi.org/10.1046/j.1523-1747.2002.19532.x
13. Yadav, P., Yadav, R., Jain, S., Vaidya, A.: Caspase-3: a primary target for natural and synthetic compounds for cancer therapy. Chem. Biol. Drug Des. **00**, 1–22 (2021). https://doi.org/10.1111/cbdd.13860
14. Crawford, E.D., Seamant, J.E., Agard, N., Hsu, G.W., et al.: The DegraBase: a database of proteolysis in healthy and apoptotic human cells. Mol. Cell. Proteomics. **12**, 813–824 (2013). https://doi.org/10.1074/mcp.O112.024372
15. Kothakota, S., Azuma, T., Reinhard, C., Klippel, A., Tang, J., Chu, K., et al.: Caspase-3-generated fragment of gelsolin: effector of morphological change in apoptosis. Science. **278**, 294–298 (1997). https://doi.org/10.1126/science.278.5336.294
16. Harrington, H.A., Ho, K.L., Ghosh, S., Tung, K.C.: Construction and analysis of a modular model of caspase activation in apoptosis. Theor. Biol. Med. Model. **5**(1), 26 (2008). https://doi.org/10.1186/1742-4682-5-26
17. Bhadra, K.A.: Mini review on molecules inducing caspase-independent cell death: a new route to cancer therapy. Molecules. **27**, 6401 (2022). https://doi.org/10.3390/molecules27196401
18. Eskandari, E., Eaves, C.J.: Paradoxical roles of caspase-3 in regulating cell survival, proliferation, and tumorigenesis. J. Cell Biol. **221**(6), e202201159 (2022). https://doi.org/10.1083/jcb.202201159
19. Dou, H., Yu, P.Y., Liu, Y.Q., Zhu, Y., Li, F.C., Wang, Y.Y., Xiao, M.: Recent advances in caspase-3, breast cancer, and traditional Chinese medicine: a review. J. Chemother. **36**, 1–19 (2023). https://doi.org/10.1080/1120009X.2023.2278014
20. Srivastava, N., Saxena, A.: Caspase-3 activators as anticancer agents. Curr. Protein Pept. Sci. **24**(10), 783–804 (2023). https://doi.org/10.2174/1389203724666230227115305
21. Hu, S., Xu, Y., Meng, L., Huang, L., Sun, H.: Curcumin inhibits proliferation and promotes apoptosis of breast cancer cells. Exp. Ther. Med. **16**(2), 1266–1272 (2018). https://doi.org/10.3892/etm.2018.6345
22. Kuttikrishnan, S., Siveen, K.S., Prabhu, K.S., Khan, A.Q., Ahmed, E.I., Akhtar, S., Ali, T.A., Merhi, M., Dermime, S., Steinhoff, M., Uddin, S.: Curcumin induces apoptotic cell death via inhibition of PI3-kinase/AKT pathway in B-precursor acute lymphoblastic leukemia. Front. Oncol. **9**, 484 (2019). https://doi.org/10.3389/fonc.2019.00484
23. Song, W., Ren, Y.Z., Liu, L.L., Zhao, Y.Y., Li, Q.F., Yang, H.B.: Curcumin induced the cell death of immortalized human keratinocytes (HaCaT) through caspase-independent and caspase-dependent pathways. Food Funct. **12**, 8669–8680 (2021). https://doi.org/10.1039/D1FO01560E
24. Li, P., Pu, S., Lin, C.: Curcumin selectively induces colon cancer cell apoptosis and S cell cycle arrest by regulating Rb/E2F/p53 pathway. J. Mol. Struct. **1263**, 133180 (2022). https://doi.org/10.1016/j.molstruc.2022.133180
25. Ali, M.S., Smiley, R.: Curcumin-induced apoptosis pathways in positive and negative human epidermal growth factor receptor breast cancer cells. FASEB J. **34**(S1), 1 (2020). https://doi.org/10.1096/fasebj.2020.34.s1.02326
26. Chaudhary, M., Kumar, N., Baldi, A., Chandra, R., Babu, M.A., Madan, J.: 4-Bromo-4′-chloro pyrazoline analog of curcumin augmented anticancer activity against human cervical cancer, HeLa cells: in silico-guided analysis, synthesis, and in vitro cytotoxicity. J. Biomol. Struct. Dyn. **38**(5), 1335–1353 (2020). https://doi.org/10.1080/07391102.2019.1604266
27. Razali, N.C., Lam, K.W., Rajab, N.F., et al.: Curcumin piperidone derivatives induce caspase-dependent apoptosis and suppress miRNA-21 expression in LN-18 human glioblastoma cells. Genes Environ. **46**, 4 (2024). https://doi.org/10.1186/s41021-023-00297-y

28. Li, W., Chen, Y., He, K., Cao, T., Song, D., Yang, H., Li, L., Lin, J.: The apoptosis of liver cancer cells promoted by curcumin/TPP-CZL nanomicelles with mitochondrial targeting function. Front. Bioeng. Biotechnol. **10**, 804513 (2022). https://doi.org/10.3389/fbioe.2022.804513
29. Zou, P., Zhang, J., Xia, Y., Kanchana, K., Guo, G., Chen, W., Huang, Y., Wang, Z., Yang, S., Liang, G.: ROS generation mediates the anti-cancer effects of WZ35 via activating JNK and ER stress apoptotic pathways in gastric cancer. Oncotarget. **6**, 5860–5676 (2015). https://doi.org/10.18632/oncotarget.3333
30. Dai, X., Zhang, J., Guo, G., Cai, Y., Cui, R., Yin, C., Liu, W., Vinothkumar, R., Zhang, T., Liang, G., Zhang, X.: A mono-carbonyl analog of curcumin induces apoptosis in drug-resistant EGFR-mutant lung cancer through the generation of oxidative stress and mitochondrial dysfunction. Cancer Manag. Res. **10**, 3069–3082 (2018). https://doi.org/10.2147/CMAR.S159660
31. Utaipan, T., Boonyanuphong, P., Chuprajob, T., et al.: A trienone analog of curcumin, 1,7-bis(3-hydroxyphenyl)-1,4,6-heptatrien-3-one, possesses ROS- and caspase-mediated apoptosis in human oral squamous cell carcinoma cells in vitro. Appl. Biol. Chem. **63**, 7 (2020). https://doi.org/10.1186/s13765-020-0491-8
32. Rosanò, L., Cianfrocca, R., Spinella, F., Di Castro, V., Nicotra, M.R., Lucidi, A., Ferrandina, G., Natali, P.G., Bagnato, A.: Acquisition of chemoresistance and EMT phenotype is linked with activation of the endothelin A receptor pathway in ovarian carcinoma cells. Clin. Cancer Res. **17**, 2350–2360 (2011). https://doi.org/10.1158/1078-0432.ccr-10-2325
33. Zhou, Y., Shu, F., Liang, X., Chang, H., Shi, L., Peng, X., Zhu, J., Mi, M.: Ampelopsin induces cell growth inhibition and apoptosis in breast cancer cells through ROS generation and endoplasmic reticulum stress pathway. PLoS One. **9**, e890212014 (2018). https://doi.org/10.1371/journal.pone.0089021
34. Ni, F., Gong, Y., Li, L., Abdolmaleky, H.M., Zhou, J.R.: Flavonoid ampelopsin inhibits the growth and metastasis of prostate cancer in vitro and in mice. PLoS One. **7**, e388022012 (2018). https://doi.org/10.1371/journal.pone.0038802
35. Guo, Z., Guozhang, H., Wang, H., Li, Z., Liu, N.: Ampelopsin inhibits human glioma through inducing apoptosis and autophagy dependent on ROS generation and JNK pathway. Biomed. Pharmacother. **116**, 108524 (2019). https://doi.org/10.1016/j.biopha.2018.12.136
36. Huang, T., Zhang, H., Wang, X., Xu, L., Jia, J., Zhu, X.: Gambogenic acid inhibits the proliferation of small-cell lung cancer cells by arresting the cell cycle and inducing apoptosis. Oncol. Rep. **41**(3), 1700–1706 (2019). https://doi.org/10.3892/or.2018.6950
37. Sun, X., Yu, M. W., Ma, X. M., Zhang, G.L.: *In vitro* and *in vivo* study of the regulation of fisetin in PI3K/Akt/mTOR pathway to induce apoptosis of breast cancer cells, in The 15th National Conference of Integrated Traditional Chinese and Western Medicine oncology (2018)
38. Afroze, N., Pramodh, S., Shafarin, J., Bajbouj, K., et al.: Fisetin deters cell proliferation, induces apoptosis, alleviates oxidative stress and inflammation in human cancer cells, HeLa. Int. J. Mol. Sci. **23**(3), 1707 (2022). https://doi.org/10.3390/ijms23031707
39. Al-oudat, B.A., Alqudah, M.A., Audat, S.A., et al.: Design, synthesis, and biologic evaluation of novel chrysin derivatives as cytotoxic agents and caspase-3/7 activators. Drug Des. Dev. Ther. **13**, 423–433 (2019). https://doi.org/10.2147/DDDT.S189476
40. Qiu, J., Zhang, T., Zhu, X., Yang, C., Wang, Y., Zhou, N., et al.: Hyperoside induces breast cancer cells apoptosis via ROS-mediated NF-Kb signaling pathway. Int. J. Mol. Sci. **21**(1), 131 (2019). https://doi.org/10.3390/ijms21010131
41. Cheng, X., Tan, S., Duan, F., Yuan, Q., Li, Q., Deng, G.: Icariin induces apoptosis by suppressing autophagy in tamoxifen-resistant breast cancer cell line MCF-7/TAM. Breast Cancer. **26**(6), 766–775 (2019). https://doi.org/10.1007/s12282-019-00980-5

42. Lan, L., Wang, Y., Pan, Z., Wang, B., Yue, Z., Jiang, Z., et al.: Rhamnetin induces apoptosis in human breast cancer cells via the miR-34a/Notch-1 signaling pathway. Oncol. Lett. **17**(1), 676–682 (2019). https://doi.org/10.3892/ol.2018.9575
43. Xu, Q.N., Zhu, D., Wang, G.H., Lin, T., et al.: Phenolic glycosides and flavonoids with antioxidant and anticancer activities from Desmodium caudatum. Nat. Prod. Res. **35**(22), 4534–4541 (2020). https://doi.org/10.1080/14786419.2020.1739044
44. Moradi, M., Gholipour, H., Sepehri, H., Attari, F., Delphi, L., Arefian, E., et al.: Flavonoid calycopterin triggers apoptosis in triple-negative and ER-positive human breast cancer cells through activating different patterns of gene expression. Naunyn Schmiedeberg's Arch. Pharmacol. **393**(11), 2145–2156 (2020). https://doi.org/10.1007/s00210-020-01917-y
45. Lee, H.L., Lin, C.S., Kao, S.H., Chou, M.C.: Gallic acid induces G1 phase arrest and apoptosis of triple-negative breast cancer cell MDA-MB-231 via P38 mitogen-activated protein kinase/p21/p27 axis. Anti-Cancer Drugs. **28**(10), 1150–1156 (2017). https://doi.org/10.1097/CAD.0000000000000565
46. Yang, S., Si, L., Jia, Y., et al.: Kaempferol exerts antiproliferative effects on human ovarian cancer cells by inducing apoptosis, G0/G1 cell cycle arrest and modulation of MEK/ERK and STAT3 pathways. J. BUON. **24**, 975–981 (2019)
47. Jiang, D.C., Jin, L., Chen, H.X., Zhang, L.: Quercetin promotes apoptosis of breast cancer cells by targeting GAS5/Notch1 signaling pathway. J. Chin. Pharmacol. Bull. **37**(05), 637–644 (2021)
48. Wang, R., Wang, J., Dong, T., Shen, J., Gao, X., Zhou, J.: Naringenin has a chemoprotective effect in MDA-MB-231 breast cancer cells via inhibition of caspase-3 and -9 activities. Oncol. Lett. **17**, 1217–1222 (2019). https://doi.org/10.3892/ol.2018.9704
49. Noori, S., Tavirani, M.R., Deravi, N., Rabbani, M.I.M., Zarghi, A.: Naringenin enhances the anti-cancer effect of cyclophosphamide against MDA-MB-231 breast cancer cells via targeting the STAT3 signaling pathway. Iran. J. Pharm. Res. **19**(3), 122–133 (2020). https://doi.org/10.22037/ijpr.2020.113103.14112
50. Shafiee, G., Saidijam, M., Tavilani, H., Ghasemkhani, N., Khodadadi, I.: Genistein induces apoptosis and inhibits proliferation of HT29 colon cancer cells. Int. J. Mol. Cell Med. **5**, 178–191 (2016)
51. Xu, H., Ma, H., Zha, L., Li, Q., Pan, H., Zhang, L.: Genistein promotes apoptosis of lung cancer cells through the IMPDH2/AKT1 pathway. Am. J. Transl. Res. **14**(10), 7040–7051 (2022)
52. Wang, C.J., Sheng, Y., Zhou, Q., He, H.X.: Anti-proliferation and pro-apoptotic activities of diosmetin on MCF-7 cells via mitochondrial apoptosis pathway. J. Tradit. Chin. Drug Res. Clin. Pharmacol. **31**(02), 149–155 (2020)
53. Huang, W.C., Gu, P.Y., Fang, L.W., Huang, Y.L., Lin, C.F., Liou, C.J., et al.: Sophoraflavanone G from Sophora flavescens induces apoptosis in triple-negative breast cancer cells. Phytomedicine. **61**, 152852 (2019). https://doi.org/10.1016/j.phymed.2019.152852
54. Ivashkevich, A.: The role of isoflavones in augmenting the effects of radiotherapy. Front. Oncol. **12**, 800562 (2023). https://doi.org/10.3389/fonc.2022.800562
55. Fang, K., Liu, J.C., Qu, M.L., Chang, N.D.: Calycosin promotes apoptosis of breast cancer cells MCF-7 through inhibiting SIRT1. J. Prog. Mod. Biomed. **15**(23), 4431–4434 (2015)
56. Jia, S.H., Liu, L.N., Yan, T.H.: Mechanism of apoptosis and oxidative stress in MCF–7 cells induced by formononetin. West China J. Pharm. Sci. **35**(04), 385–391 (2020)
57. Ranjithkumar, R., Saravanan, K., Balaji, B., Hima, S., et al.: Novel daidzein molecules exhibited anti-prostate cancer activity through nuclear receptor ERβ modulation, in vitro and in vivo studies. J. Chemother. **33**(8), 582–594 (2021). https://doi.org/10.1080/1120009X.2021.1924935

58. Youssef, M., Tolba, M., Badawy, N., et al.: Novel combination of sorafenib and biochanin-A synergistically enhances the anti-proliferative and pro-apoptotic effects on hepatocellular carcinoma cells. Sci. Rep. **6**, 30717 (2016). https://doi.org/10.1038/srep30717
59. Aslan, A., Gok, O., Beyaz, S., Uslu, H., Erman, F., Erman, O., Baspinar, S.: Ellagic acid inhibits proinflammatory intermediary manufacture by suppressing NF-κB/Akt, VEGF and activating Nrf-2/Caspase-3 signaling pathways in rat testicular damage: a new way for testicular damage cure and in silico approach. Toxicol. Mech. Methods. **32**(6), 463–476 (2022). https://doi.org/10.1080/15376516.2022.2046668
60. Peña-Morán, O.A., Villarreal, M.L.: Álvarez-Berber, et al.: Cytotoxicity, post-treatment recovery, and selectivity analysis of naturally occurring podophyllotoxins from Bursera fagaroides var. fagaroides on breast cancer cell lines. Molecules. **21**, 1013 (2016)
61. Zhang, X., Rakesh, K., Shantharam, C., Manukumar, H., Asiri, A., Marwani, H., Qin, H.-L.: Podophyllotoxin derivatives as an excellent anticancer aspirant for future chemotherapy: a key current imminent needs. Bioorg. Med. Chem. **26**(2), 340–355 (2018)
62. Nitiss, J.L.: Targeting DNA topoisomerase II in cancer chemotherapy. Nat. Rev. Cancer. **9**, 338 (2009). https://doi.org/10.1016/j.bmc.2017.11.026
63. Hyder, I., Yedlapudi, D., Kalivendi, S.V., et al.: Synthesis and biological evaluation of novel 4β-[(5-substituted)-1,2,3,4-tetrazolyl] podophyllotoxins as anticancer compounds. Bioorg. Med. Chem. Lett. **25**, 2860–2863 (2015)
64. Mottaghi, S., Abbaszadeh, H.: A comprehensive mechanistic insight into the dietary and estrogenic lignans, arctigenin and sesamin as potential anticarcinogenic and anticancer agents, Current status, challenges, and future perspectives. Crit. Rev. Food Sci. Nutr. **62**(26), 7301–7318 (2021). https://doi.org/10.1080/10408398.2021.1913568
65. Siao, A.C., Hou, C.W., Kao, Y.H., Jeng, K.C.: Effect of sesamin on apoptosis and cell cycle arrest in human breast cancer Mcf-7 cells. Asian Pac. J. Cancer Prev. **16**(9), 3779–3783 (2015). https://doi.org/10.7314/apjcp.2015.16.9.3779
66. Li, Q., Liang, Y., Tian, Y., Hu, G.: Arctigenin induces apoptosis in colon cancer cells through ROS/p38MAPK pathway. J. BUON. **21**(1), 87–94 (2016)
67. Feng, T., Cao, W., Shen, W., Zhang, L., Gu, X., Guo, Y., et al.: Arctigenin inhibits STAT3 and exhibits anticancer potential in human triple-negative breast cancer therapy. Oncotarget. **8**(1), 329–344 (2017). https://doi.org/10.18632/oncotarget.13393
68. Lu, Z., Cao, S., Zhou, H., Hua, L., Zhang, S., Cao, J.: Mechanism of arctigenin-induced specific cytotoxicity against human hepatocellular carcinoma cell lines: Hep G2 and SMMC7721. PLoS One. **10**(5), e0125727 (2015). https://doi.org/10.1371/journal.pone.0125727
69. He, Y., Fan, Q., Cai, T., Huang, W., Xie, X., Wen, Y., Shi, Z.: Molecular mechanisms of the action of arctigenin in cancer. Biomed. Pharmacother. **108**, 403–407 (2018). https://doi.org/10.1016/j.biopha.2018.08.158
70. Tang, D., Kang, R., Berghe, T.V., Vandenabeele, P., Kroemer, G.: The molecular machinery of regulated cell death. Cell Res. **29**, 347–364 (2019). https://doi.org/10.1038/s41422-019-0164-5
71. Yadav, A.K., Jang, B.C.: Anti-survival and pro-apoptotic effects of 6-shogaol on SW872 human liposarcoma cells via control of the intrinsic caspase pathway, STAT-3, AMPK, and ER stress. Biomol. Ther. **10**(10), 1380 (2020). https://doi.org/10.3390/biom10101380
72. Hsu, C.M., Su, H.C., Yang, M.Y., Tsai, Y.T., Tsai, M.S., Yang, Y.H., Wu, C.Y., Chang, S.F.: 6-Shogaol is a potential treatment for head and neck squamous cell carcinoma. Int. J. Med. Sci. **20**(2), 238–246 (2023). https://doi.org/10.7150/ijms.80542
73. Peng, S., Yu, S., Zhang, J., Zhang, J.: 6-Shogaol as a novel thioredoxin reductase inhibitor induces oxidative-stress-mediated apoptosis in HeLa cells. Int. J. Mol. Sci. **24**(5), 4966 (2023). https://doi.org/10.3390/ijms24054966

74. Tailor, D., Going, C.C., Resendez, A., Kumar, V., et al.: Novel aza-podophyllotoxin derivative induces oxidative phosphorylation and cell death via AMPK activation in triple-negative breast cancer. Br. J. Cancer. **124**, 604–615 (2021). https://doi.org/10.1038/s41416-020-01137-4
75. Hossain, F., Andreana, P.R.: Developments in carbohydrate-based cancer therapeutics. Pharmaceuticals (Basel). **12**(2), 84 (2019). https://doi.org/10.3390/ph12020084
76. Lee, H.M., Moon, A.: Amygdalin regulates apoptosis and adhesion in Hs578T triple-negative breast cancer cells. Biomol. Ther. (Seoul). **24**(1), 62–66 (2016). https://doi.org/10.4062/biomolther.2015.172
77. Al-Khafaji, K., Tok, T.T.: Understanding the mechanism of amygdalin's multifunctional anti-cancer action using computational approach. J. Biomol. Struct. Dyn. **39**(5), 1600–1610 (2020). https://doi.org/10.1080/07391102.2020.1736159
78. Lee, M.S., Yuet-Wa, J.C., Kong, S.K., et al.: Effects of polyphyllin D, a steroidal saponin in Paris polyphylla, in growth inhibition of human breast cancer cells and in xenograft. Cancer Biol. Ther. **4**, 1248–1254 (2005)
79. Kong, M., Fan, J., Dong, A., Cheng, H., Xu, R.: Effects of polyphyllin I on growth inhibition of human non-small lung cancer cells and in xenograft. Acta Biochim. Biophys. Sin. **42**, 827–833 (2010)
80. Han, Y.H., Kee, J.Y., Kim, D.S., Mun, J.G., Jeong, M.Y., Park, S.H., et al.: Arctigenin inhibits lung metastasis of colorectal cancer by regulating cell viability and metastatic phenotypes. Molecules. **21**(9), 1135 (2016). https://doi.org/10.3390/molecules21091135
81. Watanabe, S., Suzuki, T., Hara, F., Yasui, T., et al.: Polyphyllin D, a steroidal saponin in Paris polyphylla, induces apoptosis and necroptosis cell death of neuroblastoma cells. Pediatr. Surg. Int. **33**(6), 713–719 (2017). https://doi.org/10.1007/s00383-017-4069-4
82. Kim, J.H., Choi, J.S.: Effect of ginsenoside Rh-2 via activation of caspase-3 and Bcl-2-insensitive pathway in ovarian cancer cells. Physiol. Res. **65**, 1031–1037 (2016). https://doi.org/10.33549/physiolres.933367
83. Li, C., Gao, H., Feng, X., Bi, C., Zhang, J., Yin, J.: Ginsenoside Rh2 impedes proliferation and migration and induces apoptosis by regulating NF-κB, MAPK, and PI3K/Akt/mTOR signaling pathways in osteosarcoma cells. J. Biochem. Mol. Toxicol. **34**(12), e22597 (2020)
84. Hong, Y., Fan, D.: Ginsenoside Rk1 induces cell cycle arrest and apoptosis in MDA-MB-231 triple negative breast cancer cells. Toxicology. **418**, 22–31 (2019). https://doi.org/10.1016/j.tox.2019.02.010
85. Liu, Y., Fan, D.: The preparation of ginsenoside Rg5, its antitumor activity against breast cancer cells and its targeting of PI3K. Nutrients. **12**, 246 (2020). https://doi.org/10.3390/nu12010246
86. Kim, S.N., Kim, D.H., Lee, et al.: Effects of ginsenoside Rg3 on inhibiting differentiation, adipogenesis, and ER stress-mediated cell death in brown adipocytes. Evid. Based Complement. Alternat. Med. **2021**, 6668665 (2021). https://doi.org/10.1155/2021/6668665
87. Zhang, F., Li, M., Wu, X., Hu, Y., Cao, Y., Wang, X., Xiang, S., Li, H., Jiang, L., Tan, Z., Lu, W., Weng, H., Shu, Y., Gong, W., Wang, X., Zhang, Y., Shi, W., Dong, P., Gu, J., Liu, Y.: 20(S)-ginsenoside Rg3 promotes senescence and apoptosis in gallbladder cancer cells via the p53 pathway. Drug Des. Dev. Ther. **9**, 3969–3987 (2015). https://doi.org/10.2147/DDDT.S84527
88. Tian, Y.Z., Liu, Y.P., Tian, S.C., Ge, S.Y., Wu, Y.Z., Zhang, B.L.: Antitumor activity of ginsenoside Rd in gastric cancer via up-regulation of caspase-3 and caspase-9. Pharmazie. **75**, 147–150 (2020). https://doi.org/10.1691/ph.2020.9931
89. Oh, J.-M., Lee, J., Im, W.-T., Chun, S.: Ginsenoside Rk1 induces apoptosis in neuroblastoma cells through loss of mitochondrial membrane potential and activation of caspases. Int. J. Mol. Sci. **20**, 1213 (2019). https://doi.org/10.3390/ijms20051213

90. Salimi, A., Bahiraei, T., Ahdeno, S., Vatanpour, S., Pourahmad, J.: Evaluation of cytotoxic activity of betanin against U87MG human glioma cells and normal human lymphocytes and its anticancer potential through mitochondrial pathway. Nutr. Cancer. **73**(3), 450–459 (2021). https://doi.org/10.1080/01635581.2020.1764068
91. Kamran, S., Sinniah, A., Abdulghani, M.A.M., Alshawsh, M.A.: Therapeutic potential of certain terpenoids as anticancer agents: a scoping review. Cancers (Basel). **14**(5), 1100 (2022). https://doi.org/10.3390/cancers14051100
92. Hu, J., Zhang, H., Li, J., Jiang, X., Zhang, Y., Wu, Q., et al.: ROCK1 activation-mediated mitochondrial translocation of Drp1 and cofilin are required for arnidiol-induced mitochondrial fission and apoptosis. J. Exp. Clin. Cancer Res. **39**(1), 37 (2020). https://doi.org/10.1186/s13046-020-01545-7
93. Abu-Izneid, T., Rauf, A., Shariati, M.A., et al.: Sesquiterpenes and their derivatives-natural anticancer compounds: an update. Pharmacol. Res. **161**, 105165 (2020). https://doi.org/10.1016/j.phrs.2020.105165
94. Chan, M.L., Liang, J.W., Hsu, L.C., Chang, W.L., Lee, S.S., Guh, J.H.: Zerumbone, a ginger sesquiterpene, induces apoptosis and autophagy in human hormone-refractory prostate cancers through tubulin binding and crosstalk between endoplasmic reticulum stress and mitochondrial insult. Naunyn Schmiedeberg's Arch. Pharmacol. **388**, 1223–1236 (2015)
95. Wang, S.D., Wang, Z.H., Yan, H.Q., Ren, M.Y., Gao, S.Q., Zhang, G.Q.: Chemotherapeutic effect of zerumbone on melanoma cells through mitochondria-mediated pathways. Clin. Exp. Dermatol. **41**, 858–863 (2016). https://doi.org/10.1111/ced.12986
96. Wang, D., Li, Y., Cui, P., Zhao, Q., Tan, B.B., Zhang, Z.D., Liu, Y., Jia, N.: Zerumbone induces gastric cancer cells apoptosis: involving cyclophilin A. Biomed. Pharmacother. **83**, 740–745 (2016). https://doi.org/10.1016/j.biopha.2016.07.034
97. Rajan, I., Jayasree, P.R., Kumar, P.R.: Zerumbone induces mitochondria-mediated apoptosis via increased calcium, generation of reactive oxygen species and upregulation of soluble histone H2AX in K562 chronic myelogenous leukemia cells. Tumour Biol. **36**, 8479–8489 (2015). https://doi.org/10.1007/s13277-015-3583-z
98. Urla, C., Stagno, M.J., Fuchs, J., et al.: Anticancer bioactivity of zerumbone on pediatric rhabdomyosarcoma cells. J. Cancer Res. Clin. Oncol. **149**, 3313–3323 (2023). https://doi.org/10.1007/s00432-022-04237-1
99. Al-Fatlawi, A.A., Al-Fatlawi, A.A., Irshad, M., Rahisuddin, Ahmad, A.: Effect of parthenolide on growth and apoptosis regulatory genes of human cancer cell lines. Pharm. Biol. **53**(1), 104–109 (2015). https://doi.org/10.3109/13880209.2014.911919
100. Han, Y.Y., Yang, J.F., Sun, Y., Li, L.H.: Parthenolide on proliferation and apoptosis of breast cancer. Northwest J. Defense Med. **40**(04), 205–211 (2019)
101. Jorge, J., Neves, J., Alves, R., Geraldes, C., Gonçalves, A.C., Sarmento-Ribeiro, A.B.: Parthenolide induces ROS-mediated apoptosis in lymphoid malignancies. Int. J. Mol. Sci. **24**(11), 9167 (2023). https://doi.org/10.3390/ijms24119167
102. Halmurat, O., Gulibositan, A., Abudoukaher, A., Yu, L., Shaobin, D.: Parthenolide induces gallbladder cancer cell apoptosis via MAPK signalling. Ann. Med. Surg. **86**(4), 1956–1966 (2024). https://doi.org/10.1097/MS9.0000000000001828
103. Ma, Q., Chen, J., Xiong, S., Li, G.L.: Effects of dehydrocostus lactone on the apoptosis of human breast cancer SK-BR-3 cells and its mechanism. J. Tradit. Chin. Drug Res. Clin. Pharmacol. **32**(02), 200–207 (2021)
104. Egbujor, M.C., Petrosino, M., Zuhra, K., Saso, L.: The role of organosulfur compounds as Nrf2 activators and their antioxidant effects. Antioxidants (Basel). **11**(7), 1255 (2022). https://doi.org/10.3390/antiox11071255

105. Li, G., Zhou, J., Budhraja, A., Hu, X., Chen, Y., Cheng, Q., et al.: Mitochondrial translocation and interaction of cofilin and Drp1 are required for erucin-induced mitochondrial fission and apoptosis. Oncotarget. **6**(3), 1834–1849 (2015). https://doi.org/10.18632/oncotarget.2795
106. Rosas-González, V.C., Téllez-Bañuelos, M.C., et al.: Differential effects of alliin and allicin on apoptosis and senescence in luminal A and triple-negative breast cancer: caspase, ΔΨm, and pro-apoptotic gene involvement. Fundam. Clin. Pharmacol. **34**(6), 671–686 (2020). https://doi.org/10.1111/fcp.12559
107. Shoaib, S., Tufail, S., Sherwani, M.A., Yusuf, N., Islam, N.: Phenethyl isothiocyanate induces apoptosis through ROS generation and caspase-3 activation in cervical cancer cells. Front. Pharmacol. **12**, 673103 (2021). https://doi.org/10.3389/fphar.2021.673103
108. Francois, R.A., Zhang, A., Husain, K., Wang, C., Hutchinson, S., Kongnyuy, M., Batra, S.K., Coppola, D., Sebti, S.M., Malafa, M.P.: Vitamin E δ-tocotrienol sensitizes human pancreatic cancer cells to TRAIL-induced apoptosis through proteasome-mediated down-regulation of c-$FLIP_s$. Cancer Cell Int. **19**, 189 (2019). https://doi.org/10.1186/s12935-019-0876-0
109. Zu, C., Zhang, M., Xue, H., Cai, X., Zhao, L., He, A., et al.: Emodin induces apoptosis of human breast cancer cells by modulating the expression of apoptosis-related genes. Oncol. Lett. **10**(5), 2919–2924 (2015). https://doi.org/10.3892/ol.2015.3646
110. Ren, L., Li, Z., Dai, C., Zhao, D., Wang, Y., Ma, C., et al.: Chrysophanol inhibits proliferation and induces apoptosis through NF-κB/cyclin D1 and NF-κB/Bcl-2 signaling cascade in breast cancer cell lines. Mol. Med. Rep. **17**(3), 4376–4382 (2018). https://doi.org/10.3892/mmr.2018.8443
111. Chandrashekar, K.R., Prabhu, A., Rekha, P.D.: Tetrandrine isolated from Cyclea peltata induces cytotoxicity and apoptosis through ROS and caspase pathways in breast and pancreatic cancer cells. In Vitro Cell Dev. Biol. Anim. **55**(5), 331–340 (2019). https://doi.org/10.1007/s11626-019-00332-9
112. Dhyani, P., Quispe, C., Sharma, E., et al.: Anticancer potential of alkaloids: a key emphasis to colchicine, vinblastine, vincristine, vindesine, vinorelbine and vincamine. Cancer Cell Int. **22**, 206 (2022). https://doi.org/10.1186/s12935-022-02624-9
113. Asokan, S.M., Mariappan, R., Muthusamy, S., Velmurugan, B.K.: Pharmacological benefits of neferine – a comprehensive review. Life Sci. **199**, 60–70 (2018). https://doi.org/10.1016/j.lfs.2018.02.032
114. Zhang, Y.J., Qian, L., Chen, C.J.: Effect of berberine on proliferation and apoptosis of MCF-7 cells. J. Acta Universitatis Medicinalis Anhui. **50**(09), 1276–1280 (2015)
115. Yang, M., Zhu, L., Li, L., Li, J., Xu, L., Feng, J., et al.: Digital gene expression analysis provides insight into the transcript profile of the genes involved in aporphine alkaloid biosynthesis in lotus (Nelumbo nucifera). Front. Plant Sci. **8**, 80 (2017)
116. Sivalingam, K.S., Paramasivan, P., Weng, C.F., Viswanadha, V.P.: Neferine potentiates the antitumor effect of cisplatin in human lung adenocarcinoma cells via a mitochondria-mediated apoptosis pathway. J. Cell. Biochem. **118**, 2865–2876 (2017)
117. Zhang, X., Li, D., Xue, X., Zhang, Y., et al.: First total synthesis of a novel amide alkaloid derived from Aconitum taipeicum and its anticancer activity. Nat. Prod. Res. **32**(2), 128–132 (2017). https://doi.org/10.1080/14786419.2017.1340283
118. Zhang, H.J., Zhang, G.R., Piao, H.R., Quan, Z.H.: Synthesis and characterization of celastrol derivatives as potential anticancer agents. J. Enzyme Inhib. Med. Chem. **33**(1), 190–198 (2018). https://doi.org/10.1080/14756366.2017.1404590
119. Chen, M., Xiao, C., Jiang, W., Yang, W., Qin, Q., Tan, Q., et al.: Capsaicin inhibits proliferation and induces apoptosis in breast cancer by down-regulating FBI-1-mediated NF-Kb pathway. Drug Des. Dev. Ther. **15**, 125–140 (2021). https://doi.org/10.2147/DDDT.S269901

120. Wei, J., Li, K.J., Wei, J.B., Lv, C.P.: Influence of gambogic acid lysinate in proliferation and apoptosis in breast cancer MCF 7 cells and its mechanism. J. J. Jilin Univ. (Medical Sci. Edition). **41**(03), 476–480 (2015)
121. Wang, D., Zhang, Y., Lu, J., Wang, Y., Wang, J., Meng, Q., et al.: Cordycepin, a natural anti-neoplastic agent, induces apoptosis of breast cancer cells via caspase-dependent pathways. Nat. Prod. Commun. **11**(1), 63–68 (2016). https://doi.org/10.1177/1934578x1601100119
122. Liu, C., Qi, M., Li, L., Yuan, Y., Wu, X., Fu, J.: Natural cordycepin induces apoptosis and suppresses metastasis in breast cancer cells by inhibiting the hedgehog pathway. Food Funct. **11**(3), 2107–2116 (2020). https://doi.org/10.1039/c9fo02879j
123. Wei, M., Wu, Y., Liu, H., Xie, C.: Genipin induces autophagy and suppresses cell growth of oral squamous cell carcinoma via PI3K/AKT/MTOR pathway. Drug Des. Dev. Ther. **4**, 395–405 (2020). https://doi.org/10.2147/DDDT.S222694

Modelling of the Cyclic Guanidines as 5 - HT_{5A} Receptor Ligands Using Chemometric Tools

Manju Choudhary[1], Brij K. Sharma[2], Sisir Nandi[3], and Pradeep Pilania[1] (✉)

[1] Department of Chemistry, Government Science College, Sikar, India
[2] Government College, Talera, Rajasthan, India
[3] Global Institute of Pharmaceutical Education & Research (GIPER), Kashipur, Uttarakhand, India

Abstract Depression affects over 250 million people worldwide. The 5 - HT_{5A} receptors play a great role in mood-related disorders. The cyclic guanidine derivatives having binding affinities on 5 - HT_{5A} receptors have been quantitatively modelled taking Dragon computed molecular descriptors to formulate the quantitative structure-activity relationship (QSAR) models. These models can capture various parameters such as Moran autocorrelation MATS3m, MATS3v, and MATS4v, BEHp1, Me, and JGI4 responsible for explaining the binding affinity of the studied compounds utilizing combinatorial protocol in multiple linear regression (CP-MLR). These parameters suggested that the substituents of guanidine moiety for these congeners are responsible for the 5 - HT_{5A} receptor antagonism activity.

Keywords QSAR · cyclic guanidines · 5 - HT_{5A} ligands · binding affinity · PLS · CP-MLR

1 Introduction

Depression is a major mental problem that engulfs around 250 million persons worldwide [1]. The leading cause of disability in the general population is this disease, which is characterized by emotional symptoms like sad mood, irritability, and anhedonia, as well as somatic symptoms affecting sleep, appetite, and libido. Additionally, abnormal activities such as suicidal tendency, delayed speech, lack of movement, and deficiencies in learning, attention, and memory are all associated with this condition. Drugs that activate

Supplementary Information The online version contains supplementary material available at https://doi.org/10.1007/978-3-031-75457-9_10.

A. K. Saxena, S. Nandi (eds.), *Global Trends in Health, Technology and Management*, https://doi.org/10.1007/978-3-031-75457-9_10

or inhibit specific subtypes of 5-HTR are now being used to treat various ailments. Identifying specific ligands for several subtypes of the 5-hydroxytryptamine receptor (5-HTR) has made these receptors a viable target for treating various illnesses [2]. The 5 - HT_{5A} receptor subtype is located in the cerebral cortex, amygdala, thalamus, hippocampus, hypothalamus, brain stem, basal ganglia, and cerebellum [3], and its association with schizophrenia through genetic variations [4] underscores its crucial role in the regulation of mood disorders [5]. The potential of antipsychotic effects in animal models has also been established using selective 5 - HT_{5A} receptor antagonists [6]. Based on these discoveries, the 5 - HT_{5A} receptor emerged as a promising candidate for therapeutic targeting. In recent years, a new group of guanidine type 5 - HT_{5A} receptor antagonists with strong binding and functional properties were discovered and documented [7, 8].

Due to the significance 5 - HT_{5A} antagonists in the proven treatment of various illnesses, QSAR modelling is being explored using 5 - HT_{5A} target binding affinities of these cyclic guanidines. The primary objective of this study is to analyse and justify the many changes of substituents in these analogues, with the intention of providing valuable knowledge for future attempts.

2 Materials and Methods

2.1 Compound Structure and Database

The present research includes a published structural database consisting of fifty-six cyclic guanidine derivatives. The affinities of these compounds for the recombinant human 5 - HT_{5A} receptor were evaluated in a radioligand binding assay using [3H]LSD expressed in cells of HEK-293-EBNA. Table 1 provides information on the chemical structural differences and binding affinities. The activity data supplied on a molar scale have been utilized as the response variable for further QSAR generation. In order to facilitate modelling, the 56 analogues were divided into separate training and test sets. Among the total 56 analogues, subsets of 15 molecules were selected for the purpose of validating the model. The chemicals used for training and testing are also presented in Table 1.

2.2 Theoretical Molecular Descriptors

The compounds' structures were depicted in 2D using ChemDraw Ultra 6.0 and Chem3D Ultra software developed by Cambridge Soft Corporation. The standard conversion method in CS Chem3D Ultra then turned these 2D structures into 3D. The AM1 technique for closed-shell systems in the MOPAC module helped to minimize the energy of these 3D constructions. This ensures that the investigated chemicals have well-defined conformer relationships. After that, molecular descriptors (Table 1) were obtained by analysing the energy-minimized structures with the Dragon program [9]. This program provides many more descriptors such as 0D-, 1D-, and 2D-types. There are ten separate classes such as topological, constitutional, molecular walk counts, GALVEZ topological charge indices, 2D autocorrelations, BCUT descriptors, atom-centred fragments, functional groups, empirical descriptors, and properties descriptive of nature among the modules described

Table 1 Structures[a] observed and modelled 5 - HT$_{5A}$ binding affinities of the cyclic guanidines

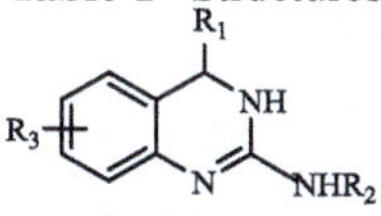

Cpd.	**R$_1$**	**R$_2$**	**R$_3$**	**pK$_i$(M)**[b]					
				Obsd.	Eq. (10)	Eq. (11)	Eq. (12)	Eq. (13)	PLS
1[c]	H	H	5-Cl	7.00	6.67	6.80	6.81	6.71	7.02
2[c]	H	H	5,6-di-Cl	7.37	7.61	7.57	7.43	7.58	7.64
3	H	H	H	6.18	5.78	5.90	5.65	5.83	5.79
4	Me	H	H	7.42	6.78	6.78	6.86	6.69	6.68
5[c]	Et	H	H	7.62	6.91	6.96	6.66	6.87	6.59
6	Pr	H	H	6.09	6.74	6.76	6.51	6.81	6.57
7	Me	H	H	5.93	6.39	6.26	6.23	6.16	5.89
8	H	H	H	5.00	5.35	5.26	5.51	5.22	5.23
9	H	Me	H	5.75	5.93	6.02	5.50	5.98	5.88
10[c]	H	Me	H	5.55	5.06	4.89	4.89	4.93	4.93
11	H	H	5-Me	6.84	6.58	6.49	6.96	6.51	6.63
12	H	H	5-OMe	-[d]	6.31	6.71	6.48	6.39	6.50
13[c]	H	H	5-F	6.26	6.20	6.23	6.15	6.25	6.19
14	H	H	6-Cl	6.09	5.98	5.99	6.16	6.04	6.12
15	H	H	6-Me	6.05	6.02	5.83	6.05	5.96	5.73
16	H	H	6-OMe	-[d]	6.19	6.49	6.50	6.26	6.31
17[c]	H	H	6-F	6.04	6.11	6.13	6.07	6.17	5.91
18	H	H	7-Cl	6.46	5.98	5.99	6.33	6.04	6.24
19[c]	H	H	7-Me	6.56	6.02	5.83	6.20	5.96	5.83
20	H	H	7-OMe	5.86	6.19	6.49	6.54	6.26	6.39
21	H	H	7-F	6.19	6.11	6.13	6.09	6.17	5.98
22	H	H	8-Cl	6.05	6.36	6.42	6.05	6.41	6.19
23	H	H	8-Me	6.33	6.32	6.19	6.31	6.26	6.04
24[c]	H	H	8-OMe	7.01	6.22	6.61	6.39	6.31	6.30
25	H	H	8-F	5.70	6.16	6.18	6.04	6.20	5.93
26	Et	H	5-Cl	7.07	7.81	7.70	7.37	7.76	7.44
27[c]	Me	H	5-Cl	8.29	7.54	7.55	7.61	7.46	7.66
28	Me	Me	5-Cl	7.97	7.57	7.52	7.44	7.50	7.52
29[c]	Et	H	5,6-di-Cl	7.58	8.52	8.42	8.02	8.47	8.16
30	Me	H	5,6-di-Cl	8.70	8.42	8.26	8.22	8.31	8.41
31	Me	Me	5,6-di-Cl	8.20	8.27	8.22	8.10	8.18	8.17
32[c]	Me	H	8-OMe	8.20	7.17	7.40	7.51	7.12	7.13
33	Me	H	8-O*i*Pr	8.00	7.75	7.77	7.62	7.63	7.43
34	Me	Me	8-O*i*Pr	7.60	7.75	7.72	7.59	7.64	7.43
35[c]	Me	H	5-Me,6-Cl, 8-OMe	8.33	8.45	8.55	9.21	8.38	8.63
36	Me	H	5-OMe,6-Cl	7.64	7.52	7.74	8.01	7.51	7.87

(continued)

Table 1 (continued)

Cpd.	R_1	R_2	R_3	$pK_i(M)$[b]					
				Obsd.	Eq. (10)	Eq. (11)	Eq. (12)	Eq. (13)	PLS
37[c]	Me	H	5,8-di-Cl	8.19	8.11	7.96	8.16	8.04	8.19
38	Me	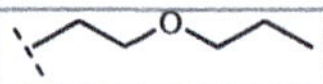	H	6.62	6.64	6.48	6.37	6.66	6.72
39	Me		H	6.75	6.76	6.62	6.49	6.78	6.91
40	Me	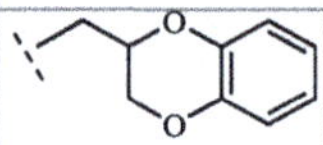	H	7.43	7.57	7.65	7.30	7.92	7.46
41	Me	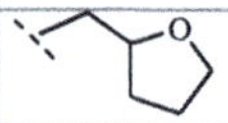	H	6.79	6.70	6.78	7.16	6.84	7.02
42	Me	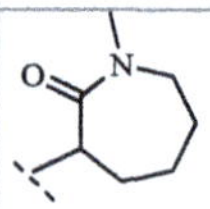	H	6.55	6.73	6.70	7.33	6.66	6.90
43	Me	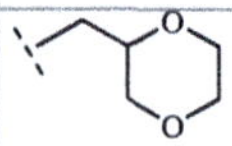	H	7.02	6.59	7.08	7.25	6.76	7.13
44	Me	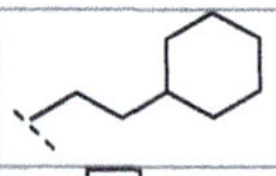	H	6.91	6.55	6.56	6.44	6.62	6.67
45	Me		H	6.91	7.07	7.03	7.13	7.23	7.17
46	Me	Me	H	6.99	6.68	6.76	6.73	6.61	6.66
47	Me	Et	H	6.71	6.97	7.07	6.63	7.00	6.98
48[c]	Me	Pr	H	6.82	6.72	6.67	6.37	6.75	6.69
49	Me	Bu	H	6.72	6.82	6.50	6.40	6.81	6.72
50	Me	CH_2 - CH_2F	H	6.40	6.52	6.75	6.65	6.62	6.72
51	Me	CH_2 - CHF_2	H	7.06	6.64	6.56	6.66	6.60	6.89
52	Me	CH_2 - CF_3	H	6.31	6.64	6.40	6.67	6.52	6.77
53[c]	Me	CH_2 - CHF_2	5-Cl	8.00	7.52	7.40	7.40	7.48	7.44
54	Me	CH_2 - CHF_2	5,6-di-Cl	8.24	8.20	8.21	8.02	8.20	8.03
55	Me	CH_2 - CHF_2	8-OMe	7.55	6.98	7.21	7.35	7.02	7.18
56	Me	CH_2 - CHF_2	8-O*i*Pr	7.65	7.51	7.50	7.56	7.45	7.78
57[c]	Me	CH_2 - CHF_2	5-OMe, 6-Cl	7.65	7.54	7.70	7.85	7.59	7.69
58	Me	CH_2 - CHF_2	5,8-di-Cl	7.48	7.97	7.98	7.96	7.99	7.90

[a]Reference9,10 (H = Me at C-4 in cpd. 7, Me at N-3 in cpd. 8 and Me in cpd. 10)
[b]On molar basis
[c]The test set
[d]Activity not reported

here. The molecular descriptors are computed using the DRAGON program, which can generate many descriptors unique to the molecules for every one of these classes [10]. 'The constitutional descriptors indicate the molecular basic constitution. The topological descriptors indicate topostructural and topochemical indices. The topostructural indices represent the molecular topology and structural features of the compounds, such as size, shape, symmetry, branching and cyclicity, whereas topochemical indices encode chemical information concerning atom type and bond multiplicity. The molecular walk counts represent self-returning walk counts of different lengths. The BCUT represents the eigenvalues including positive and negative values of the adjacency matrix weights of the diagonal elements and atoms of the chemical compound. The GALVEZ focuses the first 10 eigenvalues of the corrected adjacency matrix, whereas the 2D-autocorrelations find the relationship between pars of atoms in a specified position of the molecules. The functional group descriptors can be signified by Ghose-Crippen to explain the properties of different groups and the atom-centred fragments count of 120 fragments as given. The empirical (EMP) descriptors represent the counts of hydrophilic groups, non-single bonds, and the ratio of the number of aromatic bonds and total bonds in an H-depleted molecule'.

The present work develops QSAR models using the combinatorial technique in multiple linear regression [11, 12] and partial least squares [13–15]. Before running the CP-MLR operation, descriptors that correlated less than 0.1 with the biological endpoints and were highly intercorrelated (above 0.90) were removed. This step reduced the total number of descriptors from 481 to 87 for the binding activity analysis of the complete dataset of chemicals. The computing process is succinctly described here.

2.3 Model Development

In multiple linear regression (CP-MLR), the combinatorial protocol is a 'filter'-based variable selection method for model generation in QSARs. It uses specific subset regressions. This approach has interfaced a combinatorial technique of 'filters' coupled with MLR to screen many structure-activity models, each containing a unique combination of descriptors calculated from the dataset. The contents and number of variables to be examined are mixed according to the set limits. Here, important assessors of the variables in regression at several phases of model construction are the 'filters'. 'Filter-1 has been assigned as the inter-parameter correlation cut-off criteria for variables with a default value of 0.3 and upper limit ≤ 0.79. The corresponding variable combination is banned and denied if two variables are linked higher than a given cut-off value. The second filter is based on the t-values of regression coefficients of variables connected with a subset (filter-2, default value 2.0). Here, the variable combination will be denied if the ratio of the regression coefficient and related standard error of any variable is less than a predefined cut-off value. The square root of the adjusted multiple correlation coefficient of the regression equation, called r-bar, has been used to compare the internal explanatory power of models with different numbers of variables since successive additions of variables to multiple regression equations will increase successive multiple correlation coefficient (r) values. To determine the variables' "merit" in the model building, a filter has thus been set in terms of a specified threshold level of r-bar (filter-3, default value 0.71). Ultimately, the external consistency of the variables of the model has been addressed in terms of cross-validated R^2 or Q^2 criteria from the leave-one-out (LOO) cross-validation procedure as the default option (filter-4, the default threshold value $0.3 \leq Q^2 \leq 1.0$). This helps to exclude false or artificial correlations. These filters screen and ensure the variable selection process. Using the r-bar value of the previous optimum model as the new threshold for the next generation, the threshold of filter-3 was successively raised with an increasing number of descriptors (per equation), collecting the descriptors with higher information content and explanatory power' [11].

A partial least squares (PLS) study on descriptors found by CP-MLR helps to validate the results. The work helps classify some descriptors' potential in describing the binding activity profiles of drugs aiming at 5 - HT_{5A}. It promotes the construction of a 'single window' of the structure-activity model. It also allows one to evaluate the significance of the corresponding descriptors. The normalized regression coefficients of the descriptors offer fraction contributions that allow comparison within the modelled activity.

2.4 Model Validation

The present data set is divided into two subsets for this work: a training set for model building and a test set for the validation of the developed training model predicting the test compound's biological activities. The multiple correlation coefficient (r), the standard deviation (s), and the F-ratio between the variances of the calculated and observed activities (F) were used to assess how well the models fit the data. To assess the optimal model, several additional statistical parameters have also been derived, including the Kubinyi function (FIT) [16, 17], Friedman's lack of fit (LOF) [18], and Akaike's information criterion (AIC) [19, 20], represented in Eqs. 1–3.

$$AIC = \frac{RSS \times (n + p')}{(n - p')^2} \tag{1}$$

$$FIT = \frac{r^2 \times (n - k - 1)}{(n + k^2) \times (1 - r^2)} \tag{2}$$

$$LOF = \frac{RSS/n}{\left[1 - \frac{k(d+1)}{n}\right]^2} \tag{3}$$

where k denotes the number of variables, p' is the adjustable parameter, d is the smoothing parameter in the model, and RSS is the sum of the squared deviation between the observed and the calculated activity levels. The AIC considers the statistical quality of fit and the number of parameters that must be estimated to achieve that level of fit.

The Fisher ratio, or FIT, which is close to the F-value has been shown to be a useful metric for evaluating the model quality. The primary drawback of the F-value is that if k (the number of variables in the equation that characterizes the model) is small, it will be more sensitive to changes in k, while if k is big, it will be less sensitive. When k-values are modest, the FIT criterion is not very sensitive to changes in them; nevertheless, when k-values are large, the sensitivity increases significantly. The model that yields the highest FIT value and the lowest AIC value is thought to be the best and most potentially useful. The LOF is not biased towards large numbers of parameters and considers the number of terms in the equation, in contrast to other indicators. The statistical soundness of the generated model is implied by a minimum LOF value.

Using the cross-validated index, Q^2, from the leave-one-out (LOO) and leave-five-out (L5O) methods, the internal validity of the derived model was determined. The LOO method creates changed datasets by deleting each observation from the parent dataset exactly once. Next, a model is created for every reduced data set, and these models are used to forecast the response values of the deleted observations. The predictive residual sum of squares, or PRESS, is calculated by adding the squared discrepancies between the actual and anticipated values. PRESS will thus have a single contribution from every observation. The cross-validated Q^2_{LOO} is calculated as

$$Q^2_{LOO} = 1 - \frac{PRESS}{SSY} \tag{4}$$

where SSY represents the variation of the molecules' measured activity around the mean value. Using the leave-five-out method, five compounds are randomly excluded from analysis each time, ensuring that, for the first time, all of the compounds end up in the prediction groupings. A Q^2-index value higher than 0.5 suggests a moderately robust model.

Test set compounds are the foundation for the generated model's external validation or predictive ability. The following is the computed squared correlation coefficient between the observed and predicted values of the substances in test set r^2_{Test}:

$$r^2_{Test} = 1 - \frac{\sum\left(Y_{Pred(Test)} - Y_{(Test)}\right)^2}{\sum\left(Y_{(Test)} - \bar{Y}_{(Training)}\right)^2} \tag{5}$$

where $\bar{Y}_{(Training)}$ represents mean activity of the training set; $Y_{Pred(Test)}$ and $Y_{(Test)}$ are the predicted and observed activity values of the test compounds. The squared correlation coefficient (r^2_{Test}) between the test sets observed and projected data is what's known as this. When the r^2_{Test} value is higher than 0.5, it indicates that the model derived from the training set possesses dependable prediction ability.

2.5 Y-Randomization

By repeatedly scrambling the biological response in the randomization test [21, 22], chance correlations, if any, related to the CP-MLR models were identified. Multiple regression analysis has been used to re-evaluate the data sets containing response vectors that are jumbled (MRA). Regression equations that were produced, if any, and had correlation coefficients greater than or equal to the response data that had been unscrambled were tallied. These simulation runs have been performed 100 times on each model. This has been employed as a metric to convey the model's per cent chance association that is being examined.

3 Results and Discussion

Computed structural descriptor-based QSAR models have been developed for the cyclic guanidines. The data set was broken into training set (41) and test (15) set as shown in Table 1. Eighty-seven descriptors were filtered from 0D-, 1D, and 2D-classes of the descriptor pool using the default 'filters' set in it. Many statistical models based on the one, two, and three descriptor(s) have been developed successively to achieve the best QSAR utilizing the CP-MLR program. Higher descriptor models were achieved by incrementing the filter-3 with increasing descriptors per equation. The new threshold for filter-3 in the next generation is determined by using the optimal r-bar value from the previous level model.

The selected models in one, two, and three descriptors are presented below.

$$\begin{aligned} & pK_i = -2.849(0.409)X5A + 8.513 \\ & `n = 41, r = 0.744, s = 0.558, F = 48.516, FIT = 1.155, LOF = 0.327, AIC = 0.343, \\ & Q^2_{LOO} = 0.507, Q^2_{L5O} = 0.513, r^2_{randY}(sd) = 0.024(0.036), r^2_{Test} = 0.568' \end{aligned} \quad (6)$$

$$\begin{aligned} & pK_i = 2.417(0.303)MATS3v + 1.366(0.276)nCaR + 5.128 \\ & `n = 41, r = 0.864, s = 0.425, F = 56.288, FIT = 2.501, LOF = 0.206, AIC = 0.209, \\ & Q^2_{LOO} = 0.708, Q^2_{L5O} = 0.691, r^2_{randY}(sd) = 0.053(0.044), r^2_{Test} = 0.598' \end{aligned} \quad (7)$$

$$\begin{aligned} & pK_i = 1.207(0.288)PW2 + 2.933(0.314)MATS3v + 4.470 \\ & `n = 41, r = 0.846, s = 0.451, F = 48.002, FIT = 2.133, LOF = 0.231, AIC = 0.235, \\ & Q^2_{LOO} = 0.667, Q^2_{L5O} = 0.682, r^2_{randY}(sd) = 0.050(0.044), r^2_{Test} = 0.551' \end{aligned} \quad (8)$$

$$\begin{aligned} & pK_i = 1.361(0.208)C\text{-}008 + 1.313(0.200)C\text{-}026 + 5.881 \\ & `n = 41, r = 0.831, s = 0.470, F = 42.576, FIT = 1.892, LOF = 0.252, AIC = 0.256, \\ & Q^2_{LOO} = 0.639, Q^2_{L5O} = 0.639, r^2_{randY}(sd) = 0.040(0.036), r^2_{Test} = 0.537' \end{aligned} \quad (9)$$

$$\begin{aligned} & pK_i = 1.418(0.405)JGI4 + 2.458(0.340)MATS3v + 4.899 \\ & `n = 41, r = 0.828, s = 0.474, F = 41.642, FIT = 1.850, LOF = 0.255, AIC = 0.260, \\ & Q^2_{LOO} = 0.636, Q^2_{L5O} = 0.657, r^2_{randY}(sd) = 0.053(0.048), r^2_{Test} = 0.587' \end{aligned} \quad (10)$$

$$\begin{aligned} & pK_i = -0.847(0.307)Me + 2.615(0.288)MATS3v + 1.859(0.311)nCaR + 5.201 \\ & `n = 41, r = 0.889, s = 0.392, F = 46.595, FIT = 2.795, LOF = 0.191, AIC = 0.187, \\ & Q^2_{LOO} = 0.750, Q^2_{L5O} = 0.745, r^2_{randY}(sd) = 0.070(0.049), r^2_{Test} = 0.624' \end{aligned} \quad (11)$$

$$\begin{aligned} & pK_i = -0.810(0.296)Lop + 2.991(0.350)MATS3v + 1.200(0.263)nCaR + 5.143 \\ & `n = 41, r = 0.888, s = 0.393, F = 46.382, FIT = 2.782, LOF = 0.191, AIC = 0.188, \\ & Q^2_{LOO} = 0.752, Q^2_{L5O} = 0.745, r^2_{randY}(sd) = 0.075(0.063), r^2_{Test} = 0.559' \end{aligned} \quad (12)$$

$$\begin{aligned}&pK_i = 1.681(0.398)\text{MATS3v} + 1.544(0.266)\text{nCaR} + 0.648(0.247)\text{C-08} + 5.232\\&\text{`}n = 41, r = 0.887, s = 0.396, F = 45.598, FIT = 2.735, LOF = 0.194, AIC = 0.190,\\&Q^2_{LOO} = 0.742, Q^2_{L5O} = 0.741, r^2_{randY}(sd) = 0.069(0.055), r^2_{Test} = 0.603'\end{aligned} \quad (13)$$

$$\begin{aligned}&pK_i = -0.873(0.262)nF + 1.427(0.264)PW2 + 3.330(0.304)\text{MATS3v} + 4.227\\&\text{`}n = 41, r = 0.884, s = 0.400, F = 44.225, FIT = 2.653, LOF = 0.199, AIC = 0.195,\\&Q^2_{LOO} = 0.721, Q^2_{L5O} = 0.698, r^2_{randY}(sd) = 0.071(0.060), r^2_{Test} = 0.543'\end{aligned} \quad (14)$$

The values provided within the brackets in the regression equations represent the standard errors associated with the regression coefficients. The $r^2_{randY}(sd)$ is the average squared multiple correlation coefficient of the regressions in the activity (Y) randomization study, together with its standard deviation obtained from 100 simulations. In this analysis, where a hundred simulations were conducted for each model, none of the found models exhibited any instances of chance correlation. In the models, the signs of the regression coefficients show the direction of the effect of the explanatory variables.

The descriptors 'Me and 'nF' are part of the CONST class of Dragon descriptors. The classification of constitutional classes is determined solely by basic constitutional facts and is not influenced by molecular connections or conformations. Both the 'Me' descriptor (which represents the mean atomic Sanderson electronegativity scaled on a carbon atom) and the 'nF' descriptor (which represents the number of fluorine atoms) show a negative correlation with the activity. This suggests that molecules with fewer electronegative atoms and a lower number of fluorine atoms will have a higher binding affinity. The descriptors 'X5A', 'PW2', and 'Lop' are members of the TOPO class of Dragon descriptors. 'The TOPO class descriptors are numerical quantifiers of molecular topology. They are derived by applying algebraic operators to matrices representing molecular networks, which are based on a graph representation of the molecule. The values of these descriptors are independent of vertex numbering or labelling. They may exhibit sensitivity to several structural characteristics of the molecule, including but not limited to size, shape, symmetry, branching, and cyclicity. Additionally, they have the ability to encode chemical information pertaining to atom type and bond multiplicity'. The presence of a positive correlation between the descriptor 'PW2' (path/walk 2—Randic shape index) and negative correlations between the descriptors 'X5A' (average connectivity index chi-5) and 'Lop' (Lopping centric index) suggests that a higher value of path/walk 2—Randic shape index and lower values of average connectivity index and Lopping centric index would be advantageous for the activity.

In the models above, 'MATS3v' is the only Dragon descriptor from the 2D-AUTO class. 'The 2D-AUTO descriptors are derived from the autocorrelation of the topological structure of Broto-Moreau (ATS), Moran (MATS), and Geary (GATS). The computation of these descriptors entails adding several autocorrelation functions that correspond to different fragment lengths, resulting in distinct autocorrelation vectors that correspond to the lengths of the structural fragments. Furthermore, these descriptors have been enhanced by incorporating a weighting component based on a physicochemical feature. Consequently, these descriptors specifically consider the arrangement of the structure or its components in relation to a chosen physicochemical characteristic. In the naming system of these descriptors, the second-to-last character, which is a number, represents the number of consecutive edges that are taken into account while calculating the descriptor. This number is referred to as the autocorrelation vector of lag k, which corresponds to the number of edges in the unit fragment. The final character in the descriptor's name denotes the specific physicochemical feature that is considered when calculating the weighting component'. The 'MATS3v' descriptor, which measures the Moran autocorrelation weighted by atomic van der Waals volumes at a lag of 3, shows a positive connection with the activity. This indicates that favourable conditions are related with the lag 3 weighted by atomic van der Waals volumes. The descriptors 'C-008' and 'C-026' that have been identified in the above models belong to the ACF (atom-centred fragments) class. 'ACF are molecular descriptors which are derived from the enumeration of 120 atom-centred pieces, as specified by Ghose and Crippen. Molecular descriptors are basic metrics that quantify the number of specific types of atoms present in a molecule. Their calculation involves knowledge of the molecular composition and atom connectivities'. Descriptors 'C-008' and 'C-026' have exhibited a favourable association with the activity. Therefore, the inclusion of CHR2X (descriptor C-008) and R—CX—R (descriptor C-026) structural fragments in a molecular structure would enhance its activity.

The descriptors 'JGI4' and 'nCaR', mentioned in the previous models, are classified as GALVEZ and FUNC descriptors, respectively. 'The GALVEZ descriptors refer to the Galvez topological charge indices, which are derived from the first ten eigenvalues of the polynomial of the rectified adjacency matrix of the compounds. All the descriptors of the GALVEZ class can be classified into two groups such as topological charge index of order n (GGIn), and the mean topological charge index of order n (JGIn), where "n" represents the order of eigenvalue'. The average topological charge index of order 4 (represented by the descriptor JGI4) has had a beneficial impact on the activity and indicates that a greater value of this index will enhance the binding activity of named substances to the 5-HT5A receptor. 'FUNC class molecular descriptors are derived from the enumeration of chemical functional groups. The descriptor "nCaR" represents the number of substituted sp2 hybridized aromatic carbon atoms'. The sign of 'nCaR' in above equations suggests that having a higher number of these carbon atoms will increase the binding affinity of the named compounds.

The three descriptor models were able to accurately predict 79.03% of the variation in the measured activity of the chemicals. Given the dataset's quantity of observations, we examined models with a maximum of four descriptors. CP-MLR yielded a collection of 91 models, each of which shared 32 descriptors. Table 2 contains a list of 32 descriptors, together with their concise definitions, average regression coefficients, and total occurrence. This table serves as a metric for estimating these descriptors across the models.

The following models bear four-descriptor QSAR having 83.72% variance in the observed activities of the studied compounds.

Table 2 Descriptors[a] predicted for modelling the binding affinity of cyclic guanidines along with the average regression coefficient[b], standard deviation, and total incidence

'Descriptor'	'Avg reg coeff(sd) total incidence'	'Descriptor'	'Avg reg coeff(sd) total incidence'	'Descriptor'	'Avg reg coeff(sd) total incidence'
Mv	1.195(0.292)5	PW3	0.742(0.000)1	MATS3v	2.527(0.655) 67
Me	1.190(0.000)1; −0.993(0.215)5	BAC	1.303(0.416)6	MATS4v	−1.308(0.295)26
nF	−0.703(0.157)5	Lop	−0.991(0.268)21	MATS5v	−1.148(0.061)3
nCL	1.034(0.301)8	T(F..Cl)	−0.743(0.040)2	MATS5e	−0.639(0.144)6
HNar	−0.919(0.117)3	BEHe1	1.474(0.660)4; −3.592(0.000)1	GATS1e	0.640(0.090)5
MAXDP	−0.539(0.087)8	BELe1	1.446(1.197)6; −1.582(0.000)1	nCaR	1.379(0.271)43
TIE	0.963(0.159)16	BEHp1	1.580(0.663)12	C-008	1.343(0.418)32
X1A	0.686(0.000)1; −0.867(0.151)12	JGI3	0.675(0.198)4	C-026	1.024(0.579)12
X5A	−1.377(0.000)1	JGI4	1.107(0.171)3	H-052	0.740(0.275)4
X0Av	1.145(0.489)9	MATS3m	−1.042(0.229)12	PSA	−0.870(0.162)4
PW2	1.063(0.224)13	MATS4m	−0.693(0.144)2		

[a]The four-parameter models that evolved from the CP-MLR procedure have identified these 32 descriptors, with filter-1 being 0.79; filter-2 as 2.0; filter-3 as 0.87; filter-4 as $0.3 \leq Q^2 \leq 1.0$; The standard definitions of the descriptors are as follows: '**CONST**: Mv, mean atomic van der Waals volume (scaled on carbon atom); Me, mean atomic Sanderson electronegativity (scaled on carbon atom); nF, number of fluorine atoms; nCL, number of chlorine atoms; **TOPO**: HNar, Narumi harmonic topological index; MAXDP, maximal electrotopological positive variation; TIE, E-state topological parameter; X1A, average connectivity index chi-1; X5A, average connectivity index chi-5; X0Av, average valence connectivity index chi-0; PW2, path/walk 2—Randic shape index; PW3, path/walk 3—Randic shape index; BAC, Balaban centric index; Lop, Lopping centric index; T(F..Cl), sum of topological distances between F..Cl; **BCUT**: BEHe1, highest eigenvalue n. 1 of Burden matrix/weighted by atomic Sanderson electronegativities; BELe1, lowest eigenvalue n. 1 of Burden matrix/weighted by atomic Sanderson electronegativities; BEHp1, highest eigenvalue n. 1 of Burden matrix/weighted by atomic polarizabilities; **GALVEZ**: JGI3 and JGI4, mean topological charge index of order 3 and 4, respectively; **2D-AUTO**: MATS3m and MATS4m, Moran autocorrelation—lag 3 and 4, respectively/weighted by atomic masses; MATS3v, MATS4v, and MATS5v, Moran autocorrelation—lag 3, 4, and 5, respectively/weighted by atomic van der Waals volumes; MATS5e, Moran autocorrelation—lag 5/ weighted by atomic Sanderson electronegativities; GATS1e, Geary autocorrelation—lag 1/weighted by atomic Sanderson electronegativities; **FUNC**: nCaR, number of substituted aromatic C(sp2); **ACF**: C-008,CHR2X; C-026, R—CX—R; H-052, H attached to C0(sp3) with one X attached to next C atom; **PROP**: PSA, fragment-based polar surface area

[b]The average regression coefficient of the descriptor corresponds to all models and the total number of its incidences; the arithmetic sign represents the sign of the regression coefficient in the models' [9, 10]

$$pK_i = -1.161(0.291)\,Me + 1.001(0.311)\,PW2 + 2.993(0.283)\,MATS3v + 1.388(0.314)\,nCaR + 4.641$$
$$n = 41, r = 0.915, s = 0.350, F = 46.379, FIT = 3.254, LOF = 0.166, AIC = 0.157,$$
$$Q^2_{LOO} = 0.788, Q^2_{L5O} = 0.775, r^2_{randY}(sd) = 0.110(0.078), r^2_{Test} = 0.619 \quad (15)$$

$$pK_i = -0.933(0.271)\,Lop - 0.875(0.287)\,MATS3m + 3.516(0.360)\,MATS3v + 1.161(0.238)\,nCaR + 5.412$$
$$n = 41, r = 0.912, s = 0.355, F = 44.891, FIT = 3.150, LOF = 0.171, AIC = 0.161,$$
$$Q^2_{LOO} = 0.791, Q^2_{L5O} = 0.787, r^2_{randY}(sd) = 0.104(0.066), r^2_{Test} = 0.669 \quad (16)$$

$$pK_i = 1.729(0.387)\,BEHp1 - 1.327(0.396)\,MATS4v + 1.401(0.213)\,C\text{-}008 + 0.858(0.181)\,C\text{-}026 + 6.093$$
$$n = 41, r = 0.908, s = 0.363, F = 42.538, FIT = 2.985, LOF = 0.179, AIC = 0.169,$$
$$Q^2_{LOO} = 0.772, Q^2_{L5O} = 0.778, r^2_{randY}(sd) = 0.106(0.065), r^2_{Test} = 0.634 \quad (17)$$

$$pK_i = -0.887(0.286)\,Me - 0.765(0.298)\,X1A + 2.950(0.299)\,MATS3v + 1.581(0.310)\,nCaR + 5.413$$
$$n = 41, r = 0.907, s = 0.366, F = 41.860, FIT = 2.937, LOF = 0.181, AIC = 0.171,$$
$$Q^2_{LOO} = 0.776, Q^2_{L5O} = 0.768, r^2_{randY}(sd) = 0.098(0.065), r^2_{Test} = 0.598 \quad (18)$$

In the present randomization, with a hundred simulations per model, no chance correlation is found. The $Q^2 > 0.5$ indicates robust QSAR model. These models are derived from the training set and validated with the test set. The test set compounds were evaluated using external validation, and the activity predictions were found acceptable, which is reflected by the test set r2 values and anticipated activity values given in Table 1. The goodness of fit between observed versus calculated activities for the training and test set compounds has been plotted in Fig. 1.

In previous models, the descriptors such as 'Me', 'PW2', 'MATS3v', 'nCaR', 'Lop', 'C-008', and 'C-026' were important and have shown consistent implications for the activity. In the new models, descriptors 'MATS3m', 'MATS4v', 'X1A', and 'BEHp1' have emerged from 2D-AUTO, TOPO, and BCUT classes of Dragon descriptors. 'The BCUT descriptors are modified Burden adjacency matrix represented as BEHwk and BELwk, respectively. Here, w refers to some specific atomic property, and k refers to the rank of eigenvalue from 1 to 8'. The ordered sequence of the highest and lowest eigenvalues reflects the pertinent characteristics of molecule structure, which is significantly meaningful for the purpose of similarity searching. In the above-selected model Eq. 17, the BCUT descriptor, 'BEHp1', reflects the modified Burden adjacency matrix for the first highest absolute eigenvalue weighted for atomic polarizability. The sign of the regression coefficient of 'BEHp1' descriptor delineates that a higher value is advantageous for activity. A smaller value of the descriptors 'MATS3m' (Moran autocorrelation—lag 3 weighted by atomic masses), 'MATS4v' (Moran autocorrelation—lag 4 weighted by atomic van der Waals volumes), and 'X1A' (average connectivity index, chi-1) would be supportive of improving the activity, as they were all detrimental to it.

The 32 CP-MLR identified descriptors (Table 3) have been subjected to a partial least squares (PLS) analysis to help develop a 'single window' structure-activity model and determine the potential of these descriptors in explaining the binding affinities of cyclic guanidine derivatives to the 5-HT5A receptor. It provides an opportunity to compare the descriptors' relative significance. This comparison within the modelled activity is made possible by the proportion contributions obtained from the normalized regres-

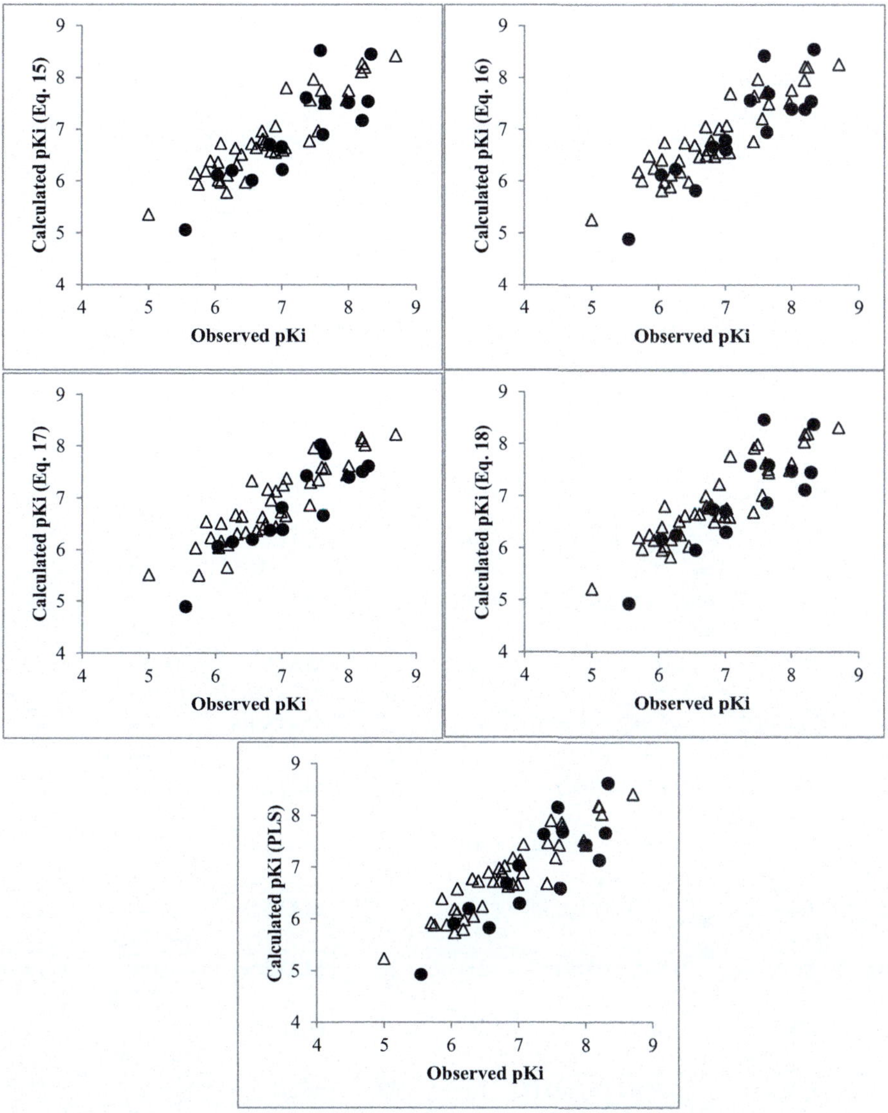

Fig. 1 Plots of the observed versus calculated pK_i values of cyclic guanidines' training (Δ) and test sets (●)

sion coefficients of the descriptors. To give each descriptor equal weight in the study, the PLS descriptors have been auto-scaled (zero mean and unit standard deviation). Four components were determined to be the best fit for these 32 descriptors in the PLS cross-validation, accounting for 86.11% of the variance in the activity.

Table 3 gives the MLR-like PLS coefficients of the concerned descriptors. The computed activity of the test and training set compounds is very close to the corresponding observed activity listed in Table 1. Figure 1 provides a plot of observed versus calcu-

Table 3 PLS and MLR-like PLS models from the descriptors of four-parameter CP-MLR models for the 5 - HT_{5A} binding affinity

A: PLS equation	
PLS components	PLS coefficient (s.e.)[a]
Comp.-1	−0.221(0.016)
Comp.-2	−0.092(0.022)
Comp.-3	0.102(0.028)
Comp.-4	0.075(0.026)
Constant	6.863

B: MLR-like PLS equation

S. No.	'Descriptor'	'MLR-like coefficient(f.c.)[b]'	'Order'	S. No.	'Descriptor'	'MLR-like coefficient (f.c.)[b]'	'Order'
1	Mv	0.045 (0.020)	18	17	BELe1	0.026 (0.011)	25
2	Me	0.019 (0.008)	26	18	BEHp1	0.091 (0.040)	7
3	nF	0.010 (0.004)	29	19	JGI3	−0.031 (−0.014)	23
4	nCL	0.057 (0.025)	14	20	JGI4	−0.017 (−0.007)	28
5	HNar	0.041 (0.018)	22	21	MATS3m	0.128 (0.056)	6
6	MAXDP	−0.082 (−0.036)	9	22	MATS4m	−0.091 (−0.040)	8
7	TIE	0.198 (0.087)	4	23	MATS3v	0.221 (0.097)	2
8	X1A	−0.010 (−0.004)	30	24	MATS4v	−0.223 (−0.098)	1
9	X5A	−0.068 (−0.030)	12	25	MATS5v	−0.045 (−0.020)	17
10	X0Av	0.044 (0.019)	21	26	MATS5e	−0.044 (−0.020)	19
11	PW2	0.005 (0.002	31	27	GATS1e	−0.018 (−0.008)	27
12	PW3	−0.047 (−0.021)	16	28	nCaR	0.167 (0.073)	5
13	BAC	0.044 (0.019)	20	29	C-008	0.207 (0.091)	3
14	Lop	−0.052 (−0.023)	15	30	C-026	0.074 (0.033)	10
15	T(F..Cl)	−0.031 (−0.014)	24	31	H-052	0.071 (0.031)	11
16	BEHe1	0.061 (0.027)	13	32	PSA	−0.004 (−0.002)	32
	Constant	6.198					

(continued)

Table 3 (continued)

C: PLS regression statistics	Values
N	41
R	0.928
S	0.322
F	56.478
Q^2_{LOO}	0.828
Q^2_{L5O}	0.816
r^2_{Test}	0.597

[a]Regression coefficient of PLS factor and its standard error

[b]Coefficients of MLR-like PLS equation in terms of descriptors for their original values; f.c. represents the fraction contribution of the regression coefficient computed from the normalized regression coefficients derived from the auto-scaled (zero mean and unit standard deviation) data

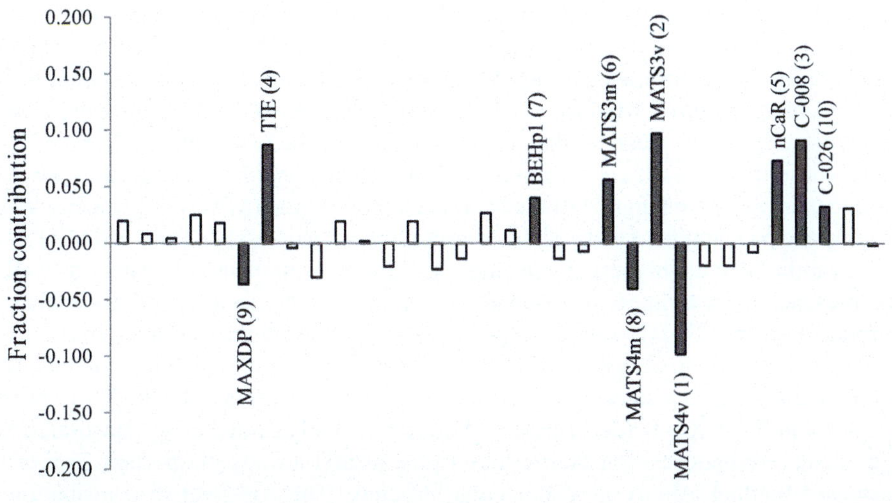

Fig. 2 Fraction contribution plot of MLR-like PLS coefficients (normalized) against 32 descriptors associated with cyclic guanidines binding affinity

lated activities (using PLS analysis) for the training and test set compounds, which demonstrates the goodness of fit. Figure 2 shows the fraction contribution plot of normalized regression coefficients of these descriptors to the activity (Table 3).

Additionally, the PLS analysis has recommended 'MATS4v' descriptor as the most important descriptor for modelling the chemicals' activities. The remaining nine significant descriptors (Table 3; Fig. 2) are 'MATS3v', 'C-008', 'TIE', 'nCaR', 'MATS3m', 'BEHp1', 'MATS4m', 'MAXDP', and 'C-026'. They are listed in decreasing order of significance. All descriptors listed in Table 3 are part of Eqs. 6–18 and convey the same inference in the PLS model as well. The 'TIE' (E-state topological parameter) and 'MAXDP' (maximal electrotopological positive variation) are TOPO class descriptors. A higher E-state topological parameter value and a smaller maximal electrotopological positive variation value would benefit the activity. The negative regression coefficient of Moran autocorrelation—lag 4 weighted by atomic masses (descriptor MATS4m) recommends that a greater positive value is unfavourable to the activity. The other descriptors appear to have a lesser degree of significance in relation to these ten descriptors when it comes to influencing the compounds' activity (Table 4; Fig. 2). Additionally, it is noted that the PLS model from the dataset (Table 3) lacking 32 descriptors performs worse in elucidating the analogues' activity.

4 Conclusions

Overall, this study has established the link between the binding affinities of cyclic guanidines to 5-HT_{5A} receptors and their structural characteristics. The binding affinity now depends on the combined impact of many structural characteristics that have been established as individual descriptors. The descriptor nF suggests that reducing the number of fluorine atoms in the chemical structure is recommended to enhance the binding affinity of a drug to the 5 - HT_{5A} receptor. A decrease in the values of the molecular topology and symmetry accounting parameters, specifically the average connectivity indices chi-1 and chi-5 (descriptors X1A and X5A), as well as the Lopping centric index (descriptor Lop), and an increase in the value of the path/walk 2—Randic shape index (descriptor PW2), are advantageous for the activity. The relationship between atomic mass and path length 3 of the Moran autocorrelation (MATS3m), van der Waals volume and path lengths 3 and 4 of the Moran autocorrelations (MATS3v and MATS4v), polarizability and the highest eigenvalue n.1 of the Burden matrix (BEHp1), as well as the presence of descriptors Me (mean atomic Sanderson electronegativity scaled on carbon atom) and JGI4 (fourth order mean topological charge index) have demonstrated the significance of atomic properties and charge content in explaining the binding affinity. In addition, the existence of structural fragments, specifically CHR2X (descriptor C-008) and R—CX—R (descriptor C-026), and a higher number of substituted sp2 hybridized aromatic carbon atoms (descriptor nCaR) are also important factors for increased binding affinity in a molecular structure. The constructive contribution of information content descriptors determined by CP-MLR has been validated by PLS analysis. The generated models and the descriptors involved in them have indicated ample potential for further modification of the substituents of the guanidine molecule.

Acknowledgements The authors thank their respective institutions for furnishing the essential resources to conduct this study successfully.

Supplementary Data The complete dataset of molecular descriptors corresponding to all the 58 compounds included in Table 1.

References

1. Ślifirski, G., Król, M., Turło, J.: 5-HT receptors and the development of new antidepressants. IJMS. **22**(16), 9015 (2021). https://doi.org/10.3390/ijms22169015
2. Barnes, N.M., Sharp, T.: A review of central 5-HT receptors and their function. Neuropharmacology. **38**(8), 1083–1152 (1999). https://doi.org/10.1016/S0028-3908(99)00010-6
3. Oliver, K.R., Kinsey, A.M., Wainwright, A., Sirinathsinghji, D.J.S.: Localization of 5-HT$_{5A}$ receptor-like immunoreactivity in the rat brain. Brain Res. **867**(1–2), 131–142 (2000). https://doi.org/10.1016/S0006-8993(00)02273-3
4. Arias, B., Collier, D.A., Gastó, C., Pintor, L., Gutiérrez, B., Vallès, V., Fañanás, L.: Genetic variation in the 5-HT5A receptor gene in patients with bipolar disorder and major depression. Neurosci. Lett. **303**(2), 111–114 (2001). https://doi.org/10.1016/S0304-3940(01)01701-3
5. Thomas, D.: 5-Ht5A receptors as a therapeutic target. Pharmacol. Ther. **111**(3), 707–714 (2006). https://doi.org/10.1016/j.pharmthera.2005.12.006
6. Kassai, F., Schlumberger, C., Kedves, R., Pietraszek, M., Jatzke, C., Lendvai, B., Gyertyán, I., Danysz, W.: Effect of 5-HT$_{5A}$ antagonists in animal models of schizophrenia, anxiety and depression. Behav. Pharmacol. **23**(4), 397–406 (2012). https://doi.org/10.1097/FBP.0b013e3283565248
7. Peters, J.-U., Lübbers, T., Alanine, A., Kolczewski, S., Blasco, F., Steward, L.: Cyclic guanidines as dual 5-HT$_{5A}$/5-HT$_7$ receptor ligands: optimising brain penetration. Bioorg. Med. Chem. Lett. **18**(1), 262–266 (2008). https://doi.org/10.1016/j.bmcl.2007.10.078
8. Peters, J.-U., Lübbers, T., Alanine, A., Kolczewski, S., Blasco, F., Steward, L.: Cyclic guanidines as dual 5-HT$_{5A}$/5-HT$_7$ receptor ligands: structure–activity relationship elucidation. Bioorg. Med. Chem. Lett. **18**(1), 256–261 (2008). https://doi.org/10.1016/j.bmcl.2007.10.080
9. Dragon software (version 1.11-2001) by Todeschini, R, Consonni, V, Milano, Italy. http://disat.unimib.it/chm/Dragon.htm
10. Todeschini, R., Consonni, V.: Handbook of molecular descriptors. In: Methods and Principles in Medicinal Chemistry, 1st edn. Wiley (2000). https://doi.org/10.1002/9783527613106
11. Prabhakar, Y.S.: A combinatorial approach to the variable selection in multiple linear regression: analysis of Selwood *et al.* Data set—a case study. QSAR Comb. Sci. **22**(6), 583–595 (2003). https://doi.org/10.1002/qsar.200330814
12. Sharma, B.K., Singh, P., Pilania, P., Shekhawat, M., Prabhakar, Y.S.: QSAR of 2-(4-methyl sulphonylphenyl) pyrimidine derivatives as Cyclooxygenase-2 inhibitors: simple structural fragments as potential modulators of activity. J. Enzyme Inhib. Med. Chem. **27**(2), 249–260 (2012). https://doi.org/10.3109/14756366.2011.587414
13. Wold, S.: Cross-validatory estimation of the number of components in factor and principal components models. Technometrics. **20**(4), 397–405 (1978). https://doi.org/10.1080/00401706.1978.10489693
14. Ståhle, L., Wold, S.: Multivariate data analysis and experimental design in biomedical research. In: Progress in Medicinal Chemistry, vol. 25, pp. 291–338. Elsevier (1988). https://doi.org/10.1016/S0079-6468(08)70281-9
15. Kettaneh, N., Berglund, A., Wold, S.: PCA and PLS with very large data sets. Comput. Stat. Data Anal. **48**(1), 69–85 (2005). https://doi.org/10.1016/j.csda.2003.11.027
16. Kubiny, H.: Variable selection in QSAR studies. I. An evolutionary algorithm. Quant. Struct. Act. Relat. **13**(3), 285–294 (1994). https://doi.org/10.1002/qsar.19940130306
17. Kubinyi, H.: Variable selection in QSAR studies. II. A highly efficient combination of systematic search and evolution. Quant. Struct. Act. Relat. **13**(4), 393–401 (1994). https://doi.org/10.1002/qsar.19940130403

18. Friedman, J.: Multivariate Adaptive Regression Splines, Technical Report No. 102, Laboratory for Computational Statistics, Department of Statistics, Stanford University, November 1988 (revised August 1990)
19. Akaike, H.: A new look at the statistical model identification. IEEE Trans. Autom. Control. **19**(6), 716–723 (1974). https://doi.org/10.1109/TAC.1974.1100705
20. Akaike, H.: Information theory and an extension of the maximum likelihood principle. In: Parzen, E., Tanabe, K., Kitagawa, G. (eds.) *Selected Papers of Hirotugu Akaike* Springer Series in Statistics, pp. 199–213. Springer New York, New York, NY (1998). https://doi.org/10.1007/978-1-4612-1694-0_15
21. So, S.-S., Karplus, M.: Three-dimensional quantitative structure−activity relationships from molecular similarity matrices and genetic neural networks. 1. Method and validations. J. Med. Chem. **40**(26), 4347–4359 (1997). https://doi.org/10.1021/jm970487v
22. Prabhakar, Y.S., Solomon, V.R., Rawal, R.K., Gupta, M.K., Katti, S.B.: CP-MLR/PLS directed structure-activity modeling of the HIV-1 RT inhibitory activity of 2, 3-Diaryl-1,3-thiazolidin-4-ones. QSAR Comb. Sci. **23**(4), 234–244 (2004). https://doi.org/10.1002/qsar.200330854

Computer-Aided Drug Design of Potential Anti-Tubercular Drug Targeting the Mycobacterial F-Type ATP Synthase Enzyme

Ankur Kumar[1] and Kuldeep K. Roy[1] (✉)

[1] Department of Pharmaceutical Sciences, School of Health Sciences and Technology, UPES, Dehradun, Uttarakhand, India

Abstract Tuberculosis (TB) remains one of the leading infectious diseases with a significant impact on morbidity and mortality worldwide. *Mycobacterium tuberculosis* (Mtb), the causative agent of TB, is highly dependent on the electron transport chain (ETC) pathway for the generation of energy. Mycobacteria opportunistically move to another site for energy generation if any of their ETC sites are inhibited by any drug. In this study, we aim to design and evaluate novel scaffolds as potential anti-tubercular agents targeting the inhibition of ATP synthase. Recently, an Amiloride derivative HM2-16F was reported as an anti-tubercular agent targeting the inhibition of Mtb ATP synthase. Using HM2-16F as template compound, a library of the novel compounds was generated and subjected to docking analysis against the mycobacterial F-type ATP synthase (PDB-ID: 4V1F). The designed new compounds were found to have optimal ADMET properties and predicted binding affinity comparable to Bedaquiline. The benzofuran ring of compounds 4, 5, and 11 showed π-π interactions with Phe69, whereas compound 5 formed hydrogen bond with Gly62 (backbone) and with Glu65. Furthermore, compounds 4 and 11 showed hydrogen bond interaction with Glu65. Further synthesis and biological evaluation of the designed compounds may lead to a potential drug candidate against TB.

Keywords Tuberculosis · ATP synthase · Electron Transport Chain · Inhibitor · Scaffold design

Abbreviations

ADMET	Absorption, Distribution, Metabolism, Excretion, and Toxicity
ATP	Adenosine triphosphate
BBB	Blood–brain barrier
BDQ	Bedaquiline
DILI	Drug-induced liver injury
ETC	Electron transport chain
HBA	Hydrogen bond acceptor

A. K. Saxena, S. Nandi (eds.), *Global Trends in Health, Technology and Management*, https://doi.org/10.1007/978-3-031-75457-9_11

HBD	Hydrogen bond donor
hERG	Human ether-a-go-go-related gene
HIA	Human intestinal absorption
logD7.4	logarithm of the *n*-octanol/water distribution coefficient at pH = 7.4
logP	logarithm of the *n*-octanol/water distribution coefficient
logS	logarithm of aqueous solubility value
MIC	Minimum inhibitory concentration
Mtb	*Mycobacterium tuberculosis*
MW	Molecular weight
OSCP	Oligomycin sensitivity conferring protein
P-gp	P-glycoprotein
PAINS	Pan assay interference compounds
RMSD	Root mean square deviation
TB	Tuberculosis
TPSA	Topological polar surface area
WHO	World Health Organization

1 Introduction

Tuberculosis (TB) remains a significant global health challenge, with a profound impact on morbidity and mortality worldwide. TB is caused by *Mycobacterium tuberculosis*, which mainly affects the lungs but is capable of infecting other organs such as brain, kidneys, and bones. Despite being a curable disease, TB continues to afflict millions, with an estimated 10 million new cases and 1.4 million deaths annually according to the global TB 2023 report by the World Health Organization (WHO) [1]. The burden of TB is exacerbated by factors such as poverty, overcrowding, HIV/AIDS co-infection, latent TB, and the emergence of drug-resistant strains.

The standard treatment for drug-susceptible TB involves a combination of antibiotics, typically isoniazid, rifampicin, pyrazinamide, and ethambutol, administered over 6 to 9 months. However, the rise of multidrug-resistant TB (MDR-TB) and extensively drug-resistant TB (XDR-TB) poses formidable challenges to TB control efforts. Treating MDR-TB and XDR-TB requires prolonged, complex regimens with second-line drugs that are less effective, more toxic, and often inaccessible in resource-limited settings. The incidence of latent TB infection (23% of the global population) and the advent of multi-drug resistant TB (MDR-TB) (82%) toward the first line of therapy, particularly against Rifampicin, constitute a challenge to global health throughout time [1].

Bedaquiline (BDQ) has emerged as a beacon of hope in the fight against tuberculosis (TB), particularly MDR-TB and extensively drug-resistant tuberculosis (XDR-TB). Bedaquiline exerts its potent anti-tubercular activity by selectively inhibiting mycobacterial ATP synthase, a crucial enzyme involved in cellular energy production [2]. This unique mechanism disrupts the energy metabolism of *Mycobacterium tuberculosis*, leading to bacterial death. Its long half-life allows for once-daily dosing, enhancing patient adherence to treatment regimens [3]. Bedaquiline, also known as "Sirturo" is approved by US-FDA for its use in combination therapy to treat adults with MDR-TB [4].

The mycobacterial F-type ATP synthase is a multi-subunit complex consisting of two main regions: the F_1 catalytic region, located in the mitochondrial matrix or bacterial cytoplasm, and the F_O proton-translocating region, embedded in the inner mitochondrial membrane or bacterial plasma membrane [5]. The F_1 region typically consists of multiple subunits, including α, β, γ, δ, and ε subunits, which form the catalytic core responsible for ATP synthesis. The F_O region comprises multiple subunits, including a, b, c, and Oligomycin Sensitivity Conferring Protein (OSCP) in bacteria. The γ subunit acts as a rotor, rotating within the $\alpha_3\beta_3$ hexamer of the F_1 region, while the *c* subunits of the F_O region form a ring-like structure that rotates within the membrane in response to proton translocation. Protons are pumped across the membrane from the intermembrane space (or periplasm in bacteria) to the matrix (or cytoplasm in bacteria) through the F_O region during ATP synthesis. This proton movement drives the rotation of the γ subunit, leading to conformational changes in the catalytic sites of the β subunits, ultimately resulting in ATP synthesis [6]. The structure of F-type ATP synthase is now available that can be used to design and discover novel candidates against TB such as chloroquinoline [7]. BDQ is reported to bind to subunit *c* of this enzyme and thereby, disrupts the proton translocation and ATP synthesis leading to killing of mycobacteria. Figure 1 shows the binding of BDQ to the subunit *c* (rotary ring) of *M. phlei* ATP synthase [8, 9].

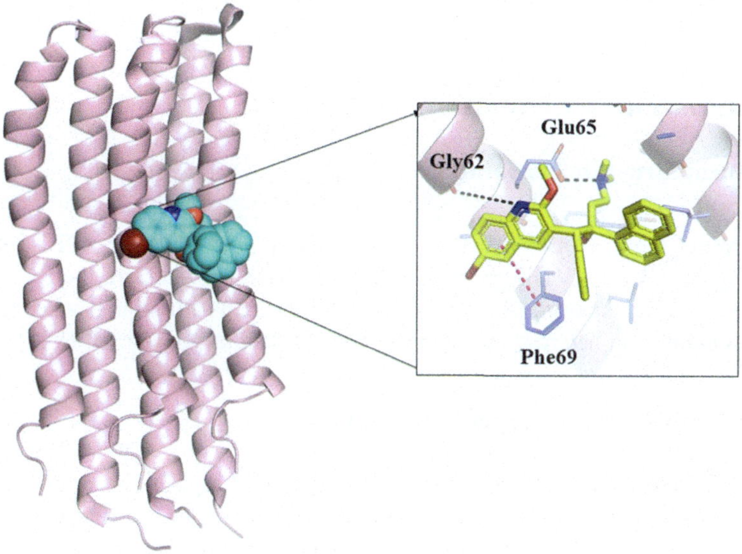

Fig. 1 Depiction of BDQ bound to subunit *c* of *M. phlei* ATP synthase (PDB-ID: 4V1F) with surrounding interacting residues. Subunit *c* is shown in purple cartoon, while BDQ is shown in cyan spheres (left side) and yellow sticks (right side). Selected amino acids are shown in thick lines and dotted lines representing direct or indirect interactions

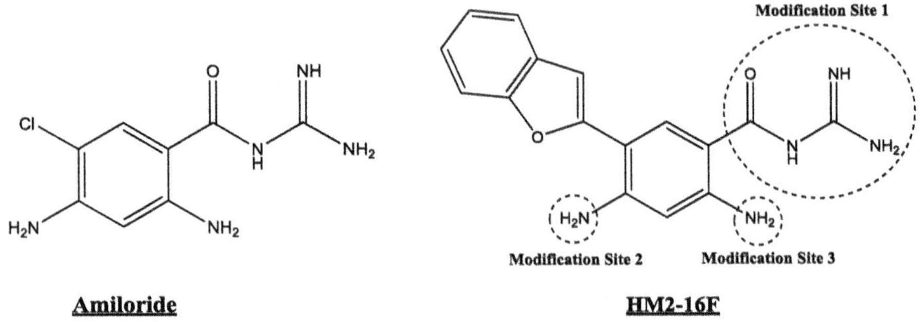

Fig. 2 The chemical structure of Amiloride, HM2-16F, and the scheme of the library design based on HM2-16F

Amiloride, an approved drug for diuresis, has shown moderate anti-tubercular activity with minimum inhibitor concentration (MIC) of 256 μM (H37Rv strain). Recently, few analogues of Amiloride have been explored against mycobacterial ATP synthase and a lead candidate HM2-16F (Fig. 2) has been reported to potently inhibit this enzyme and kill mycobacteria (MIC = 4 μM) [10]. Since it is known that the dimethylamino group of BDQ acts as an arginine mimic in its interaction with the subunit-c (rotary ring) of F_1F_O ATP synthase, we hypothesized that the acylguanidine of HM2-16F could make it an inhibitor of Mtb F_1F_O ATP synthase [11, 12]. By taking it into our consideration, we aimed to conduct the library design and screening using docking studies, followed by absorption, distribution, metabolism, excretion, and toxicity (ADMET) property calculations to discover novel potential compounds for further exploration.

2 Materials and Methods

2.1 Protein Structure Preparation

In the present work, the X-ray structure of BDQ-bound *M. phlei* ATP synthase (F_O rotary ring) (PDB-ID: 4V1F) resolved at 1.70 Å was retrieved from the RCSB Protein Data Bank (www.rcsb.org). The protein was prepared by removing water molecules, another subunits and heteroatoms using Python Molecular Viewer (PMV).

2.2 Library Design and Ligand Preparation

The 2D structures of compounds were sketched using ChemDraw and then converted into 3D structures by the addition of Lamarckian charge. The 3D structures were then subjected to ligand preparation, followed by energy minimization using the Open Babel tool, considering the Amber force field. The resulting ligands structures were exported in the "pdbqt" format for further use in the molecular docking using the AutoDock Vina.

2.3 Molecular Docking Tool Validation and Library Screening

Molecular docking is among the most widely used methods in drug discovery, which has driven the development of several docking algorithms and scoring functions. However, their performances can vary greatly depending on the ligand–protein complex under study. The reliability of the docking tool can be measured by root mean square deviation (RMSD) between a docked (predicted) structure and an experimentally determined one. In order to validate the AutoDock Vina tool, the newly optimized 3D structure of Bedaquiline was docked into the reported site of crystal structure of ATP synthase and compared with the experimentally determined one in terms of RMSD. Docking was done with the default settings considering the grid center (X = −172, Y = 57, Z = 33.98) and grid size (X = 24, Y = 14 and Z = 16). Afterwards, the designed library of several compounds was docked at this site to find out the top-ranked compounds for further ADMET analysis and optimal compounds' selection.

2.4 ADMET Analysis

Predictions of the ADMET properties of the designed compounds were performed through the ADMETlab server (https://admetmesh.scbdd.com) [13]. ADMET parameters include the aqueous solubility value logarithm (−4 to 0.5 log mol/L), the n-octanol/water distribution coefficient logarithm (0 to 3 log mol/L), the inhibitor of P-glycoprotein (0–0.3: excellent (green); 0.3–0.7: medium (yellow); 0.7–1.0(++): poor (red); and so on. The number of alerts for Pan-Assay Interference Compounds (PAINS) should be zero. Human ether-a-go-go related gene (hERG) encodes the inward rectifying voltage-gated potassium channel in the heart (I_{Kr}) which is involved in cardiac repolarization: 0–0.3: excellent (green); 0.3–0.7: medium (yellow); 0.7–1.0 (++): poor (red). Drug-induced liver injury (DILI) (0–0.3: excellent (green); 0.3–0.7: medium (yellow); 0.7–1.0 (++): poor (red); Carcinogenicity (0–0.3: excellent (green); 0.3–0.7: medium (yellow); 0.7–1.0 (++): poor (red)). Based on the calculated ADMET parameters, relevant compounds were identified as potential hits for further exploration.

3 Results and Discussion

3.1 Docking Tool Assessment

The trustworthiness of any docking tool can be measured by measuring the deviation between a docked pose and an experimentally determined one of a particular ligand at the binding site of a biomolecule (target). In order to validate the AutoDock Vina tool, the newly optimized 3D structure of Bedaquiline was docked into the reported site (subunit c) in the ATP synthase and the predicted pose was compared with the experimentally determined one. The overlay and comparison of Bedaquiline's co-crystallized conformation (green carbon) and best predicted posture (cyan carbon) are depicted in Fig. 3. The co-crystallized and anticipated conformations of Bedaquiline were observed

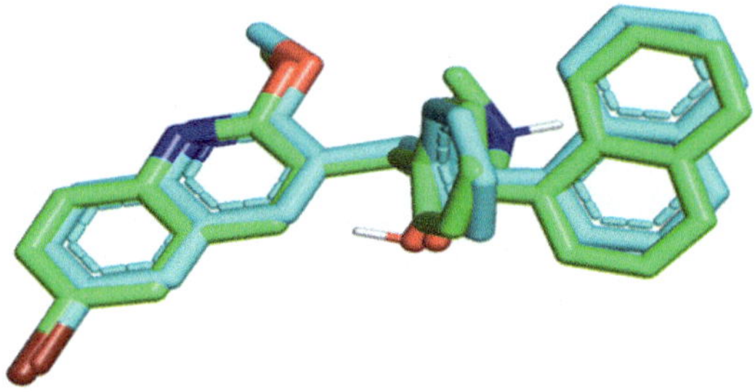

Fig. 3 Overlay and comparison of the best predicted pose (cyan carbon) with co-crystallized conformation (green carbon) of Bedaquiline. Structures are represented in sticks

to match precisely. The binding affinity of Bedaquiline was −6.1 kcal/mol. This allowed us to draw the conclusion that since AutoDock Vina could accurately predict the recognized ATP synthase inhibitor, its prediction of new molecules would also be trustworthy.

3.2 Library Design and Screening

The ligands were designed based on the substitution of Amiloride derivative HM2-16F with various suitable rings and function groups to mimic the guanidine moiety. Figure 2 depicts the scheme of the library design based on HM2-16F. A total of 30 compounds were rationally created and analyzed for drug-likeness based on Lipinski's rule of five. The designed drug-like library was afterward used in the docking-based screening using the AutoDock Vina tool. All compounds were docked at the BDQ binding site in the subunit c (F_O rotary ring) of mycobacterial ATP synthase (Fig. 1). It led to the top 10 compounds with predicted binding affinity similar to or comparable with BDQ. The details are summarized in Table 1. Figure 4 depicts predicted binding modes of these top 10 compounds at the BDQ binding site in the F_O rotary ring (subunit c) of the mycobacterial ATP synthase.

3.3 ADMET Properties

The ADMET properties can be predicted with the help of ADMETlab server. The ADMETLAB online server was used to calculate the physicochemical, ADME, and toxicity properties of the all the top-ranked compounds. Table 2 gives a summary of the calculated ADMET properties of the selected top-ranked compounds. A total of 13 hit compounds exhibited the optimal and acceptable range of ADMET profiles which needs further exploration in the in vitro setting.

Table 1 Summary of the top 10 compounds along with reference compound

S.N.	Compound name	Chemical structure	Binding Energy (kcal/mol)
1.	Bedaquiline or BDQ (reference molecule)	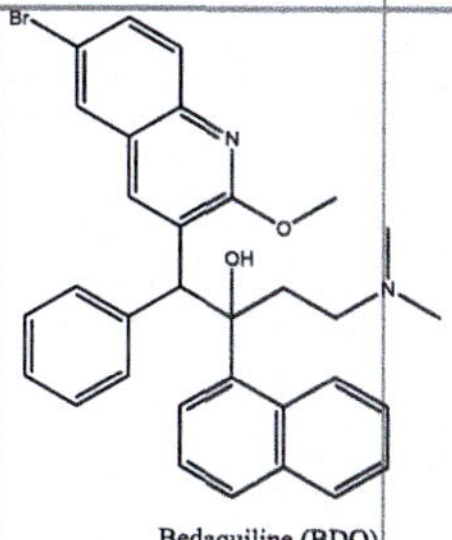 Bedaquiline (BDQ)	−6.1
2.	Compound 5	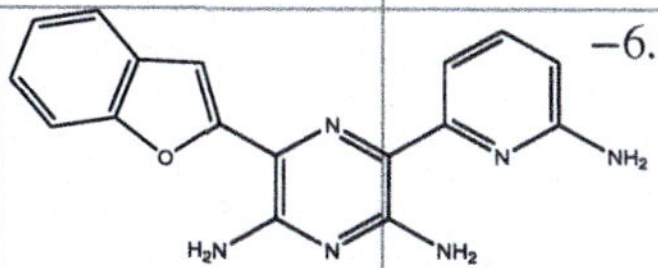	−6.2
3.	Compound 4	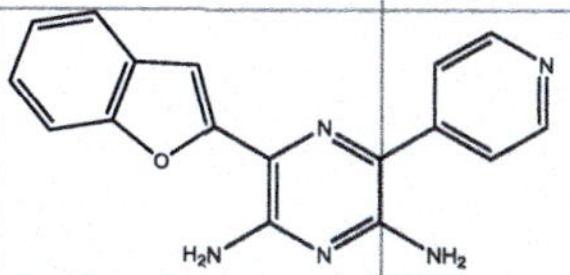	−6.0
4.	Compound 14	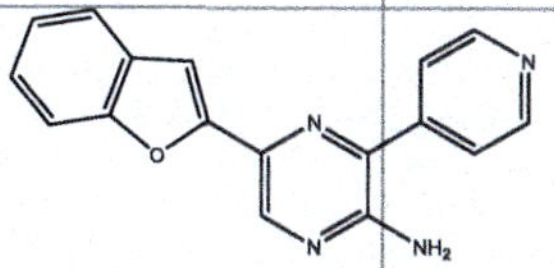	−6.0
5.	Compound 15	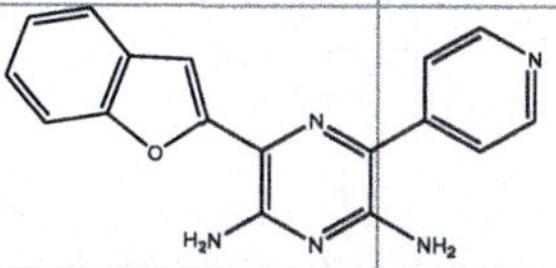	−6.0
6.	Compound 2	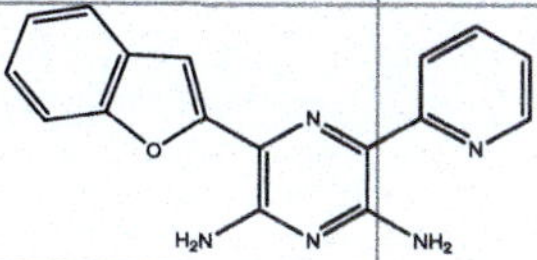	−6.0
7.	Compound 3	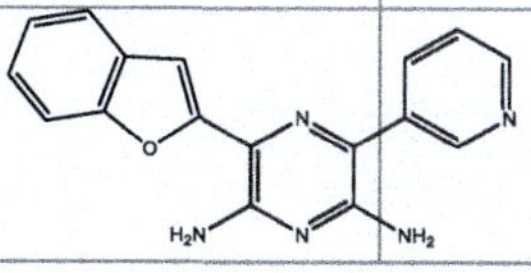	−5.7

(continued)

Table 1 (continued)

S.N.	Compound name	Chemical structure	Binding Energy (kcal/mol)
8.	Compound 6	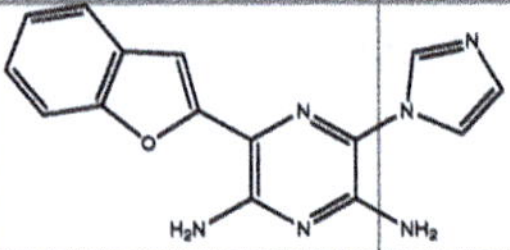	−5.3
9.	Compound 11	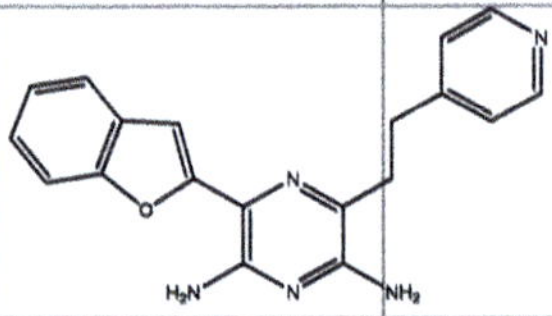	−6.1
10.	Compound 9	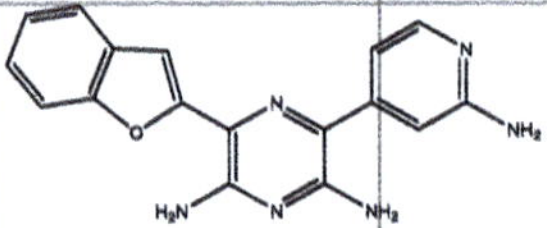	−5.6
11.	Compound 8	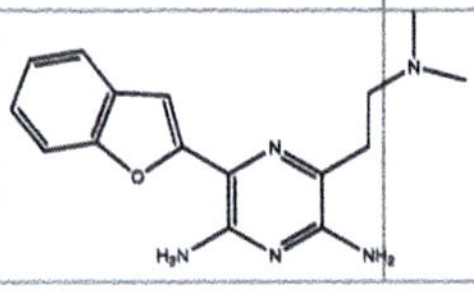	−5.2

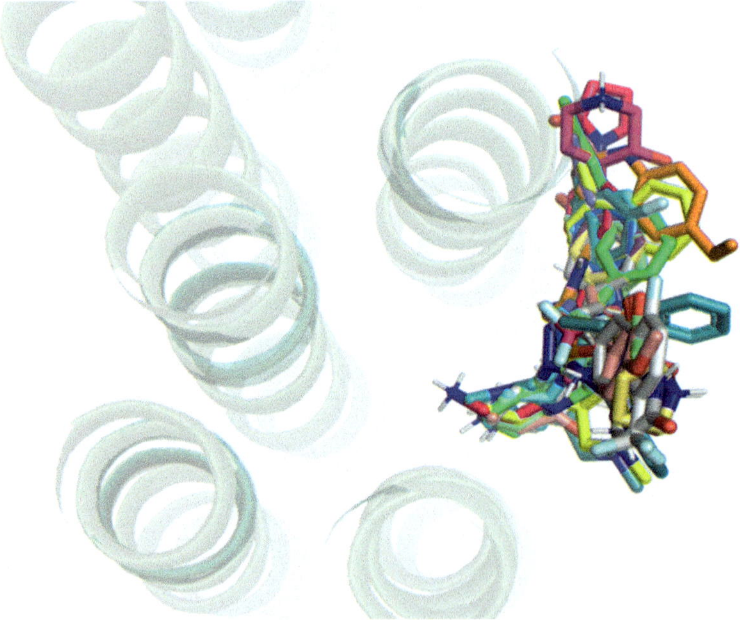

Fig. 4 Depiction of the designed molecules docked at same pocket as BDQ at subunit *c* (rotary ring) of the mycobacterial ATP synthase

Table 2 ADMET properties of the top 10 hits[a]

Compound number	Log S	Log D	Log P	Pgp-inh	Pgp-sub	HIA	BBB	hERG	DILI	Carcinogenicity	MW	HBA	HBD	TPSA	PAINS
Reference molecule (Bedaquiline)	−6.907	4.619	6.489	0.999	0.001	0.006	0.871	0.984	0.854	0.032	554.16	4	1	45.59	0
Compound 2	−3.88	2.4	2.6	0.994	0.57	0.01	0.16	0.111	0.982	0.109	318.1	7	6	130	0
Compound 3	−3.82	2.2	1.9	0.038	0.002	0.01	0.12	0.077	0.995	0.068	292.1	7	4	109	0
Compound 4	−3.68	2.6	2.7	0.994	0.012	0.01	0.38	0.037	0.989	0.047	303.1	6	4	104	0
Compound 5	−3.53	1.3	1.8	0.956	0.849	0.01	0.06	0.052	0.995	0.108	334.1	9	8	169	0
Compound 6	−2.86	1.6	1.7	0.646	0.943	0.01	0.76	0.352	0.975	0.274	283.1	6	4	94.2	0
Compound 8	−3.29	0.4	1.3	0.901	0.979	0.01	0.32	0.303	0.985	0.462	311.1	9	8	170	0
Compound 9	−3.63	2	2.2	0.992	0.653	0.01	0.17	0.083	0.987	0.094	318.1	7	6	130	0
Compound 11	−2.86	1.7	2	0.941	0.959	0.01	0.78	0.53	0.968	0.25	297.2	6	4	94.2	0
Compound 14	−3.52	2.2	2.4	0.996	0.03	0.01	0.23	0.061	0.99	0.041	303.1	6	4	104	0
Compound 15	−3.64	2.2	2.4	0.995	0.016	0.01	0.28	0.072	0.991	0.039	303.1	6	4	104	0

[a] Ideal Range of ADMET properties: LogS = −4 to 0.5; LogP = 0 to 3; Pgp-inhibitor = Excellent 0–0.3, medium 0.3–0.7 poor 0.7–1.0; Pgp-substrate = Excellent 0–0.3, medium 0.3–0.7, poor 0.7–1.0; HIA = Excellent 0–0.3, medium 0.3–0.7, poor: 0.7–1.0; BBB = Excellent 0–0.3, Medium 0.3–0.7, poor 0.7–1.0; hERG = Excellent 0–0.3, medium 0.3–0.7: poor 0.7–1.0(++); DILI = Excellent 0–0.3, Medium 0.3–0.7, poor 0.7–1.0; Carcinogenicity = Excellent 0–0.3, medium 0.3–0.7, poor 0.7–1.0

3.4 Molecular Interaction

The structure of the *M. phlei* *c*-ring bound to BDQ revealed that BDQ mimics the *a*-subunit arginine, which temporarily interacts with the *c*-ring Glu65 carboxylate side chain, while the phenyl ring of BDQ forms π-π interaction with Phe69 in the subunit *c* (rotary ring) of ATP synthase. In comparison with reference ligand (BDQ), we have selected three best compounds as potential hits based on their binding affinity, ADMET properties, and the interactions at the binding site as depicted in Fig. 5 The benzofuran ring of compounds 4, 5, and 11 showed π-π interactions with Phe69, whereas the compound 5 formed hydrogen bond with Gly62 (backbone) and with Glu65. Furthermore, compounds 4 and 11 showed hydrogen bond interaction with Glu65.

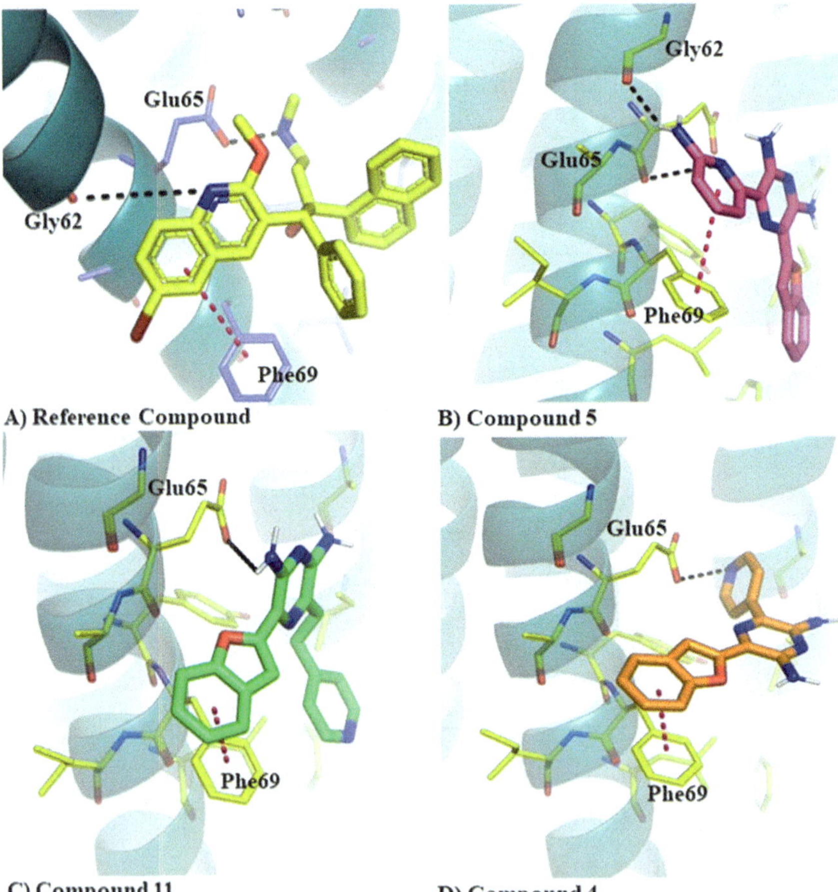

Fig. 5 Protein–ligand interactions of the selected best compounds and the reference compound BDQ

4 Conclusions

Mycobacterium tuberculosis, the causative agent of tuberculosis, is highly dependent on the ETC pathway for the generation of energy. In this study, we have designed and evaluated novel scaffolds as potential anti-tubercular agents targeting the inhibition of ATP synthase. Recently, few analogues of Amiloride have been explored against mycobacterial ATP synthase and a lead candidate HM2-16F has been reported to potently inhibit this enzyme and kill mycobacteria (MIC = 4 μM). Considering HM2-16F as a template compound, a library of the novel compounds has been generated and subjected to docking analysis against the mycobacterial F-type ATP synthase (PDB-ID: 4V1F). The designed new compounds are found to have optimal ADEMT properties and predicted binding affinity comparable to Bedaquiline. The benzofuran ring of compounds 4, 5, and 11 showed π-π interactions with Phe69 whereas compound 5 formed hydrogen bond with Gly62 (backbone) and with Glu65. Furthermore, compounds 4 and 11 showed hydrogen bond interaction with Glu65. Further synthesis and biological evaluation of the designed compounds may lead to a potential drug candidate against TB.

Acknowledgments A. K. is thankful to UPES, Dehradun for providing support in terms of doctoral fellowship and infrastructure. A.K. also expresses thankfulness to his colleague Ms. Divya Pandey for giving her support, as and when needed, in completing this work.

References

1. WHO Tuberculosis Report. https://www.who.int/teams/global-tuberculosis-programme/tb-reports (2023). Accessed 15 Apr 2024
2. Roy, K.K., Wani, M.A.: Emerging opportunities of exploiting mycobacterial electron transport chain pathway for drug-resistant tuberculosis drug discovery. Expert Opin. Drug Discov. **15**(2), 231–241 (2020)
3. Luo, M., Zhou, W., Patel, H., Srivastava, A.P., Symersky, J., Bonar, M.M., Faraldo-Gómez, J.D., Liao, M., Mueller, D.M.: Bedaquiline inhibits the yeast and human mitochondrial ATP synthases. Commun. Biol. **3**, 452 (2020)
4. Cox, E., Laessig, K.: FDA approval of Bedaquiline-the benefit-risk balance for drug-resistant tuberculosis. N. Engl. J. Med. **371**, 689–691 (2014)
5. Haagsma, A.C., Driessen, N.N., Hahn, M.M., Lill, H., Bald, D.: ATP synthase in slow- and fast-growing mycobacteria is active in ATP synthesis and blocked in ATP hydrolysis direction. FEMS Microbiol. Lett. **313**(1), 68–74 (2010)
6. Guo, H., Rubinstein, J.L.: Structure of ATP synthase under strain during catalysis. Nat. Commun. **13**, 2232 (2022)
7. Kalia, D., KS, A.K., Meena, G., Sethi, K.P., Sharma, R., Trivedi, P., Khan, S.R., Verma, A.S., Singh, S., Sharma, S., Roy, K.K., Kant, R., Krishnan, M.Y., Singh, B.N., Sinha, S., Chaturvedi, V., Saxena, A.K., Dikshit, D.K.: Synthesis and anti-tubercular activity of conformationally constrained and bisquinoline analogs of TMC207. MedChemComm. **6**(8), 1554–1563 (2015)

8. Singh, S., Roy, K.K., Khan, S.R., Kashyap, V.K., Sharma, A., Jaiswal, S., Sharma, S.K., Krishnan, M.Y., Chaturvedi, V., Lal, J., Sinha, S., Dasgupta, A., Srivastava, R., Saxena, A.K.: Novel, potent, orally bioavailable, and selective mycobacterial ATP synthase inhibitors that demonstrated activity against both replicating and non-replicating M. Tuberculosis. Bioorg. Med. Chem. **23**(13), 742 (2015)
9. Singh, S., Roy, K. K., Khan, S. R., Kashyap, V. K., Sharma, S. K., Krishnan, M. Y., Chaturvedi, V., Sinha, S. K., Shrivastava, R., Saxena, A. K.: N-(3-((diethylamino) methyl)-4-hydroxyphenyl)-n-(quinolin-4-yl) sulfonamides for the treatment of tuberculosis and process of preparation thereof, World Intellectual Property Organization International Bureau WO2013/102936 Al (2013)
10. Hards, K., Cheung, C.Y., Waller, N., Adolph, C., Keighley, L., Tee, Z.S., Harold, L.K., Menorca, A., Bujaroski, R.S., Buckley, B.J., Tyndall, J.D.A., McNeil, M.B., Rhee, K.Y., Opel-Reading, H.K., Krause, K., Preiss, L., Langer, J.D., Meier, T., Hasenoehrl, E.J., Berney, M., Kelso, M.J., Cook, G.M.: An amiloride derivative is active against the F_1F_O-ATP synthase and cytochrome *bd* oxidase of *Mycobacterium tuberculosis*. Commun. Biol. **5**, 166 (2022)
11. Gomes, A.R., Varela, C.L., Pires, A.S., Tavares-da-Silva, E.J., Roleira, F.M.F.: Synthetic and natural guanidine derivatives as antitumor and antimicrobial agents: A review. Bioorg. Chem. **138**, 106600 (2023)
12. Preiss, L., Langer, J.D., Yildiz, O., Eckhardt-Strelau, L., Guillemont, J.E.G., Koul, A., Meier, T.: Structure of the mycobacterial ATP synthase Fo rotor ring in complex with the anti-TB drug bedaquiline. Sci. Adv. **1**(4), 1–8 (2015)
13. Xiong, G., Wu, Z., Yi, J., Fu, L., Yang, Z., Hsieh, C., Yin, M., Zeng, X., Wu, C., Lu, A., Chen, X., Hou, T., Cao, D.: ADMETlab 2.0: an integrated online platform for accurate and comprehensive predictions of ADMET properties. Nucleic Acids Res. **49**(W1), W5–W14 (2021)

Natural Drugs to Combat Dengue Fever

Archana Sagar[1], Maheshwari Agari[1], and Sisir Nandi[1] (✉)

[1] Global Institute of Pharmaceutical Education and Research (GIPER), Affiliated to Veer Madho Singh Bhandari Uttarakhand Technical University, Kashipur, Uttarakhand, India

Abstract Dengue is a dangerous mosquito-borne, viral infection caused by the dengue virus. The female mosquitoes of the *Aedes* genus are the carriers of this virus. Dengue is a neglected tropical disease that can kill many lakhs of people around the world because there are no particular small molecules or specific chemotherapeutics discovered yet. Major symptoms such as high fever, vomiting, headache, body ache, redness of the skin, and head moving happen after 14 to 21 days of the infection. Only symptomatic treatment is given by acetaminophen, ivermectin, doxycycline, zanamivir, eltrombopag, multivitamins, and multiminerals to combat dengue. Early diagnosis followed by treatment is urgent. Yet, the road of success for the specific synthetic drug has not been achieved. To date, only the Dengvaxia vaccine has been developed to protect against dengue. The Mother Nature has a lot of medicine. The Indian traditional medicines such as *Azadirachta indica (Neem)*, *Carica papaya (Papaya)*, *Andrographis paniculata* (Kalmegh), *Euphorbia hirta*, *Cissampelos pareira* (velvet leaf), and animal-sourced goat milk are given to combat the dreadful attack of dengue virus.

Keywords Dengue · DENV · Synthetic drugs · Natural medicines

1 Introduction

Dengue virus (DENV), a member of the genus *Flavivirus* of the family Flaviviridae, a positive single-stranded RNA virus is a causative agent of Dengue fever. The virus is transmitted by the female *Aedes* mosquito. An infected mosquito bites a healthy person and the genome of the virus is transferred to the healthy person and becomes infected. This transmission and infection stage requires several processes. This infection depends on the transmission of the viral genome and some accessory proteins [1] as we know the dengue virus has an endemic impact that results in high morbidity and mortality rates. There are no specific treatments for this disease. Early reorganization and medical recommendations reduce the mortality rate a little bit [2]. Annually more than millions of people are affected by this virus. DENV has 4 serotypes—DENV 1, DENV 2, DENV 3, and DENV 4. Recently fifth serotype was discovered in the year 2013. DENV 2 is the most prevalent serotype in India over the past 50 years. In more than 110 countries dengue is endemic. There is no such treatment or vaccine for dengue. Nowadays, it has

A. K. Saxena, S. Nandi (eds.), *Global Trends in Health, Technology and Management*, https://doi.org/10.1007/978-3-031-75457-9_12

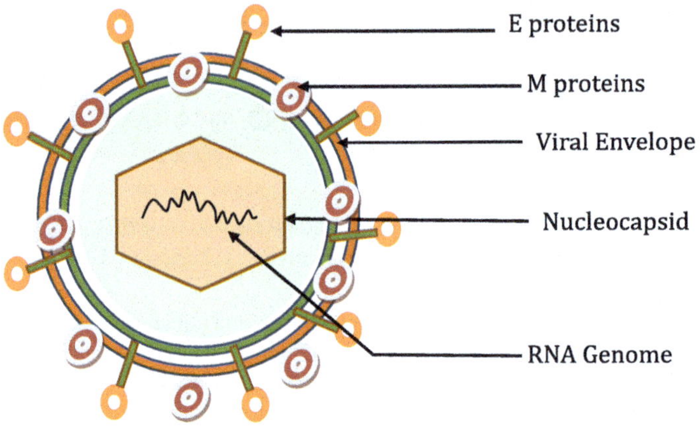

Fig. 1 Various targets of dengue virus

become a threatening disease for society. The mosquito-borne disease, dengue is caused by a positive sense RNA virus. It spreads in humans by *Aedes* mosquito, mainly *Aedes aegypti* [3]. Nowadays, dengue is a major global threat. Every year, more than 390 million viral infections and 20,000 deaths occur worldwide. The mortality rate of severe dengue disease is up to 5% [4]. Infections may be asymptomatic or symptomatic. Symptoms include fever, severe headache, macular rash, abdominal pain, vomiting, etc. The severe clinical manifestation of dengue is dengue hemorrhagic fever (DHF) or dengue shock syndrome (DSS) [5]. The asymptomatic dengue is a silent killer; therefore, the exploration of various targets of dengue virus is urgent need for the design of more potent drugs. Various targets are structural proteins and non-structural proteins. Capsid (C), Envelope (E), pre-membrane (prM) proteins are structural proteins and the non-structural proteins are NS1, NS2A, NS2B, NS3, NS4A, NS4B, and NS5 [6]. Both structural and non-structural proteins play different role which is essential for the survival of viruses.

Structural proteins are involved in viral capsid formation, membrane formation and maturation, and the formation of envelope. One of the largest non-structural proteins NS5 has RNA-dependent RNA polymerase and MTase activity. NS3 has helicase activity. Along with NS2B, NS3 has protease activity. During its infectious life cycle, it has to make large poly-protein which is translated from the viral RNA genome (~11 kb). The poly-protein is processed into three structural proteins, namely capsid, membrane protein {PrM/M}, and envelope protein and seven non-structural proteins (NS1, NS2A, NS2B, NS3, NS4A, NS4B, and NS5) by cellular and viral proteases [7] (Fig. 1).

Various targets have been taken into consideration to design and report various potential inhibitors in a recent review paper [8].

2 Synthetic Drugs to Combat the Dengue

Clinically dengue can be managed by fluid administration along with paracetamol as antipyretics at a dose of 10 mg/kg/dose. The maximum dose for an adult patient is 4 g/ day [9]. The JNJ-64281802) is an analog of JNJA07 developed by Janssen Pharmaceuticals. It has been registered for the phase II clinical trials to show patient efficacy for the prophylaxis is of dengue-injected individuals (NCT05201794) and for the treatment against dengue confirmed patient (NCT04906980). This drug can inhibit the interaction between NS4B and NS3 proteins responsible for the DENV viral replication [10]. The antiparasitic Ivermectin was shown to inhibit DENV1–4 serotypes in an in vitro experiment explaining inhibitory action against NS5 proteins [11]. In an in vivo model, it was given at a dose of 400 g/kg for 2–3 days in adult dengue-infected patients. A phase 2/3 clinical trial reported that this drug is effective against NS1 positive dengue patient [12].

Doxycycline is a broad-spectrum tetracycline congener that has shown inhibition toward N52B-N53 protease in vitro activity [13]. This drug was tested in randomized clinical trials or hospitalized dengue-positive patients in Brazil. It was shown to reduce viremia induced by proinflammatory cytokines IL-6 and TNF and promote faster recovery of platelet and leukocyte counts reducing the hospital stay [14].

Zanamivir, a neuraminidase inhibitor, was clinically trialed in randomized groups of DENV-infected patients. It has been shown to reduce DENV NS1-induced endothelial hyper permeability and vascular leakage (NCT04597437) [15]. Platelets reduction is a major issue for dengue patients. 25 mg eltrombopag, a thrombopoietin receptor agonist, was shown to increase platelet count for the 3 days of treatment. Natural medicines have been proven as effective measures to increase the platelet count [16].

3 Natural Drugs to Combat Dengue

The phytobioactives were reported to inhibit different serotypes of DENV and are simultaneously also being scientifically validated by researchers. The natural plant resources such as *Azadirachta indica (Neem)*, *Carica papaya (Papaya)*, *Andrographis paniculata* (Kalmegh), *Euphorbia hirta*, *Cissampelos pareira* (velvet leaf), and animal-sourced goat milk have been found effective and demonstrated improvement in clinical symptoms and direct inhibitory effect on dengue virus.

3.1 Neem

The active component of neem leaf extracts is nimbin which shows inhibitory activity against the envelope proteins of the dengue virus in all 4 serotypes. This study primarily reports on the inhibitory potential of pure neem chemical (azadirachtin) and crude aqueous extract of neem leaves on the reproduction process of dengue virus type-2, both in vitro and in vivo. Using a virus inhibition experiment, the antiviral efficacy of this aqueous neem leaf extract was evaluated in C(6/36) cells, which are cloned cells of

Aedes albopictus larvae, and the inhibition was dose-dependent. The absence of cytopathic effects demonstrated that the aqueous extract of neem leaves, at its maximum non-toxic dosage of 1.897 mg/ml, totally inhibited 100–10,000 TCID (50) of virus. Neem leaf extract at its highest non-toxic dosages of 120–130 mg/ml was used in in vivo protection experiments. The absence of clinical signs associated with Dengue in nursing mice and the absence of a virus-specific 511 bp amplicon in RT-PCR both demonstrated the inhibition of virus multiplication. In both in vitro and in vivo systems, pure neem, or azadirachtin, did not show any suppression on Dengue virus type-2 replication. Although neem is being studied for its possible efficacy in managing dengue viruses, no conclusive data from clinical trials have been obtained to yet. Neem may have antiviral qualities, according to several studies, but further investigation is required to fully understand its safety and efficacy in treating dengue virus infections, particularly through clinical trials [17].

3.2 Papaya

Carica papaya leaf extract having two significant active compounds such as papain and chymopapain is an anti-inflammatory drug that promotes megakaryocyte development and platelet creation. It reduces the generation of proinflammatory cytokines and increases platelet count in vivo. Three randomized controlled studies found that platelet counts improved.

Carica papaya (CP) extract is gaining popularity as an unregulated herbal medicine that claims to speed up recovery from dengue infection, mostly due to studies that it may enhance platelet counts. This systematic review and meta-analysis will critically examine the evidence from controlled clinical studies on the efficacy and safety of CP extract in the treatment of dengue illness. The inclusion criteria were met by nine studies (India-6, Pakistan-1, Indonesia-1, and Malaysia 1). In seven investigations, patients who received CP extract had higher platelet counts; in one, there was no significant difference between the two groups; in the remaining three, direct comparison was not possible. No reports of serious adverse events were made. A study with 580 participants found that the mean difference in hospital stay duration (mean difference −1.98 days, 95% confidence interval −1.83 to −2.12) and mean platelet counts (mean difference 35.45, 95% confidence interval 23.74 to 47.15, 3 studies, 129 participants, low-quality evidence) may be improved by CP extract.

There was no information available about any other clinical outcome. The early phases of clinical trials are being conducted to examine the potential of papaya in treating dengue. More thorough clinical trials are required to validate the efficacy and safety of papaya leaf extract in treating dengue fever, despite some evidence suggesting that it may have benefits, such as raising platelet count [18].

Carica papaya is one of the several therapeutic plants with anti-DENV potential that has undergone extensive clinical testing. It was discovered that giving patients 25 milliliters of plant aqueous leaf extract increased the levels of platelets, white blood cells, and neutrophils in their blood, which lessened the severity of dengue fever. Similarly, two doses (5 mL) of C. papaya leaf extract given at 8 h intervals along with standard dengue symptomatic care led to an increase in platelet count and white blood cell count within 24 h of treatment in a clinical study on 12 patients with dengue fever and a platelet count less than 130,000/mm3.

In addition, compared to the control group, the mean platelet count of 228 patients with dengue fever and dengue hemorrhagic fever increased after receiving 30 mL of Carica papaya leaf juice orally for 3 days in a row. As a result, it might minimize these patients' hospital stays and avoid the consequences of thrombocytopenia [19].

3.3 Kalmegh

Kalmegh contains major active components such as quinic acid, diterpenoids, and flavonoids that act as anti-dengue agents. The antiviral activities of andrographolide against DENV-2 infected HepG2 and HeLa cells as well as DENV-4 with HepG2 cells were assessed. Significant anti-DENV serotype 2 and 4 effects were demonstrated by andrographolide on both HepG2 and HeLa, resulting in decreased viral output and cellular infection with 50% effective concentrations (EC50) of 22.739 μM and 21.304 μM for HepG2 and HeLa, respectively. However, it has been observed that andrographolide dramatically lowers the amount of virus produced, the number of copies of the DENV-4 genome, and the expression of the DENV-4 E protein in HepG2 cells from the first day after infection (Panraksa et al. Citation 2017). Time addition studies carried out by treating andrographolide to: (i) pre-treat cells before infection for 1 h, 30 min, or 15 min, (ii) post-treat infected cells at 0, 1, 2, 3, 4, 6, 9, 12 h after infection showed that addition of andrographolide post-infection stage produced marked effects as compared to insignificant results showed in pre-treat cells (Panraksa et al. Citation2017). It was evidenced that the disruptive effects of Andrographis paniculata extracts and andrographolide were via binding of DENV-2 and DENV-4 E protein to actin and several cellular pathways (Jitoboam et al. Citation2016; Panraksa et al. Citation2017). A follow-up proteomic study carried out by Paemanee et al. found that the glucose-regulated protein 78 (GRP78) was significantly down-regulated in DENV-2 infected andrographolide-treated HepG2 cells as compared to control [20]. GRP78 functioned as a chaperone in protein folding in ER and a mediator of unfolded protein response which was manipulated by DENV to benefit its viral production (Pena and Harris Citation2011). This suggested that GRP78 and unfolded protein response played an important role in mediating andrographolide anti-DENV activity which may help in explaining the broad antiviral pathway of andrographolide. Andrographis paniculata, commonly known as Kalmegh, has shown promise in preclinical studies for its potential antiviral activity against dengue virus. However, the clinical stage of Kalmegh specifically in dengue treatment can vary. Some clinical trials may be ongoing or planned to evaluate its efficacy and safety in treating dengue fever, but specific information about the current clinical stage would require up-to-date research or trial registries [19].

3.4 Euphorbia hirta

Different parts of *Euphorbia hirta (E. hirta)* usage can cure dengue fever and it is used as a natural drug across different regions. Interim uses of barks and roots alleviate fever, headache, skin blister, and mouth bleeding which are allied with dengue fever.

Somewhere the leaves and stems of *E. hirta* are used by boiling down [19]. One of the major clinical problems in dengue fever is a variation in platelet count. Leaf extract of *E. hirta* contains phenolics and flavonoids responsible for enhancing the platelet count in several models in animal that are affected in dengue fever.

In addition to lowering the thrombocytopenic rats' clotting time, when *E. hirta* leaves were administered orally to AG129 dengue-infected mice, they also raised plasma monocyte chemoattractant protein-1 (MCP-1) levels during the peak of viremia, indicating the potential immunomodulatory ability of this plant during DENV infection. Animal models could be used to study the impact of medicinal plants on the pathophysiology of dengue with respect to the mechanisms governing the causes of capillary leakage and a drop in platelet count. A variety of animal models, including transgenic or knockout mice, can be used to assess the anti-DENV efficacy of treatment regimens. Oral administration of E. hirta aqueous extract to dengue fever patients increased platelet and total leukocyte counts, but had no discernible effect on hematocrit, according to a clinical investigation. There is not much clinical research on the anti-DENV properties of medicinal herbs. To verify whether extracts and/or the bioactive chemicals they have isolated are effective against DENV, more research of this kind is required [21].

3.5 *Cissampelos pareira* (Velvet Leaf)

To study the native herbal bio-resource and find plants with pan-DENV inhibitory activity, we created a systematic bioassay-guided screening approach, drawing on our knowledge of traditional Indian medicine, or Ayurveda. The alcoholic extract of *Cissampelos pareira* (Cipa extract) was a powerful inhibitor of all four DENVs, as measured by viral replication as determined by plaque assays and viral NS1 antigen secretion as determined by ELISA. Assays for virus production reduction demonstrated that Cipa extract could significantly lower viral titers. Using the AG129 mouse model, the extract provided statistically significant protection against DENV infection. Velvet leaf was also used to raise the platelet count and erythrocyte count in DENV-infected mice [22].

3.6 Goat Milk

Goat milk contains selenium as a microcomponent that produces T cells and interleukins responsible for immunity and can inhibit DENV infectivity by interfering with its replication. To understand the mechanisms of GME-mediated inhibition of DENV infection, we first evaluated its effect on DENV replication. Vero cells were infected with DENV followed by treatment with 50 μg/ml of GME or HiGME for 36 h. Subsequently, cells were fixed and stained with J2 antibody for visualizing the accumulation of dsRNA intermediate formed during the DENV replication cycle. Our immunofluorescence results indicate a significant reduction in the accumulation of DENV dsRNA intermediate in the cytoplasm of infected cells treated with GME when compared to only DENV-infected cells [23].

4 Conclusion

Since there are no particular anti-dengue viral drugs developed yet, to save the patient's life, we are solely dependent on the mother natural resources along with the repurposed synthetic compounds. For the betterment of treatment in case of drug-resistant strain, the natural medicines might be very helpful. The natural medicines such as *Azadirachta indica (Neem)*, *Carica papaya (Papaya)*, *Andrographis paniculata* (Kalmegh), *Euphorbia hirta, Cissampelos pareira* (velvet leaf) and animal-sourced goat milk have been shown as potent anti-dengue treatment measurement. The natural components isolated from the above medicines could be taken as scaffold for the design and development of further congeneric potent compounds to overcome all the difficulties against dengue virus. By this study we can establish a molecular tool for the treatment of dengue fever which will be easiest and cheapest treatment of the patients.

References

1. Shimizu, H., Saito, A., Mikuni, J., Nakayama, E.E., Koyama, H., Honma, T., Shirouzu, M., Sekine, S., Shioda, T.: Discovery of a small molecule inhibitor targeting dengue virus NS5 RNA-dependent RNA polymerase. PLoS Negl. Trop. Dis. **13**(11), e0007894 (2019)
2. Item. https://www.who.int/emergencies/disease-outbreak-news/item/2022-DON412 (2022, October 10)
3. Chagas, G.C.L., Rangel, A.R., Noronha, L.M., Veloso, F.C.S., Kassar, S.B., Oliveira, M.J.C., Meneses, G.C., da Silva Junior, G.B., Daher, E.F.: Risk factors for mortality in patients with dengue: A systematic review and meta-analysis. Trop. Med. Int. Health. **27**(8), 656–668 (2022)
4. World Health Organization: WHO & World Health Organization: WHO. *Dengue and severe dengue*. https://www.who.int/news-room/fact-sheets/detail/dengue-and-severe-dengue (2024, April 23)
5. Norazharuddin, H., Lai, N.S.: Roles and prospects of dengue virus nonstructural proteins as antiviral targets: an easy digest. Malays. J. Med. Sci. **25**(5), 6–15 (2018)
6. Nanaware, N., Banerjee, A., Bagchi, S.M., Bagchi, P., Mukherjee, A.: Dengue virus infection: a tale of viral exploitations and host responses. Viruses. **13**(10), 1967 (2021)
7. Xu, T., Sampath, A., Chao, A., Wen, D., Nanao, M.H., Chene, P., Vasudevan, S.G., Lescar, J.: Structure of the dengue virus helicase/nucleoside triphosphatase catalytic domain at a resolution of 2.4 A. J. Virol. **79**(16), 10278–10288 (2005)
8. Nath, S., Malakar, P.K., Biswas, B., Das, S., Shabnam, N., Nandi, S., Samadder, A.: Exploring the targets of dengue virus and designs of potential inhibitors. Comb. Chem. High Throughput Screen. **27**(17), 2485–2524 (2023). https://doi.org/10.2174/011386207324768923103015305 4
9. Vasikasin, V., Rojdumrongrattana, T., Chuerboonchai, W., Siriwiwattana, T., Thongtaeparak, W., Niyasom, S., Lertliewtrakool, N., Jitsiri, S., Changpradub, D.: Effect of standard dose paracetamol versus placebo as antipyretic therapy on liver injury in adult dengue infection: a multicentre randomised controlled trial. Lancet Glob. Health. **7**(5), e664–e670 (2019)
10. Kaptein, S.J.F., Goethals, O., Kiemel, D., Marchand, A., Kesteleyn, B., Bonfanti, J., Bardiot, D., Stoops, B., Jonckers, T.H.M., Dallmeier, K., Geluykens, P., Thys, K., Crabbe, M., Chatel-Chaix, L., Münster, M., Quérat, G., Touret, F., De Lamballerie, X., Raboisson, P., Simmen,

K., Chaltin, P., Bartenschlager, R., Loock, M.V., Neyts, J.: A pan-serotype dengue virus inhibitor targeting the NS3–NS4B interaction. Nature. **598**(7881), 504–509 (2021)
11. Tay, M.Y.F., Fraser, J.E., Chan, W.K., Moreland, N.J., Rathore, A.P.S., Wang, C., Vasudevan, S.G., Jans, D.A.: Nuclear localization of dengue virus (DENV) 1–4 non-structural protein 5; protection against all 4 DENV serotypes by the inhibitor Ivermectin. Antivir. Res. **99**(3), 301–306 (2013)
12. Suputtamongkol, Y., Avirutnan, P., Mairiang, D., Angkasekwinai, N., Niwattayakul, K., Yamasmith, E., Saleh-Arong, F.A., Songjaeng, A., Prommool, T., Tangthawornchaikul, N., Puttikhunt, C., Hunnangkul, S., Komoltri, C., Thammapalo, S., Malasit, P.: Ivermectin accelerates circulating nonstructural protein 1 (NS1) clearance in adult dengue patients: a combined phase 2/3 randomized double-blinded placebo controlled trial. Clin. Infect. Dis. **72**(10), e586–e593 (2021)
13. Rothan, H.A., Zeriouh, M., Paydar, M., Rahman, N.A., Yusof, R.: Inhibitory effect of doxycycline against dengue virus replication in vitro. Arch. Virol. **159**(4), 711–718 (2013)
14. Fredeking, T., Zavala-Castro, J., González-MartíNez, P., Moguel-Rodríguez, W., Sánchez, E.S., Foster, M.D., Díaz-Quijano, F.A.: Dengue patients treated with doxycycline showed lower mortality associated to a reduction in IL-6 and TNF levels. Recent Pat. Antiinfect. Drug Discov. **10**(1), 51–58 (2015)
15. Glasner, D.R., Ratnasiri, K., Puerta-Guardo, H., Espinosa, D.A., Beatty, P.R., Harris, E.: Dengue virus NS1 cytokine-independent vascular leak is dependent on endothelial glycocalyx components. PLoS Pathog. **13**(11), e1006673 (2017)
16. Chakraborty, S., Alam, S., Sayem, M., Sanyal, M., Das, T., Saha, P., Sayem, M., Byapari, B.K., Tabassum, C.T., Kabir, A., Amin, M.A., Nabi, A.N.: Investigation of the efficacy and safety of eltrombopag to correct thrombocytopenia in moderate to severe dengue patients – a phase II randomized controlled clinical trial. EClinicalMedicine. **29–30**, 100624 (2020)
17. Parida, M., Upadhyay, C., Pandya, G.A., Jana, A.M.: Inhibitory potential of neem (AzadirachtaindicaJuss) leaves on dengue virus type-2 replication. J. Ethnopharmacol. **79**(2), 273–278 (2002)
18. Rajapakse, S., De Silva, N.L., Weeratunga, P., Rodrigo, C., Sigera, C., Fernando, S.D.: Carica papaya extract in dengue: a systematic review and meta-analysis. BMC Complement. Altern. Med. **19**, 265 (2019)
19. Saleh, M.S.M., Kamisah, Y.: Potential medicinal plants for the treatment of dengue fever and severe acute respiratory syndrome-coronavirus. Biomol. Ther. **11**(1), 42 (2020)
20. Lim, S.Y.M., Chieng, J.Y., Pan, Y.: Recent insights on anti-dengue virus (DENV) medicinal plants: review on in vitro, in vivo and in silico discoveries. All Life. **14**(1), 1–33 (2021)
21. De Guzman, G.Q., Dacanay, A.T.L., Andaya, B.A., Alejandro, G.J.D.: Ethnopharmacological studies on the uses of Euphorbia hirta in the treatment of dengue in selected indigenous communities in Pangasinan (Philippines). J. Intercult. Ethnopharmacol. **5**(3), 239 (2016)
22. Sood, R., Raut, R., Tyagi, P., Pareek, P.K., Barman, T.K., Singhal, S., Shirumalla, R.K., Kanoje, V., Ramesh, S., Rajerethinam, R., Sharma, N., Kanaujia, A., Shukla, G., Gupta, Y.K., Katiyar, C.K., Bhatnagar, P.K., Upadhyay, D.J., Swaminathan, S., Khanna, N.: Cissampelospareira Linn: natural source of potent antiviral activity against all four dengue virus serotypes. PLoS Negl. Trop. Dis. **9**(12), e0004255 (2015)
23. Yenuganti, V.R., Afroz, S., Khan, R.A., Bharadwaj, C., Nabariya, D.K., Nayak, N.S., Subbiah, M., Chintala, K., Banerjee, S., Reddanna, P., Khan, N.: Milk exosomes elicit a potent antiviral activity against dengue virus. J. Nanobiotechnol. **20**(1), 317 (2022)

Development and Assessment of Nanostructured Lipid Carriers Loaded with *Calotropis gigantea* Extract for the Enhancement of Therapeutic Effect

Disha Dutta[1] (✉), Ramsha Aslam[1], and Anjali Toliya[2]

[1] Department of Pharmaceutics, Devsthali Vidyapeeth College of Pharmacy, Affiliated to Veer Madho Singh Bhandari Uttarakhand Technical University, Rudrapur, India
[2] Department of Pharmacology, School of Pharmaceutical Sciences, Sardar Bhagwan Singh University, Dehradun, India

Abstract The aim of the present study was to formulate and evaluate nanostructured lipid carriers (NLC) loaded with *Calotropis gigantea* (giant milkweed) extract prepared by a modified micro-emulsification technique followed by ultrasonication. Nanocarriers have several advantages, such as enhanced solubility and bioavailability of hydrophobic molecules, reduced dose by controlled and sustained release of therapeutic agents, and attenuated drug toxicity. NLC was developed using stearic acid as a solid lipid, oleic acid as a liquid lipid, Tween 80 as a surfactant, and Phospholipon 80H as a stabilizer. The formulated NLC was characterized for particle size, polydispersibility index (PDI), zeta potential, morphology, drug content, entrapment efficiency, and in-vitro drug release. The optimized NLC formulation exhibited a particle size of 210.46 nm with PDI of 0.256 and zeta potential of −20.3 mV. Scanning electron microscopy showed smooth spherical morphology. Optimized formulation demonstrated drug content and drug loading capacity of 90.52% and 98.29%, respectively. In-vitro drug release exhibited a sustained release profile, with 75.64% cumulative drug release over 28 h following Higuchi release kinetics. Stability studies showed insignificant changes in evaluated parameters over 90 days demonstrating good stability of optimized NLC. The study demonstrates the successful development of stable *Calotropis gigantea* loaded NLC with suitable physicochemical properties for topical application.

Keywords *Calotropis gigantea* · Nanostructured lipid carriers · Wound healing · Micro-emulsification · Topical delivery

A. K. Saxena, S. Nandi (eds.), *Global Trends in Health, Technology and Management*, https://doi.org/10.1007/978-3-031-75457-9_13

1 Introduction

Calotropis gigantea, commonly known as the crown flower or giant milkweed, is a medicinal plant native to tropical regions of Asia and Africa. Traditionally, various parts of *Calotropis gigantea* have been utilized in folk medicine for their diverse therapeutic properties, including anti-inflammatory, analgesic, antimicrobial, and anticancer effects. One of the key challenges in harnessing the medicinal potential of *Calotropis gigantea* lies in optimizing its delivery system to enhance bioavailability, stability, and therapeutic efficacy [1].

Nanostructured lipid carriers (NLC) have emerged as promising drug delivery systems due to their ability to encapsulate both hydrophilic and hydrophobic compounds, improve drug solubility and bioavailability of poorly water-soluble drugs, and facilitate sustained release [2]. The rational design and formulation of NLC offer a strategic approach to address the limitations associated with conventional dosage forms as well as other novel formulations and maximize the therapeutic benefits of natural products. In the present study, we aimed to develop and evaluate NLC loaded with *Calotropis gigantea* leaves extract and its characterization [3], with a focus on enhancing its pharmacokinetic profile and therapeutic potential. The rationale behind this work stems from keeping considerations like enhanced bioavailability: The bioactive compounds present in Calotropis gigantea extract often exhibit poor aqueous solubility, limiting their absorption and bioavailability upon oral administration. Formulating the extract into NLC can improve its solubility, leading to enhanced absorption and systemic exposure when given through the topical route [4]. Natural products, including plant extracts, are prone to degradation, oxidation, and loss of potency during storage and transportation. Encapsulation of *Calotropis gigantea* extract within NLC can protect it from environmental factors and enhance its stability, thereby prolonging shelf life and ensuring consistent therapeutic efficacy [5]. NLC offer the advantage of targeted drug delivery, allowing for site-specific accumulation of the therapeutic agent and minimizing off-target effects. By incorporating Calotropis gigantea extract into NLC, it may be possible to achieve targeted delivery to specific tissues or cells, enhancing therapeutic outcomes while reducing systemic toxicity [6]. *Calotropis gigantea* contains a complex mixture of bioactive compounds, including flavonoids, alkaloids, and glycosides, which may exert synergistic effects when delivered together. Formulating the extract into NLC enables the preservation of these natural synergies and may potentiate the therapeutic effects observed in traditional medicine practices [7, 8]. Moreover, NLC can provide sustained and controlled release of herbal compounds, leading to prolonged exposure and improved absorption over time. This can result in higher and more consistent plasma levels, enhancing the bioavailability and therapeutic efficacy of the compounds [9, 10]. NLC can enhance the solubility and dissolution rate of poorly water-soluble herbal compounds. The lipid matrix of NLC can solubilize these hydrophobic compounds, increasing their effective surface area and facilitating better dissolution in the gastrointestinal fluids [11].

2 Materials and Methods

The leaves of *Calotropis gigantea* were collected from the Haldwani city growing in the wild. The plant was identified and authenticated by the Botanical Survey of India 192, Kaulagarh, Dehradun. This has a voucher specimen number 1275/2023 Calotropis gigantea. The leaves were collected from the plant. It was cleaned, dried under shade, and crushed to get coarse powder. The extract of dried leaves powder was prepared at Devsthali Vidyapeeth College of Pharmacy, Rudrapur.

Materials: Stearic acid, oleic acid, and Phospholipon 80H were procured from HiMedia and Yarrow Chem. Tween 80 from Finar.

2.1 Extraction Process

Fresh and healthy leaves of the *Calotropis gigantea* plant were chosen for extraction. All of the gathered leaves were properly cleansed with distilled water to eliminate any dirt or pollutants. The leaves were dried in the shade and crushed into a coarse powder. For the maceration procedure, 50 gm of dried leaf powder was put in a beaker holding 500 ml of a solvent combination of 80% ethanol and 20% distilled water and left at room temperature for 1 week. The beaker was closed with foil paper. The mixture was stirred periodically (Figs. 1 and 2). After 7 days, the ethanolic extract was filtered through a muslin cloth, the marc was pressed, and the filtrate was stored in a vacuum oven at 40 °C for evaporation [12, 7].

Fig. 1 *Calotropis gigantea* plant and dried crushed leaves (Self)

Fig. 2 Extract of *Calotropis gigantea* leaves (Self)

2.2 Physicochemical Evaluation of Crude Drug

2.2.1 Loss on Drying

Three China plates were collected and weighed. Each China dish received 1.5 gm of powdered drugs. The weight of each China dish was then measured again, and the plates were placed in an oven set to 40–45 °C. After 20 min in the oven, the China plates were taken out and placed on a desiccator for 5 min. The China plates were weighed again, and the loss from drying was determined. Again, the China plates were maintained in the oven at 40–45 °C. After 10 min, the same method was repeated, and the loss on drying was determined [13].

2.2.2 pH Measurement

Two different concentrations of solutions were prepared as 1% w/v and 10% w/v and the pH was checked using a pH meter [13].

2.2.3 Swelling Index

1 gm of crushed drug was dissolved in 25 ml of distilled water with continuous shaking for 30 min, then set aside for 24 h, and the height was measured to determine swelling [13] (Fig. 3).

2.2.4 Foaming Index

1 gm of crushed drug was added to 100 ml of boiling water and held at the optimal temperature. After 30 min, the mixture was cooled and filtered. The filtrate was diluted in increments of 1 ml, 2 ml, 3 ml, and so on until a volume of 10 ml was reached, shaken for 15 s, and set aside for 15 min. The foam's length was then examined [13].

Fig. 3 Determination of swelling index (Self)

2.2.5 Ash Value

Three silica crucibles were collected and weighed. Each silicon crucible received 2 gm of the drug and was weighed again. The silica crucibles were then placed in a muffle furnace. After 45 min, the silica crucibles were removed and stored in a desiccator. The silica crucibles were then weighed once again. The ash produced was then used to compute the water and acid-soluble ash values [13].

- **Acid-Soluble Ash Value** The ash obtained from above was then dissolved in 25 ml of conc. HCl and boiled for 5 min. The boiled mixture was then filtered using the ash-less filter paper. Then the boiled water was added to the residue. The residue was then again transferred to the silica crucible and the silica crucible was again kept in the muffle furnace for 45 min after then it was weighed [13].
- **Water-Soluble Ash Value** The ash obtained was then dissolved in 25 ml of distilled water and boiled for 5 min. The boiled mixture was then filtered using the ash-less filter paper. Then the boiled water was added to the residue (Fig. 4). The residue was then again transferred to the silica crucible and the silica crucible was again kept in the muffle furnace for 45 min after then it was weighed [13].

2.3 Fabrication of NLC

The liquid phase was created by melting the solid lipid at 10 °C above its melting point. The determined amount of liquid lipid was then added to the melted solid lipid, followed by gradually adding the medication while stirring constantly to produce a

Fig. 4 Silica crucible with crude drug in the muffle furnace (Self)

Table 1 Ratio of five different formulations

S.no.	Drug	Solid lipid: liquid lipid	SPAN 20	TWEEN 80	Water
1.	100 mg	4:6	2%	2%	50 ml
2.	100 mg	5:5	2%	2%	50 ml
3.	100 mg	6:4	2%	2%	50 ml
4	100 mg	7:3	2%	2%	50 ml
5	100 mg	8:2	2%	2%	50 ml

transparent solution. The lipid and water phases were heated to the same temperature. The water phase contained Tween 80 as the detergent and Phospholipon® 80H (1 percent w/v) as a stabilizer. To create a pre-emulsion, the watery phase was progressively mixed with the oily phase while shaking at the same temperature. The heated pre-emulsion was then shaken using a sonicator. The heated-up, homogeneous pre-emulsion was ultrasonically treated with a probe sonicator (PCI Analytics) using a 20:2 on/off cycle [14, 15]. The prepared o/w nanoemulsion was allowed to cool to room temperature so that the lipid phase may recrystallize and contribute to the formation of NLC. Using a high-speed cooling centrifuge (REMI), the unbound medication in the NLC aqueous mixture was removed. Five different formulations were prepared by changing the ratio of solid lipid and liquid lipid as shown in Table 1.

2.4 Characterization of *Calotropis gigantea* Loaded NLC

2.4.1 Drug Content

From the prepared formulation, 1 ml was taken and the volume was made up to 10 ml with methanol. Stir it for 15 min into the magnetic stirrer. Filter it with 0.2 filter paper. Draw 1 ml from this solution and make up the volume with methanol.

2.4.2 Entrapment Efficiency

From the prepared formulation, 1 ml was taken and volume was made up to 10 ml with methanol and refrigerated-centrifuged for 30 min at 1500 rpm 2–4 ° C. Filter it with 0.2 filter paper. Draw 1 ml from this solution and make up the volume up to 10 ml with methanol.

2.4.3 In-Vitro Drug Release

In-vitro drug release tests were performed on the optimized batch at a temperature of 35.5 °C using a dialysis bag technique with a dialysis membrane in phosphate buffer pH 7.4 [16, 17]. After being accurately weighed out, *Calotropis gigantea*-loaded NLC dispersions were put into a dialysis bag and then the bag was put into use. Following that, the bag was placed inside a beaker that contained 250 ml of a physiological solution [18, 19]. At various time intervals of 2, 3, 4, 5, 6, 24, and 28 h, 1 ml of the receptor phase was removed and replaced with an equal volume of new fluid. A UV spectrophotometer set at 261 nm was used to evaluate each of the samples [20].

2.4.4 Size Distribution and Polydispersity Index Study

Photon correlation spectroscopy (PCS) with a zeta sizer at a set angle was used to figure out the average diameter and PDI of the made formulation. The mean diameter and PDI of the samples under study were found by taking the average of three measures. All of the tests were done when it was 25 °C outside [21].

2.4.5 Zeta Potential Analysis

The electrophoretic mobility of the *Calotropis gigantea*-loaded NLC dispersion was evaluated with the help of the Malvern Zetasizer, and the zeta potential of the dispersion was computed [21, 22]. The measurements of each sample were performed three times.

2.4.6 Shape and Surface Morphology

The morphology of the optimized *Calotropis gigantea*-loaded NLC formulation was analyzed using SEM [23, 24].

2.4.7 Drug Release Kinetics

The data on drug release was mathematically processed using the zero-order model, the first-order model, the Higuchi diffusion model, and the Korsmeyer-Peppas model so that the drug release mechanism could be described [25].

2.4.8 Stability

The stability tests were carried out for 3 months at temperatures ranging from 2 to 8 °C and at room temperature with 60% RH [25].

3 Results and Discussion

Calotropis gigantea has various components demonstrating a wide spectrum of pharmacological activities, including anti-inflammatory, analgesic, anticonvulsant, and antioxidant effects. To harness the full potential of these bioactive compounds, the

Table 2 Physicochemical evaluation of crude drug

S.no	Evaluation parameters		Result
1	Loss on drying		>1%
2	pH		6.8
3	Foaming index		No foaming found
4	Swelling index		2.5 ml
5	Ash value	Water soluble	10%
		Acid soluble	12%

development of efficient and targeted drug delivery systems is crucial. Nanostructured lipid carriers (NLCs) emerge as a promising platform for this purpose. NLCs can accommodate lipophilic and hydrophilic compounds, improving the solubility of *Calotropis gigantea* extracts, which often contain a complex mixture of compounds with varying polarities. Moreover, the unique lipid composition of NLCs can enhance the permeability of the drug across biological barriers, leading to better absorption and distribution. NLCs can accommodate a wide range of drug loadings, enabling optimization of therapeutic effects. The versatility of NLCs allows for their application in various drug delivery systems, including oral, topical, and parenteral formulations. A number of evaluation tests were performed and results of individual results were discussed below along with the respective results which confirm that the extract was efficiently entrapped into the formulation and release in a controlled manner (Table 2).

3.1 Physicochemical Evaluation of Crude Drug

3.2 Characterization of NLC

The optimized NLC formulation exhibited a particle size of 210.86 nm with a PDI of 0.256, indicating a narrow size distribution. The zeta potential was −20.3 mV, suggesting good stability of the NLC dispersion. This indicates the successful formation of nanosized particles, which is beneficial for drug delivery applications. Particles in this size range have been shown to improve drug delivery efficiency, increase cellular uptake, and enhance bioavailability due to their small size and increased surface area as shown in Figs. 5 and 6.

3.3 Determination of Drug Content & Drug Entrapment

The drug content of the above formulation was found to be in the range of 82.57–98.29% and drug entrapment was found to be in the range of 82.36–90.52% in which Formulation F3 (Solid Lipid Ratio 6:4) shows optimum drug content 98.29% with maximum entrapment 90.52%. The highest drug entrapment of F3 formulation may be due to the perfect blend of solid and liquid lipids (Table 3).

			Size (d.n...	% Intensity:	St Dev (d.n...
Z-Average (d.nm):	210.86	Peak 1:	210.86	100.0	40.67
PdI:	0.256	Peak 2:	0.000	0.0	0.000
Intercept:	0.928	Peak 3:	0.000	0.0	0.000
Result quality	Good				

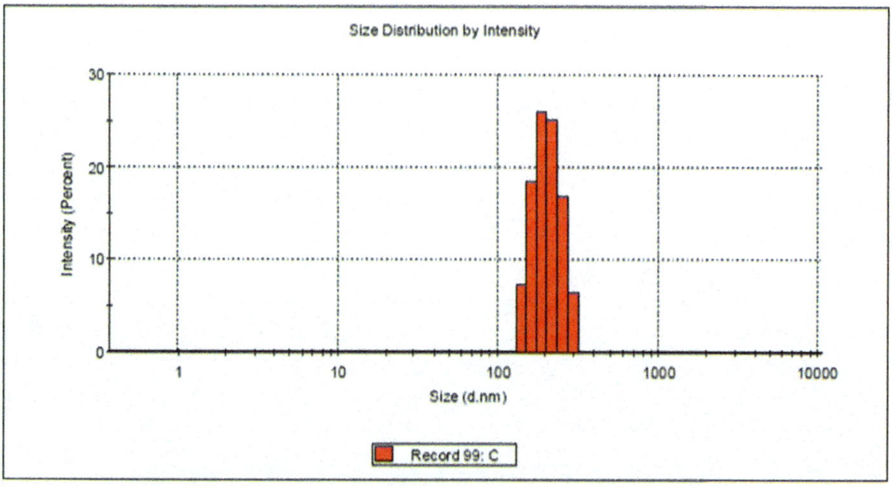

Fig. 5 Polydispersity index of the optimized formulation

			Mean (mV)	Area (%)	Width (mV)
Zeta Potential (mV):	-20.3	Peak 1:	-20.3	99.8	4.54
Zeta Deviation (mV):	4.73	Peak 2:	0.00	0.2	0.00
Conductivity (mS/cm):	0.105	Peak 3:	0.00	0.0	0.00
Result quality :	See result quality report				

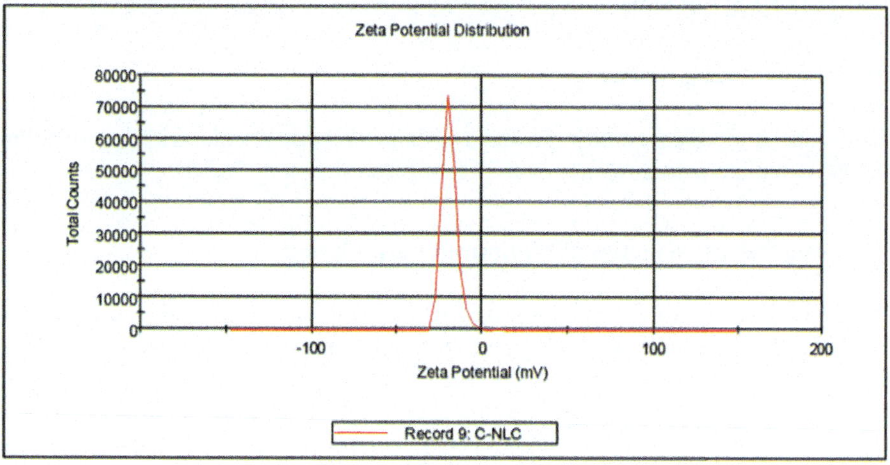

Fig. 6 Zeta potential of optimized formulation

Table 3 Determination of drug content & drug entrapment

S.no.	Formulation ratio	Drug content	Drug entrapment
1.	F1	82.57%	82.36%
2.	F2	91.32%	80.54%
3.	F3	98.29%	90.52%
4	F4	98.23%	88.34%
5	F5	92.34	78.93%

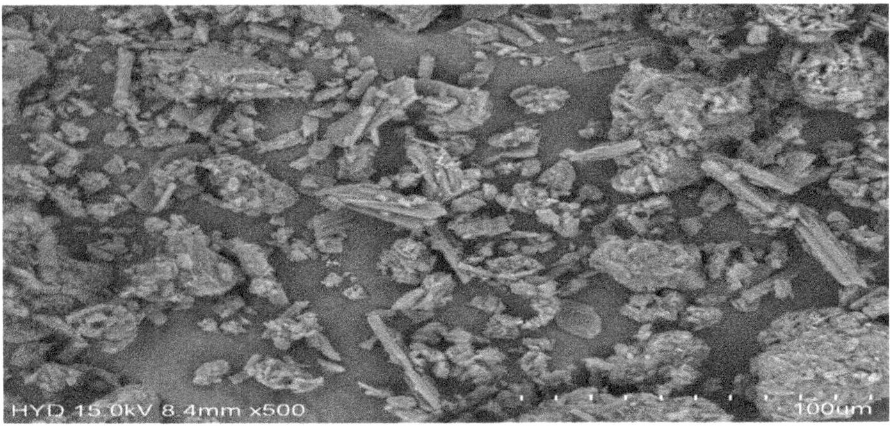

Fig. 7 Scanning electron microscopy of the optimized formulation

3.4 Scanning Electron Microscopy

The optimized batch of NLC was lyophilized and used for the SEM analysis. The morphology of NLC was determined as shown in Fig. 7.

3.5 In-Vitro Drug Release

The in-vitro drug release study showed a sustained release profile, with a maximum of 75.64% cumulative drug release over 28 h of F3 as shown in Fig. 8.

3.6 Drug Release Kinetics of Optimized Formulation

The drug release profile of the optimized NLC formulation was analyzed using many different kinetic models, including the zero-order model, the first-order model, the Korsmeyer-Peppas model, the Higuchi model, and the Hixson-Crowell model (Fig. 9). Based on the value of R^2, the Higuchi model (Table 4 and 5) is most suited to reflect the release. According to extant research, the Higuchi model depicts the release process as a diffusion mechanism regulated by a matrix. The release kinetics followed the Higuchi model, indicating diffusion-controlled drug release from the NLC.

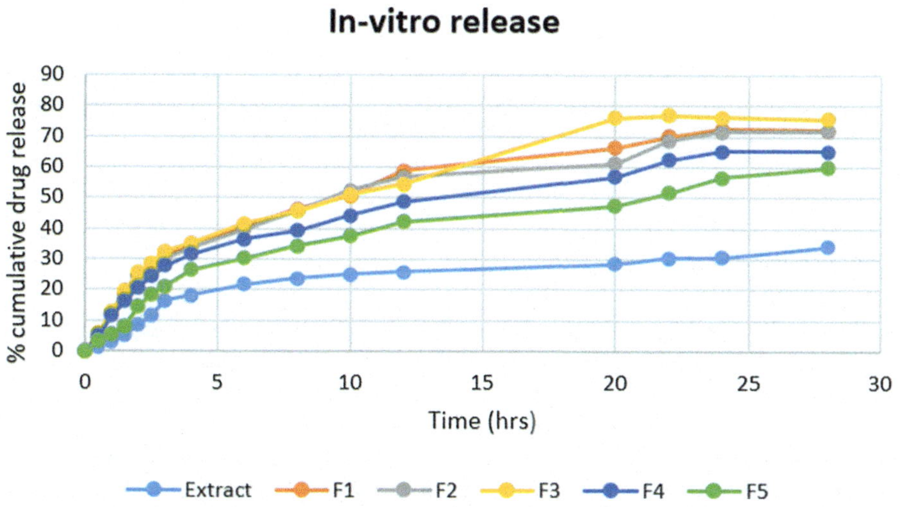

Fig. 8 In-vitro drug release profile of different formulations

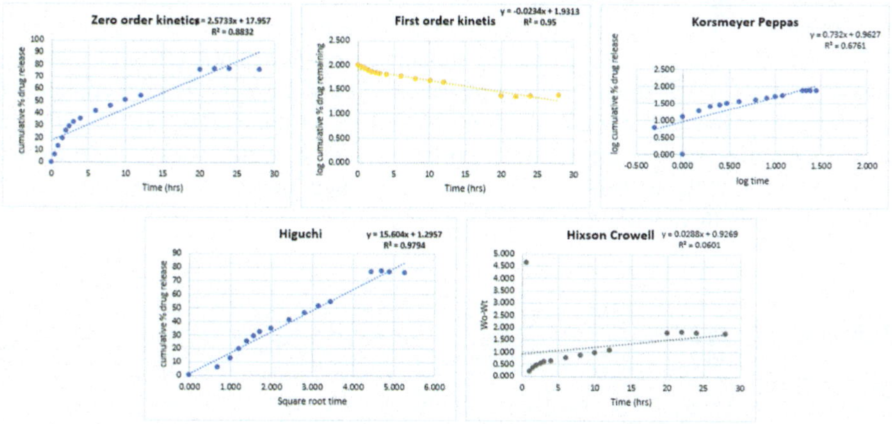

Fig. 9 Drug release kinetics of optimized formulation

Table 4 R^2 value of different release kinetics model

Models	Zero-order	first-order	Korsmeyer-Peppas	Higuchi	Hixson-Crowell
R2	0.8832	0.95	0.6761	0.9794	0.0601

3.7 Stability Studies

As shown in Table 5, no significant changes were observed in 3-month stability studies of optimized formulation (F3).

Table 5 Three months of stability studies of optimized formulation (F3)

Storage condition	Particle size(nm)			Entrapment efficiency (%)			Release profile (%)		
	Initial	30d	90d	Initial	30d	90d	Initial	30d	90d
2–8 ° C	210.46	210.89	212.22	90.52	90.02	89.68	75.64	74.33	74.02
25 °C/60% RH	210.46	211.32	213.65	90.52	89.36	88.98	75.64	76.32	75.32

4 Conclusion

NLC have proven to be suitable carriers for active pharmaceuticals, especially for the topical route. Nanocarriers have several advantages, such as enhanced solubility and bioavailability of hydrophobic molecules, etc. In this study, the nanostructured solid lipid carrier of *Calotropis gigantean* leaves extract was successfully developed by using the sonication method. The F_3 formulation was found to be optimal because it shows high drug content and drug entrapment which was 98.29% and 90.52%. The developed NLC loaded with *Calotropis gigantea* extract exhibited desirable physicochemical properties including appropriate particle size, narrow size distribution, and sustained drug release profile. The high drug content and entrapment efficiency suggest efficient encapsulation of the extract within the NLC. These findings indicate the potential of NLC as an effective delivery vehicle for enhancing the therapeutic effect of *Calotropis gigantea* extract, warranting further exploration of their pharmacological applications. Furthermore, investigation may have the possibility of combining NLCs with other nanotechnologies, such as liposomes or polymeric nanoparticles, for synergistic effects could be a promising avenue including preclinical and clinical trials, are warranted to fully evaluate its therapeutic potential.

References

1. Javed, M.N., Dahiya, E.S., Ibrahim, A.M., Alam, M.S., Khan, F.A., Pottoo, F.H.: Recent advancement in clinical application of nanotechnological approached targeted delivery of herbal drugs. In: Nanophytomedicine: Concept to Clinic, pp. 151–172 (2020)
2. Padmapriya, R., Singh, R.J.A.: Milkweed plant calotropis, a review. Int. J. Trend Sci. Res. Dev. **2**, 391–399 (2018)
3. Chauhan, I., Yasir, M., Verma, M., Singh, P.A.: Nanostructured lipid carriers: a groundbreaking approach for transdermal. Drug Deliv. **10**(2), 150–165 (2020)
4. Gajanan, V.G., Yadav Adhikrao, Y.: Study of methanolic extract of leaves Calotropis gigantea (L.) as an anti-solar. Adv. Complement. Altern. Med. **1**(4), 56–58 (2018)
5. Phad, R.A., Nandgude, D.T., Ganapathy, S.R.: Emulgel: a comprehensive review for topical delivery of hydrophobic drugs. Asian J. Pharm. **12**(2), 382–393 (2018)
6. Tadić, V.M., Žugić, A., Martinović, M., Stanković, M., Maksimović, S., Frank, A., Nešić, I.: Enhanced skin performance of emulgel vs. cream as systems for topical delivery of herbal actives (immortelle extract and hemp oil). Pharmaceutics. **13**(11), 1919 (2021)
7. Verma, N.: The chemical study of Calotropis. Int. Lett. Chem. Phys. Astron. **1**, 74–90 (2014)

8. Patel, K.D., Tripathy, S., Nair, K.S., Kesharwani, R.: Nanostructured lipid carrier (Nlc) a modern approach for topical delivery: a review. World J. Pharm. Pharm. Sci. **2**(3), 921–938 (2013)
9. Kim, S.J., Puranik, N., Yadav, D., Jin, J.O., Lee, P.C.: Lipid nanocarrier-based drug delivery systems: therapeutic advances in the treatment of lung cancer. Int. J. Nanomedicine. **31**, 2659–2676 (2023)
10. Kumar, M., Tiwari, A., Asdaq, S.M., Nair, A.B., Bhatt, S., Shinu, P., Al Mouslem, A.K., Jacob, S., Alamri, A.S., Alsanie, W.F., Alhomrani, M.: Itraconazole loaded nano-structured lipid carrier for topical ocular delivery: optimization and evaluation. Saudi J. Biol. Sci. **29**(1), 1–10 (2022)
11. Ghate, M.V., Lewis, A.S., Prabhu, P., Dubey, A., Patel, N.: Nanostructured lipid carriers for the topical delivery of tretinoin. Eur. J. Pharm. Biopharm. **108**, 253–261 (2016)
12. Purohit, K.D., Nandgude, D.T., Poddar, S.S.: Nano-lipid carriers for topical application: current scenario. Asian J. Pharm. **9**(5), 1–9 (2016)
13. Falana, M.B., Nurudeen, Q.O.: Evaluation of phytochemical constituents and *in vitro* antimicrobial activities of leaves extracts of *Calotropis procera* against certain human pathogens. Notulae Scientia Biologicae. **12**(2), 208–221 (2020)
14. Phatak, A.A., Chaudhari, D.P.: Development and evaluation of Nanostructured Lipid Carrier (NLC) based topical delivery of an anti-inflammatory drug. J. Pharm. Res. **7**, 677–685 (2013)
15. Patel, D., Dasgupta, S., Dey, S., Ramani, R., Ray, S., Mazumder, B.: Nanostructured Lipid Carriers (NLC)-based gel for the topical delivery of Aceclofenac: preparation, characterization, and in vivo evaluation. Scipharm. **80**, 749–764 (2012)
16. Müller, R.H., Radtke, M., Wissing, S.A.: Solid lipid nanoparticles (SLN) and nanostructured lipid carriers (NLC) in cosmetic and dermatological preparations. Adv. Drug Deliv. Rev. **54**(1), 131–155 (2002)
17. Mei, Z., Chen, H., Weng, T., Yang, Y., Yang, X.: Solid lipid nanoparticle and microemulsion for topical delivery of triptolide. Eur. J. Pharm. Biopharm. **56**(2), 189–196 (2003)
18. Patra, J.K., Das, G., Fraceto, L.F., Campos, E.V.R., Rodriguez-Torres, M.D.P., Acosta-Torres, L.S., Diaz-Torres, L.A., Grillo, R., Swamy, M.K., Sharma, S., Habtemariam, S.: Nano based drug delivery systems: recent developments and future prospects. J. Nanobiotechnol. **16**(1), 1–33 (2018)
19. Das, S., Chaudhury, A.: Recent advances in lipid nanoparticle formulations with solid matrix for oral drug delivery. AAPS PharmSciTech. **12**(1), 62–76 (2011)
20. Singh, S., Kamal, S.S., Sharma, A., Kaur, D., Katual, M.K., Kumar, R.: Formulation and *in vitro* evaluation of solid lipid nanoparticles containing levosulpiride. Open Nanomed. J. **4**(1), 17–29 (2017)
21. Koppel, D.E.: Analysis of macromolecular polydispersity in intensity correlation spectroscopy: the method of cumulants. J. Chem. Phys. **57**(11), 4814–4820 (1972)
22. Joshi, G., Tiwari, A., Upadhyay, P.: Optimization and evaluation of piperine-loaded herbosomes for their antioxidant and hepatoprotective potential. Asian J. Chem. **34**(2), 383–388 (2022)
23. Abdelbary, G., Fahmy, R.H.: Diazepam-loaded solid lipid nanoparticles: design and characterization. AAPS PharmSciTech. **10**(1), 211–219 (2009)
24. Ghanbarzadeh, S., Hariri, R., Kouhsoltani, M., Shokri, J., Javadzadeh, Y., Hamishehkar, H.: Enhanced stability and dermal delivery of hydroquinone using solid lipid nanoparticles. Colloids Surf. B Biointerfaces. **136**, 1004–1010 (2015)
25. Kar, N., Chakraborty, S., De, A.K., Ghosh, S., Bera, T.: Development and evaluation of a cedrol-loaded nanostructured lipid carrier system for *in vitro* and in vivo susceptibilities of wild and drug-resistant Leishmania donovani amastigotes. Eur. J. Pharm. Sci. **104**, 196–211 (2017)

Identification of Novel Potent $KRAS^{G12D}$ Inhibitors Through Target-Based Virtual Screening

Divya Pandey[1] and Kuldeep K. Roy[1] (✉)

[1] Department of Pharmaceutical Sciences, School of Health Sciences and Technology, UPES, Dehradun, Uttarakhand, India

Abstract Cancer, a complex and multifaceted group of diseases, is a leading cause of death worldwide. Cancer leads to uncontrolled cell growth and has the capacity to spread to nearby tissues. There are different approaches reported for cancer treatment. Notably, 25% of cancer cases are because of mutations in KRAS (Kirsten rat sarcoma viral oncogene homolog) protein that lead to cancer cell proliferation and angiogenesis. Recently, the Food and Drug Administration (FDA) has granted approval for Adagrasib and Sotorasib for use in adult patients in metastatic $KRAS^{G12C}$-mutated non-small cell lung cancer (NSCLC) [Lanman, B. A. et al., J. Med. Chem. **63**, 52 (2020), Li, Y., Han, L., Zhang, Z., Comput. Struct. Biotechnol. J. **20**, 1056 (2022)]. Here, we have used the AutoDock Vina tool to perform a docking-based virtual screening of the 7,00,000 commercially available compounds in the ZINC database against the $KRAS^{G12D}$ (PDB ID: 6GJ5). The top-ranked 50 compounds, identified from the ZINC database, were analyzed for binding modes, binding affinities, and drug-likeness properties. We found three hit compounds whose binding affinities were comparable to the reported reference compound 15 (a known $KRAS^{G12D}$ inhibitor). In conclusion, using an in silico approach, we have identified novel hit compounds that need further in vitro analysis to confirm their $KRAS^{G12D}$ inhibitory activities.

Keywords KRAS · In silico · Virtual screening · Molecular docking · Mutation

1 Introduction

Cancer is one of the major causes of deaths globally. An estimated 20 million new cases of cancer and 9.7 million fatalities were reported for the year 2022 [1]. Mutations in the RAS gene family include NRAS, HRAS, and KRAS. KRAS is the most frequently mutated oncogene, which is associated with several human cancer forms, including those found in lung, colon, and pancreatic cancers [2]. KRAS mutations have been

A. K. Saxena, S. Nandi (eds.), *Global Trends in Health, Technology and Management*, https://doi.org/10.1007/978-3-031-75457-9_14

found in three main hotspots: 83% of KRAS mutations are attributed to G12, with G13 accounting for 14% and Q61 for 2% [3]. Although mutations in the RAS gene are linked with development of malignant tumors, the G12D mutation is one of the most prevalent KRAS mutations and has long been thought of as a possible therapeutic target for the treatment of cancers [4]. The KRAS protein functions as a molecular switch that regulates cell proliferation, differentiation, and survival by the upstream activation of EGFR and the downstream MAPK and PI3K/mTOR pathways [5]. Mutation in the KRAS leads to the stabilization of the RAS gene into its active state which leads to uncontrolled cell proliferation and tumor development [6].

Direct and indirect approaches can be used to roughly categorize the common strategies used for the development of RAS-targeted anti-cancer drugs. Compounds that bind to RAS directly and stop RAS from interacting with GEFs/effectors are examples of direct methods. Some indirect methods are as follows: (1) blocking the action of proteins such as farnesyltransferase, which encourage RAS binding to the plasma membrane, (2) targeting proteins involved in downstream signaling of the RAS, (3) preventing mutant RAS from interacting with artificially lethal partners, and (4) focusing on RAS-regulated mechanisms including autophagy and micropinocytosis. Since the mutant cysteine would facilitate irreversible binding to small molecules, targeting RAS genes directly has been considered viable. A considerable amount of study has been done to identify inhibitors for $KRAS^{G12C}$. With the FDA approval of Sotorasib and Adagrasib, a promising future for KRAS researchers is presented [7].

Computer-aided drug design is essential for anticipating possible targets and compounds, assessing biological capabilities, and maximizing therapeutic activity [8]. A pocket known as the switch I/II (SI/II) interface pocket has recently been discovered on the surface of both the active and inactive forms of KRAS; Compound 15, $KRAS^{G12D}$ inhibitor, has been reported to bind in this pocket and is one of the potent non-covalent $KRAS^{G12D}$ inhibitors. Compound 15 has an antiproliferative effect on KRAS mutant cells and modifies the levels of pERK and pAKT. But because of its high molecular weight, other scaffolds must be explored to continue the investigation on non-covalent inhibitors of $KRAS^{G12D}$. In this study, an *in-silico* approach including docking-based virtual screening and pose scoring was used as outlined in Fig. 1. We aimed to find novel potential $KRAS^{G12D}$ inhibitors by screening the commercial small-molecules databases at the SI/II interface pocket of the $KRAS^{G12D}$ protein (Fig. 1).

2 Materials and Methods

2.1 Protein Structure Download and Preparation

The X-ray crystal structure of the human $KRAS^{G12D}$ mutant bound to compound 15 (PDB ID: 6GJ5) [9] with a resolution of 1.50 Å was retrieved from the RSCB protein data bank (PDB) (www.rcsb.org). The protein–ligand complex was selected because the ligand has better binding affinity of 5- to 10-fold and has antiproliferative effect which modifies the level of the downstream protein pERK and pAKT levels. Furthermore, the structure was pre-processed in terms of addition of hydrogens, energy

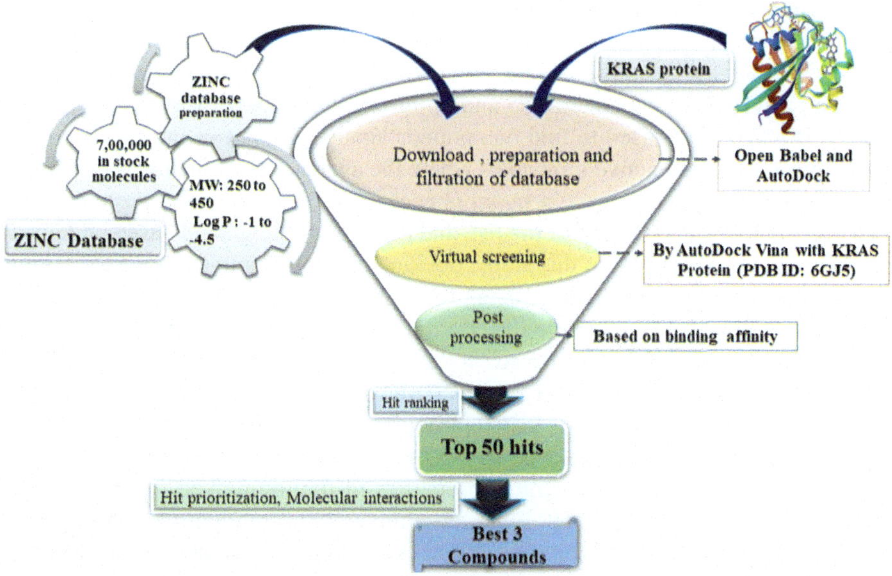

Fig. 1 Workflow of the target-based virtual screening used in the present study

minimization (optimization), and the removal of co-crystal ligand and water molecules which were subsequently subjected to optimization using the Python Molecular Viewer (PMV).

2.2 Drug-Likeness Filtering and Preparation of ZINC Database

A database of 700,000 in stock drug-like molecules were downloaded in sdf file format from the online ZINC database (https://zinc20.docking.org) after applying the following custom criteria of drug-likeness rule (molecular weight of $250 \leq MW \geq 450$; logP $-1 \leq LogP \geq 4.5$). The molecules were subjected to ligand preparation followed by energy minimization using the Open Babel tool considering the Amber force field [10]. The resulting ligand structures were exported in the "pdbqt" format for further use in the TBVS process.

2.3 Docking Tool Assessment/Validation

The reliability of the docking tool can be measured by root mean square deviation (RMSD) between a docking structure and an experimentally determined one. In order to validate the AutoDock Vina tool, the newly optimized 3D structure of Compound15 was docked into the reported site (SI/II) of the KRAS^{G12D} (PDB ID: 6GJ5) and compared with the experimentally determined one in terms of RMSD.

2.4 Target-Based Virtual Screening (TBVS)

Virtual screening (VS) represents one of the promising in-silico methods for facilitating quick and affordable lead generation and optimization. Target-based virtual screening (TBVS) is a technique used to find compounds against a target of interest for the purpose of discovering bioactive molecules where the availability of the three-dimensional structure of the respective protein target is necessary. Thus, the most promising compounds are displayed at the top of the list, and the ligands are graded based on how well they bind to the target [11]. In this study, docking of ligands into the $KRAS^{G12D}$ mutant (target) (PDB-ID: 6GJ5) was done using the AutoDock Vina tool, where the grid box size was kept default (19 Å) while the coordinate of the grid center was $X = -27.701$, $Y = -26.007$, and $Z = -14.244$ based on the co-crystalized ligand. A total of 700,000 compounds from the ZINC library (https://zinc.docking.org/) were screened against $KRAS^{G12D}$ mutant. A total of 10 different conformations of the ligand fitted into the target site were generated, ranked, and analyzed. Afterward, the "ADMETlab" tool was used to calculate the physicochemical properties of the top-ranked 50 compounds as detailed below in another section. Final hit compounds were identified based on multiple criteria such as scoring (binding affinity), structural diversity, and drug-likeness properties.

2.5 Drug Likeliness Properties Prediction

The drug likeliness properties, which include molecular weight (MW), log S, log P, hydrogen bond donor (HBD), and hydrogen bond acceptor (HBA), are essential for small molecule drug development and therapy. Utilizing a web server called ADMET LAB [11] (admetmesh.scbdd.com) one can predict the physicochemical properties of particular substances. These parameters include the aqueous solubility value logarithm (−4 to 0.5 log mol/L), molecular weight of $250 \leq MW \geq 450$; logP $-1 \leq LogP \geq 4.5$, hydrogen bond donor not more than 5, and hydrogen bond acceptor not more than ten. Based on the calculated physiochemical parameters, relevant compounds were identified as potential hits for further exploration.

3 Results and Discussion

3.1 Molecular Docking Tool Assessment/Validation

To find out whether the AutoDock tool could replicate the same orientation and position of the ligand as it was reported in the crystal structure, the optimized 3D structure of the same ligand was re-docked and scored at the original target site. The re-docked ligand's pose was then compared with the experimental one. Figure 2 shows the comparison of the predicted and experimentally determined one (co-crystalized conformation). As shown, the docking tool was able to correctly reproduce the experimental pose of compound 15 in the same binding pocket of the KRAS protein (Fig. 2). This result indicated

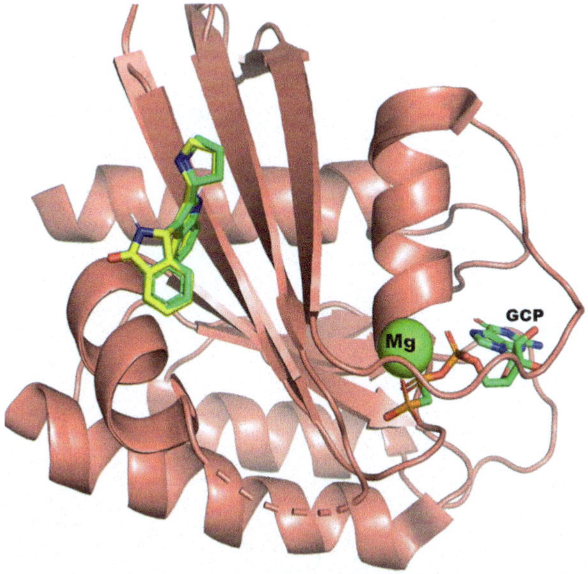

Fig. 2 An overlay of X-ray-bound (green) and re-docked (yellow) conformations of compound15 at SI/II site of the KRAS^{G12D} protein (PDB ID: 6GJ5)

that the docking tool was able to correctly predict the ligand binding orientation in the KRAS protein binding pocket and therefore can be used to predict new compound.

3.2 Target-Based Virtual Screening and Post-Processing

The commercial ZINC database of 700,000 compounds was screened using docking and scoring at SI/II site of KRAS^{G12D} protein (PDB ID: 6GJ5) considering compound 15 as reference control. Post-processing of the molecules was done on the basis of binding affinity and key interactions with reference to the control ligand. In comparison with reference compound 15 (known KRAS^{G12D} inhibitor), top 50 hits were identified which were subsequently subjected to the calculation of physicochemical properties. Based on hit prioritization and diversity analysis as well interactions, a total of 16 compounds were identified. Table 1 summarizes the ZINC ID, chemical structures, and binding affinity values of the top-ranked compounds (Table 2).

3.3 Drug Likeliness Properties Analysis and Hit Selection

The ADMETlab online server [11] was utilized to ascertain the physicochemical characteristics of the 16 hits that were found top-ranked binders at the active site residues of the SI/II interface of KRAS^{G12D}. This method made it possible to forecast the compounds' characteristics with accuracy. All 16 compounds were predicted for physicochemical properties; however, of these, nine hit molecules exhibited the best and

Table 1 List of top-ranked 16 compounds identified through TBVS process

S.no.	Compound ID	Structure	Binding affinity (kcal/mol)	References
1	Compound 15 (reference compound)	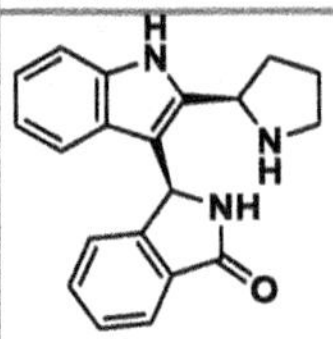	−8.6 (reported $K_D = 440\ \mu M$)	[12]
2	ZINC000008673103	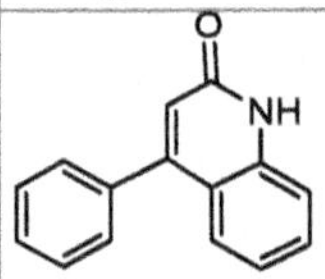	−8.3	https://zinc20.docking.org
3	ZINC000002529013	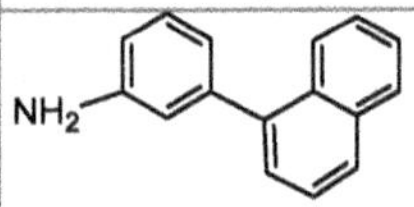	−8.3	https://zinc20.docking.org
4	ZINC000071715221	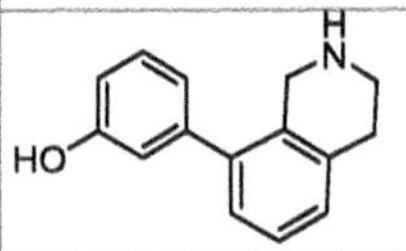	−8.2	https://zinc20.docking.org
5	ZINC000014358823	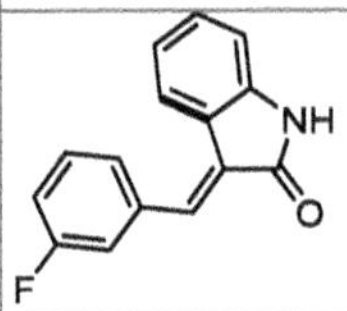	−8.2	https://zinc20.docking.org
6	ZINC000072164556	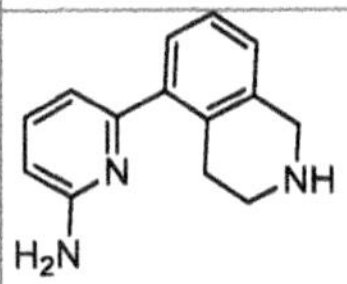	−8.1	https://zinc20.docking.org
7	ZINC000065372225	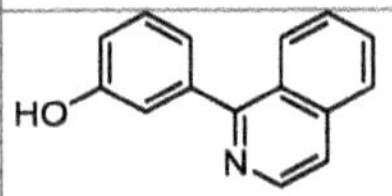	−8.1	https://zinc20.docking.org
8	ZINC000091364317	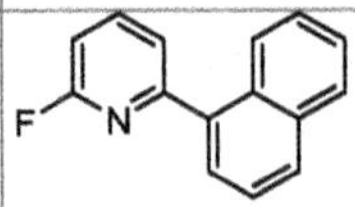	−8.1	https://zinc20.docking.org
9	ZINC000015773550	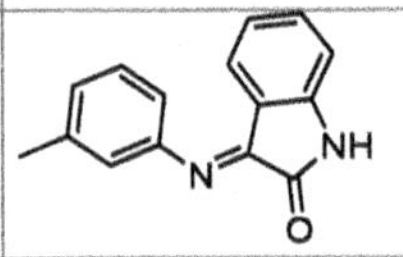	−8.1	https://zinc20.docking.org
10	ZINC000000224430	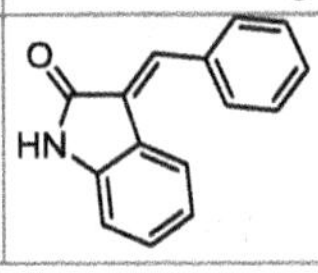	−8.1	https://zinc20.docking.org

(continued)

Table 1 (continued)

S.no.	Compound ID	Structure	Binding affinity (kcal/mol)	References
11	ZINC000015773548	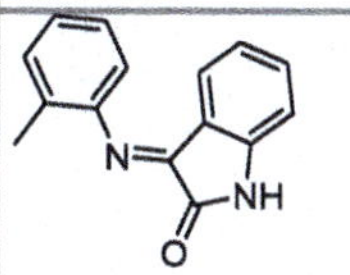	−8.1	https://zinc20.docking.org
12	ZINC000015446030	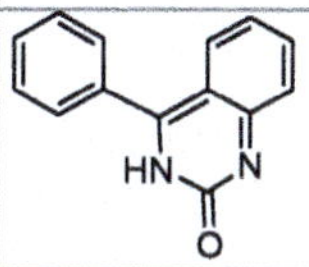	−8.1	https://zinc20.docking.org
13	ZINC000038875445	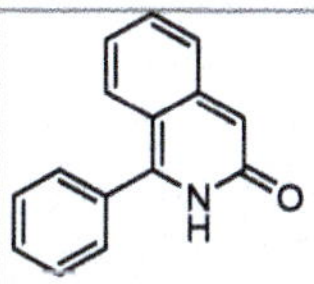	−8	https://zinc20.docking.org
14	ZINC000019281077	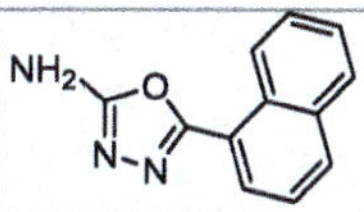	−8	https://zinc20.docking.org
15	ZINC000001564626	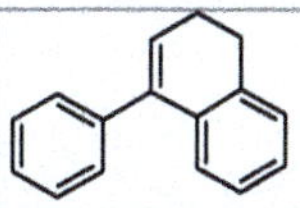	−8	https://zinc20.docking.org
16	ZINC000028523322	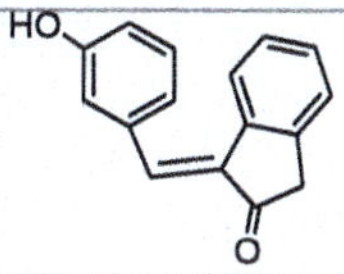	−8	https://zinc20.docking.org
17	ZINC000000154863	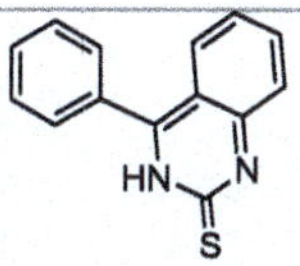	−8	https://zinc20.docking.org

acceptable range of drug likeliness profile which needs further exploration in the in vitro setting.

3.4 Molecular Interactions

The purpose of molecular docking and scoring is to determine a ligand's capacity to favorably occupy the target's active site and form hydrogen bonds as well as other electrostatic and hydrophobic interactions with amino acids underlying the active site. The reference compound's molecular interaction pattern was examined and compared with the identified hits. The reference compound, compound 15 showed that its methylamino at position 7 and the N1 nitrogen of indole form hydrogen bonds with Asp54, while the carbonyl oxygen of the dihydroisoindolinone forms H-bond with Thr74. The N-benzyl imidazole displaced water molecules to fit in the pocket and created polar

Table 2 Physicochemical characteristics of the top nine compounds

Parameters[a]	ZINC000008673103	ZINC000002529013	ZINC000071715221	ZINC000014358823	ZINC000072164556	ZINC000015773550	ZINC000015446030	ZINC000019281077	ZINC000091364317
Binding affinity (kcal/mol)	−8.3	−8.3	−8.2	−8.2	−8.1	−8.1	−8.1	−8.0	−8.1
MW	221.08	219.1	225.12	239.07	225.13	236.09	222.08	213.764	223.08
HBD	2	1	2	2	3	3	3	4	1
BHA	1	2	2	1	3	1	1	2	0
Log S	−3.577	−4.079	2.373	4.175	−1.923	−2.873	−4.004	−3.373	−5.752
Log P	3.284	2.193	2.59	3.246	1.749	3.652	2.285	2.355	4.39

[a]Ideal Range of Drug likeliness properties: LogS = −4 to 0.5; 250 ≤ MW ≥ 450; logP −1 ≤ LogP≥4.5, HBD not more than 5, HBA not more than 10

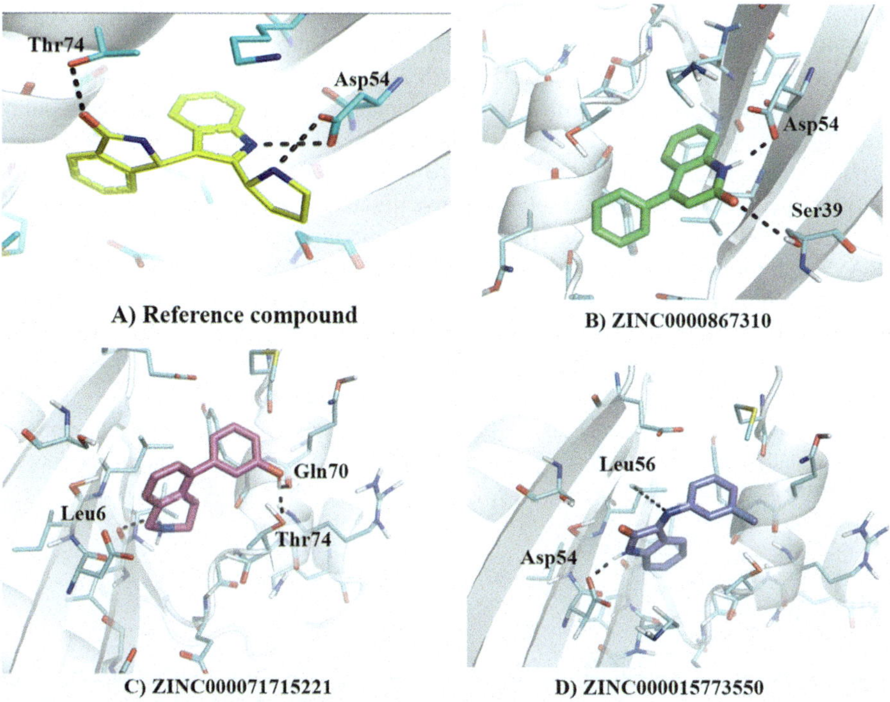

Fig. 3 Depiction of the protein–ligand interactions of: (**a**) Reference compound; (**b-d**) Hit compounds docked at SI/II interface pocket of KRAS^{G12D}

contact with them. The physicochemical properties revealed three top hit molecules as depicted in Fig. 3. ZINC000015773550 showed polar hydrogen bonds with Thr74 and Asp54. Compound ZINC000071715221 exhibited hydrogen bonds with Thr74, Leu6, and Gln70, while the compound ZINC0000867310 showed polar contacts with Asp54 and Ser39. All of the chosen hit compounds were fitted well at the target's Switch I and II interface pocket and interacted with crucial amino acids in the primary active site.

4 Conclusion

The RAS gene had been a challenging target in drug discovery initiatives since it was long thought to be an undruggable target. Since RAS protein surfaces lack a clearly defined hydrophobic pocket, much of the early research centered on indirect strategies such as blocking downstream RAS effectors or blocking farnesyltransferase to inhibit RAS. But recently, several molecules that directly bind to the KRAS variant and impair its function have been found. Compound 15 is one such molecule that was found to be a direct and non-covalent inhibitor of KRAS^{G12D}. All GEFs, GAP, and effector contacts are blocked by compound 15 when it binds to a pocket between switches I and II, which inhibits downstream signaling and has antiproliferative effects. Considering the compound 15 as reference ligand with binding energy (−8.6 kcal/mol, reported $K_D = 440\,\mu M$), the binding energies of the identified hit compounds are ZINC0000867310 (−8.3 kcal/mol), ZINC000071715221 (−8.2 kcal/mol), and ZINC000015773550 (−8.1 kcal/mol). By binding with Asp54, a crucial amino acid in the catalytic subunit of the $KRAS^{G12D}$ active site, these hit compounds shared the same pocket of the target protein. All the chosen hit compounds were with excellent drug likeliness properties. The identified compound may be further investigated to create a novel lead candidate that targets KRAS mutations. The chosen compounds are potential non-covalent $KRAS^{G12D}$ inhibitors and during the Hit-to-Lead optimization process, their binding affinities could still be enhanced through targeted manipulations of their chemical structures. We anticipate that our work may be useful to researchers, particularly those who specialize in drug design and exploring medicinal chemistry.

Acknowledgments D. P. is thankful to UPES, Dehradun for providing doctoral fellowship. The authors would like to thank UPES Dehradun for providing SEED funding and other necessary infrastructural support.

References

1. World Health Organization. Global Cancer Burden Growing, amidst mounting need for services. https://www.who.int/news/item/01-02-2024-global-cancer-burden-growing%2D%2Damidst-mounting-need-for-services. Accessed 24 Apr 2024
2. Yang, W., Ge, J., Yuan, M., Li, J., Pan, L., Ren, J., Dou, G., Yang, L., Zhou, Y., Xie, H., Wang, X., Hu, H.: Computational study of novel natural inhibitors targeting Kirsten rat sarcoma viral oncogene homolog G12C. Anti-Cancer Drugs. **34**, 609 (2023)
3. Hobbs, G.A., Der, C.J., Rossman, K.L.: RAS isoforms and mutations in cancer at a glance. J. Cell Sci. **129**, 1287 (2016)
4. Alam, S., Khan, F.: Virtual screening, docking, ADMET and system pharmacology studies on garcinia caged xanthone derivatives for anticancer activity. Sci. Rep. **8**, 5524 (2018)
5. Waters, A.M., Der, C.J.: KRAS: the critical driver and therapeutic target for pancreatic cancer. Cold Spring Harb. Perspect. Med. **8**, a031435 (2018)
6. Kulkarni, A.M., Kumar, V., Parate, S., Lee, G., Yoon, S., Lee, K.W.: Identification of new KRAS inhibitors through computer-aided drug discovery methods. Int. J. Mol. Sci. 23, 1309 (2022)
7. Kulkarni, A.M., Kumar, V., Parate, S., Lee, G., Yoon, S., Lee, K.W.: Identification of new KRAS G12D inhibitors through computer-aided drug discovery methods. Int. J. Mol. Sci. **23**, 1309 (2022)
8. Hakeem, I.J.: Molecular docking analysis of KRAS inhibitors. Bioinformation. **19**, 411 (2023)
9. Trott, O., Olson, A.J.: AutoDock Vina: improving the speed and accuracy of docking with a new scoring function, efficient optimization, and multithreading. J. Comput. Chem. **31**, 455 (2010)
10. Liu, S., Alnammi, M., Ericksen, S.S., Voter, A.F., Ananiev, G.E., Keck, J.L., Hoffmann, F.M., Wildman, S.A., Gitter, A.: Practical model selection for prospective virtual screening. J. Chem. Inf. Model. **59**, 282 (2019)

11. Xiong, G., Wu, Z., Yi, J., Fu, L., Yang, Z., Hsieh, C., Yin, M., Zeng, X., Wu, C., Lu, A., Chen, X., Hou, T., Cao, D.: ADMETlab 2.0: an integrated online platform for accurate and comprehensive predictions of ADMET properties. Nucleic Acids Res. **49**, W5 (2021)
12. Kessler, D., Gmachl, M., Mantoulidis, A., Martin, L.J., Zoephel, A., Mayer, M., Gollner, A., Covini, D., Fischer, S., Gerstberger, T., Gmaschitz, T., Goodwin, C., Greb, P., Häring, D., Hela, W., Hoffmann, J., Karolyi-Oezguer, J., Knesl, P., Kornigg, S., Koegl, M., et al.: Drugging an undruggable pocket on KRAS. Proc. Natl. Acad. Sci. USA. **116**, 15823–15829 (2019)

Oncolytic Viral Therapy: A Promising Approach in Cancer Treatment

Mohd Javed[1] and Sisir Nandi[1] (✉)

[1] Global Institute of Pharmaceutical Education and Research (GIPER), Affiliated to Veer Madho Singh Bhandari Uttarakhand Technical University), Kashipur, Uttarakhand, India

Abstract Oncolytic viral (OV) therapy has become an emergent tool to treat the cancer which exploits the specific replication and oncolysis of viruses in tumor cells keeping safeguard the surrounding healthy tissues. It can directly kill cancerous cells inducing an immune response against tumor antigens. Literature search was done in PubMed taking the keywords of oncolytic viral (OV) therapy to combat cancer and current literatures were focused to present different types of oncolytic viruses along with their precise mechanisms and application of OVs in clinical field. Furthermore, an attempt has been made to highlight the challenges and limitations associated with this therapy, such as host immune response and potential side effects. Finally, we discuss future directions to enhance the efficacy and optimize the clinical utility of oncolytic viral therapy, including combination strategies with other treatment modalities. The growing body of evidence supports the prospective of oncolytic viral therapy as a valuable addition to armamentarium of cancer therapy, offering new hope for improved patient outcomes and ultimately, the possibility of eradicating cancer.

Keywords Oncolytic viruses · Oncotropism · GM-CSF gene · Cancer treatment

1 Introduction

Viruses are microscopic pathogens which attack humans, plants, animals, bacteria, and fungi [1]. When a healthy human get viral infections, it may cause diseases like respiratory illnesses, gastrointestinal problems, diarrhea, vomiting, sexually transmitted infections, and skin conditions [2]. Viruses are microscopic having genetic contents enclosed in a protective coating called a capsid. Unlike cells, viruses lack the necessary equipment to replicate on their own. They rely on host cells, hijacking their genetic machinery to produce copies of themselves. It's like uninvited guests breaking into your house to use your kitchen and cooking with your utensils. Unfortunately, they often leave a

A. K. Saxena, S. Nandi (eds.), *Global Trends in Health, Technology and Management*, https://doi.org/10.1007/978-3-031-75457-9_15

mess behind when they are finally eliminated [3]. Sometimes referred to as "virions," viruses invade cells and utilize the host cell's resources to reproduce themselves. Some pathological condition produces cancer in liver caused by "hepatitis B virus (HBV)" and the "human papillomavirus (HPV)" causes cervical cancer, head and neck cancer. However, it's important to note that cancer vaccines play a leading role in preventing these viral infections and reducing the risk of associated cancers [4].

The viruses have been shown to target and attack existing tumors. These modified viruses are called as oncolytic viruses, which offer a potential approach to combat the cancer [5]. There are several reasons why oncolytic viruses hold promise. Cancer cells often have weakened antiviral defenses, making them more vulnerable to viral infection. These viruses can be engineered to reduce their ability to infect healthy cells while selectively targeting and attacking tumor cells. They can also be modified to carry out the therapeutic actions directly to tumors and stimulate the immune system. Oncolytic viruses can cause infected cancer cells to rupture, leading to their destruction and the release of cancerous antigens which then trigger immune responses, allowing the immune system to target and eliminate remaining tumor cells, not just in the initial site but potentially throughout the body [5].

2 Emergence of Oncolytic Virotherapy

Saint Peregrine is a historical figure associated with the phenomenon of spontaneous tumor regression. Peregrine Laziosi, an Italian priest from the thirteenth and fourteenth centuries, experienced with the miraculous disappearance of a cancerous growth on his leg while in prayer in front of a crucifix. This event led to his connection with spontaneous tumor regression and his subsequent canonization as the patron saint of cancer patients [6].

These observations have prompted researchers to investigate various hypotheses to shed light on the reasons behind spontaneous tumor regression. Some studies have proposed that infections, both pyrogenic and non-pyrogenic, may trigger an immune response that contributes to tumor regression. However, further scientific research is necessary to validate these claims in understanding the potential role of infections and the immune system in combating cancer. It underscores the importance of ongoing scientific research in unraveling the complexities of cancer and improving our strategies for its treatment and prevention [7].

The basic virology and the oncolytic properties of different viruses have greatly expanded. These developments have provided researchers with opportunities to delve deeper into the concept of viral oncotropism, referring to viruses that selectively target cancer cells. As early as 1922, physician, cytologist, and immunologist Levaditi observed that viruses display a preference for replicating in cancer cells, highlighting their susceptibility to viral infection [7].

These technological and scientific advancements have paved the way for a better understanding of viral oncotropism mechanisms and the potential applications of oncolytic viruses in cancer treatment. Researchers continue to study the oncolytic activity of various viruses, further advancing our knowledge and offering potential new therapeutic avenues in the fight against cancer.

3 Oncotropism in Virus

The success of oncolytic virotherapy depends on targeting the specific malignant cells. Multiple strategies have been taken to enhance the ability of naturally occurring specific viruses which can exhibit selective destruction of malignant cells [8]. These strategies include genetic modification of viral envelopes or capsids to make them specifically target cells with tumor-associated surface markers (transductional targeting) [9], expression of viral genes under the control of "promoters" that only function in transcriptional targeting cells, such as "tissue-specific" and "hypoxia-responsive promoters" [10], insertion of essential viral genes downstream of internal ribosome entry sites (IRESs) that cannot be used as translational origins in specific tissues (translational targeting) [11], utilization of "destabilization domains" that make viral products unstable (post-translational targeting) [12], inactivation of virulence factors (attenuated viral strains) that only replicate in cells due to abnormal signaling and over-expression of RAS or inactive p53 (oncogenetic targeting) [13], and integration of tissue-specific microRNA-binding elements at the 3'-UTR of essential viral genes to allow expression of those microRNAs for the productive infection targeting the microRNA [14].

The cytotoxic effects of oncolytic viruses have been analyzed on cancer cells beyond the typical cytopathic effect caused by replication-competent particles. Various strategies have been attempted for the incorporation of sequences into the viral genome that encode enzymes capable of converting the non-toxic prodrugs into lethal cytotoxic chemicals. These enzymes such as thymidine kinase (TK) of herpes simplex viruses (HSVs) can convert ganciclovir into deoxyguanosine triphosphate [15], and cytosine deaminase catalyzes the conversion of 5-fluorocytosine into 5-fluorouracil [16]. Additionally, proteins that have tumor-restricted cytotoxic functions, such as wild-type p53 or death receptor ligands like FASL, have been integrated into the viral genome [17]. Short-hairpin RNAs targeting proteins specifically required for the survival of malignant cells, such as survivin, have also been utilized [18].

It is worth noting that some viral particles can cause the death of malignant cells, even without viral gene expression, upon uncoating. The precise molecular mechanisms are not yet clear, but the efficacy appears to be influenced by the number of virions that enter the target cells and undergo uncoating [19]. Furthermore, a number of viral proteins have shown tumor-selective pro-apoptotic functions [20]. For example, apoptin, a 15 kDa protein encoded by the chicken anemia virus, and the polypeptide coded by the adenoviral E4 open reading frame have been shown to preferentially induce apoptotic cell death in malignant cells rather than non-transformed cells [21].

4 Mechanism of Action of OVs

The oncolytic viruses can attach and remove the specific tumor cells without harming the normal cells. The mechanism has been graphically plotted in Fig. 1.

They have particular targets such as human telomerase reverse transcriptase, nuclear transcription factors, prostate-specific antigen, osteocalcin, cyclooxygenase-2, and surface markers such as prostate-specific membrane antigen, folate receptor, endothelial growth factor receptor, CD20, and Her2/neu produced by the tumor cells [22]. In laboratory, scientists have been able to test the OVs which can specifically target the pathogenic viral genes of tumor cells without affecting the normal tissues. Therefore, this has been promising under preclinical and clinical research as a novel method for cancer treatment [23].

The route of administration of oncolytic viruses influences the effectiveness of the therapy. The choice of administration route depends on the type of tumor being treated, the availability of the virus on-site, and the natural barrier of the organism to fight the antigens [24]. OVs can be administered through various routes such as intraperitoneal, intrathecal, subcutaneous, intratumoral, and intravenous. Intraperitoneal administration involves injecting the virus into the peritoneal cavity, while intrathecal administration involves injecting the virus into the cerebrospinal fluid. Subcutaneous administration involves injecting the virus under the skin, while intratumoral administration involves injecting the virus directly into the tumor. Intravenous administration involves injecting the virus into the bloodstream associated with the treatment of distant metastases. Intratumoral administration provides greater control of viral quantity in the tumor environment and less adverse effects [25].

When it comes to evading the immune system, tumor cells can employ various strategies like abnormal expression and activation of signaling pathways of protein kinase R and interferon 1. These modifications interfere with the response to viral infections, programmed cell death, and the maturation of inflammatory cells. These changes in the antiviral response, coupled with viral factors that inhibit apoptosis, enable oncolytic

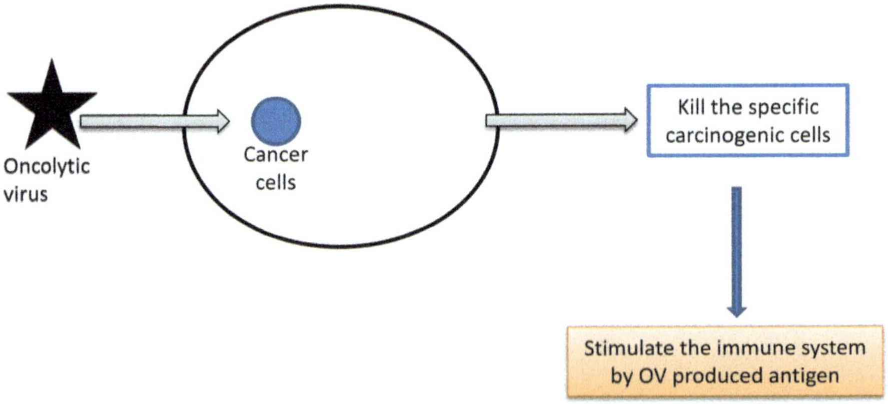

Fig. 1 Graphical representation of mechanism of action of oncoviral therapy

viruses to persist for longer periods within cancer cells. As a result, OVs can complete their life cycle and mature into the lytic phase [26].

When viruses enter the human body, they trigger the immune system to recognize specific markers associated with the virus's structure. These markers include viral proteins, RNA, DNA, and the viral capsid, collectively known as "pathogen-associated molecular patterns (PAMPs)" [27]. Dendritic cells play a crucial role in this recognition process by using pattern recognition receptors called "toll-like receptors (TLRs)." Once dendritic cells detect PAMPs through TLRs, they initiate the production of inflammatory molecules that possess antiviral properties. These molecules include type 1 interferons, "tumor necrosis factor alpha (TNF-alpha)," and cytokines storms such as interleukin 2 (IL-2). These molecules are important for recruiting immune cells to the site of infection and maintaining an inflammatory environment [28].

TNF-alpha plays a crucial role in the immune response to viral infections by positively regulating the expression of class 1 major histocompatibility complex (MHC) on cell membranes [29]. It also influences caspase enzyme action, promoting cell apoptosis in certain tumors. TNF-alpha stimulates cancer cell death through necrosis and apoptosis, which can cause thrombotic events via its antiangiogenic effects, leading to the destruction of tumor blood vessels [30]. Moreover, TNF-alpha is involved in stimulating T helper cells type 1 (Th1), enhancing natural killer cell cytotoxicity, and maturing antigen-presenting cells [29].

Various studies have demonstrated the role of IL-2 in stimulating cytotoxic lymphocytes and activating T cell response. This leads to the maturation and expansion of CD8+ T cells and natural killer cells, while also positively regulating CD4+ T cells [31]. Additionally, IL-2 plays a crucial role in regulating the action and maintaining the balance of T regulatory cells, thereby creating an inflammatory environment that aids in combating tumors. Furthermore, the Th1 inflammatory profile has been associated with a decrease in T regulatory cells, increased levels of effector TCD4 and TCD8 cells, and the stimulation and differentiation of T lymphocytes and maturation of dendritic cells. Ultimately, these processes contribute to reversing the immunosuppressive state of tumors and promoting an inflammatory response [32].

Apart from the detrimental effects of the inflammatory response, the action of the virus within the infected cell plays a significant role in the lysis and demise of aberrant cells. The presence of oncolytic viruses can induce dysfunction in various organelles, such as the endoplasmic reticulum, mitochondria, and lysosomes, which compromises normal cellular functions. This disruption triggered by the virus can also result in the generation of oxidative stress by producing reactive nitrogen species and causing endoplasmic reticulum stress. Consequently, intracellular calcium levels rise, further contributing to the stabilization and reduction of the tumor [28].

The synergistic use of cell checkpoint blockers and oncolytic viruses plays a pivotal role in enhancing the survival rates of viruses within the human body. This is primarily due to their ability to instigate an inflammatory response against tumors, thereby contributing to their elimination. By inhibiting the negative regulation of PD-L1, an immune evasion strategy employed by tumors, the maturation of T cells can be hindered. In turn, this inhibition stimulates a robust Th1 response, promoting the emergence of TCD8 cells to combat tumors and enhancing the activity of natural killer cells [33]. Additionally, research has showcased the potential of combining OVs with monoclonal antibodies targeting "cytotoxic T lymphocyte-associated antigen 4," as this approach leads to a significant boost in the immunotherapy effectiveness [32].

5 Oncolytic Viruses

Various oncolytic viruses cause elimination of tumor cells through various pathways, including death of autophagic cell, pyroptosis, apoptosis, and necrosis. This elimination process leads to the generation of immune signals associated with cellular damage, known as "damage-associated molecular patterns (DAMPs)." High mobility group box 1 protein and ATP are examples of DAMPs that are released during this process. DAMPs are essential in promoting the dendritic cells maturation and facilitate the tumor-associated antigens to immune cells through cross-presentation. This, in turn, perpetuates the inflammatory response [34]. Furthermore, the lysis of tumor cells allows the release of viruses into the extracellular environment, enabling the succeeding infection of other tumor cells. This creates a chain reaction, amplifying the antitumor response [27]. Additionally, cell death results in the release of tumor antigens that can be recognized by immune cells in the inflammatory milieu. This further stimulates a response against tumor cells, even those that are not directly infected by OVs [26].

5.1 Herpes Simplex Virus

OVs, including wild-type and genetically modified live viruses, offer a promising approach to cancer immunotherapy. One example is T-VEC, which is a genetically engineered herpes simplex virus type 1 (HSV-1) that selectively infects tumor tissue and damages tumor cells, promoting antitumor immunity [35].

T-VEC contains the human GM-CSF gene, this gene leads to generation of GM-CSF in the tumor. This increases the entry and activation of dendritic cells (DCs), crucial for initiating an antitumor response in immune system [36]. In addition, the deletion of neurovirulence factor ICP34.5 and ICP47 genes reduces viral pathogenicity, suppressing antigen presentation. The deletion of ICP47 also enhances expression of the US11 gene of HSV, promoting virus growth in tumor cells and not compromising with tumor selectivity [37].

Cancer immunotherapy works by recognizing of TAA (tumor-associated antigens) by immune cells and involves components of immunity and proinflammatory cytokines such as TNF α (tumor necrosis factor) and interferon-γ [29]. T-VEC can be administered intratumorally, which leads to direct tumor cell death and produces antigens, leading to the influx of immune cells into the microenvironment of the cancer cell. Primed T cells then induce antitumor response in the body which addresses intra- and intertumoral heterogeneity [38].

Herpes virus glycoproteins are present on the cell surface of the cancer cell that helps in the entry of T-VEC, allowing favorable replication within the cells and triggering cell lysis. This releases tumor-derived antigens (TDA) along with, GM-CSF, and new viral particles, the surrounding tumor cells get infected and increases the spread of

TDA. Furthermore, activation of danger-associated molecular pathways leads to an antitumor effect in the body and a complete immune response is generated [39]. Overall, OVs like T-VEC hold a great potential for cancer immunotherapy by causing cell death, releasing antigens, and creating a tumor killing immune response in the body.

5.2 Adenovirus

Adenoviruses are a double-stranded linear DNA containing virus. The capsid of adenovirus has an icosahedral shape that lacks an envelope. The capsid contains three major proteins known as penton base, fiber, and hexon. These proteins help in triggering the immune response against the virus. The Adenoviridae family consists of over 80 different types of adenoviruses, all of which affect humans [40]. Adenoviruses have a strong affinity for various tissues in the body, including eyes, respiratory system, intestines, kidneys, and lymphoid tissues. To penetrate host cells, they can utilize different receptors like the HCA (human coxsackie-adenovirus) receptor, CD86, CD46, and CD80 [41].

In the initial phase of replication of virus that occurs in the nucleus of the tumor cell and triggers the production of specific proteins, like E1a and E1b, into the cytoplasm, these proteins act as catalyst in activating the process of autophagy, whereby the cell digests and eliminates its own components. As a result, autophagosomes are generated, which later fuse with lysosomes, leading to the degradation of organelles and, in severe cases, the death of the entire cell [42]. Recent studies have also indicated that the production of E1a protein in tumor cells may promote the production of autophagic complexes, while E1b potentially enhances the effectiveness of these complexes. Both mechanisms contribute toward stabilizing and reducing tumor growth [43].

When the human body detects and interacts with various proteins present in the viral capsid of adenoviruses, it triggers a surge of immune responses. This encompasses the generation of proinflammatory cytokines, including IL-12 and TNF-alpha [44], which are pivotal in activating cytotoxic cells, notably natural killer cells and TCD8 cells. Moreover, these cytokines facilitate the maturation of immune cells and booster the body's defense mechanisms against tumors. Among the different adenovirus types, Adenovirus type 5 (Ad5) is commonly utilized in oncolytic therapy because it can be recognized by "Toll-like receptors (TLRs)" located both on the cell membrane (specifically, TLR-2) and within the cell (notably, TLR-9). This detection process initiates a series of mechanisms that together promote an inflammatory response characterized by the inflammation mediated by the Th1 profile [44].

Furthermore, adenoviruses can also activate different immune pathways, like the complement system. This activation triggers opsonization, causing inflammatory cells to migrate more rapidly and produce extra-inflammatory cytokines [33]. Together, these immune responses help destroy infected cells, thereby strengthening the body's defense against viral infections.

Ultimately, the viral infection-induced cellular stress and subsequent inflammatory responses result in the death of tumor cells through various mechanisms such as necrosis, autophagy, or apoptosis. This process also leads to the release of DAMPs (danger-associated molecular patterns) or PAMPs (pathogen-associated molecular patterns) into the inflammatory environment [40]. These released molecules further stimulate the maturation and migration of inflammatory cells, as well as the production of cytokines.

Importantly, besides the direct elimination of tumor cells, adenoviruses have the remarkable ability to initiate the development of an antitumor immune memory. This immune memory helps in combating tumor growth in metastatic sites [45]. The immune memory generated by adenoviruses provides long-lasting protection against future tumor growth and contributes to the overall effectiveness of adenovirus-based therapies [45].

5.3 Protoparvovirus

Protoparvoviruses, which belong to the Parvoviridae family, are non-enveloped, single-stranded DNA viruses. Protoparvoviruses have the capability to enter mammalian cells, including human ones, by using fixation agents like the transferrin receptor or glycosidic substances such as N-acetylneuraminic acid found on the cell membrane. These molecules help create a favorable environment for the virus to attach to the cell [46].

The VP1 protein, a primary component of the viral capsid, is essential for helping protoparvoviruses enter host cells via endocytosis. It also facilitates the breakdown of the endocytic vesicle within the cell, which leads to the release of viral proteins into the cytoplasm. Moreover, VP1 contains nuclear localization signals that help transport these viral proteins to the cell's nucleus [28]. Once in the nucleus, the virus can stay inactive until the cell transitions into the S/G2 phases of the cell cycle. At this point, the NS1 protein plays a role by inhibiting the replication of the cell's genome, allowing the viral DNA to integrate with the host's genetic material, thereby ensuring its persistence [47].

Through the activity of the NS1 protein, H-1PV can generate an oxidative stress condition inside the cell by boosting the levels of reactive oxygen and nitrogen species. The NS1 protein is also involved in controlling RNA viral replication, which eventually leads to the degradation of cellular genetic material. This mechanism then initiates apoptosis pathways, resulting in cell death. Furthermore, the virus can prompt the release of proteases from the lysosome into the cytoplasm, leading to the necrosis of cancer cells [48].

Furthermore, protoparvoviruses have the tendency to produce an inflammatory response that exhibits antitumor properties. This response leads to the release of cytokines, such as IL-2 and TNF-alpha, with a Th1 profile [49]. This inflammatory environment created by the protoparvoviruses is capable of effectively combating tumor cells. H-1PV, specifically, helps in the proliferation of T lymphocytes, including cytotoxic TCD8 cell and auxiliary TCD4 cells. This immune response contributes to the formation of immunological memory against the tumor, further enhancing the antitumor effects of H-1PV [49].

During the lytic phase, viral activity leads to an increase in the permeability of lysosomal membranes. This increased permeability allows cathepsin enzymes to pass into the cytoplasm, while also reducing the inhibitory effects of agents that normally suppress these proteases. Both elements play a crucial role in amassing cathepsins in the cytoplasm, enhancing their function, and aiding in apoptosis pathways, ultimately leading to the demise of tumor cells [50]. Moreover, the presence of NS1 further facilitates cellular apoptosis by damaging genetic material, activating and intensifying caspase activity, and triggering oxidative stress processes. These mechanisms efficiently circumvent the apoptotic evasion strategies used by tumor cells [51].

5.4 Vaccinia Virus

These are unique enveloped viruses with their remarkable history dates back to 1796 when they were utilized for smallpox vaccination. With the successful eradication of smallpox, their scientific utility has shifted toward the development of innovative vaccines and therapies for various diseases [52]. Notably, among the VACVs, the exceptional Pexa-Vec (pexastimogene devacirepvec, JX-594) stands out. Pexa-Vec is a genetically altered virus, designed with the granulocyte-macrophage colony-stimulating factor (GM-CSF) and a specific deletion of the thymidine kinase (TK) gene. These precise genetic alterations enable Pexa-Vec to exhibit enhanced tropism toward tumor cells while confining its replication exclusively to cells expressing unusual levels of TK [53].

The application of VACVs within the tumor microenvironment has been linked to the activation and production of key elements such as GM-CSF and IL-24, which work together to stabilize and destroy tumor cells. GM-CSF is crucial for the growth and development of immune cells, including dendritic cells and neutrophils. This process creates an inflammatory environment that effectively targets the tumor. Conversely, IL-24 hampers tumor blood vessel formation, thereby enhancing apoptosis pathways and fostering an antitumor response. Moreover, IL-24 prevents the spread of tumor metastases, further boosting its antitumor capabilities [54].

VACV strains exhibit viral activity that triggers various cell death pathways, and associated molecules like ATP and high mobility group box 1 protein create an immunogenic environment. These molecules, which are classified as DAMPs, aid in cross-presenting tumor antigens, thereby enhancing the activation of TCD8 cells and contributing to antitumor responses [55]. Furthermore, Pexa-Vec specifically aims at endothelial cells, which are critical for tumor growth because of their expression of vascular endothelial growth factor or fibroblast growth factor. By targeting these cells, Pexa-Vec disrupts the tumor vasculature, leading to tissue necrosis and a decrease in tumor size [56].

6 Potential Applications of OVs in Clinical Field

6.1 Melanoma

The incidence of melanoma, a life-threatening form of skin cancer, is continuously rising worldwide, in stark contrast to the decreasing trend observed in all the other types of cancer [57]. Within various Western nations, the risk of developing skin cancer stands at approximately one in every 50 individuals [58], with a higher prevalence observed among populations with lighter skin tones [59].

In the field of oncolytic virus (OV) therapy, the vaccinia virus stands out as a prototypical poxvirus with substantial clinical importance, primarily due to its potential for genetic customization. By deleting genes responsible for virulence and inserting therapeutic genes [60], the vaccinia virus can be readily weakened. JX-594, a modified oncolytic vaccinia virus, was specifically engineered to activate local macrophages and dendritic cells [61]. This modified virus was the centerpiece of two phase I trials involving 17 patients with inoperable cutaneous melanoma. The studies proved that JX-594 successfully replicated within the tumor microenvironment, causing local oncolysis. Additionally, the trials showed that increasing doses of JX-594 were both safe and effective in combating the disease [62, 63].

In the same vein, two other phase I clinical trials utilized a vaccinia virus encoding B7.1 T cell co-stimulating molecules in 25 patients with non-resectable melanoma [64]. Hearteningly, these trials resulted in a 20% rate of full objective response, with minimal toxicity and mild-grade reactions [64, 65].

The herpes simplex virus has gained attention as a viable candidate for oncolytic virotherapy in melanoma due to its large genome, which includes non-essential genes that can be deleted to lower its pathogenicity and facilitate the insertion of therapeutic genes [66]. The United States Food and Drug Administration has approved T-VEC as the first oncolytic virus for melanoma treatment [67]. Numerous clinical trials, encompassing Phase I, II, and III studies, have shown encouraging results for T-VEC in treating melanoma [68]. Biopsies during Phase I revealed significant tumor destruction in the injected lesions caused by T-VEC [69]. In Phase II, the overall objective response rate was 26%, with a 1-year survival rate of 58% for all patients, and minimal adverse effects were observed in 85% of the participants [69]. In Phase III, the objective response rate for the T-VEC arm remained stable at 26%, with 11% of patients experiencing complete responses [70]. Unfortunately, the final survival data are still awaited. Nonetheless, this pioneering randomized clinical trial demonstrated the therapeutic benefits of oncolytic virotherapy for patients with advanced or inoperable melanoma [66].

HF10, a strain of herpes simplex virus type 1 (HSV-1), which has experienced spontaneous mutations resulting in the loss of certain viral genes [71], was employed in an in vitro study that showed notable cytolytic effects on both murine and human melanoma cells upon HF10 infection [72]. In the same research, immunocompetent mice bearing advanced melanomas were treated with intratumoral injections of HF10. Interestingly, there was a reduction in tumor growth not only in the injected tumors but also in the non-injected ones, suggesting both direct oncolysis and the initiation of a systemic antitumor immune response [72]. To assess the oncolytic effectiveness of HF10, it was combined with dacarbazine in mouse models with subcutaneous melanoma. The combination therapy of dacarbazine and HF10 showed a rapid and potent cytotoxic effect compared to single therapy. This was attributed to the strong induction of a systemic antitumor immune response, resulting in prolonged survival [73].

Other viruses that have garnered less attention have also shown promising results in research. For example, Coxsackievirus A21 (CVA21) has exhibited oncolytic properties in melanoma cells during preclinical studies, while remaining tolerable and presenting low viral pathogenicity [74]. CVA21, the marketable form of Coxsackievirus A21, advanced through phase I and II clinical trials involving patients with advanced, unresectable melanoma. These patients received intratumoral injections of the virus over the span of 15 weeks. The treatment was generally well tolerated, showing only mild reactions, and even resulted in complete therapeutic responses, indicating a favorable safety profile [75, 76]. Additionally, in a separate phase II trial, the oncolytic efficacy of Reolysin® was assessed in 21 patients with metastatic melanoma who received intravenous injections of the virus [77]. All patients handled the injections well, and viral replication was confirmed in biopsy samples from 2 out of the 13 patients who were evaluated after treatment. Although the trial did not meet its primary efficacy endpoint nor show objective responses, the collected data support the potential use of reovirus in combination with other therapies for treating malignant melanoma [77].

6.2 Hepatic Cancer

Hepatocellular carcinoma (HCC) is an extremely aggressive and difficult-to-treat type of tumor, presenting significant challenges in the medical field [78]. HCC stands as the most prevalent form of liver cancer [79, 80], ranking among the six most frequent and four deadliest cancer types globally [80]. Alarmingly, the incidence of HCC has been escalating in recent years [81]. This rise in cases is particularly pronounced in developing nations, where liver cancer is exceptionally widespread [82].

A significant challenge in treating liver cancer is the substantial recurrence rate following standard therapies. Existing treatment methods have limited efficacy, leading to a 5-year survival rate of just above 10% for patients [81].

These disheartening statistics emphasize the critical need for more efficient and groundbreaking methods to fight this aggressive type of cancer. Medical scientists and experts around the world are committed to discovering hopeful solutions and enhancing the prognosis for individuals with liver cancer [81].

Numerous studies in the literature have underscored the potential of oncolytic viruses (OVs) in targeting hepatocellular carcinoma (HCC), particularly in advanced stages. These studies prioritize safety and aim to minimize the risk of recurrence [81, 83]. Among OVs, adenovirus is noteworthy for its high affinity for liver cells [80]. Researchers have explored various strategies to augment viral adaptation to overcome challenges posed by tumor cells. For example, one approach involves integrating the Ad5 viral vector with GP73 and SphK1-shRNA promoters [82]. In preclinical trials, this modified adenovirus showed the ability to induce apoptosis in tumor cells and significantly curb tumor growth, thereby enhancing survival rates in mice [83]. In another study, scientists modified the adenovirus ZD55 vector to address the high resistance of HCC cells to tumor necrosis factor-related apoptosis-inducing ligand (TRAIL). By combining ZD55-TRAIL with ZD55-Smac, a variant containing a second mitochondrial-derived caspase activator, they were able to effectively reduce tumor size [80]. These encouraging results highlight the potential of these modified adenoviruses in overcoming the hurdles posed by HCC. Continued research and development of tailored OV treatments promise significant improvements in outcomes for HCC patients, offering renewed hope in the fight against this aggressive form of cancer.

Another virus that has demonstrated potential in treating hepatocellular carcinoma (HCC) is the vaccinia virus. Specifically, the JX-594 strain of the vaccinia virus has been extensively researched and found to be safe and effective in preclinical studies using rabbits. This strain successfully eradicated lung metastases and liver tumors in these animals [78, 80]. Besides the JX-594 strain, the vaccinia virus can also be paired with cytokines to boost its therapeutic effectiveness. For instance, recombinant VV-IL-37, which integrates interleukin 37 into the virus genome, has shown the ability to impede the growth of liver tumors [82].

Another noteworthy treatment option for hepatocellular carcinoma (HCC) involves the use of the herpes simplex virus (HSV). In a research study using mice, a more precise and oncolytic vector named Ld0-GFP was crafted specifically for targeting liver cells. This vector has shown both safety and effectiveness by inducing cell death (apoptosis) in cancer cells and encouraging the release of damage-associated molecular patterns (DAMPs) linked with immunogenic cell death. These results underscore the potential of vaccinia virus variants, as well as the application of HSV, in developing innovative therapeutic strategies for HCC. Ongoing research and clinical trials are crucial for further investigating and refining these approaches to benefit HCC patients [81].

6.3 Mammary Carcinoma

Mammary carcinoma or breast cancer (BC) is a complex and diverse carcinoma influenced by various factors, including history of family members, living conditions, and hormonal components [84, 85]. Around the world, the incidence of mammary carcinoma is rising beyond the imagination. Additionally, the advent of improved early detection techniques has contributed to the increased detection rate of BC cases. Currently, it is estimated that approximately in every 8 women 1 woman is at risk of being diagnosed with mammary carcinoma worldwide, making it one of the most common form of cancer among women [86]. The multifactorial nature of BC highlights the importance of comprehensive research, prevention strategies, and effective early detection methods. Continued efforts in raising awareness, promoting healthy lifestyles, and advancing medical technologies are essential in combating this significant health challenge impacting women across the globe.

There is significant potential for treating advanced breast cancer through oncolytic virotherapy, which has demonstrated a variety of cellular and molecular mechanisms [87]. Among the promising options, double-stranded DNA viruses have drawn attention as they replicate and transcribe within the cell nucleus without integrating their genetic material into the host cell's DNA [87]. It is critical for oncolytic viruses (OVs) to show high specificity in replicating within cancer cells, which has been verified through tests demonstrating the tolerability and selectivity of genetically engineered viruses like the vaccinia virus [88]. Adenovirus, another thoroughly investigated OV for breast cancer treatment, remains somewhat contentious. Preclinical studies suggest its effectiveness in reducing tumor size and controlling metastases in mice [87]. However, phase I trials have shown limited effectiveness for breast cancer, either as a standalone therapy or in combination with other drugs [88]. In addition to these OVs, T-VEC, which is presently approved for certain types of melanomas [89] in the United States and Europe, has also been clinically tested for breast cancer. Patient tolerability has been favorable, and there has been moderate success in inducing tumor necrosis and immune response [88, 90]. These advancements indicate the ongoing exploration of oncolytic virotherapy for breast cancer treatment. Further research, clinical trials, and optimization of treatment strategies are essential to fully harness the potential of OVs in combating advanced breast cancer.

6.4 Bowl Cancer

Bowel cancer, also known as colorectal cancer, stands as one of the most common types of cancer in Western countries and is a leading cause of cancer-related deaths [91]. Unfortunately, current effective treatments for this cancer are limited, pushing researchers to explore alternative methods such as oncolytic viruses (OVs). In certain studies involving heavily pretreated colorectal cancer patients, administration of oncolytic vaccinia virus either alone or combined with elevated levels of granulocyte-macrophage colony-stimulating factor (GM-CSF), a hematopoietic growth factor, resulted in stable disease in 67% of patients [92, 93]. Another study examined the use of oncolytic herpes simplex virus type 2 (HSV2). In vitro analysis demonstrated that oncolytic HSV2 effectively inhibited the growth of CT-26 colorectal cancer cells. Mouse model experiments with xenograft tumors also showed a reduction in liver metastasis following treatment with oncolytic HSV2 [94]. These findings suggest that OVs could be a promising option for colorectal cancer treatment. While further research and clinical trials are needed, these preliminary studies underscore the potential of oncolytic viruses in meeting the urgent demand for effective therapies for colorectal cancer patients.

7 Merits of Oncolytic Virus Therapy

The selective destruction of malignant cell is the biggest merit of the oncolytic virus therapy. Oncolytic viruses can infect cancer cells while sparing the healthy cells, which is not in the case of other therapies like chemotherapy or radiotherapy. These viruses are genetically engineered to replicate predominantly in the tumor environment, which reduces the damage to normal cells [8]. As the viruses begin their action and destroy the cancer cells, they release some tumor-related markers and antigen which generates an immune response [11]. This can potentially result in the stimulation of a systemic anti-tumor immune response that can target metastatic sites. Oncolytic viruses can also modify the host genome which acts as a viral vaccine in long run [32]. Oncolytic viruses are genetically engineered viruses, thus a large variety of gene combinations can be incorporated in the viral genome. This makes OVs very versatile vassals. For instance, they can be modified to carry and express immune stimulating molecules or enzymes that can convert non-toxic prodrugs into cytotoxic prodrugs. This ability enhances the oncolytic activity of the virus [14]. Some cancer cells develop resistance to conventional drugs and therapies; in such cases, virotherapy can perform very well since oncolytic viruses use various mechanisms to damage a cancer cell, this makes very hard for the cell to become resistant to oncolytic viruses [95]. Advancements in genetic engineering will provide a continuous improvement in the OVs therapy. In the future, viruses that are more advance and safe can be designed and developed which will provide better horizons in the treatment of cancer.

8 Demerits of Oncolytic Virus Therapy

The efficient delivery and introduction of oncolytic viruses into the tumor cell is still a space which has not been explored much. The intravenous administration of OVs can be hindered by the immune system and can cause serious complications. For now only intratumoral administration seems the most feasible way of delivery [25]. The excretion of OVs after the therapy is a part of open debate. The immune system might not allow the OVs to excrete out and start attacking the OVs and also those cells which have recently interacted with OVs [96]. Although mild selective but OV therapy can cause immune responses and produce side effects like inflammation, chills, fever, fatigue, and in some cases tissue damage [97].

9 Clinical Trials of OVs Therapy

Very recently clinical trials on the OVs therapy have been carried out utilizing "herpes simplex virus-1" and "adenovirus" for the design of novel dosing strategies. Friedman et al. tested a phase 1 clinical trial of OVs taking children and adolescents having 7–18 years of age with progressive supratentorial brain tumor after surgery, radiotherapy, and chemotherapy with intratumoral administration of the herpesvirus (HSV-1) G207 [98]. Next, Gállego Pérez-Larraya et al. finished a phase 1 clinical trial of adenovirus DNX-2401 taking children and adolescents (ages 1–18) having "diffuse intrinsic pontine glioma (DIPG)" [99]. Todo et al. carried out a phase 2 trial investigation of intratumoral administration of the herpesvirus G47Δ to the adults having more than 18 years of age suffering from supratentorial glioblastoma [100]. A significant reduction of the cancer cells resulted in the above three trials.

10 Current Breakthrough in Antiviral Therapies

The cervical cancer is most prevalent nowadays and is caused by sexually transmitted human papillomavirus (HPV), and it was reported that the HPV16 and HPV 18 caused 84% of HPV-related cancer lesions in cervix. Therefore, attempt has been made to design, develop, and formulate three types of HPV vaccines such as quadrivalent HPV vaccine, bivalent HPV vaccine, and nonavalent HPV vaccine which are commercially available for the prophylaxis to protect 90% against the HPV infection.

11 Conclusion

Oncolytic viral therapy, an innovative strategy for cancer treatment, hinges on the use of viruses that can selectively target and destroy cancer cells. The foundational concept, known as oncotropism, allows these viruses to infect and replicate within tumor cells, minimizing harm to normal tissues.

Historically, the notion of using viruses for cancer treatment has fascinated researchers for over a century. Several types of viruses, such as adenoviruses, herpes simplex viruses, and vaccinia viruses, have showcased potential in this therapeutic area. Scientists have modified these viruses to enhance their safety and oncolytic properties, leading to numerous clinical trials and some FDA-approved therapies.

This therapy not only directly kills cancer cells but also activates an antitumor immune response, making it a dual-action treatment. However, challenges remain. Enhancing tumor selectivity by making viruses more precise in targeting cancer cells, improving delivery methods to ensure viruses reach the tumor site, and managing pre-existing neutralizing antibodies that can diminish therapy effectiveness are crucial areas for future research.

Moreover, combining oncolytic viral therapy with other treatments such as chemotherapy or immunotherapy could amplify its effectiveness. Combinatorial approaches might offer more comprehensive anti-cancer strategies by leveraging multiple mechanisms of action.

Looking ahead, advancements in viral engineering and deeper insights into tumor biology are expected to yield more potent, specifically tailored viral agents. Ongoing clinical trials and expanded knowledge on how these viruses modulate the immune system will be instrumental in optimizing this therapeutic approach.

In summary, oncolytic viral therapy holds significant promise as a safe and effective cancer treatment. Clinical investigation has been going on and yet to come to the global market to combat cancer. With continued research and innovation, it has the potential to improve patient outcomes and offer new hope in the battle against various types of cancer.

References

1. Louten, J.: Virus structure and classification. Ess. Hum. Viro., 19–29 (2016)
2. Louten, J.: Detection and diagnosis of viral infections. Ess. Hum. Viro., 111–132 (2016)
3. Louten, J.: Virus replication. Ess. Hum. Viro., 49–70 (2016)
4. Zapatka, M., Borozan, I., Brewer, D.S., Iskar, M., Grundhoff, A., Alawi, M., Desai, N., Sültmann, H., Moch, H., PCAWG Pathogens, Cooper, C.S., Eils, R., Ferretti, V., Lichter, P., PCAWG Consortium: The landscape of viral associations in human cancers. Nat. Genet. **52**(3), 320–330 (2020)
5. Manton, K., Akushevich, I., Kravchenko, J.: Cancer Mortality and Morbidity Patterns in the U.S. Population: An Interdisciplinary Approach. Springer Science & Business (2008)
6. Hewitt, E.W.: The MHC class I antigen presentation pathway: strategies for viral immune evasion. Immunology. **110**(2), 163–169 (2003)

7. Jackson, R.: Saint Peregrine.: O.S.M.--the patron saint of cancer patients. Can. Med. Assoc. J. **111**(8), 824 (1974)
8. Bifulco, M., Zazzo, E.D., Napolitano, F., Malfitano, A.M., Portella, G.: History of how viruses can fight cancer: from the miraculous healings to the approval of oncolytic viruses. Biochimie. **206**, 89–92 (2023)
9. Vähä-Koskela, M.J., Heikkilä, J.E., Hinkkanen, A.E.: Oncolytic viruses in cancer therapy. Cancer Lett. **254**, 178–216 (2007)
10. Ayala-Breton, C., Barber, G.N., Russell, S.J., Peng, K.W.: Retargeting vesicular stomatitis virus using measles virus envelope glycoproteins. Hum. Gene Ther. **23**, 484–491 (2012)
11. Passer, B.J., Cheema, T., Zhou, B., Wakimoto, H., Zaupa, C., Razmjoo, M., et al.: Identification of the ENT1 antagonists dipyridamole and dilazep as amplifiers of oncolytic herpes simplex virus-1 replication. Cancer Res. **70**, 3890–3895 (2010)
12. Yang, X., Chen, E., Jiang, H., Muszynski, K., Harris, R.D., Giardina, S.L., et al.: Evaluation of IRES-mediated, cell-type-specific cytotoxicity of poliovirus using a colorimetric cell proliferation assay. J. Virol. Methods. **155**, 44–54 (2009)
13. Banaszynski, L.A., Sellmyer, M.A., Contag, C.H., Wandless, T.J., Thorne, S.H.: Chemical control of protein stability and function in living mice. Nat. Med. **14**, 1123–1127 (2008)
14. Stojdl, D.F., Lichty, B.D., ten Oever, B.R., Paterson, J.M., Power, A.T., Knowles, S., et al.: VSV strains with defects in their ability to shutdown innate immunity are potent systemic anti-cancer agents. Cancer Cell. **4**, 263–275 (2004)
15. Sugio, K., Sakurai, F., Katayama, K., Tashiro, K., Matsui, H., Kawabata, K., et al.: Enhanced safety profiles of the telomerase-specific replication-competent adenovirus by incorporation of normal cell-specific micro RNA-targeted sequences. Clin. Cancer Res. **17**, 2807–2818 (2011)
16. Wildner, O., Blaese, R.M., Morris, C.: Therapy of colon cancer with oncolytic adenovirus is enhanced by the addition of herpes simplex virus-thymidine kinase. Cancer Res. **59**, 410–413 (1999)
17. Foloppe, J., Kintz, J., Futin, N., Findeli, A., Cordier, P., Schlesinger, Y., et al.: Targeted delivery of a suicide gene to human colorectal tumors by a conditionally replicating vaccinia virus. Gene Ther. **15**, 1361–1371 (2008)
18. Shinoura, N., Yoshida, Y., Asai, A., Kirino, T., Hamada, H.: Adenovirus-mediated transfer of p 53 and Fas ligand drastically enhances apoptosis in gliomas. Cancer Gene Ther. **7**, 732–738 (2000)
19. Jiang, G., Li, J., Zeng, Z., Xian, L.: Lentivirus-mediated gene therapy by suppressing survivin in BALB/c nude mice bearing oral squamous cell carcinoma. Cancer Biol. Ther. **5**(4), 435–440 (2006)
20. Labrada, L., Bodel6n, G., Viñuela, J., Benavente, J.: Avian reoviruses cause apoptosis in cultured cells: viral uncoating, but not viral gene expression, is required for apoptosis induction. J. Virol. **76**, 7932–7941 (2002)
21. Galluzzi, L., Brenner, C., Morselli, E., Touat, Z., Kroemer, G.: Viral control of mitochondrial apoptosis. PLoS Pathog. **4**, e1000018 (2008)
22. Olijslagers, S., Dege, A.Y., Dinsart, C., Voorhoeve, M., Rommelaere, J., Noteborn, M.H., et al.: Potentiation of a recombinant oncolytic parvovirus by expression of Apoptin. Cancer Gene Ther. **8**, 958–965 (2001)
23. Russell, S.J., Peng, K.W., Bell, J.C.: Oncolytic virotherapy. Nat. Biotechnol. **30**, 658–670 (2012)
24. Thorne, S.H., Hwang, T.H., O'Gorman, W.E., Bartlett, D.L., Sei, S., Kanji, F., Brown, C., Werier, J., Cho, J.H., Lee, D.E., Wang, Y., Bell, J., Kirn, D.H.: Rational strain selection and engineering creates a broad-spectrum, systemically effective oncolytic poxvirus, JX-963. J. Clin. Invest. **117**, 3350–3358 (2007)

25. Li, L., Liu, S., Han, D., Tang, B., Ma, J.: Delivery and biosafety of oncolytic Virotherapy. Front. Oncol. **10**, 475 (2020)
26. Kaufman, H.L., Kohlhapp, F.J., Zloza, A.: Oncolytic viruses: a new class of immunotherapy drugs. Nat. Rev. Drug Discov. **14**, 642–662 (2015)
27. Chiocca, E.A., Rabkin, S.D.: Oncolytic viruses and their application to cancer immunotherapy. Cancer Immunol. Res. **2**, 295–300 (2014)
28. Marchini, A., Daeffler, L., Pozdeev, V.I., Angelova, A., Rommelaere, J.: Immune conversion of tumor microenvironment by oncolytic viruses: the Protoparvovirus H-1PV case study. Front. Immunol. **10**, 1848 (2019)
29. Conlon, K.C., Miljkovic, M.D., Waldmann, T.A.: Cytokines in the treatment of cancer. J. Interf. Cytokine Res. **39**, 6–21 (2019)
30. Shen, J., Xiao, Z., Zhao, Q., Li, M., Wu, X., Zhang, L., Hu, W., Cho, C.H.: Anti-cancer therapy with TNFα and IFNγ: a comprehensive review. Cell Prolif. **51**, e12441 (2018)
31. Kim, J.H., Lee, K.J., Lee, S.W.: Cancer immunotherapy with T-cell targeting cytokines: IL-2 and IL-7. BMB Rep. **54**, 21–30 (2021)
32. Zamarin, D., Holmgaard, R.B., Subudhi, S.K., Park, J.S., Mansour, M., Palese, P., Merghoub, T., Wolchok, J.D., Allison, J.P.: Localized oncolytic virotherapy overcomes systemic tumor resistance to immune checkpoint blockade immunotherapy. Sci. Transl. Med. **6**, 226–232 (2014)
33. Khare, R., May, S.M., Vetrini, F., Weaver, E.A., Palmer, D., Rosewell, A., Grove, N., Ng, P., Barry, M.A.: Generation of a Kupffer cell-evading adenovirus for systemic and liver-directed gene transfer. Mol. Ther. **19**, 1254–1262 (2011)
34. Varghese, S., Rabkin, S.D.: Oncolytic herpes simplex virus vectors for cancer virotherapy. Cancer Gene Ther. **9**(12), 967–978 (2002)
35. Rehman, H., Silk, A.W., Kane, M.P., et al.: Into the clinic: talimogene laherparepvec (T-VEC), a first-in-class intratumoral oncolytic viral therapy. J. Immunother. Cancer. **4**, 53 (2016)
36. Kohlhapp, F.J., Kaufman, H.L.: Molecular pathways: mechanism of action for talimogene laherparepvec, a new oncolytic virus immunotherapy. Clin. Cancer Res. **22**(5), 1048–1054 (2011)
37. Harrington, K.J., Puzanov, I., Hecht, J.R., et al.: Clinical development of talimogene laherparepvec (T-VEC): a modified herpes simplex virus type-1-derived oncolytic immunotherapy expert. Rev. Anticancer. Ther. **15**(12), 1389–1403 (2015)
38. Cody, J.J., Hurst, D.R.: Promising oncolytic agents for metastatic breast cancer treatment. Oncolytic Virother. **4**, 63–73 (2015)
39. Lasswitz, L., Chandra, N., Arnberg, N., Gerold, G.: Glycomics and proteomics approaches to investigate early adenovirus-host cell interactions. J. Mol. Biol. **430**, 1863–1882 (2018)
40. Uusi-Kerttula, H., Hulin-Curtis, S., Davies, J., Parker, A.L.: Oncolytic adenovirus: strategies and insights for vector design and immuno-oncolytic applications. Viruses. **7**, 6009–6042 (2015)
41. Hendrickx, R., Stichling, N., Koelen, J., Kuryk, L., Lipiec, A., Greber, U.F.: Innate immunity to adenovirus. Hum. Gene Ther. **25**, 265–284 (2014)
42. Tazawa, H., Kuroda, S., Hasei, J., Kagawa, S., Fujiwara, T.: Impact of autophagy in oncolytic adenoviral therapy for cancer. Int. J. Mol. Sci. **18**, 1479 (2017)
43. Rodriguez-Rocha, H., Gomez-Gutierrez, J.G., Garcia-Garcia, A., Rao, X.M., Chen, L., McMasters, K.M., Zhou, H.S.: Adenoviruses induce autophagy to promote virus replication and oncolysis. Virology. **416**, 9–15 (2011)
44. Wold, W.S., Toth, K.: Adenovirus vectors for gene therapy, vaccination and cancer gene therapy. Curr. Gene Ther. **13**, 421–433 (2013)

45. Ros, C., Bayat, N., Wolfisberg, R., Almendral, J.M.: Protoparvovirus cell entry. Viruses. **9**(11), 313 (2017)
46. Marchini, A., Bonifati, S., Scott, E.M., Angelova, A.L., Rommelaere, J.: Oncolytic parvoviruses: from basic virology to clinical applications. Virol. J. **12**, 6 (2015)
47. Geletneky, K., Leoni, A.L., Pohlmeyer-Esch, G., Loebhard, S., Baetz, A., Leuchs, B., Roscher, M., Hoefer, C., Jochims, K., Dahm, M., Huber, B., Rommelaere, J., Krebs, O., Hajda, J.: Pathology, organ distribution, and immune response after single and repeated intravenous injection of rats with clinical-grade parvovirus H1. Comp. Med. **65**, 23–35 (2015)
48. Moralès, O., Richard, A., Martin, N., Mrizak, D., Sénéchal, M., Miroux, C., Pancré, V., Rommelaere, J., Caillet-Fauquet, P., de Launoit, Y., Delhem, N.: Activation of a helper and not regulatory human CD4+ T cell response by oncolytic H-1 parvovirus. PLoS.One. **7**, 32197 (2012)
49. Di Piazza, M., Mader, C., Geletneky, K., Herrero, Y., Calle, M., Weber, E., Schlehofer, J., Deleu, L., Rommelaere, J.: Cytosolic activation of cathepsins mediates parvovirus H-1-induced killing of cisplatin and TRAIL-resistant glioma cells. J. Virol. **81**, 4186–4198 (2007)
50. Nüesch, J.P., Lacroix, J., Marchini, A., Rommelaere, J.: Molecular pathways: rodent parvoviruses--mechanisms of oncolysis and prospects for clinical cancer treatment. Clin. Cancer Res. **18**, 3516–3523 (2012)
51. Smith, G.L., Benfield, C.T.O., Maluquer de Motes, C., Mazzon, M., Ember, S.W.J., Ferguson, B.J., Sumner, R.P.: Vaccinia virus immune evasion: mechanisms, virulence and immunogenicity. J. Gen. Virol. **94**, 2367–2392 (2013)
52. Chon, H.J., Lee, W.S., Yang, H., Kong, S.J., Lee, N.K., Moon, E.S., Choi, J., Han, E.C., Kim, J.H., Ahn, J.B., Kim, C.: Tumor microenvironment remodeling by intratumoral oncolytic vaccinia virus enhances the efficacy of immune-checkpoint blockade. Clin. Cancer Res. **25**, 1612–1623 (2019)
53. Deng, L., Yang, X., Fan, J., Ding, Y., Peng, Y., Xu, D., Huang, B., Hu, Z.: An oncolytic vaccinia virus armed with GM-CSF and IL-24 double genes for cancer targeted therapy. Onco. Targets. Ther. **13**, 3535–3544 (2020)
54. Guo, Z.S., Lu, B., Guo, Z., Giehl, E., Feist, M., Dai, E., Liu, W., Storkus, W.J., He, Y., Liu, Z., Bartlett, D.L.: Vaccinia virus-mediated cancer immunotherapy: cancer vaccines and oncolytics. J. Immunother. Cancer. **7**, 6 (2019)
55. Breitbach, C.J., Arulanandam, R., De Silva, N., Thorne, S.H., Patt, R., Daneshmand, M., Moon, A., Ilkow, C., Burke, J., Hwang, T.H., Heo, J., Cho, M., Chen, H., Angarita, F.A., Addison, C., McCart, J.A., Bell, J.C., Kirn, D.H.: Oncolytic vaccinia virus disrupts tumor-associated vasculature in humans. Cancer Res. **73**, 1265–1275 (2013)
56. Mac Kie, R.M., Hauschild, A., Eggermont, A.M.: Epidemiology of invasive cutaneous melanoma. Ann. Oncol. **20**(6), 1–7 (2009)
57. Duncan, L.M.: The classification of cutaneous melanoma. Hematol. Oncol. Clin. North. Am. **23**, 501–513 (2009)
58. Caini, S., Gandini, S., Sera, F., Raimondi, S., Fargnoli, M.C., Boniol, M., Armstrong, B.K.: Meta-analysis of risk factors for cutaneous melanoma according to anatomical site and clinico-pathological variant. Eur. J. Cancer. **45**, 3054–3063 (2009)
59. Tartaglia, J., Pincus, S., Paoletti, E.: Poxvirus-based vectors as vaccine candidates. Crit. Rev. Immunol. **10**, 13–30 (1990)
60. Kaufman, H.L., Ruby, C.E., Hughes, T., Slingluff Jr., C.L.: Current status of granulocyte-macrophage colony-stimulating factor in the immunotherapy of melanoma. J. Immunother. Cancer. **2**, 11 (2014)
61. Hwang, T.H., Moon, A., Burke, J., Ribas, A., Stephenson, J., Breitbach, C.J., Daneshmand, M., De Silva, N., Parato, K., Diallo, J.S., Lee, Y.S., Liu, T.C., Bell, J.C., Kirn, D.H.: A

mechanistic proof-of-concept clinical trial with JX-594, a targeted multi-mechanistic oncolytic poxvirus, in patients with metastatic melanoma. Mol. Ther. **19**, 1913–1922 (2011)
62. Mastrangelo, M.J., Maguire Jr., H.C., Eisenlohr, L.C., Laughlin, C.E., Monken, C.E., McCue, P.A., Kovatich, A.J., Lattime, E.C.: Intratumoral recombinant GM-CSF-encoding virus as gene therapy in patients with cutaneous melanoma. Cancer Gene Ther. **16**, 409–422 (1999)
63. Kaufman, H.L., Deraffele, G., Mitcham, J., Moroziewicz, D., Cohen, S.M., Hurst-Wicker, K.S., Cheung, K., Lee, D.S., Divito, J., Voulo, M., Donovan, J., Dolan, K., Manson, K., Panicali, D., Wang, E., Hörig, H., Marincola, F.M.: Targeting the local tumor microenvironment with vaccinia virus expressing B7.1 for the treatment of melanoma. J. Clin. Invest. **115**, 1903–1912 (2005)
64. Kaufman, H.L., Cohen, S., Cheung, K., DeRaffele, G., Mitcham, J., Moroziewicz, D., Schlom, J., Hesdorffer, C.: Local delivery of vaccinia virus expressing multiple costimulatory molecules for the treatment of established tumors. Hum. Gene Ther. **17**, 239–244 (2006)
65. Dharmadhikari, N., Mehnert, J.M., Kaufman, H.L.: Oncolytic virus immunotherapy for melanoma. Curr. Treat. Options in Oncol. **16**, 326 (2015)
66. Poh, A.: First oncolytic viral therapy for melanoma. Cancer Discov. **6**, 6 (2016)
67. Hu, J.C., Coffin, R.S., Davis, C.J., Graham, N.J., Groves, N., Guest, P.J., Harrington, K.J., James, N.D., Love, C.A., McNeish, I., Medley, L.C., Michael, A., Nutting, C.M., Pandha, H.S., Shorrock, C.A., Simpson, J., Steiner, J., Steven, N.M., Wright, D., Coombes, R.C.: A phase I study of Onco VEXGM-CSF, a second-generation oncolytic herpes simplex virus expressing granulocyte macrophage colony-stimulating factor. Clin. Cancer Res. **12**, 6737–6747 (2016)
68. Senzer, N.N., Kaufman, H.L., Amatruda, T., Nemunaitis, M., Reid, T., Daniels, G., Gonzalez, R., Glaspy, J., Whitman, E., Harrington, K., Goldsweig, H., Marshall, T., Love, C., Coffin, R., Nemunaitis, J.J.: Phase II clinical trial of a granulocyte-macrophage colony-stimulating factor-encoding, second-generation oncolytic herpesvirus in patients with unresectable metastatic melanoma. J. Clin. Oncol. **27**, 5763–5771 (2009)
69. Kaufman, H.L., Bines, S.D.: OPTIM trial: a Phase III trial of an oncolytic herpes virus encoding GM-CSF for unresectable stage III or IV melanoma. Future Oncol. **6**, 941–949 (2010)
70. Sahin, T.T., Kasuya, H., Nomura, N., Shikano, T., Yamamura, K., Gewen, T., Kanzaki, A., Fujii, T., Sugae, T., Imai, T., Nomoto, S., Takeda, S., Sugimoto, H., Kikumori, T., Kodera, Y., Nishiyama, Y., Nakao, A.: Impact of novel oncolytic virus HF10 on cellular components of the tumor microenviroment in patients with recurrent breast cancer. Cancer Gene Ther. **19**, 229–237 (2012)
71. Watanabe, D., Goshima, F., Mori, I., Tamada, Y., Matsumoto, Y., Nishiyama, Y.: Oncolytic virotherapy for malignant melanoma with herpes simplex virus type 1 mutant HF10. J. Dermatol. Sci. **50**, 185–196 (2008)
72. Tanaka, R., Goshima, F., Esaki, S., Sato, Y., Murata, T., Nishiyama, Y., Watanabe, D., Kimura, H.: The efficacy of combination therapy with oncolytic herpes simplex virus HF10 and dacarbazine in a mouse melanoma model. Am. J. Cancer Res. **7**, 1693–1703 (2017)
73. Au, G.G., Lindberg, A.M., Barry, R.D., Shafren, D.R.: Oncolysis of vascular malignant human melanoma tumors by Coxsackievirus A21. Int. J. Oncol. **26**, 1471–1476 (2005)
74. Shafren, D.R., Au, G.G., Nguyen, T., Newcombe, N.G., Haley, E.S., Beagley, L., Johansson, E.S., Hersey, P., Barry, R.D.: Systemic therapy of malignant human melanoma tumors by a common cold-producing enterovirus, coxsackievirus a21. Clin. Cancer Res. **10**, 53–60 (2004)
75. Andtbacka, R.H.I., Curti, B.D., Kaufman, H., Daniels, G.A., Nemunaitis, J.J., Spitler, L.E., Hallmeyer, S., Lutzky, J., Schultz, S., Whitman, E.D., Zhou, K., Weisberg, J.I., Shafren, D.: CALM study: a phase II study of intratumoral coxsackievirus A21 in patients with stage IIIc and stage IV malignantmelanoma. J. Clin. Oncol. **32**, 3031–3031 (2014)

76. Galanis, E., Markovic, S.N., Suman, V.J., Nuovo, G.J., Vile, R.G., Kottke, T.J., Nevala, W.K., Thompson, M.A., Lewis, J.E., Rumilla, K.M., Roulstone, V., Harrington, K., Linette, G.P., Maples, W.J., Coffey, M., Zwiebel, J., Kendra, K.: Phase II trial of intravenous administration of Reolysin(®) (Reovirus Serotype-3-Dearing strain) in patients with metastatic melanoma. Mol. Ther. **20**, 1998–2003 (2012)
77. Xie, Y., Xiang, Y., Sheng, J., Zhang, D., Yao, X., Yang, Y., Zhang, X.: Immunotherapy for hepatocellular carcinoma: current advances and future expectations. J Immunol Res. **2018**, 8740976 (2018)
78. Johnston, M.P., Khakoo, S.I.: Immunotherapy for hepatocellular carcinoma: current and future. World J. Gastroenterol. **25**, 2977–2989 (2019)
79. Yoo, S.Y., Badrinath, N., Woo, H.Y., Heo, J.: Oncolytic virus-based immunotherapies for hepatocellular carcinoma. Mediat. Inflamm. **2017**, 5198798 (2017)
80. Luo, Y., Lin, C., Ren, W., Ju, F., Xu, Z., Liu, H., Yu, Z., Chen, J., Zhang, J., Liu, P., Huang, C., Xia, N.: Intravenous injections of a rationally selected oncolytic herpes virus as a potent Virotherapy for hepatocellular carcinoma. Mol. Ther. Oncolytics. **15**, 153–165 (2019)
81. Chen, Z., Xie, H., Hu, M., Huang, T., Hu, Y., Sang, N., Zhao, Y.: Recent progress in treatment of hepatocellular carcinoma. Am. J. Cancer Res. **10**, 2993–3036 (2020)
82. Bai, Y.H., Yun, X.J., Xue, Y., Zhou, T., Sun, X., Gao, Y.J.: A novel oncolytic adenovirus inhibits hepatocellular carcinoma growth. J. Zhejiang Univ. Sci. B. **20**, 1003–1013 (2019)
83. De Cicco, P., Catani, M.V., Gasperi, V., Sibilano, M., Quaglietta, M., Savini, I.: Nutrition and breast cancer: a literature review on prevention, treatment and recurrence. Nutrients. **11**, 1514 (2019)
84. Armstrong, N., Ryder, S., Forbes, C., Ross, J., Quek, R.G.: A systematic review of the international prevalence of BRCA mutation in breast cancer. Clin. Epidemiol. **11**, 543–561 (2019)
85. Yedjou, C.G., Sims, J.N., Miele, L., Noubissi, F., Lowe, L., Fonseca, D.D., Alo, R.A., Payton, M., Tchounwou, P.B.: Health and racial disparity in breast cancer. Adv. Exp. Med. Biol. **1152**, 31–49 (2019)
86. Bryan, S.M., Mathis, J.M.: Oncolytic Virotherapy for breast cancer treatment. Curr. Gene Ther. **18**, 192–205 (2018)
87. Eissa, I.R., Bustos-Villalobos, I., Ichinose, T., Matsumura, S., Naoe, Y., Miyajima, N., Morimoto, D., Mukoyama, N., Zhiwen, W., Tanaka, M., Hasegawa, H., Sumigama, S., Aleksic, B., Kodera, Y., Kasuya, H.: The current status and future prospects of oncolytic viruses in clinical trials against melanoma, glioma, pancreatic, and breast cancers. Cancers (Basel). **10**, 356 (2018)
88. Cao, G.D., He, X.B., Sun, Q., Chen, S., Wan, K., Xu, X., Feng, X., Li, P.P., Chen, B., Xiong, M.M.: The oncolytic virus in cancer diagnosis and treatment. Front. Oncol. **10**, 1786 (2020)
89. Kimata, H., Imai, T., Kikumori, T., Teshigahara, O., Nagasaka, T., Goshima, F., Nishiyama, Y., Nakao, A.: Pilot study of oncolytic viral therapy using mutant herpes simplex virus (HF10) against recurrent metastatic breast cancer. Ann. Surg. Oncol. **13**, 1078–1084 (2006)
90. Siegel, R.L., Miller, K.D., Jemal, A.: Cancer statistics, 2017. C.A. Cancer. J. Clin. **67**, 7–30 (2017)
91. Park, S.H., Breitbach, C.J., Lee, J., Park, J.O., Lim, H.Y., Kang, W.K., Moon, A., Mun, J.H., Sommermann, E.M., Maruri Avidal, L., Patt, R., Pelusio, A., Burke, J., Hwang, T.H., Kirn, D., Park, Y.S.: Phase 1b Trial of Biweekly Intravenous Pexa-Vec (JX-594), an Oncolytic and Immunotherapeutic Vaccinia Virus in Colorectal Cancer. Mol. Ther. **23**, 1532–1540 (2015)
92. Morelli, M.P., Xie, C., Brar, G., Floudas, C.S., Fioravanti, S., Walker, M., Mabry-Hrones, D., Greten, T.F.: A phase I/II study of pexa-vec oncolytic virus in combination with immune checkpoint inhibition in refractory colorectal cancer: safety report. J. Clin. Oncol. **37**, 646 (2019)

93. Zhang, W., Wang, F., Hu, X., Liang, J., Liu, B., Guan, Q., Liu, S.: Inhibition of colorectal cancer liver metastasis in BALB/c mice following intratumoral injection of oncolytic herpes simplex virus type 2 for the induction of specific antitumor immunity. Oncol. Lett. **17**, 815–822 (2019)
94. Vacchelli, E., Eggermont, A., Sautès-Fridman, C., Galon, J., Zitvogel, L., Kroemer, G., Galluzzi, L.: Trial watch: oncolytic viruses for cancer therapy. Oncoimmunology. **2**, e24612 (2013)
95. Liu, S., Li, F., Ma, Q., Du, M., Wang, H., Zhu, Y., Deng, L., Gao, W., Wang, C., Liu, Y., Zhao, Z., Liu, H., Wang, R., Tian, Y., Hu, M., Wan, Y., Lu, W., Zhang, M., Zhao, M., Cao, Y., Zhang, H., Wang, W., Wang, H., Wang, Y.: OX40L-armed oncolytic virus boosts T-cell response and remodels tumor microenvironment for pancreatic cancer treatment. Theranostics. **13**(12), 4016–4029 (2023)
96. Fröhlich, A., Niebel, D., Fietz, S., Egger, E., Buchner, A., Sirokay, J., Landsberg, J.: Talimogene laherparepvec treatment to overcome loco-regional acquired resistance to immune checkpoint blockade in tumor stage IIIB-IV M1c melanoma patients. Cancer Immunol. Immunother. **69**(5), 759–769 (2020)
97. Zhang, J., He, Q., Mao, D., Wang, C., Huang, L., Wang, M., Zhang, J.: Efficacy and adverse reaction management of oncolytic viral intervention combined with chemotherapy in patients with liver metastasis of gastrointestinal malignancy. Front. Oncol. **13**, 1159802 (2023)
98. Friedman, G.K., Johnston, J.M., Bag, A.K., et al.: Oncolytic HSV-1 G207 immunovirotherapy for pediatric high-grade gliomas. N. Engl. J. Med. **384**(17), 1613–1622 (2021)
99. Gállego Pérez-Larraya, J., Garcia-Moure, M., Labiano, S., et al.: Oncolytic DNX-2401 virus for pediatric diffuse intrinsic pontine glioma. N. Engl. J. Med. **386**(26), 2471–2481 (2022)
100. Todo, T., Ito, H., Ino, Y., et al.: Intratumoral oncolytic herpes virus G47Δ for residual or recurrent glioblastoma: a phase 2 trial. Nat. Med. **28**(8), 1630–1639 (2022)

QSAR of Repurposed Anti-COVID-19 Chemotherapeutics Utilizing Theoretical Structural Descriptors

Laxmi Kotwal[1], Priyanka Maindoliya[1], and Sisir Nandi[1] (✉)

[1] Global Institute of Pharmaceutical Education and Research (GIPER), Kashipur, Uttarakhand, India

Abstract By the end of December 2019, the novel coronavirus (nCoV) caused COVID-19 to erupt from Wuhan, China. It was speeded by leaps and bounds and very shortly nCoV attacked more than 180 countries around the world. It killed mil lions of million people. The lives had been forcefully kept under the complete lock-down. The economy and lives of the country had been devastated and people were crying but there is no small molecule therapy developed to date except a few approved vaccines such as COVISHIELD, COVAXIN, CORONAVAC, etc. Natural medicines and a few synthetic small molecules such as remdesivir, favipiravir, ribavirin, penciclovir, nitazoxanide, nafamostat, chloroquine, hydroxychloroquine, umifenovir, daclatasvir, ivermectin, etc. were repurposed to combat COVID-19. The chemobioinformatic scientists tried to screen new molecules to combat the nCoV. The quantitative structure-activity relationship (QSAR) is a chemoinformatic tool for the ligand-based drug design and screening of bioactive molecules. In the present attempt computed descriptor-based QSAR models have been developed to capture the promising features that would focus on biochemical mechanisms for the designing of new therapeutics.

Keywords Novel coronavirus · COVID-19 · Calculated theoretical descriptors · QSAR modeling

1 Introduction

COVID-19 is a highly contagious viral disease caused by a novel strain of severe acute respiratory syndrome coronavirus 2 (nCoV or SARS-CoV-2). Out of 704,753,890 COVID-19 active cases, 7,010,681 deaths occurred reported as of April 13, 2024. In late December 2019 the first case was reported in Wuhan, China, followed by the rapid spread of this virus in a short span caused the pandemic, which was declared as COVID-19 by World Health Organization (WHO) on March 11, 2020 [1]. Symptoms in humans are SARS-associated high fever or chills, dry cough, throat pain, shortness

A. K. Saxena, S. Nandi (eds.), *Global Trends in Health, Technology and Management*, https://doi.org/10.1007/978-3-031-75457-9_16

of breath, lungs effusion, bronchitis, pneumonia, muscle and body pain, pharyngitis, sinusitis, and others such as loss of taste and smell, and occasional diarrhea. The dangerous condition occurs by happy hypoxia, circulatory arrests, and multiple organ failure [2].

The novel coronavirus attaches the host target, after entering through the mouth or nostrils, with the help of spike protein–host ACE-2 interaction. The fusion of corona viral-host target proteins occurs through hemagglutinin esterase, then the viral replication in the host cell is started by nCoV main protease responsible for the corona viral replication and transcription. The virus had been neuro-invasive because the ACE-2 receptor is expressed in brain neuronal cells [3].

The general public had been forcefully kept under the complete lockdown situation. The economy and lives of countries had been devastated. People were under threat due to the dreadful attack of nCoV because there is no specific antiviral treatment to date. The doctor scientists gave symptomatic treatment by repurposing a few small molecule inhibitors. Computational scientists screened few natural and synthetic drugs for the treatment of COVID-19 utilizing chemometric simulation [4]. One of the most useful chemometric tools is ligand-based quantitative structure-activity relationship (QSAR) based on the structural descriptors that are numerical quantification of a certain structure to generate information about the property. The molecular descriptor-based QSAR model is often used as an optimization tool in drug design and development research. In this connection, the quantitative structure-activity relationship models for 15 repurposed anti-COVID-19 compounds have been developed taking theoretical descriptors solely computed from the structures of these studied compounds to assign the crucial properties responsible for the anti-corona viral action.

2 Materials and Methods

2.1 Compounds' Biological Activity Data

A number of 15 drugs such as remdesivir, favipiravir, ribavirin, penciclovir, nitazoxanide, nafamostat, chloroquine (RAC), R-chloroquine, S-chloroquine, hydroxychloroquine (RAC), R-hydroxychloroquine, S-hydroxychloroquine, ivermectin, daclatasvir, and umifenovir along with their biological activity to combat COVID-19 have been taken as data set. Table 1 contains molecular structures along with activity data in terms of 50% effective concentration (EC_{50}) of these compounds. The EC_{50} means median effective concentration (EC_{50}) which is the concentration of a substance to produce a certain effect in 50% of test organisms. The pEC_{50} is the negative logarithm of EC_{50}. A higher pIC50 value means higher inhibitory activity of the corresponding compound. The pEC_{50} is taken as the dependent variable for the QSAR modeling. Wang et al. carried out a standard assay to measure the effect of these compounds such as remdesivir, favipiravir, ribavirin, penciclovir, nitazoxanide, and nafamostat utilizing Vero E6 cells (ATCC-1586) of nCoV-2019BetaCoV/Wuhan/WIV04/2019 strain [5]. To quest the anti-corona viral activity of antiparasitic ivermectin, the Vero/hSLAM cells infected with SARS-CoV-2 clinical isolate strain of Australia/VIC01/2020 were taken [6]. The

Table 1 Biological activity data of the studied 15 drugs

Sl. No	Drug name	Structure	**EC$_{50}$** uM	pEC50
1.	Remdesivir	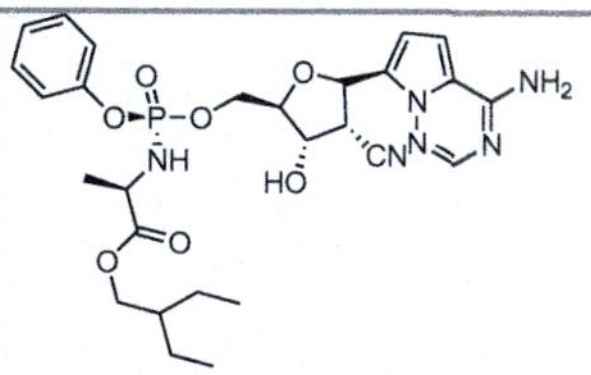	0.77	0.113
2.	Favipiravir	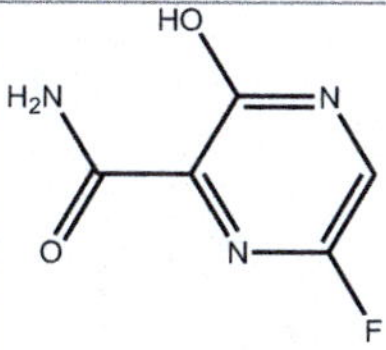	61.88	1.791
3.	Ribavirin	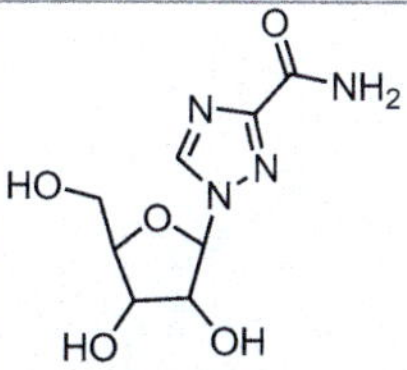	109.50	−2.039
4.	Penciclovir	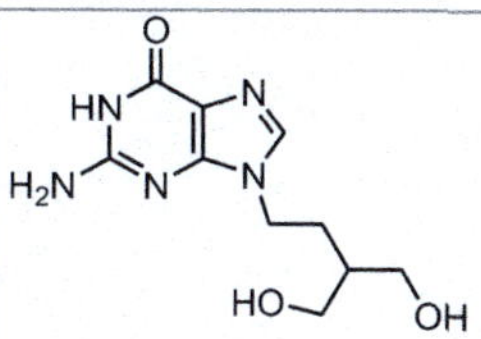	95.56	−1.980
5.	Nitazoxanide	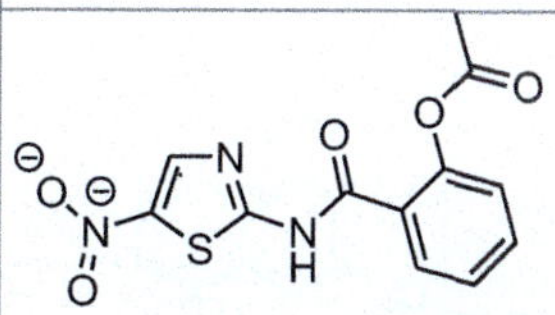	2.12	−0.326
6.	Nafamostat	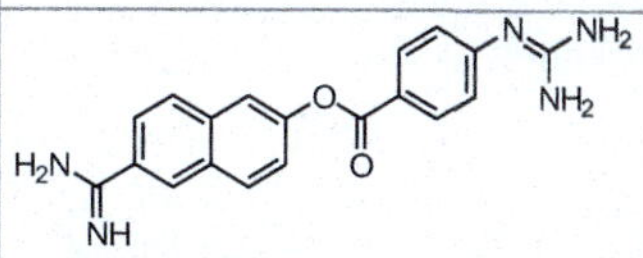	22.50	−1.352
7.	Ivermectin	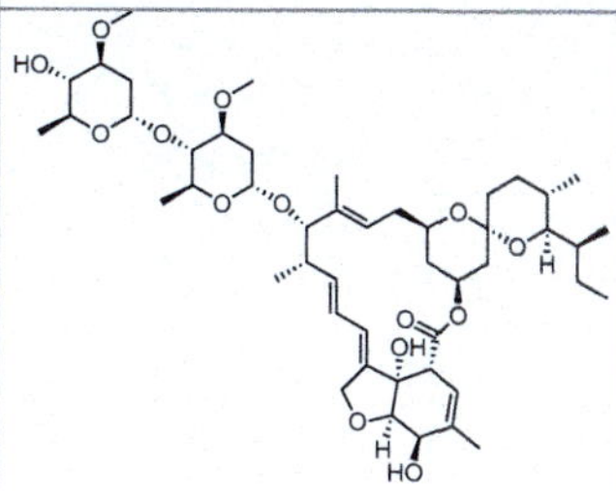	2.4	−0.380

(continued)

Table 1 (continued)

Sl. No	Drug name	Structure	**EC$_{50}$** uM	pEC50
8.	Chloroquine (Rac)	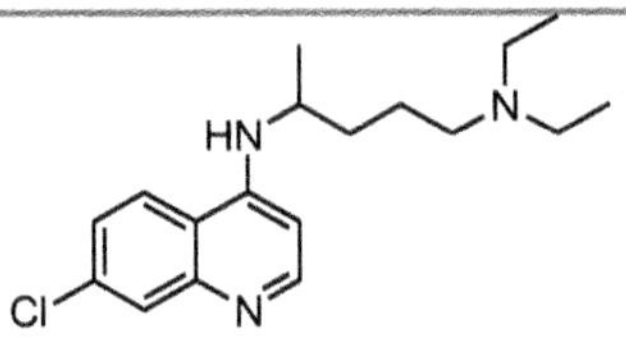	1.801	−0.255
9.	S-chloroquine	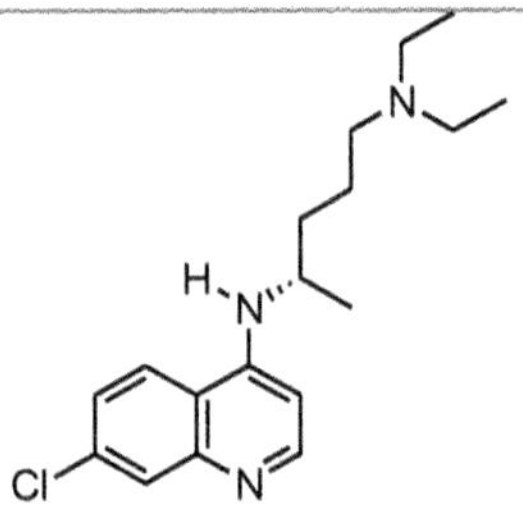	1.975	−0.295
10.	R-chloroquine	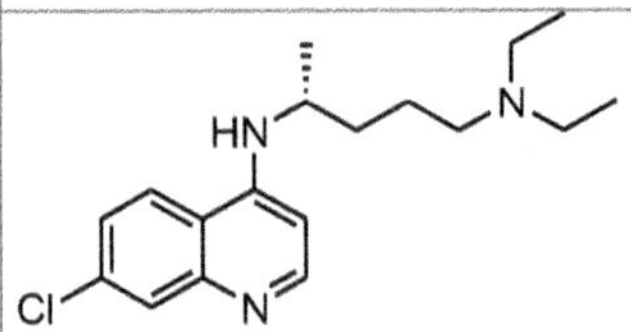	1.752	−0.245
11.	Hydroxychloroquine (RAC)	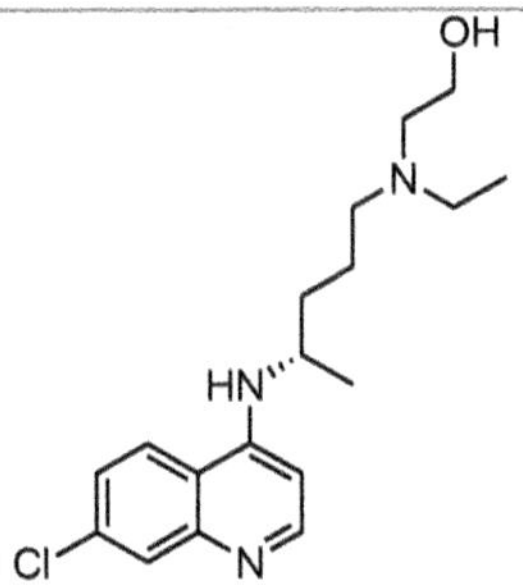	1.752	−0.243
12.	S-hydroxychloroquine	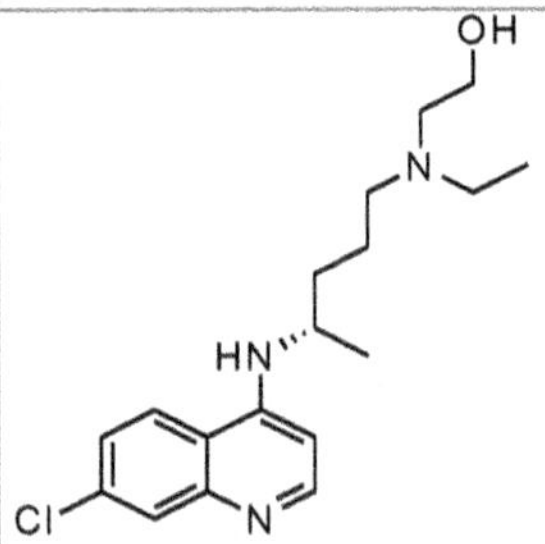	2.445	−0.386
13.	R-hydroxychloroquine	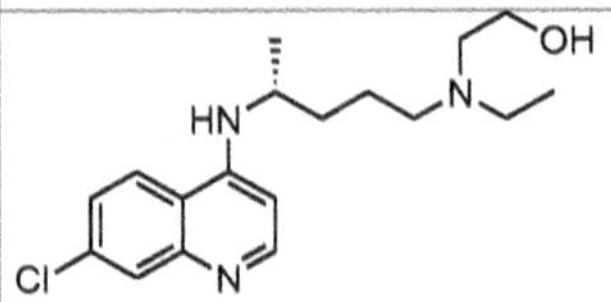	1.444	−0.159

(continued)

Table 1 (continued)

Sl. No	Drug name	Structure	EC_{50} uM	pEC50
14.	Daclatasvir	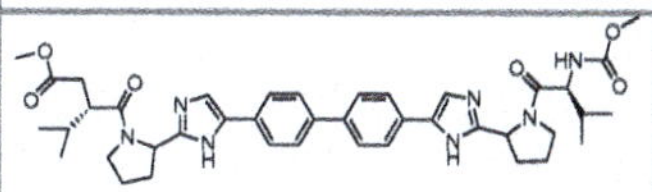	0.8	0.096
15.	Umifenovir	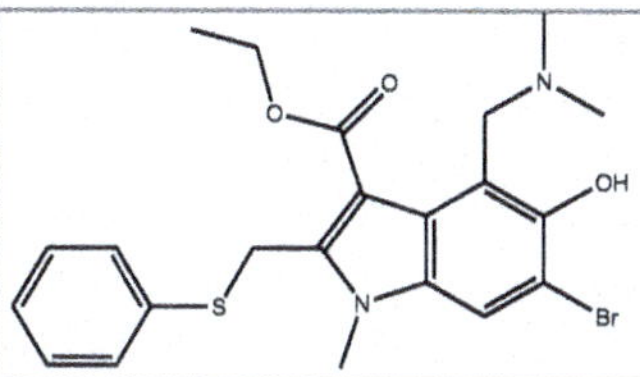	4.1	−0.612

chloroquine and hydroxychloroquine, well-known antimalarials, had been reused to combat nCoV. The racemate chloroquine (CQ), hydroxychloroquine (HCQ), and their corresponding enantiomers have been tested on SARS-CoV-2 main protease to check the in vitro anti-COVID-19 activities of these compounds [7]. The anti-hepatitis C viral drug daclatasvir [8] and anti-influenza umifenovir were shown to be effective against COVID-19 [9].

2.2 Structure Drawing, Molecular Optimization, and Descriptor Computation

The 2D structure of all studied 15 compounds was drawn using 2D ChemDraw 8.0 software. The drawn 2D structures were transformed into 3D modules on the ChemUltraDraw. The molecular optimization is done for the energy minimizations using Merck molecular force field (MMFF) and MMFF charge-taking distance-dependent dielectric constant of 1.0 and the convergence criterion of 0.01 kcal/mol values, respectively [10].

This will ensure a well-stable 3D conformer having minimal surface energy. All these optimized structures were incorporated in the PaDEL software for the computation of various 1D, 2D, and 3D theoretical descriptors including physicochemical, electrostatic, constitutional, geometrical, and topological [11].

Total structural independent variables were calculated as 1875 which is more than enough and may contain highly correlated descriptors. Therefore, descriptor thinning is done by highly intercorrelated descriptors considering variance and correlation coefficient cut-off values of 0.0001 and 0.99 using V-WSP algorithmic NanoBRIDGES Software [12]. Therefore, 1875 number of structural descriptors were reduced into 311. The abbreviated descriptor data are given in Table 2.

2.3 QSAR Modeling

The reduced theoretical molecular descriptors data of 311 were taken as independent structural invariants while the negative logarithm of effective concentration 50 of the molecules (Biological Activity, pEC50) of the studied 15 compounds was taken as the

Table 2 Abbreviated structural descriptor data after thinning

ALogP, ALogp2, AMR, nN, nH, nO, ATSOm, ATS3m, AATS0m, AATS5m, AATS1m, AATS8m, AATS6m, AATS2v, AATS0v, AATS5v, AATS3v, AATS7v, AATS6v, AATS6e, AATS0e, AATS8e, AATS7e, AATS0p, AATS2p, AATS5p, AATS3p, AATS0i, AATS7p, AATS2i, AATS1i, ATSC1c, AATS3i, ATSC3c, ATSC2c, ATSC4c, ATSC6c, ATSC5c, ATSC8c, ATSC7c, ATSC1m, ATSC0m, ATSC2m, ATSC4m, ATSC3m, ATSC5m, ATSC6m, ATSC8m, ATSC7m, ATSC2v, ATSC1v, ATSC3v, ATSC5v, ATSC4v, ATSC6v, ATSC1e, ATSC8v, ATSC2e, ATSC4e, ATSC3e, ATSC6e, ATSC5e, ATSC8e, ATSC7e, ATSC1p, ATSC3p, ATSC2p, ATSC4p, ATSC5p, ATSC7p, ATSC6p, ATSC8p, ATSC2i, ATSC1i, ATSC3i, ATSC5i, ATSC4i, ATSC7i, ATSC6i, ATSC8i, AATSC0m, AATSC0c, AATSC2m, AATSC1m, AATSC4m, AATSC3m, AATSC6m, AATSC5m, AATSC8m, AATSC2v, AATSC1v, AATSC4v, AATSC3v, AATSC6v, AATSC5v, AATSC1e, AATSC8v, AATSC2e, AATSC5e, AATSC3e, AATSC0p, AATSC8e, AATSC3p, AATSC1p, AATSC6p, AATSC5p, AATSC7p, AATSC0i, AATSC8p, AATSC1i, AATSC2i, AATSC4i, AATSC6i, AATSC3i, AATSC8i, AATSC5i, AATSC7i, MATS2c, MATS1c, MATS4c, MATS3c, MATS6c,MATS5c, MATS8c, MATS7c, MATS2m, MATS4m, MATS1e, MATS6m, MATS3e, MATS2e, MATS5e, MATS4e, MATS2p, MATS7e, MATS1i, MATS1s, MATS3s, MATS2s, MATS5s, MATS4s, MATS7s, MATS6s, GATS1c, MATS8s, GATS4c, GATS3c, GATS5c, GATS6c, GATS7c, GATS8c, GATS1m, GATS2m, GATS3m, GATS4m, GATS5m, GATS7m, GATS8m, GATS2v, GATS1v, GATS4v, GATS3v, GATS8v, GATS6v, GATS1e, GATS3e, GATS2e, GATS5e, GATS4e, GATS8e, GATS2p, GATS1p, GATS4p, GATS3p, GATS6p, GATS5p, GATS1i, GATS8p, GATS3i, GATS2i, GATS4i, GATS5i, GATS6i, GATS8i, GATS1s, GATS2s, GATS3s, GATS4s, GATS5s, GATS6s, GATS8s, GATS7s, SpAbsDzZ, SM1_DzZ, SpMAD_DzZ, VE3_DzZ, VE1_DzZ, VR1_DzZ, VE1_Dzv, VR2_DzZ, VR1_Dzv, VE1_Dze, VR2_Dzv, VE1_Dzp, VR2_Dze, VR1_Dzp, VE3_Dzp, VR1_Dzi, VR2_Dzp, VR2_Dzi, SpMAD_Dzs, SM1_Dzs, VE3_Dzs, VE1_Dzs, VR1_Dzs, BCUTw-1h, BCUTw-1l, BCUTc-1h, BCUTc-1l, BCUTp-1h, BCUTp-1l, NBondsD, C3SP2, C2SP2, C2SP3, C1SP3, SCH-6, SCH-7SC-5, ASP-0, ASP-1, ASP-2, AVP-0, CrippenLogP, VE1_Dt, VE3_Dt, VR2_Dt, NHBd, NHsOH, NHssNH, NHdsCH, ndssC, SHBd, MinHsOH, MaxHBd, MinHCsats, MaxwHBa, MaxHBa, maxHBint5, maxHssNH, maxHBint6, maxHCsats, maxHdsCH, maxdsN, ETA_Epsilon_5, ETA_AlphaP, ETA_Shape_P, ETA_dEpsilon_B, ETA_Shape_Y, ETA_BetaP, ETA_dBetaP, ETA_BetaP_s, ETA_Eta_F_L, ETA_EtaP_F, FMF, nHBDon_Lipinski, HybRatio, IC0, IC1, IC2, IC3, SIC2, SIC0, SIC4, CIC0, SIC3, BIC4, BIC3, MIC0, MIC1, MIC2, ZMIC2, ZMIC4,NAtomLC, NAtomP, NAtomLAC, MDEC-33, MDEN-23, MDEC-22, MLFER_A, MDEN-22, MLFER_S, MLFER_BH, MLFER_E, piPC3, piPC5, R_TpiPCTPC, PetitjeanNumber,n6Ring, NRotB, RotBtFrac, RotBFrac, Lipinski Failures, JGI1,JGI2, JGI3, JGTVE1_D, VR2_D, TopoPSA

dependent variable for QSAR modeling. Since the thinned descriptors number (311) is much greater than the number of compounds (15), descriptor selection is done using the genetic algorithm (GA) method which gives the best solution for stochastic optimization. The GA method creates many new combinations of descriptors that are used as independent data for QSAR modeling [13]. In the present study, many QSARs have been formulated on anti-COVID-19 compounds, the best model has been reported to predict the important molecular properties for further optimization.

3 Results and Discussion

To formulate a quantitative model between these computed structural descriptors and the anti-COVID-19 activity of the studied compounds, QSAR for the whole observation data of (N = 15) the repurposed drugs has been developed using the genetic algorithm-multiple linear regression (GA-MLR) method.

The developed model is given below:

$$\begin{aligned} \text{pEC50} = -14.18 + (0.271)\text{Crippen LogP} - (0.650)\text{GATS5s}(0.381)\text{BCUTp-ll} + (2.80) \\ \text{MATS6c} - (2.86)\text{SIC4} + (1.48)\text{BCUTw-ll} - (2.80)\text{AATSC8e} \end{aligned} \tag{1}$$

$$N = 15, R^2 = 0.998, S = 0.039$$

N = 15, $R^2 = 0.998$, S = 0.039, where N is the number of observations, R is the square root of multiple R^2 for regression, and S is the standard error of estimation.

The above QSAR Eq. (1) with 7 parameters can explain 99.8% of the variances of the studied compounds. It produced an R^2 value of 0.998 and a standard error of 0.039. This is a good model but a simple R^2 does not represent the goodness-of-fit and also know that the number of modeled parameters is calculated by N/5 = 15/5 = 3. Therefore, the model should be refined by reducing the parameter up to 3 with validation. For statistical validation, many random training and test sets were generated. The optimal model quality parameters were obtained while 80% of the total data were taken as training and the 20% of the remaining data were taken as a test set. So the training data consist of 12 molecules, whereas the test set contains 3 molecules such as compound numbers 4, 12, and 14. The training data have been used to build the training QSAR which is given as

$$\text{pEC50} = 0.783(+/-0.512) - 0.525(+/-0.335)\text{IC0} + 5.331(+/-0.835)\text{MATSls} \quad (2)$$

$$N = 12, R^2 = 0.918, Q^2_{Loo} = 0.786, R^2_{pred} = 0.596, SEE = 0.303, RMSEP = 0.401, F = 34(DF: 2{,}9)$$

The model quality parameter values were greatly improved. The selected model can explain and predict 91.8% and 78.6% of variances of the nCoV inhibitory activity of the studied compounds. This model can also produce 59.6% external predictability. The standard error of estimation (SEE), root mean square error of prediction (RMSEP), F statistics, and degree of freedom are given as 0.303, 0.401, 34, and (2,9) respectively. To resolve the predictability of the developed model, Eq. 2 was used to predict the anti-COVID-19 activities of the test compounds. The square correlation coefficient (r2) between observed and predicted activities was calculated as 0.918 which carries significant model predictability. Therefore, this developed model can be acceptable.

4 Conclusion

The QSAR model (Eq. 2) contains two important parameters such as IC0 and MATS1s. The IC0 is an Information Content index signifying neighborhood symmetry of 0-order and the MATS1s is a 2D autocorrelations characterized by Moran autocorrelation of lag 1 weighted by I-state. These represent 2D autocorrelation descriptors encoding ionization potential and Sanderson electronegativity of the studied compounds. The modeled captured predictors have capability of charge distribution, producing the electron cloud of the studied molecules and enhancing the affinity toward target binding by the charge or hydrogen bond interactions.

References

1. Cui, J., Li, F., Shi, Z.L.: Origin and evolution of pathogenic coronaviruses. Nat. Rev. Microbiol. **17**, 181–192 (2019)
2. Yang, Y., Xiao, Z., Ye, K., He, X., Sun, B., Qin, Z., Yu, J., Yao, J., Wu, Q., Bao, Z., Zhao, W.: SARS-CoV-2: characteristics and current advances in research. Virol. J. **17**, 1–17 (2020)

3. Whittaker, A., Anson, M., Harky, A.: Neurological manifestations of COVID-19: a systematic review and current update. Acta Neurol. Scand. **142**, 14–22 (2020)
4. Nandi, S., Nayak, B.S., Khede, M.K., Saxena, A.K.: Repurposing of chemotherapeutics to combat COVID-19. Curr. Top. Med. Chem. **22**, 2660–2694 (2022)
5. Wang, M., Cao, R., Zhang, L., Yang, X., Liu, J., Xu, M., Shi, Z., Hu, Z., Zhong, W., Xiao, G.: Remdesivir and chloroquine effectively inhibit the recently emerged novel coronavirus (2019-nCoV) in vitro. Cell Res. **30**(3), 269–271 (2020)
6. Caly, L., Druce, J.D., Catton, M.G., Jans, D.A., Wagstaff, K.M.: The FDA-approved drug ivermectin inhibits the replication of SARS-CoV-2 in vitro. Antivir. Res. **178**, 104787 (2020)
7. Tang, L., Jiang, Y., Zhu, M., Chen, L., Zhou, X., Zhou, C., Ye, P., Chen, X., Wang, B., Xu, Z., Zhang, Q., Xu, X., Gao, H., Wu, X., Li, D., Jiang, W., Qu, J., Xiang, C., Li, L.: Clinical study using mesenchymal stem cells for the treatment of patients with severe COVID-19. Front. Med. **14**, 664 (2020)
8. Sacramento, C.Q., Fintelman-Rodrigues, N., Temerozo, J.R., Silva, A. D. P.D. D., Dias, S. D. S.G., Silva, C. D. S. D., Ferreira, A.C., Mattos, M., Pão, C.R.R., Freitas, C.S.D, Soares, V.C., Hoelz, L., Fernandes, T. V., Branco, F., Bastos, M., Boechat, N., Saraiva, F., Ferreira, M., Rajoli, R., Pedrosa, C., Vitória, G., Souza, L.R.Q., Goto-Silva, L., Guimarães, M., Rehen, S., Owen, A., Bozza, F., Bou-Habib, D., Bozza, P., Souza, T.M.: The in vitro antiviral activity of the anti-hepatitis C virus (HCV) drugs daclatasvir and sofosbuvir against SARS-CoV-2. bioRxiv. (2020)
9. Wang, X., Cao, R., Zhang, H., Liu, J., Xu, M., Hu, H., Li, Y., Zhao, L., Li, W., Sun, X., Yang, X., Shi, Z., Deng, F., Hu, Z., Zhong, W., Wang, M.: The anti-influenza virus drug, arbidol is an efficient inhibitor of SARS-CoV-2 in vitro. Cell Discov. **6**, 28 (2020)
10. Allinger, N.L., Yuh, Y.H., Lii, J.H.: Molecular mechanics. The MM3 force field for hydrocarbons. 1. J. Am. Chem. Soc. **111**, 8551–8566 (1989)
11. Yap, C.W.: PaDEL-descriptor: an open source software to calculate molecular descriptors and fingerprints. J. Comput. Chem. **32**, 1466–1474 (2011)
12. Ambure, P., Aher, R.B., Gajewicz, A., Puzyn, T., Roy, K.: "NanoBRIDGES" software: open access tools to perform QSAR and nano-QSAR modeling. Chemom. Intell. Lab. Syst. **147**, 1–13 (2015)
13. Ellis, D.I., Broadhurst, D., Kell, D.B., Rowland, J.J., Goodacre, R.: Rapid and quantitative detection of the microbial spoilage of meat by fourier transform infrared spectroscopy and machine learning. Appl. Environ. Microbiol. **68**, 2822–2828 (2002)

Nanosuspension Formulation—Activity Relationship of Anticancer Compounds to Enhance Therapeutic Efficacy of Gastric Cancer

Mayank Kumar Khede[1], Bhabani Shankar Nayak[2] (✉), Harekrishna Roy[3], and Biswaranjan Mohanty[4]

[1] BPUT NCR, Institute of Pharmacy and Technology, Cuttack, Odisha, India
[2] Faculty in Pharmacy, KIIT School of Pharmacy, KIMS, KIIT Deemed to be University, Bhubaneswar, Odisha, India
bhabanishankar.nayak@kims.ac.in
[3] Faculty in Pharmacy, Nirmala College of Pharmacy, Guntur, Andhra Pradesh, India
[4] Faculty in Pharmacy, School of Pharmacy, DRIEMS University, Cuttack, Odisha, India

Abstract Cancer continues to be a major worldwide health concern due to its high mortality rate and limited treatment options. The therapeutic choices available to treat cancer are generally characterized by low bioavailability, poor solubility, and non-specificity. Nanosuspension (NS) formulations have emerged as a viable strategy to address these difficulties, offering a number of advantages such as better drug solubility, increased bioavailability, and tailored drug administration. In this review work sincere efforts have been made to provide a comprehensive detail of the application of NS formulations in gastric cancer treatment and to explore their potential to address the shortcomings of conventional formulations. We also cover the anticancer drugs along with their mechanism of action and brand names available for the treatment of gastric cancer approved by Food and Drug Administration and their NS (nanosuspension) formulation case studies. These include polymeric NS, lipid-based NS, and biopolymeric NS. The review also discusses how the reduced particle size of these NS can affect drug performance in treating and targeting gastric cancer. The review concludes that nanosuspension formulations hold great promise in the future of cancer treatment especially gastric cancer treatment therapy, with significant implications for gastric cancer treatment and the potential to offer a promising alternative to conventional dosage form.

Keywords Nanosuspension · Cancer statistics · Gastric cancer · Anti-drugs · Treatment · SAR · Bioavailability and targeted drug delivery

A. K. Saxena, S. Nandi (eds.), *Global Trends in Health, Technology and Management*, https://doi.org/10.1007/978-3-031-75457-9_17

1 Introduction

The term "cancer," which originates from Latin and was previously known as "karkinoma" in Greek, has been identified for thousands of years, with several kinds of cancer being described by Hippocrates over 2300 years ago [1]. Cancer is one of the prime reasons and important causes of mortality around the world, with the World Health Organization (WHO) data suggesting it's among the top two causes of death before 70 years of age in over a hundred nations. A recent worldwide survey reported approximately one ninety-three lakh fresh instances of cancer cases reported globally and close to 10 million fatalities associated with the cancer disease. In the context of the United States of America (USA), cancer holds the second position in the list of leading causes of death, with annual figures showing over 1.7 million new cases and close to 600,000 fatalities [1]. As depicted in Fig. 1, cancer is characterized by the body's cells proliferating without restraint and invading other regions of the body. When this abnormal growth occurs, the cells exceed their usual tissue or organ boundaries and are invaded or spread into adjacent areas and/or other organs in the body. This process of spreading or invading into organs is termed metastasis, which is recognized as a significant cause of death from cancer [1, 2].

In India, cancer poses a significant challenge to public health. The country bears a substantial cancer burden, with certain types of cancer, such as lung, breast, cervical, and gastric cancer, being relatively prevalent. Conversely, other types, like prostate cancer, have a comparatively lower incidence [3]. Disability-Adjusted Life Years (DALYs), an indicator which signifies the total disease burden was approximately two crore sixty-seven lakhs, the projected impact of cancer in India in the year 2021. This number is

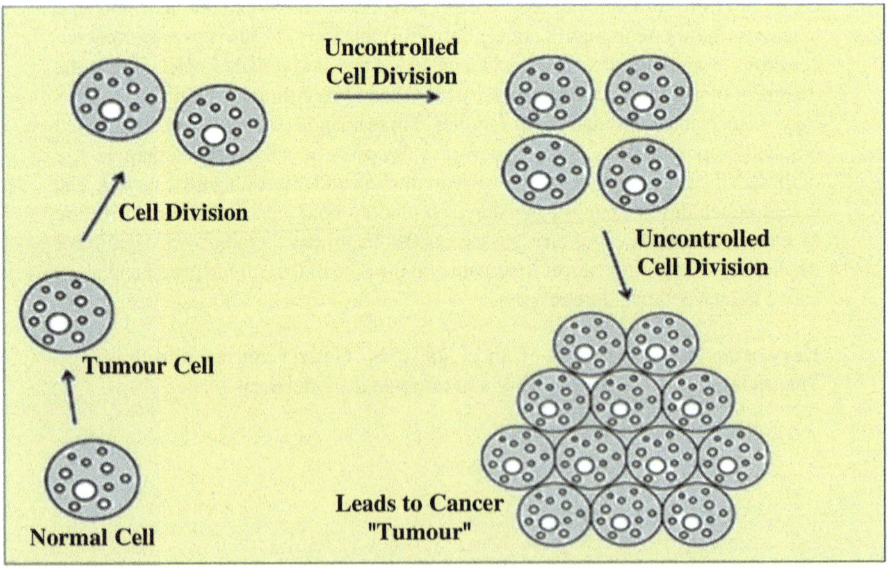

Fig. 1 Diagrammatic representation of healthy cell escalating into tumor

expected to increase to nearly two crore ninety-eight lakh DALYs by 2025 indicating an increasing trend [3]. Geographically, the northern and northeastern regions of the country bear the highest burden, with 2408 and 2177 DALYs per 100,000 populations, respectively. Furthermore, the burden of cancer is higher among males compared to females. A closer look at the data reveals that more than 40% of the total cancer burden in India is attributed to seven primary cancer sites. The cancers accounted for are lung cancer at 10.6%, breast cancer at 10.5%, esophageal cancer at 5.8%, oral cancer at 5.7%, stomach cancer at 5.2%, liver cancer at 4.6%, and cervical cancer at 4.3% [4]. Each of these types contributes significantly to the overall cancer burden, highlighting the need for targeted interventions and strategies for prevention, early detection, and treatment [3, 4].

According to the information depicted in Table 1, in terms of worldwide occurrence, gastric cancer is the fifth most common, and it stands as the fourth leading cause of death related to cancer. In the Indian context, this form of cancer, often referred to as

Table 1 Ranking of the leading cancers as per incidence and country wise load [5–7]

Incidence (Globally)	Type of cancer	New cases reported in the year 2020	% of all cancers
1	Breast cancer	2,261,419	12.5
2	Lung cancer	2,206,771	12.2
3	Colorectal cancer	1,931,590	10.7
4	Prostate cancer	1,414,259	7.8
5	Stomach cancer	1,089,103	6.0
Incidence (Top five countries)	**Name of countries**	**Number of cases reported in year 2020**	**Cancer rate per 100,000 people**
1	Denmark	39,996	334.9
2	Ireland	27,067	326.6
3	Belgium	74,162	322.8
4	Hungary	62,399	321.6
5	France	422,828	320.1
Mortality rate (Globally)	**Type of cancer**	**Number of deaths**	**Mortality rate per 100,000 people**
1	Lung cancer	1,796,144	2.4
2	Colorectal cancer	935,173	19.5
3	Liver cancer	830,180	9.5
4	Stomach cancer	768,793	47.8
5	Breast cancer	684,996	11.1
Mortality rate (Top five countries)	**Type of cancer**	**Number of deaths**	**Mortality rate per 100,000 people**
1	Mongolia	4465	175.9
2	Serbia	27,820	150.6
3	Hungary	32,637	148.1
4	Montenegro	1752	144.5
5	Slovakia	16,192	140.7

"stomach cancer," poses a major public health issue. Notably, in the male population of India, cancers of the stomach and esophagus are frequently diagnosed [8]. According to the Global Burden of Disease India study report spanning from 1990 to 2016, 9% of the total cancer-related Disability-Adjusted Life Years (DALYs) in India were due to stomach cancer. Other notable contributors included cancers of the colon and rectum (5.8%), esophagus (4.3%), liver (3.5%), and gallbladder and biliary tract (3.1%) as reported by the India State-Level Disease Burden Initiative Cancer Collaborators in 2018. The National Cancer Registry Program (NCRP) in India has been diligently gathering data on cancer since 1981, utilizing two kinds of registries: Population-Based Cancer Registries (PBCRs) and Hospital-Based Cancer Registries (HBCRs). These registries provide crucial insights into the prevalence and patterns of cancer in the country. As per the NCRP data, cancer incidence varies based on gender. Gastric cancer ranks second in terms of prevalence among males and fourth among females. The data also reveal that men are more likely to develop cancer than women, with a male-to-female ratio of approximately 2:1. This suggests that men in India are twice as likely to be diagnosed with cancer as women [8, 9].

Prevalence of stomach cancer is higher in older adults, particularly those over the age of 60. Risk factors include poor dietary habits such as consuming smoked and pickled foods, lack of awareness, and limited access to screening and early detection services [10]. The high occurrence rate can be attributed to two significant factors. Firstly, individuals with a lower socioeconomic standing are more susceptible. Secondly, a high prevalence of infections caused by the bacterium known as Helicobacter pylori (H. pylori) also plays a crucial role [10]. People in lower socioeconomic groups often lack access to quality healthcare and nutritious food, which can weaken their immune system and make them more prone to various health issues, including H. pylori infections. On the other hand, H. pylori is a type of bacteria that can live in the digestive tract and is capable of causing ulcers in the stomach or the upper part of the small intestine. A high rate of H. pylori infection means that a large proportion of the population will be in high risk or susceptible of developing gastric cancer. Diagnosis often occurs at advanced stages due to insufficient awareness and screening services. Increasing public awareness, promoting healthier dietary habits, and enhancing the availability of early diagnosis procedures and screening services will be a vital preventive strategy to lessen the impact of stomach cancer [10].

With respect to the above challenges in combating the gastric cancer intensified research efforts have targeted innovative approaches to enhance the management and treatment of gastric cancer with improved therapeutic outcomes. The development of nanosuspension-based delivery systems for anticancer drugs has shown significant potential in addressing the limitations of conventional dosage form such as poor solubility, toxicity, and limited bioavailability [11]. The main objective of the review is to thoroughly investigate the latest advancements in nanosuspension technologies for treating gastrointestinal cancer, focusing specifically on the correlation between the structure (reduced particle size) and improved activity of anticancer agents. By investigating the impact of nanoscale drug delivery on treatment efficacy and safety, potential advancements and opportunities in the field are identified. Moreover, the review highlights the mortality and morbidity trends of gastric cancer, current treatment modalities, drawbacks of existing chemotherapy, and case studies of nanosuspension formulations developed to improve the bioavailability of anticancer drugs for targeting gastric tumor [11, 12]. Through a comprehensive review of current literature and research findings, the article aims to shed light on the promising role of nanosuspension formulations in combating gastric cancer, offering hope for enhanced patient outcomes and improved therapeutic outcomes [12].

2 Mortality and Morbidity of Cancer

Cancer, recognized worldwide as a leading cause of death, imposes an uneven load. The occurrence of cancer has increased over the years in both advanced and emerging countries due to a complex array of factors. These include the aging process, population expansion, rapid socioeconomic progression, and changes in the pattern of associated risk factors. As the global population continues to grow and age, cancer has surfaced as the leading cause of premature mortality, leading to a reduction in life expectancy in numerous countries [13].

According to the Global Cancer Observatory's Cancer Today data as observed and depicted in Table 1, the five most prevalent and deadly cancers globally have been determined [5]. In the year 2020, breast cancer in females surpassed lung cancer, making it the most frequently diagnosed cancer in both men and women combined. Other prevalent cancers included those of the stomach, prostate, and rectum. When ranked by the number of deaths in both sexes in 2020, the leading causes of death were lung cancer followed up by colorectal cancer, liver cancer at second and third position, with esophageal, pancreatic, and prostate cancers trailing closely behind. Interestingly, there was a reduction of nearly 14,000 cases in stomach cancer deaths in 2020. In 1990, liver cancer was the fourth most common cause of cancer-related deaths, but by 2020, it had risen to become the third most common cause. Out of the 14 most prevalent cancers, breast cancer experienced the largest rise in mortality rates [5].

2.1 Statistics at a Glance: The Burden of Cancer Worldwide

Cancer, a disease marked by abnormal cell division leading to uncontrolled growth of cells in the body, is a major global health concern. Each year, it is estimated that between 19 and 20 million individuals worldwide are diagnosed with this condition. The disease has been responsible for a significant number of deaths globally, year after year. Recent data suggest that in 2020 alone, over 19.3 million new cases of cancer were identified, leading to nearly 10 million fatalities among which lung cancer was the leading cause of death globally and Mongolia reporting the highest mortality per one lakh population as depicted in Table 1 [6, 14, 15].

According to the latest data, Denmark reported the highest number of cancer incidence per one lakh population with incidence rate of 334.9 followed up by Ireland, Belgium, Hungary, and France [6, 7]. Globally, cancer ranks the second in disease-related deaths, following cardiovascular diseases. It's a major health issue with around 19–20 million cancer fresh cases reported every year around the globe. According to the

World Health Organization (WHO), approximately one-third of fatalities from cancer can be attributed to certain lifestyle choices and dietary habits. These include the use of tobacco products, maintaining a high body mass index (BMI), consuming alcohol, not eating enough fruits and vegetables, and leading a sedentary lifestyle with minimal physical activity. These factors significantly contribute to the global cancer mortality rate. GLOBOCAN data reveal regional differences in cancer incidence and mortality rates, with Eastern Asia having the highest number of cases and deaths. Despite healthcare advancements, globally, cancer persistently ranks as the second most common cause of deaths related to diseases. Despite advancements in medical science, it remains a significant health challenge worldwide for researchers and scientists. These statistics reveal that cancer has a significant impact on healthcare systems, economies, and individuals worldwide [7, 16].

In 2018, the data published by International Agencies doing research on cancer revealed that gastric cancer accounted for more than a million new cases [7, 17]. This constituted 5.7% of all cancer diagnoses for that year, leading to an estimated 782,685 fatalities. Consequently, gastric cancer was ranked as the fifth most commonly occurring cancer in 2018 around the globe. It was responsible for 8.2% of all deaths caused by cancer, equating to one in every twelve cancer-related deaths, thereby making it the third deadliest form of cancer. By the year 2020, gastric cancer was the fourth leading cause of deaths related to cancer globally with countries like Mongolia reporting highest cancer incidence rate per one lakh population, i.e. 32.5 followed up by Japan at 31.6 and South Korea at 31.6 closely trailed by Tajikistan at 23.4 and China at 20.6, respectively [7, 17].

The survival rate of individuals diagnosed with gastric cancer is heavily influenced by the stage of the disease at the time of diagnosis. Gastric cancer significantly contributes to the worldwide total of disability-adjusted life years, which measures the overall disease burden. Studies conducted on healthy but infected individuals in Asia who show no symptoms have indicated that eliminating H. pylori infection can decrease the occurrence of gastric cancer [17].

Even with advancements in healthcare system and treatment regimens to combat gastric cancer, international benchmarking studies have indicated that the survival rates for gastric cancer continue to be low, even in countries with high income [18]. The high mortality and morbidity associated with cancer significantly impacted global health systems, economies, and societies, placing a considerable financial burden on patients, families, and healthcare systems. Addressing the global burden of cancer involved promoting public awareness and education on cancer prevention and early detection, advancing research and development of novel therapies, and improving access to quality cancer care and treatment facilities, e.g. HPV cervical cancer vaccination. Cancer remained a major global health challenge with high mortality and morbidity rates, requiring collaborative efforts from governments, healthcare organizations, researchers, and the broader community to develop effective prevention strategies, enhance early detection, and implement innovative treatment approaches to combat this devastating disease [18, 19].

3 Prevention and Treatment of Gastric Cancer

Figure 2 depicts the representation of gastric tumor in stomach. Globally, gastric cancers significantly contribute to cancer-related health issues and deaths. Although the occurrence of stomach cancer generally increases with age, about 10% of cases are identified in individuals under 45 years of age. The progression toward stomach cancer is a complex journey involving numerous stages. This journey is characterized by an accumulation of changes in the genes, leading to mutations and modifications in the way these genes express themselves. These changes are often influenced by a variety of external elements. It's these elements that act as catalysts, setting off the onset of gastric cancer [20]. Treatment decisions are influenced by various factors, including cancer type, stage, and patient health. Chemotherapy, a systemic therapy, is commonly used to target cancer cells, while surgery aims to remove the primary tumor. Both approaches have limitations and challenges. Advances in personalized medicine and targeted therapies offer promising prospects for improved treatment outcomes. The multiprong strategy of blending preoperative (neoadjuvant) chemotherapy and surgical procedures, complemented by the incorporation of therapies specifically designed to target cancer cells, is designed with the primary objective of enhancing the effectiveness of the treatment. This approach is intended to not only suppress the growth and spread of the cancer but also to significantly enhance the survival rates of the patients. This multipronged approach allows for a comprehensive treatment plan that addresses the cancer cells from several angles, thereby increasing the chances of a successful outcome. Ongoing research into novel treatment approaches, such as immunotherapies and nanoformulation dosage form, continues to shape the future of gastric cancer treatment [20].

Multiple factors have been identified that escalate the likelihood of being diagnosed with gastric cancer. These include genetic predisposition, dietary habits, consumption of alcoholic beverages, tobacco use, and bacterial infection caused by Helicobacter pylori and viral infection by Epstein–Barr virus also contributes to its onset in a healthy human body. Important factors like a person's family history can play a significant role in their susceptibility to gastric cancer. If close relatives have had this disease, the risk may be higher due to shared genetic traits. Dietary habits, particularly the consumption

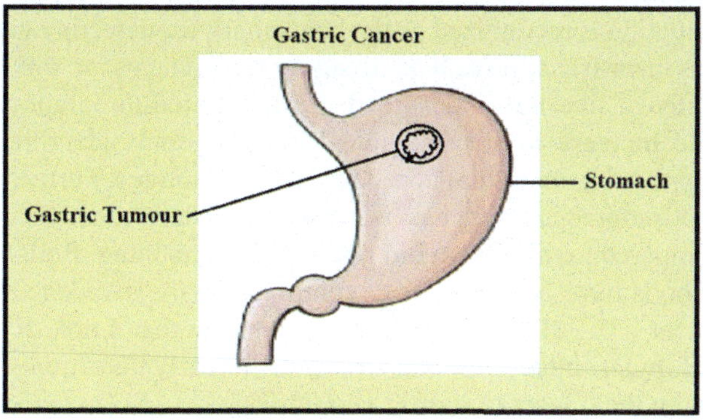

Fig. 2 Diagrammatic representation of gastric tumor in stomach

of processed foods or those high in salt, have been linked to an increased risk of gastric cancer. Conversely, a diet rich in fruits and vegetables may offer some protective benefits. Alcohol and tobacco use are well-known risk factors for various types of cancer, including gastric cancer. Both substances can cause damage to the cells lining the stomach, potentially leading to cancerous changes over time. It's important to note that while these factors can increase risk but always might end up causing cancer. Therefore, people with risk or without these risk factors should always try to maintain a healthy lifestyle and visit doctor for regular body checkups, which will play a crucial role in prevention and early detection [21, 22]. Family history is a crucial risk factor, with around 10% of gastric cancer displaying familial aggregation. Under 3% of all stomach carcinomas show a pattern of inheritance that follows Mendel's laws, with the most notable type within the family being hereditary diffuse gastric cancer (HDGC). This subtype is primarily caused by changes in the cadherin 1 gene (CDH1) [22, 23]. Individuals with a family history of gastric carcinoma are about three times more likely to develop the disease than those without such a history. The prevalence of familial gastric cancer is notably higher in Asian regions compared to Europe and North America. This suggests that environmental factors play an important role in causing familial gastric cancer in countries where the disease is more common [22–24].

3.1 Prevention of Gastric Cancer

To prevent gastric carcinoma on a broad scale, it's crucial to improve dietary habits and decrease the exposure of the body to H. pylori bacterial infection, a leading cause of this type of cancer. Dietary changes could include consuming more fresh fruits and green veggies and reducing the intake of high salt volume in the daily dietary intake and foods preserved with salt, e.g. pickles. Lifestyle changes such as engaging in regular physical activity and quitting smoking can also help lower the risk of this disease. Fruits and vegetables, which are high in folate, carotenoids, vitamin C, and phytochemicals, could potentially guard against the development of cancer [25, 26]. Helicobacter pylori eradication is another preventive strategy, with some evidence suggesting a reduction in GC incidence following eradication. However, randomized trials indicate significant limitations to the data supporting this effect [27–29]. Early detection through available resources, particularly endoscopic methods, is crucial for secondary prevention. Endoscopic evaluation is recognized as the benchmark for detecting gastric carcinoma, providing a less invasive approach to treating early stages of this type of cancer. Screening initiatives, like the widespread use of photofluorography in Japan, have shown a notable increase in survival rates for individuals identified through these screenings, as opposed to those diagnosed based on symptoms. Furthermore, the implementation of endoscopic screening has been linked to a decrease in deaths due to gastric cancer when compared to those who did not undergo screening. Both radiographic and endoscopic methods have been shown to contribute to the avoidance of gastric carcinoma development [30–32]. In conclusion we can state that a mix of dietary changes, elimination of H. pylori infections, and strategies for early detection significantly contribute to lessening the impact of gastric cancer [20].

3.2 Treatment of Gastric Cancer

In recent times, with efforts from researchers there has been great improvement in the therapies and treatment methods for gastric cancer, including chemotherapy, targeted therapy, and immunotherapy. However, due to the subtle nature of early symptoms and infrequent screening, many patients are diagnosed at advanced stages. For cases where the cancer can be surgically removed, chemotherapy around the time of surgery has become the standard care. Through ongoing advanced research in area of immunotherapy or drug targeted therapy, directed by biomarkers for advanced disease where patients are differentiated according to specific molecular markers, such as programmed cell death ligand 1 (PD-L1), microsatellite instability (MSI), and human epidermal growth factor receptor 2 (HER2), helping health care providers to identify individuals who are likely to respond positively to these therapy, remarkable progress has been made in treatment of gastric cancer [33]. These markers serve as indicators, helping us tailor treatment plans to the individual's unique genetic profile, potentially improving their prognosis. The use of diagnostic techniques like MDT also known as molecular diagnostic techniques has enabled the understanding of the genetic makeup of gastric cancer and the discovery of new potential molecular targets [33].

Neoadjuvant chemotherapy, which is therapy given before the main treatment, has been underscored as a vital part of the treatment regimen for gastric carcinoma, gastroesophageal junction, and adenocarcinoma of lower esophagus. Clinical trials have demonstrated its efficacy in improving survival outcomes. Research studies, including MAGIC and FFCD9703, have endorsed the use of perioperative chemotherapy. This treatment approach has demonstrated advantages such as shrinking the size of the tumor, lowering the cancer stage, and enhancing both the rates of survival without disease progression and overall survival [34–36]. Similarly, preoperative chemotherapy regimens like 5-fluorouracil/cisplatin have proven effective in enhancing disease-free and overall survival in advanced adenocarcinoma cases [36]. Additionally, radiation therapy, often combined with chemotherapy, has shown efficacy in treating stomach cancer. In the realm of targeted therapy, drugs like ramucirumab and trastuzumab have been identified as significant options, particularly for HER2-positive tumors. Other targeted therapies, such as lapatinib and cetuximab, have shown varying degrees of effectiveness based on molecular characteristics. Research is currently being conducted on inhibitors that focus on the fibroblast growth factor 2 receptor tyrosine kinase (FGFR-2) and the PI3K/AKT/mTOR pathway. These studies could have significant effects on a specific group of gastric cancer instances [37–39]. While some targeted therapies have shown promise in improving overall survival rates, others have yielded mixed results or shown no significant improvement. Overall, targeted therapies offer promising avenues for improving outcomes in gastric cancer treatment, particularly when tailored to molecular characteristics of individual tumors [38–41]. Table 2 depicts the list of anticancer drugs approved by the US-FDA to treat gastric cancer and their mechanisms of action, along with important brand names [42].

Table 2 Comprehensive overview of FDA-approved anticancer drugs for gastric cancer and nano-formulation availability [41, 42]

S. no.	Name of anticancer drug	Mechanism of action	Conventional dosage form brand name	Dosage form	Nanotherapeutics (Availability in market)	Nanotherapeutics (Brand name)
	Monotherapy					
1	**Capecitabine**	Prodrug converted to 5-FU in tumor cells, inhibiting DNA synthesis.	Xeloda	Tablet	No	NA
2	**Ramucirumab**	Monoclonal antibody targeting VEGFR-2(Protein), inhibiting angiogenesis	Cyramza	I.V Injection	No	NA
3	**Paclitaxel**	Inhibits microtubule disassembly, disrupting mitosis and cell division	Taxol	Injection	Yes	**Abraxane** (Nab-Paclitaxel)
4	**Docetaxel**	Inhibits microtubule disassembly, disrupting mitosis and cell division	Docetere	Injection	Yes	**Taxotere** Docetaxel Injection (Liposome-based)
5	**Doxorubicin Hydrochloride**	Intercalates DNA, inhibiting topoisomerase II, resulting in DNA damage and apoptosis	Adriamycin	Injection	Yes	**Doxil** (PEGylated Liposomal Doxorubicin)

6	**Fam-Trastuzumab Deruxtecan-nxki**	Antibody-drug conjugate targeting HER2, delivering cytotoxic payload to HER2-expressing cancer cells	Enhertu	I.V Injection	No	NA
7	**5-FU (Fluorouracil)**	Inhibits thymidylate synthase, disrupting DNA synthesis	Carac	Injection, Topical cream	No	NA
8	**Trastuzumab**	Monoclonal antibody targeting HER2, inhibiting HER2 signaling pathway	Herceptin	Injection	No	NA
9	**Pembrolizumab**	Monoclonal antibody targets PD-1, blocks its interaction with ligands, and revives the immune response against tumors	Keytruda	I.V Injection	No	NA

(continued)

Table 2 (continued)

S. no.	Name of anticancer drug	Mechanism of action	Conventional dosage form brand name	Dosage form	Nanotherapeutics (Availability in market)	Nanotherapeutics (Brand name)
10	**Trifluridine and Tipiracil Hydrochloride**	Trifluridine is incorporated into DNA, inhibiting DNA synthesis; Tipiracil increases trifluridine plasma levels by inhibiting its metabolism	Lonsurf	Tablet and Cream	No	NA
11	**Mitomycin**	Alkylates DNA, leading to interstrand crosslinks and DNA damage	Jelmyto	Injection	Yes	**MitoGel** (Nano-formulation used for Upper Tract Urothelial Carcinoma (UTUC)
12	**Nivolumab**	Nivolumab inhibits PD-1, frees immune cells from suppression, enabling them to detect and fight tumor cells	Opdivo	I.V Injection	No	NA
	Combination therapy					
13	**FU-LV**	Chemotherapy combination contains the drugs fluorouracil and leucovorin calcium	NA	NA	NA	NA

14	**TPF**	Chemotherapy combination contains the drugs docetaxel (Taxotere), cisplatin (Platinol), and fluorouracil	NA	NA	NA	NA
15	**XELIRI**	Contains the drugs capecitabine and irinotecan hydrochloride used as chemotherapy combination	NA	NA	NA	NA

3.3 Drawback of Cancer Chemotherapy Treatment

Modern innovations breakthroughs in health science have boosted cancer survival in patients through potent treatments like chemotherapy. Non-specificity is observed as traditional chemotherapy drugs, targeting rapidly dividing cells, affect healthy cells, leading to adverse effects like hair loss and compromised immune function. Drug resistance, wherein cancer cells adapt and become resistant to chemotherapy, reduces treatment effectiveness over time. Some cancers form barriers limiting drug penetration, resulting in incomplete treatment and potential recurrence. Systemic toxicity, manifesting as fatigue and increased infection risk, affects healthy tissues alongside cancer cells. Long-term health effects on organs and systems may arise post-treatment [43].

Chemotherapy, while effective, can significantly impact quality of life and is a key concern in cancer treatment. Chemotherapeutic drugs often target DNA, leading to mutations and instability, which are common in both cancer and aging. Various chemotherapy drugs are cytotoxic to healthy cells and affect normal cell cycle leading to conditions like alopecia affecting patient's mental health as well. They interfere with DNA in different ways, each with a unique distribution, kinetics, and fate. The type of DNA damage, affected organ or cell type, and exposure level can greatly influence treatment outcomes [44].

Chemotherapy treatment, while crucial in combating cancer, can unfortunately lead to substantial health complications. One such adverse effect is nephrotoxicity, which is commonly triggered by medications such as platinum compounds, alkylating drugs, antimetabolites, and anti-neoplastic antibiotics. This condition arises when the kidneys are exposed to high levels of these substances, potentially causing acute kidney damage (AKI) or long-term kidney toxicity. Another significant side effect is hepatotoxicity, which can be caused by alkylating agents, antimetabolites, topoisomerase inhibitors, anti-neoplastic antibiotics, and radiation therapy. This condition can present in various forms, including hepatitis, cholestasis, fatty liver disease (steatosis), or sinusoidal obstruction syndrome (SOS), all of which can impair liver function [43, 45]. Alkylating agents, antimetabolites, mitotic inhibitors, and proteasome inhibitors used in chemotherapy are often associated with causing neurotoxicity, primarily presented in the peripheral nervous system (PNS), leading to sensory abnormalities and motor impairments [43, 46]. Cardiotoxicity, commonly seen with anthracyclines and other chemotherapeutic agents, results in congestive heart failure and rhythm disturbances, significantly impacting cardiovascular health [47]. Hematological toxicities, induced by various chemotherapeutic agents, lead to cytopenia, increasing the risk of infection. Other toxicities include ototoxicity, gastrointestinal toxicity, and gonadal toxicity, affecting hearing, gastrointestinal function, and reproductive health, respectively. These toxicities underscore the importance of monitoring and managing side effects to optimize cancer treatment outcomes [43].

In recent years, researchers and clinicians have been working to address these drawbacks and develop more targeted and personalized approaches to cancer treatment. Targeted therapies, immunotherapies, and nanotechnology-based drug delivery systems are among the innovative strategies being explored to overcome the limitations of existing chemotherapy and improve cancer treatment outcomes [48].

4 Nanosuspension: Significance in Treatment of Gastric Cancer

4.1 Nanosuspension and Its Advantages

Figure 2 depicts the diagrammatic representation of nanosuspensions (NS) preparation, which is defined as colloidal dispersions having an ultra-fine distribution of drug particles, which enhances the drug solubility and absorption of drugs into the systemic circulation. The suspended particles are submicron in size, ranging from 10 to 50 nm in diameter, and are present in a biphasic colloid system [49]. Nanosuspensions are increasingly utilized in pharmaceutical industry due to their ability to improve dissolution of poorly water-soluble drugs, thereby improving formulation potency to target cancel cell, drug efficacy, and safety. Formulation strategies for nanosuspension involve selecting stabilizers, surfactants, solvents, and polymers critical for drug targeting to specific site. There are several techniques used to prepare colloidal dispersions. These can be broadly categorized into two approaches: top-down and bottom-up. The top-down methods include high-pressure homogenization and the milling media method, while the bottom-up techniques encompass the precipitation method and probe sonication. Various other methods such as dry-co-grinding, the supercritical fluid method, and the lipid emulsion method are also employed. Nanosuspension formulation helps to improve the physiochemical properties of drugs such as saturation solubility, dissolution characteristics of drug along with surface area, and availability of drug in the systemic circulation by reduction in particle size of drugs at the nanoscale particularly very helpful in formulating a dosage form for drugs categorized under BCS class II and class IV [50, 51].

Nanosuspensions provide a flexible approach to drug delivery, with options for oral, intravenous, topical, pulmonary, and ocular administration (Fig. 3). They can also be utilized for targeted treatment in conditions such as cancer and HIV. The production process for nanosuspensions is relatively straightforward and can be easily expanded for larger scale production [49, 50].

4.2 Significance in Treatment of Gastric Cancer

The application of both parenteral and oral nanosuspensions in gastric cancer therapy is currently in the preliminary phase of development, but the potential benefits are clear. The enhanced bioavailability and improved pharmacokinetics offered by nanosuspensions could be particularly beneficial in the treatment of gastric cancer. For instance, nanosuspensions could increase drug exposure by reducing the particle size and improving the surface area countering challenges such as limited permeability of drug into the cancer cells, reduce the number of dosing frequency, and improve drug targeting thereby increasing oral bioavailability [51, 52]. Many patients find liquid forms of medication, such as suspensions or nanosuspensions, easier to swallow than solid forms like tablets and capsules. This is particularly true for patients with gastric cancer who may need to take multiple doses of a medication like capecitabine. Additionally, nanosuspensions offer more flexibility in administering large doses of medication. Hence, there is a need to develop nanosuspension dosage form of anticancer drugs used to treat

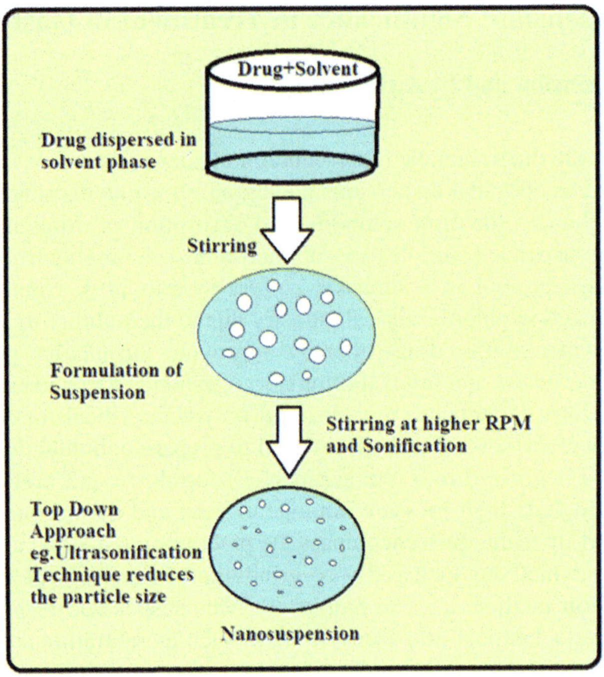

Fig. 3 Diagrammatic representation of nanosuspension preparation through a flow diagram [51]

gastrointestinal cancer which overcomes problems associated with currently available immediate release dosage form and provides better therapeutic outcome, less side effects, reduced toxicity, and a greater flexibility in dose administration [52].

5 Nanosuspension Formulations for Gastrointestinal Cancer (GI) Treatment: Case Studies

Over the past few decades, researchers have extensively explored nanosuspensions for their promising potential in targeting and treating gastrointestinal cancer (GI). It has been noted that nanosuspensions exhibit the ability to improve the bioavailability and efficacy of drug candidates employed in GC treatment. This section of the review explores significant case studies concerning the formulation of nanosuspensions tailored for targeting gastrointestinal cancer (GI).

The Chinese patent CN102961342B is related to the method used to develop a nanoscale capecitabine, utilizing silicon dioxide aerogel as a carrier, to enhance targeted treatment, bioavailability and mitigate toxic effects. The preparation method involves dissolving capecitabine in absolute ethyl alcohol, adsorbing onto silicon dioxide aerogel, emulsifying, homogenizing, and drying. The resulting nanoscale capecitabine exhibits significantly increased bioavailability for oral administration, offering a promising approach for improving gastric cancer treatment outcomes. Developed capecitabine formulation showed promising results in nude mice MGC-803 cell line testing [53].

Paclitaxel (PTX), utilized as a secondary treatment strategy for progressive or metastatic gastric cancer, encounters difficulties due to its pronounced lipophilicity, rendering it inappropriate for injection. A resolution is offered by researchers in the form of Abraxane, which is a composition of albumin nanoparticles and PTX, each particle measuring approximately 130 nm. This nanoscale formulation not only retains the therapeutic benefits of PTX but also mitigates its harmful effects. Moreover, the nanoparticles in Abraxane exploit trans-endothelial transport through an albumin-binding protein, leading to enhanced accumulation in the tumor tissue, going beyond the sole reliance on the enhanced permeability and retention (EPR) effect. The effectiveness and safety of Abraxane in the treatment of gastric cancer have been assessed through various clinical trials. These trials were crucial in determining how well the formulation works and ensuring it is safe for patient use. The clinical trials revealed encouraging results with tolerable common side effects related to taxane-based therapy which includes neutropenia (low neutrophil count), peripheral neuropathy (nerve damage), anemia (low red blood cell count), thrombocytopenia (low platelet count), nausea, vomiting, diarrhea, reduced appetite, fatigue and weakness, hypersensitivity reactions leading to its approval by the FDA [54].

In the year 2018, Bando, Hideaki, et al. conducted Phase II trial research, to evaluate combined treatment approach for advanced gastric cancer for its safety and efficacy. This approach involved the use of nanoparticle albumin-bound (nab)-paclitaxel and ramucirumab, administered to patients who had previously undergone chemotherapy. The treatment protocol involved intravenous administration of nab-paclitaxel on the first, eighth, and 15th days, complemented by ramucirumab on the first and 15th days of a 28-day cycle. The main aim of the study was to assess the overall response rate (ORR), as evaluated by the Independent Review Committee (IRC). The study achieved its goal, with an ORR of 54.8%. Additionally, the disease control rate (DCR) was found to be 92.9%. The median progression-free survival (PFS), as assessed by the IRC, was 7.6 months. The overall survival (OS) had not been determined at the time of data cut-off. These results indicate that the combined therapy of nab-paclitaxel and ramucirumab, delivered through a nanoparticle drug delivery system, could potentially serve as a promising treatment option for advanced gastric cancer. This approach appears to offer manageable side effects and favorable clinical outcomes for patients who have previously undergone chemotherapy. The most frequently observed adverse events of grade 3 or 4 included a decrease in neutrophil count, a reduction in white blood cell count, anemia, reduced appetite, febrile neutropenia, hypertension, and proteinuria. Importantly, no deaths were associated with the treatment, and there was no observed deterioration in the quality of life during the study treatment [54, 55].

Abdullah et al. in the year 2022 aimed to develop a sustained-release nanoformulation of 5-Fluorouracil (5-FU), a potent anticancer drug with a brief half-life, commonly used in the treatment of gastrointestinal (GI) cancer. The researchers developed and characterized particles coated with 5-FU-carrageenan and encapsulated them within a hydrodynamic gel composed of sodium alginate, carrageenan-lambda, or excess chitosan to create a nanosuspension. The developed suspension was evaluated for its impact on colorectal cancer cells, focusing on cell viability, apoptosis induction, and caspase-9 activity. Among the tested formulations, F5 and F6 demonstrated enhanced sustained-release profiles, forming a cohesive gel upon interaction with 0.1 N hydrochloric acid. These nanosuspensions, with particle sizes ranging from 20 to 63 nm, contained 5-FU nanoparticles encapsulated within carrageenan and sodium alginate coatings. Upon assessing their anticancer efficacy against HCT-116 cells, the 5-FU nanoparticles showcased both sensitivity and time-dependent therapeutic effects, confirming their potential for sustained-release treatment. The study suggests that this nanosuspension could be promising as a prolonged therapy option, utilizing pH-responsive, sustained-release matrices, in-situ gel formation in the gastric environment, and protective polymer coatings that shield the drug nanoparticles during release. However, potential side effects, including gastrointestinal irritation, mucosal damage, delayed gastric emptying, and localized toxicity, particularly due to extended exposure of the stomach lining to the drug can be considered for future research [56].

Taymouri *et al.* in the year 2021 developed a unique dual pH- and time-dependent polymeric-coated nanosuspension to specifically target the release of simvastatin (SIM), a drug used for treatment of colon cancer. The drug release characteristics of these coated capsules were evaluated under various simulated physiological conditions, including 0.1 N HCl, phosphate buffers of pH 6.8 and 7.4, and simulated colonic fluid with a pH of 6.8. The nanoparticles prepared exhibited a spherical shape with an average size of 261.66 nm, a zeta potential of −18.20, and a dissolution efficiency of 59.71%. Techniques such as X-ray diffraction and differential scanning calorimetry were used for analysis, revealing that the crystalline form of the drug had transformed into a more soluble amorphous form due to the nanosizing technique. Significantly, the coated capsules showed no release in gastric media, indicating a specific delivery of SIM to the colon. In vitro cytotoxicity studies were conducted on HT29 cells using SIM nanoparticles, which showed a significant increase in cytotoxic effects compared to free SIM. Based on these findings, the study suggests that the coated capsules, which employ ethyl cellulose and Eudragit S100 polymers, show promise for the targeted delivery of SIM to the colon. This innovative approach could potentially enhance the effectiveness of colon cancer treatment by simvastatin (SIM) ensuring its direct delivery to the tumor targeted area. The potential side effects of the developed simvastatin nanosuspension, including gastrointestinal irritation from pH-sensitive release and the risk of localized tissue toxicity due to targeted delivery, further research can be done to explore and fully assess these effects [57].

Curcumin, a compound renowned for its wide-ranging biological activities, has been shown to exhibit anticancer properties, particularly against gastrointestinal (GI) cancer. However, its therapeutic potential is limited by its low solubility and bioavailability. To overcome these limitations, Elbaz et al. in the year 2021 developed a nanosuspension of curcumin using a technique known as emulsion templated freeze-drying. They developed and characterized a nano-formulation that used polyethylene glycol (PEG) as an excipient, the developed nanosuspension displayed desirable characteristics such as a small size (211 nm), low polydispersity (0.06), and a zeta potential of −25 mV. By fine-tuning the parameters, they were able to optimize the stability and dissolution of the nanosuspension under simulated gastrointestinal conditions. Interestingly, while the nanosuspension enhanced the dissolution of curcumin, it exhibited reduced stability.

Cytotoxicity studies revealed that the nanosuspension of curcumin demonstrated lower cytotoxicity compared to solubilized curcumin, indicating the potential for improved oral dosing and the noted potential side effects were minimal such as gastrointestinal disturbance or hypersensitivity, which may be caused by sensitivity to the polymer, but the research work emphasizes the need for further in vivo studies to thoroughly evaluate safety. This innovative approach of creating a dispersible nanosuspension opens up new possibilities for the delivery of curcumin in the treatment of gastrointestinal (GI) cancer. It underscores the potential of nanotechnology in enhancing therapeutic efficacy of natural compounds like curcumin [58].

Ibrahim et al. in the year 2019 developed docetaxel (DOX) nanoparticles to enhance its anticancer activity. Utilizing the solvent precipitation technique, DOX-loaded nanoparticles stabilized with different sodium deoxycholate (SDC) and hydroxypropyl methylcellulose (HPMC, E5) concentrations were formulated and evaluated. The results showed that the nanoparticles stabilized with sodium deoxycholate (SDC) were smaller than those stabilized with hydroxypropyl methylcellulose. 83.97 nm was the smallest recorded particle size, achieved with a 5% SDC solution. An important observation was that an increase in the concentration of the stabilizer led to an increase in the rate of drug release. It was observed that nanoparticles stabilized with a 5% SDC solution released 33% of the drug within the first hour and 87% after 24 h and viability of FRO cells was significantly reduced when exposed to nanoparticles stabilized with a 5% SDC solution at concentrations of 100 and 1000 μM/ml. When the cells were exposed to nanoparticles stabilized with a 5% HPMC solution at concentrations of 10 and 100 μM/ml no major cytotoxic effects were observed emphasizing the enhancement of the efficacy of anticancer drugs. However potential side effects of the docetaxel nanosuspension include systemic toxicity and off-target cytotoxicity, despite improved targeting. As suggested by researchers, further studies are needed to fully assess its safety in clinical applications [59].

Patel et al. in the year 2024 formulated Mebendazole (MBZ) nanosuspension using Sodium Lauryl Sulfate (SLS) and Hydroxy Propyl Methyl Cellulose E5 (HPMC) as stabilizers aimed to enhance the solubility and effectiveness of Mebendazole (MBZ) against colorectal cancer. In the formulated MBZ NS better dissolution rates at colonic pH were observed along with excellent stability with the particle size in the range of 362.6 ± 12.3 nm. The safety of the stabilizers used in the study was confirmed through toxicity assessments on HT-29 and HT-116 cell lines. The stabilizers, HPMC E5 and SLS, were found to be non-toxic to these cell lines. The anticancer potency of MBZ-NS was highlighted by its IC50 values. The IC50 value of the formulated MBZ-NS was determined and was found to be 1.028 ± 0.030 μM and 0.884 ± 0.050 μM for HT-29 cell lines and HT-116 cell lines, respectively. Furthermore, the efficacy of MBZ-NS in inducing cell death was underscored by significant reductions in the size of 3D tumor spheroids post-treatment. Compared to the controls, there was a 1.5-fold reduction in the size of spheroids in HT-29 cell lines and a 1.8-fold reduction in HT-116 cell lines highlighting better dissolution properties and anticancer efficacy in colorectal cancer therapy. The potential side effects may include gastrointestinal distress and systemic toxicity, further clinical evaluation is recommended by researchers to optimize safety [60].

Liu et al. developed a novel nano system combining irinotecan hydrochloride and curcumin, demonstrating amplified toxicity compared to their individual components.

This effect is particularly pronounced in gastric cancer (GC) cells, wherein the nano system exhibits increased absorption and heightened toxicity under mildly acidic conditions. The increased toxicity detected in GC cells highlights the promising prospect of utilizing this nano system as a focused therapeutic approach for addressing this type of complicated malignancy. Furthermore, the enhanced absorption and toxicity of the nanoparticles under slightly acidic conditions suggest a potential mechanism for improved efficacy within the acidic tumor microenvironment characteristic of gastric cancer, the potential side effect may include toxicity from the tracking markers and possible immunogenic reactions. Further in vivo studies are needed to evaluate safety comprehensively [61].

Becit-Kizilkaya *et al.* in the year 2024 investigated the potential of daidzein nanosuspension (DZ-NS) in combination with 5-fluorouracil (5-FU) to enhance chemotherapy efficacy in human colorectal adenocarcinoma cells (Caco-2). Findings revealed that DZ-NS (particle size: 233 nm) exhibited significant cytotoxicity, surpassing that of free daidzein, and when combined with 5-FU, it synergistically inhibited cell proliferation without increasing oxidative stress. The cytotoxic effects of various concentrations of 5-fluorouracil (5-FU), daidzein (DZ), and daidzein nanosuspension (DZ-NS) on Caco-2 cells were assessed using the MTT assay. Compared to untreated control groups, which included evaluation of the solvent (DMSO 5%), treatment with 5-FU, DZ, and DZ-NS significantly reduced the viability of Caco-2 cells after 24 h. The IC50 values obtained for 5-FU, DZ, and DZ-NS were 21.27 mM, 240.3 μM, and 35.34 μM, respectively. Notably, the IC50 value for DZ-NS was lower than that of pure DZ under the same incubation conditions, indicating enhanced cytotoxicity of daidzein when formulated as a nanosuspension. Moreover, the combined treatment resulted in a rise in the levels of p53, a protein known for its tumor-suppressing properties. Concurrently, there was a decrease in the levels of the anti-apoptotic protein Bcl-2, as well as the pro-inflammatory cytokines IL-6 and TNF-α. These changes indicate a suppression of inflammation and a promotion of apoptosis, further highlighting the potential effectiveness of the treatment in combating cancer. The results suggest that DZ-NS with 5-FU could serve as a promising strategy for combating colorectal carcinoma by targeting multiple signaling pathways involved in cancer progression while minimizing oxidative damage [62]. This innovative formulation represents a promising advancement in cancer therapeutics, showing potent efficacy specifically in the context of gastric cancer (Table 3).

6 Conclusion

Nanosuspension formulations are recognized as a transformative approach to treating gastrointestinal (GI) cancer, addressing the limitations of conventional therapies. Their important properties, such as improving drug solubility, bioavailability, enhance the treatment efficacy and reduce the systemic toxicity. Nanoscale size of nanosuspension helps them to passively accumulate in tumor tissues and enable active targeting strategies, minimizing damage to healthy tissues. Moreover, nanosuspensions promise to overcome multidrug resistance, which is crucial in cancer treatment. Several key anticancer drugs are being investigated in nanosuspension formulations for GI cancer

Table 3 Nanosuspension formulation activity in the treatment of gastrointestinal cancer: drugs, structures, and clinical implications

S.no	Drug	Chemical structure	Nano particle size	Evaluation	Clinical implications	Reference
1	Capecitabine	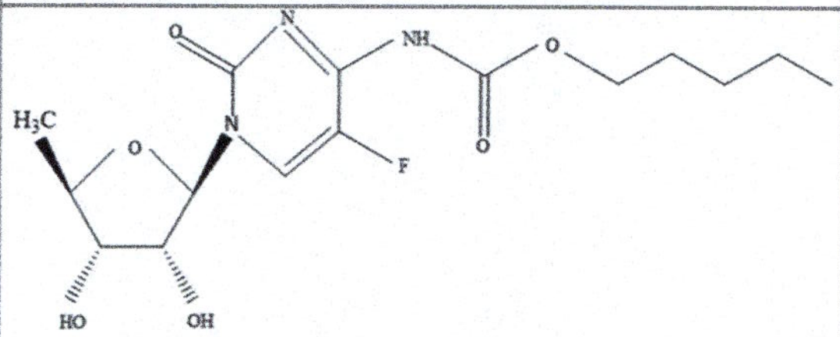	Less than ~100 nm	Improved oral bioavailability, targeted drug delivery	More convenient oral administration and potentially improved tumor targeting	[53]
2	Paclitaxel	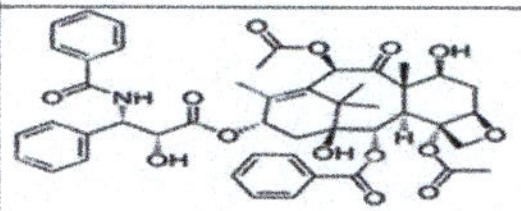	~130 nm	Improved drug solubility, enhanced cellular uptake	Potential targeted therapy for GI cancer patients with reduced side effects and improved tumor response	[55]
3	5-Fluorouracil	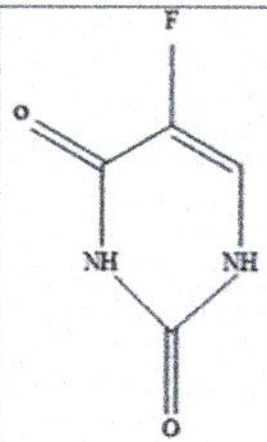	~20–63 nm	Reduced systemic toxicity	Safer alternative to the free drug, potentially reducing adverse events and improving patient tolerance	[56]
4	Simvastatin	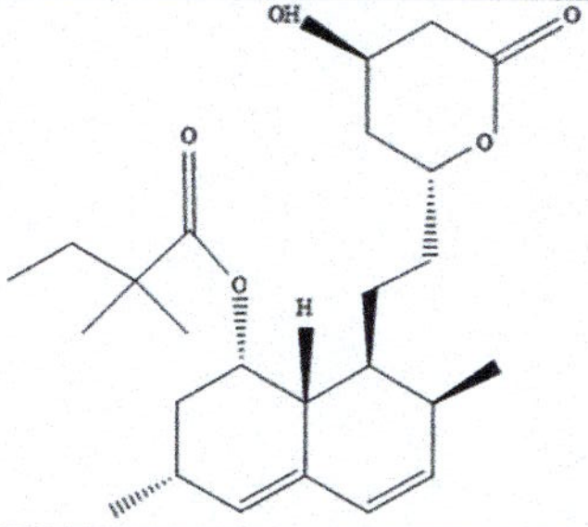	~261 nm	Significant increase in cytotoxic effects, Targeted Drug Delivery to colon	Better colon target delivery of SIM as compared to free drug	[57]

(continued)

S.no	Drug	Chemical structure	Nano particle size	Evaluation	Clinical implications	Reference
5	Curcumin	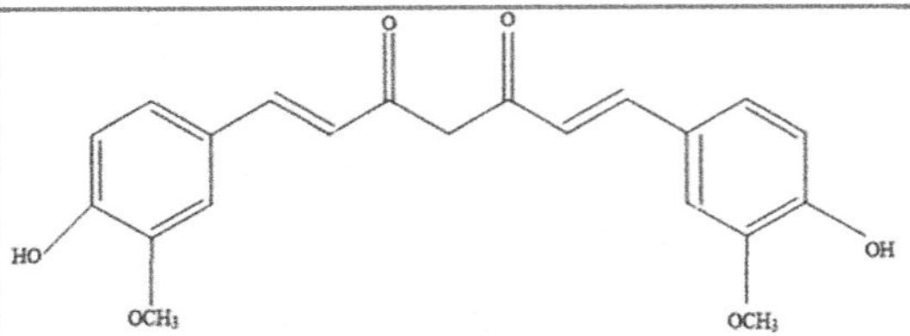	~211 nm	Increased dissolution and reduced cytotoxicity	Potential for improved oral dosing	[58]
6	Docetaxel	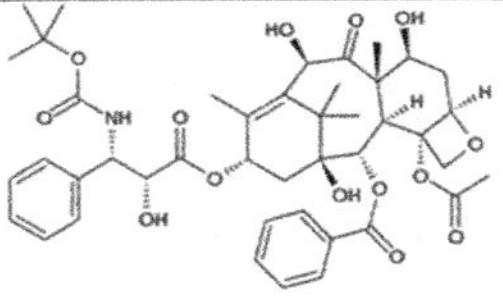	~83.97 nm	No cytotoxicity and improved bioavailability	Better drug targeting to Tumor	[59]
7	Mebendazole (MBZ)	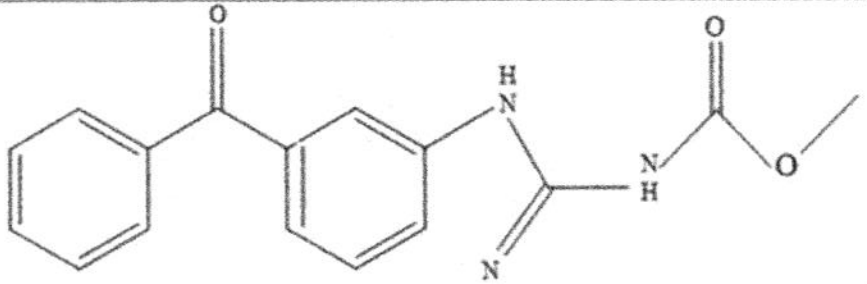	~362.6 nm	The 3D tumor spheroids had a reduction of 1.5-fold and 1.8-fold in size	improved drug solubility and efficacy of drug	[60]
8	Daidzein (Natural origin Isoflavone class compound)	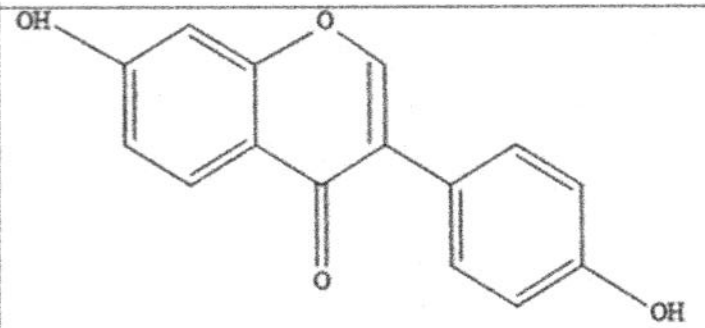	~233 nm	Targeting multiple signaling pathways involved in colorectal carcinoma progression while minimizing oxidative damage	Enhanced cytotoxicity	[62]

therapy, demonstrating improved pharmacokinetics and reduced side effects. As the field of nanotechnology and cancer treatment continues to evolve, nanosuspensions are anticipated to become a key player in transforming the treatment of gastrointestinal (GI) cancers, providing a more efficient and personalized method. The continuation of research and necessary regulatory approvals are crucial to unlock the full capabilities of nanosuspensions in clinical settings, thereby enhancing the quality of life for GI cancer patients worldwide. Current research and clinical trials will help the researchers to deep dive into the potential of nanosuspension formulations to improve drug delivery to gastric tumors, boost the effectiveness of combination therapies, and create targeted treatments designed for each patient. As advancements in nanotechnology and cancer therapeutics persist, the role of nanosuspensions in reshaping the GI cancer treatment landscape is expected to grow significantly.

Acknowledgment The authors also wish to thank GTHTM-2024 organizing team for allowing me to participate in the conference and for presentation as well.

Conflict of Interest The authors declare no conflict of interest, financial or otherwise.

Funding None.

References

1. Chanu, M.T., Singh, A.S.: Cancer disease and its' understanding from the ancient knowledge to the modern concept. World J. Adv. Res. Rev. **15**(2), 169–176 (2022)
2. Basu, D., Pal, R., Sarkar, M., Barma, S., Halder, S., Roy, H., Nandi, S., Samadder, A.: To investigate growth factor receptor targets and generate cancer targeting inhibitors. Curr. Top. Med. Chem. **23**(30), 2877–2972 (2023)
3. Kulothungan, V., Sathishkumar, K., Leburu, S., Ramamoorthy, T., Stephen, S., Basavarajappa, D., Tomy, N., Mohan, R., Menon, G.R., Mathur, P.: Burden of cancers in India-estimates of cancer crude incidence, YLLs, YLDs and DALYs for 2021 and 2025 based on National Cancer Registry Program. BMC Cancer. **22**(1), 527 (2022)
4. Sekiguchi, M., Oda, I., Matsuda, T., Saito, Y.: Epidemiological trends and future perspectives of gastric cancer in Eastern Asia. Digestion. **103**(1), 22–28 (2022)
5. Cancer Today Fact sheets. https://gco.iarc.fr/today/en/fact-sheets-cancers. Accessed 21 Jan 2024
6. Ferlay, J., Colombet, M., Soerjomataram, I., Parkin, D.M., Piñeros, M., Znaor, A., Bray, F.: Cancer statistics for the year 2020: an overview. Int. J. Cancer Res. **149**(4), 778–789 (2021)
7. Chhikara, B.S., Parang, K.: Global cancer statistics 2022: the trends projection analysis. Chem. Biol. Lett. **10**(1), 451–451 (2023)
8. Mathur, P., Sathishkumar, K., Chaturvedi, M., Das, P., Sudarshan, K.L., Santhappan, S., Nallasamy, V., John, A., Narasimhan, S., Roselind, F.S., Icmr-Ncdir-Ncrp Investigator Group: Cancer statistics, 2020: Report from national cancer registry programme, India. JCO Glob. Oncol. **6**, 1063–1075 (2020)
9. Shakuntala, T.S., Krishnan, S.K., Das, P., Sudarshan, K.L., Kotian, C.M., Santhappan, S., Vishwakarma, M.B., Sureshkumar, N., Mathur, P.: Descriptive epidemiology of gastrointes-

tinal cancers: results from National Cancer Registry Programme, India. Asian Pac. J. Cancer Prev. **23**(2), 408 (2022)
10. Yang, K., Lu, L., Liu, H., Wang, X., Gao, Y., Yang, L., Li, Y., Su, M., Jin, M., Khan, S.: A comprehensive update on early gastric cancer: defining terms, etiology, and alarming risk factors. Expert Rev. Gastroenterol. Hepatol. **15**(3), 255–273 (2022)
11. Jacob, S., Nair, A.B., Shah, J.: Emerging role of nanosuspensions in drug delivery systems. Biomater. Res. **24**(1), 3 (2020)
12. Godge, G.R., Garje, M.A., Dode, A.B., Tarkase, K.N.: Nanosuspension technology for delivery of poorly soluble drugs and its applications: a review. Int. J. Pharm. Sci. Nanotechnol. **13**(4), 4965–4978 (2020)
13. Cao, W., Chen, H.D., Yu, Y.W., Li, N., Chen, W.Q.: Changing profiles of cancer burden worldwide and in China: a secondary analysis of the global cancer statistics 2020. Chin. Med. J. **134**(07), 783–791 (2021)
14. Matthews, H.K., Bertoli, C., de Bruin, R.A.: Cell cycle control in cancer. Nat. Rev. Mol. Cell Biol. **23**(1), 74–88 (2022)
15. Hanahan, D.: Hallmarks of cancer: new dimensions. Cancer Discov. **12**(1), 31–46 (2022)
16. Glovaci, D., Fan, W., Wong, N.D.: Epidemiology of diabetes mellitus and cardiovascular disease. Curr. Cardiol. Rep. **21**, 1–8 (2019)
17. Thrift, A.P., El-Serag, H.B.: Burden of gastric cancer. Clin. Gastroenterol. Hepatol. **18**(3), 534–542 (2020)
18. Abu-Freha, N., Hizkiya, R., Abu-Abed, M., Michael, T., Jacob, B.M., Rouvinov, K., Schwartz, D., Reshef, A., Netz, U., Pinsk, I., Etzion, O.: The impact of the COVID-19 pandemic on colorectal and gastric cancer diagnosis, disease stage and mortality. Front. Med. **9**, 954878 (2022)
19. Thrift, A.P., Wenker, T.N., El-Serag, H.B.: Global burden of gastric cancer: epidemiological trends, risk factors, screening and prevention. Nat. Rev. Clin. Oncol. **20**(5), 338–349 (2023)
20. Machlowska, J., Baj, J., Sitarz, M., Maciejewski, R., Sitarz, R.: Gastric cancer: epidemiology, risk factors, classification, genomic characteristics and treatment strategies. Int. J. Mol. Sci. **21**(11), 4012 (2020)
21. Yaghoobi, M., Bijarchi, R., Narod, S.A.: Family history and the risk of gastric cancer. Br. J. Cancer. **102**(2), 237–242 (2010)
22. Lauwers, G.Y., Mullen, J.T., Schreiber, K.E.C., Chung, D.C.: Familial gastric cancers: a review with focus on hereditary diffuse gastric cancer syndrome. AJSP: Rev. Rep. **19**(2), 66–73 (2014)
23. Boland, C.R., Yurgelun, M.B.: Historical perspective on familial gastric cancer. Cell. Mol. Gastroenterol. Hepatol. **3**(2), 192–200 (2017)
24. Pinheiro, H., Oliveira, C., Seruca, R., Carneiro, F.: Hereditary diffuse gastric cancer–Pathophysiology and clinical management. Best Pract. Res. Clin. Gastroenterol. **28**(6), 1055–1068 (2014)
25. Elingarami, S., Liu, M., Fan, J., He, N.: Applications of nanotechnology in gastric cancer: Detection and prevention by nutrition. J. Nanosci. Nanotechnol. **14**(1), 932–945 (2014)
26. Vainio, H., Weiderpass, E.: Fruit and vegetables in cancer prevention. Nutr. Cancer. **54**(1), 111–142 (2006)
27. Ma, J.-L., Zhang, L., Brown, L.M., Li, J.-Y., Shen, L., Pan, K.-F., Liu, W.-D., Hu, Y., Han, Z.-X., Crystal-Mansour, S.: Fifteen-year effects of Helicobacter pylori, garlic, and vitamin treatments on gastric cancer incidence and mortality. J. Natl. Cancer Inst. **104**, 488–492 (2012)
28. Choi, J., Kim, S.G., Yoon, H., Im, J.P., Kim, J.S., Kim, W.H., Jung, H.C.: Eradication of Helicobacter pylori after endoscopic resection of gastric tumors does not reduce incidence of metachronous gastric carcinoma. Clin. Gastroenterol. Hepatol. **12**(5), 793–800 (2014)

29. Fukase, K., Kato, M., Kikuchi, S., Inoue, K., Uemura, N., Okamoto, S., Terao, S., Amagai, K., Hayashi, S., Asaka, M.: Effect of eradication of Helicobacter pylori on incidence of metachronous gastric carcinoma after endoscopic resection of early gastric cancer: An open-label, randomised controlled trial. Lancet. **372**, 392–397 (2008)
30. Choi, K.S., Suh, M.: Screening for gastric cancer: the usefulness of endoscopy. Clin. Endosc. **47**(6), 490 (2014)
31. Karimi, P., Islami, F., Anandasabapathy, S., Freedman, N.D., Kamangar, F.: Gastric cancer: descriptive epidemiology, risk factors, screening, and prevention. Cancer Epidemiol. Biomarkers Prev. **23**(5), 700–713 (2014)
32. Matsumoto, S., Ishikawa, S., Yoshida, Y.: Reduction of gastric cancer mortality by endoscopic and radiographic screening in an isolated Island: A retrospective cohort study. Aust. J. Rural Health. **21**(6), 319–324 (2013)
33. Guan, W.L., He, Y., Xu, R.H.: Gastric cancer treatment: recent progress and future perspectives. J. Hematol. Oncol. **16**(1), 57 (2023)
34. Hartgrink, H.H., Van de Velde, C.J.H., Putter, H., Songun, I., Tesselaar, M.E.T., Kranenbarg, E.K., De Vries, J.E., Wils, J.A., Van Der Bijl, J., van Krieken, J.H.J.M.: Neo-adjuvant chemotherapy for operable gastric cancer: long term results of the Dutch randomised FAMTX trial. Eur. J. Surg. Oncol. **30**(6), 643–649 (2004)
35. Cunningham, D., Allum, W.H., Stenning, S.P., Thompson, J.N., Van de Velde, C.J., Nicolson, M., Scarffe, J.H., Lofts, F.J., Falk, S.J., Iveson, T.J., Smith, D.B.: Perioperative chemotherapy versus surgery alone for resectable gastroesophageal cancer. N. Engl. J. Med. **355**(1), 11–20 (2006)
36. Boige, V., Pignon, J., Saint-Aubert, B., Lasser, P., Conroy, T., Bouché, O., Segol, P., Bedenne, L., Rougier, P., Ychou, M.: Final results of a randomized trial comparing preoperative 5-fluorouracil (F)/cisplatin (P) to surgery alone in adenocarcinoma of stomach and lower esophagus (ASLE): FNLCC ACCORD07-FFCD 9703 trial. J. Clin. Oncol. **25**(18), 4510–4510 (2007)
37. Hecht, J.R., Bang, Y.J., Qin, S.K., Chung, H.C., Xu, J.M., Park, J.O., Jeziorski, K., Shparyk, Y., Hoff, P.M., Sobrero, A., Salman, P.: Lapatinib in combination with capecitabine plus oxaliplatin in human epidermal growth factor receptor 2–positive advanced or metastatic gastric, esophageal, or gastroesophageal adenocarcinoma: TRIO-013/LOGiC—a randomized phase III trial. J. Clin. Oncol. **34**(5), 443–451 (2016)
38. Tabernero, J., Hoff, P.M., Shen, L., Ohtsu, A., Shah, M.A., Cheng, K., Song, C., Wu, H., Eng-Wong, J., Kim, K., Kang, Y.K.: Pertuzumab plus trastuzumab and chemotherapy for HER2-positive metastatic gastric or gastro-oesophageal junction cancer (JACOB): final analysis of a double-blind, randomised, placebo-controlled phase 3 study. Lancet Oncol. **19**(10), 1372–1384 (2018)
39. Wang, X., Fu, R., Hu, Y., Du, H., Li, S., Li, Z., Liu, Y., Li, Q., Zhang, L., Ji, J.: EGFR gene status predicts response and survival benefit in a preclinical gastric cancer trial treating patient-derived xenografts with cetuximab. Oncol. Rep. **38**(4), 2387–2393 (2017)
40. Lordick, F., Luber, B., Lorenzen, S., Hegewisch-Becker, S., Folprecht, G., Wöll, E., Decker, T., Endlicher, E., Röthling, N., Schuster, T., Keller, G.: Cetuximab plus oxaliplatin/leucovorin/5-fluorouracil in first-line metastatic gastric cancer: a phase II study of the Arbeitsgemeinschaft Internistische Onkologie (AIO). Br. J. Cancer. **102**(3), 500–505 (2010)
41. Macedo, F., Ladeira, K., Longatto-Filho, A., Martins, S.F.: Gastric cancer and angiogenesis: is VEGF a useful biomarker to assess progression and remission. J. Gastric Cancer. **17**(1), 1 (2017)
42. National cancer Institute, Drugs approved for stomach cancer. https://www.cancer.gov/about-cancer/treatment/drugs/stomach. Accessed 15 Feb 2024

43. Van den Boogaard, W.M., Komninos, D.S., Vermeij, W.P.: Chemotherapy side-effects: not all DNA damage is equal. Cancers. **14**(3), 627 (2022)
44. Li, L.J., Chong, Q., Wang, L., Cher, G.B., Soo, R.A.: Different treatment efficacies and side effects of cytotoxic chemotherapy. J. Thorac. Dis. **12**(7), 3785 (2020)
45. Floyd, J., Mirza, I., Sachs, B., Perry, M.C.: Hepatotoxicity of chemotherapy. Semin. Oncol. **33**(1), 50–67 (2006)
46. Cavaletti, G., Marmiroli, P.: Pharmacotherapy options for managing chemotherapy-induced peripheral neurotoxicity. Expert. Opin. Pharmacother. **19**(2), 113–121 (2018)
47. Madeddu, C., Deidda, M., Piras, A., Cadeddu, C., Demurtas, L., Puzzoni, M., Piscopo, G., Scartozzi, M., Mercuro, G.: Pathophysiology of cardiotoxicity induced by nonanthracycline chemotherapy. J. Cardiovasc. Med. **17**, e12–e18 (2016)
48. Jin, C., Wang, K., Oppong-Gyebi, A., Hu, J.: Application of nanotechnology in cancer diagnosis and therapy-a mini-review. Int. J. Med. Sci. **17**(18), 2964 (2020)
49. Maddiboyina, B., Ramaiah, Nakkala, R.K., Roy, H.: Perspectives on cutting-edge nanoparticulate drug delivery technologies based on lipids and their applications. Chem. Biol. Drug Des. **102**(2), 377–394 (2023)
50. Chakravorty, R.: Nanosuspension as an emerging Nanotechnology and Techniques for its Development. Res. J. Pharm. Technol. **15**(1), 489–493 (2022)
51. Goel, S., Sachdeva, M., Agarwal, V.: Nanosuspension technology: recent patents on drug delivery and their characterizations. Recent Pat. Drug Deliv. Formul. **13**(2), 91–104 (2019)
52. Ling, L.X., Ouyang, Y., Hu, Y.: Research trends on nanomaterials in gastric cancer: a bibliometric analysis from 2004 to 2023. J. Nanobiotechnol. **21**(1), 248 (2023)
53. Patent No. CN102961342B. Nanoscale capecitabine and preparation method thereof (2012)
54. Li, X., Ai, S., Lu, X., Liu, S., Guan, W.: Nanotechnology-based strategies for gastric cancer imaging and treatment. RSC Adv. **11**(56), 35392–35407 (2021)
55. Bando, H., Shimodaira, H., Fujitani, K., Takashima, A., Yamaguchi, K., Nakayama, N., Takahashi, T., Oki, E., Azuma, M., Nishina, T., Hironaka, S.: A phase II study of nab-paclitaxel in combination with ramucirumab in patients with previously treated advanced gastric cancer. Eur. J. Cancer. **91**, 86–91 (2018)
56. Abdullah, S., El Hadad, S., Aldahlawi, A.: The development of a novel oral 5-Fluorouracil in-situ gelling nanosuspension to potentiate the anticancer activity against colorectal cancer cells. Int. J. Pharm. **613**, 121406 (2022)
57. Taymouri, S., Ahmadi, Z., Mirian, M., Tavakoli, N.: Simvastatin nanosuspensions prepared using a combination of pH-sensitive and timed-release approaches for potential treatment of colorectal cancer. Pharm. Dev. Technol. **26**(3), 335–348 (2021)
58. Elbaz, N.M., Tatham, L.M., Owen, A., Rannard, S., McDonald, T.O.: Redispersible nanosuspensions as a plausible oral delivery system for curcumin. Food Hydrocoll. **121**, 107005 (2021)
59. Ibrahim, M.A., Shazly, G.A., Aleanizy, F.S., Alqahtani, F.Y., Elosaily, G.M.: Formulation and evaluation of docetaxel nanosuspensions: in-vitro evaluation and cytotoxicity. Saudi Pharm. J. **27**(1), 49–55 (2019)
60. Patel, R.D., Patel, A.S., Patel, H.J., Sarvepalli, S., Patel, K.: Development of rapidly soluble mebendazole nanosuspension for colorectal cancer. J. Drug Deliv. Technol. **91**, 105276 (2024)
61. Liu, H., Yuan, M., Liu, Y., Guo, Y., Xiao, H., Guo, L., Liu, F.: Self-monitoring and self-delivery of self-assembled fluorescent nanoparticles in cancer therapy. Int. J. Nanomedicine. **16**, 2487–2499 (2021)
62. Becit-Kizilkaya, M., Oncu, S., Sen, S., Kaplan, A.B.U., Celik, S., Cetin, M., Aydin Dilsiz, S.: Daidzein Nanosuspension with 5-Fluorouracil Induces Apoptosis and Alleviate Inflammation in Caco-2 Colorectal Adenocarcinoma Cells. SSRN. 4712886 (2024)

Effects of Commonly Used Pesticides on Gut Microbiota

Neha Sagar[1], Deepak Teotia[1], and Sisir Nandi[1] (✉)

[1] Global Institute of Pharmaceutical Education and Research (GIPER), Affiliated to Veer Madho Singh Bhandari Uttarakhand Technical University, Kashipur, Uttarakhand, India

Abstract The human gut microbiota can be potentially disrupted due to exposure to various environmental factors including pesticides. The widespread use of pesticides in agricultural practices has raised concerns about its potential health effects on humans. This study explores the intricate relationship between pesticide or fungicide exposure and its detrimental impact on the lungs and gastrointestinal microbiome. When pesticides are sprayed on crops, droplets can be inadvertently ingested or inhaled by individuals working in the fields without masks. Once inhaled, pesticide molecules can penetrate through the respiratory tract, gaining access to the lungs. Similarly, ingestion of pesticides through the mouth and contaminated food or water can lead to entry into the gastrointestinal tract (GIT). Upon entry into the body, pesticides can disrupt the delicate balance of the microbiota. The microbiota are mainly collection of beneficial microorganisms which include bacteria, archaea, fungi, and viruses, etc. residing in the lungs and gut. The microbiomes have a huge impact on colonization, pathogen resistance, maintain the intestinal epithelium, controlling the host immune function, and maintain a good friendship between gut and lungs. Various classes of agricultural pesticides have been shown to induce toxicity in the microbiome. This may disturb the gut environment and loss of microbiota. The patients may suffer from GIT problems (flatulence, indigestion, loss of appetite, irritable bowel syndrome, and constipation) and may have breathing difficulty and skin allergies on exposure to these pesticides. The present in vitro model is an attempt to study the detrimental effect of commonly used fungicides such as copper oxychloride and thiophanate methyl on *Lactobacillus casei* isolated from the Yakult using MRS agar media.

Keywords Pesticides · Copper oxychloride · Thiophanate methyl · Environmental exposure · Detrimental impact · Gut microbiota · *Lactobacillus casei*

A. K. Saxena, S. Nandi (eds.), *Global Trends in Health, Technology and Management*, https://doi.org/10.1007/978-3-031-75457-9_18

1 Introduction

Pesticides are chemicals used to control pests and weeds in agriculture. Pesticides are used on vegetables, wheat, rice, etc. in the fields to prevent insect infection and disease carriers. The increase in rate of population greatly demands an increase in food production [1]. The pests and insects attack the crops in the fields, causing a loss of productivity. Without the use of pesticides, there would be a 78% loss of fruit production, a 54% loss of vegetable production, and a 32% loss of cereal production. Therefore, pesticides play a critical role in reducing diseases and increasing crop yields worldwide [2]. However, their widespread use raises concerns about their impact on the environment and human health. Once sprayed, pesticides are volatilized and can be carried on air currents away from the treated surface. The high vapor pressure and higher temperature, low relative humidity, and more open air circulation in the field increase the volatilization [3].

The farmers used to spray pesticides even without masks for several hours. There is a chance to enter pesticide molecules through inhalation, consumption by mouth through volatilities, and ingested food or drinks. Lisouza et al. (2020) found that contaminated surface waters could be the major source of human exposure to organochlorine pesticides through volatilization [4]. The long exposure to pesticides may have potentially detrimental effects on the lungs and gut microbiota, the diverse community of microorganisms living in our lungs and digestive tract. These tiny organisms play crucial roles in digestion, immunity, and overall health [5]. However, the relationship between pesticides and microbiota is complex and not fully understood. Therefore, it is our target in the present study to explore the potential effects of commonly used pesticides copper oxychloride and thiophanate methyl as fungicides on gut microbiota taking *Lactobacillus casei* grown in the in vitro model as sample microbiome.

2 Materials and Methods

2.1 Isolation of *Lactobacillus casei*

Yakult is a commercially available probiotic drink that contains *Lactobacillus casei* strain Shirota. 1 ml of Yakult was dissolved into autoclaved distilled water and serial dilutions of up to 10^{-3} were made. 10^{-1}, 10^{-2}, and 10^{-3} dilutions were plated onto de Man, Rosa, Sharpe agar media (MRS agar). MRS agar is a media that has been prepared for the selective growth of *Lactobacillus* strain of bacteria [6]. The composition of MRS agar media is given in Table 1. The ingredients were poured carefully into a volumetric flask. A volume of 100 ml distilled water was added followed by shaking well to mix the ingredients properly. Plug the mouth of the conical flask using non-absorbable sterile cotton followed by the minute wrapping of the aluminum foil paper.

The agar media was autoclaved at 120° C for 1 h at 15 lb./in^2 and poured onto sterile agar plates. The agar was allowed to solidify on cooling followed by closing the lid of

Table 1 Composition of MRS agar

Composition	Quantity (gm/L)
Peptone	10.000
Dextrose	20.000
Ammonium citrate	2.000
Sodium acetate	5.000
Magnesium sulfate	0.050
Beef extract	10.000
Agar	20.000

Fig. 1 Bacterial colonies grown onto the MRS agar plate

the plates and keeping into the laminar air flow for 20–230 min. The prepared dilutions were spread onto the solid media with the help of L loop. The small and smaller circular-shaped colonies thus obtained (Fig. 1) were confirmed by Gram staining [7].

2.2 Inoculation of Colonies onto MRS Broth

The 5 ml of sterile MRS broth was transferred to each of 5 sterile test tubes. The colonies were equally selected using the flame sterilized inoculating loop and were inoculated into the MRS broth mixing well and kept for 24 to 48 h in the BOD incubator at 37–38° C. After incubation, turbidity formulated in the test tubes was observed [8]. The turbid solutions were taken as a control for further spectroscopic analyses (Fig. 2).

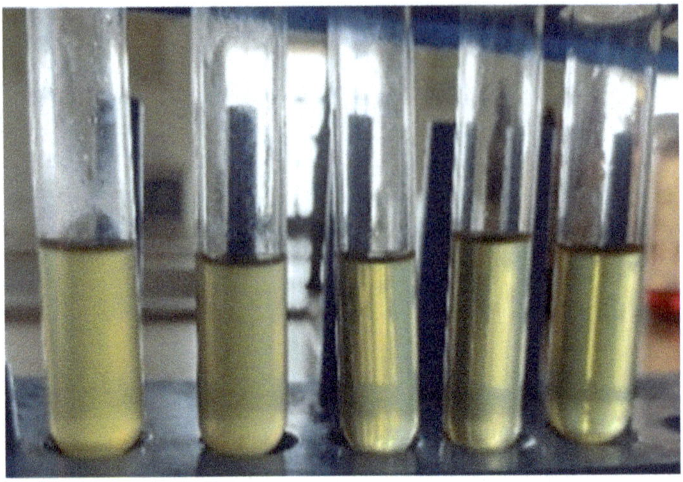

Fig. 2 Turbidity produced in the incubated MRS broth

2.3 Pesticide Compounds Data

The commonly used pesticides such as copper oxychloride and thiophanate methyl were commercially purchased and taken as biological activity data in the present attempt. The copper oxychloride was reported as a broad-spectrum fungicide used to protect potato, tomato, chili, curcuma, banana, citrus, grapes, and tea [9]. Thiophanate methyl is a fungicide used on wine grapes, tomato, aubergine, leek, fresh beans with pods and wheat (winter and durum) [10]. These pesticides stock solutions were made as 1.5 g/L for copper oxychloride, and 1 g/L for thiophanate methyl as per the guidelines written on the product label. From these stock solutions, 10^{-1} dilutions were made to study the inhibitions of the growth of *Lactobacillus*.

2.4 Growth Inhibition of *Lactobacillus* Using Turbidimetric Method

The *Lactobacillus* colonies were grown into MRS broth as stated in Sect. 2.1 and the inoculated broth was prepared as stated in Sect. 2.2. The bacterial growth was checked by measuring the optical density (OD) at 600 nm [11] using UV-Vis spectrometry. Uninoculated broth was used as a Blank. The bacterial strains were then treated with 10^{-1} dilution for each of the prepared compound solutions. OD was measured for these samples using UV-Vis spectrometry.

3 Results and Discussion

To perform a blank measurement, distilled water was taken in both the reference cuvette and sample cuvette and set the blanking as 0. Then the MRS broth (uninoculated culture media) was taken in the reference curette and the inoculated broth was taken in the

Table 2 Absorbance of the test compounds while treating

Test tube serial number	Test compound	Absorbance
1	Copper oxychloride	0.323
2	Thiophanate methyl	0.563

sample cuvette to measure absorbance. The absorbance of the inoculated broth at 600 nm was taken three times and the mean was found as 1.287 which was treated as the control. The absorbance maxima of copper oxychloride and thiophanate methyl were found as 435 nm [12] and 263 nm [13], respectively. The absorbance for the copper oxychloride and thiophanate methyl solutions at a 10^{-1} dilution, at 435 nm and 263 nm, was measured as 0.727, and 0.127 respectively.

Now 5 ml of the *L. casei* inoculated broth was taken in each of the two test tubes and 5 ml of each of the test compound solutions was added in each test tube and shaken well. 1 ml of the prepared each test solution was taken in the test cuvette and the absorbance was measured (Table 2) considering the reference of 1 ml solution mixture of the prepared 10 ml containing 5 ml uninoculated broth and 5 ml of the test compound at 10^{-1}dilution.

From the above results, it has been predicted that the test compound copper oxychloride and thiophanate methyl when treated with the *L. casei* inoculated broth negatively affected the gut microbiota showing a decrease in UV-absorbance and was found as 0.323 and 0.563 which are significantly lower than the control reading as 1.287. This observation has been corroborated by the previous experimental findings of Ma et al. who found that Cu exposure can alter the abundance of the gut microbiota [14] and Lukowicz et al. who measured the toxicity of a mixture of pesticides including thiophanate in mice, suggesting the problem in regular gut microbiota composition and the host metabolic homeostasis [15].

4 Conclusion

The harmful impact of commonly used fungicides, such as copper oxychloride and thiophanate methyl, on *Lactobacillus casei* sourced from Yakult culture was analyzed using an in vitro model. These pesticides were found to impede the growth of *Lactobacillus* in inoculated MRS broth, as evidenced by a decrease in absorbance following fungicide treatment. This research suggests the potential applicability of examining additional pesticides and conducting a structure-absorbance relationship study to pinpoint the inhibition of microbiota within the in vitro system, with potential implications for human systems. Farmers, often exposed to pesticides without proper protection, may experience gastrointestinal issues such as flatulence, indigestion, loss of appetite, breathing difficulties, and skin allergies. These pesticides could disrupt the gut microbiome followed by the lung microbiota and compromise the host's immunity. Thus, raising awareness among farmers about the importance of wearing masks while spraying pesticides is crucial.

Acknowledgments Neha Sagar thanks Sombodhi Bhattacharya for the help and advice. She also thanks the evaluators of the poster scientific session of GTHTM-2024 symposium and Dr. A. K. Saxena for the constructive suggestions.

References

1. Damalas, C.A., Eleftherohorinos, I.: Pesticide exposure, safety issues, and risk assessment indicators. Int. J. Environ. Res. Public Health. **8**, 1402–1419 (2011)
2. Tudi, M., Daniel, R.H., Wang, L., Lyu, J., Sadler, R., Connell, D., Chu, C., Phung, D.T.: Agriculture development, pesticide application and its impact on the environment. Int. J. Environ. Res. Public Health. **18**(3), 1112 (2021)
3. Connell, D.W.: Basic Concepts of Environmental Chemistry. CRC Press, Boca Raton, FL, USA (2005)
4. Lisouza, F.A., Owuor, P.O., Lalah, J.O.: Sources, distribution, and risk assessment of organochlorine pesticides in Nairobi City, Kenya. J. Environ. Sci. (China). **96**, 178–185 (2020)
5. Ali, A., AlHussaini, K.I.: Pesticides: unintended impact on the hidden world of gut microbiota. Metabolites. **14**(3), 155 (2024)
6. Renschler, M.A., Wyatt, A., Anene, N., Robinson-Hill, R., Pickerill, E.S., Fox, N.E., Griffith, J.A., McKillip, J.L.: Using nitrous acid-modified de man, Rogosa, and Sharpe medium to selectively isolate and culture lactic acid bacteria from dairy foods. J. Dairy Sci. **103**(2), 1215–1222 (2020)
7. Gram, H.C.: Über die isolierte Färbung der Schizomyceten in Schnitt- und Trockenpräparaten. Fortschritte der Medizin (in German). **2**, 185–189 (1984)
8. Nguyen, H.T., Elegado, F.B., Librojo-Basilio, N.T., Mabesa, R.C., Dizon, E.I.: Isolation and characterisation of selected lactic acid bacteria for improved processing of Nem chua, a traditional fermented meat from Vietnam. Benef. Microbes. **1**(1), 67–74 (2010)
9. Ferreira, L.C., Scavroni, J., da Silva, J.R., Cataneo, A.C., Martins, D., Boaro, C.S.: Copper oxychloride fungicide and its effect on growth and oxidative stress of potato plants. Pestic. Biochem. Physiol. **112**, 63–69 (2014)
10. Imazaki, I., Iizumi, H., Ishikawa, K., et al.: Effects of thiophanate-methyl and azoxystrobin on the composition of *Cercosporakikuchii* populations with thiophanate-methyl-resistant strains. J. Gen. Plant Pathol. **72**, 292–300 (2006)
11. Pramanick, R., Aranha, C.: Distinct functional traits of lactobacilli from women with asymptomatic bacterial vaginosis and Normal microbiota. Microorganisms. **8**(12), 1949 (2020)
12. Rao, T.N., Patrudu, T.B., Rao, N.K., Kumar, G., Rao, M.B.V.: Sissipation kinetics of copper oxychloride and kasugamycin fungicide residues in different pH waters. Anal. Chem. Indian J. **16**(10), 407–410 (2016)
13. https://pubchem.ncbi.nlm.nih.gov/compound/Thiophanate-methyl
14. Ma, F., Huo, Y., Li, H., Yang, F., Liao, J., Han, Q., Li, Y., Pan, J., Hu, L., Guo, J., Tang, Z.: New insights into the interaction between duodenal toxicity and microbiota disorder under copper exposure in chicken: involving in endoplasmic reticulum stress and mitochondrial toxicity. Chem. Biol. Interact. **366**, 110132 (2022)
15. Lukowicz, C., Simatos, S.E., Régnier, M., Polizzi, A., Lasserre, F., Montagner, A., Lippi, Y., Jamin, E.L., Martin, J.F., Naylies, C., Canlet, C., Debrauwer, L., Bertrand-Michel, J., Al Saati, T., Théodorou, V., Loiseau, N., Mselli-Lakhal, L., Guillou, H., Gamet-Payrastre, L.: Metabolic effects of a chronic dietary exposure to a low-dose pesticide cocktail in mice: sexual dimorphism and role of the constitutive and rostane receptor. Environ. Health Perspect. **126**(6), 067007 (2018)

Pharmacophore Modeling of Tricyclic Antipsychotics

Sandhya Dharkiya[1], Rukayya[1], and Sisir Nandi[1] (✉)

[1] Global Institute of Pharmaceutical Education and Research (GIPER), Kashipur, Uttarakhand, India

Abstract Psychoses are bipolar disorder characterized by complex mental illness, lack of communication/speech, abnormal thoughts, talks, behavior, and activities, irritation, loss of patience, loss of family touch, negative symptoms such as feelings of sadness, a deficit of attention, and depression. Chlorpromazine is widely available as an antipsychotic drug used in the treatment of psychosis or schizophrenia. Chlorpromazine contains a phenothiazine nucleus having activity against the D2 receptor, but it is not active against the 5HT2A receptor; therefore, it is not used in the treatment of psychoses with depression. Tricyclic antipsychotics including dibenzothiazepine, dibenzodiazepine, and thienobenzodiazepine compounds show very good effects against schizophrenia as well as depression.

Pharmacophore modeling of tricyclic antidepressants was established but pharmacophore modeling of tricyclic antipsychotics has not yet been developed. Therefore, an attempt has been made to develop the distance-based pharmacophore followed by the development of quantitative structure-pharmacophore-affinity relationship models for dibenzodiazepine, dibenzothiazepine, thienobenzodiazepines derivatives having affinity against D2 as well as 5HT2A receptors. So, the developed pharmacophore may be fit for both the receptors for further design of the congeneric higher active molecules to be used in depression and psychoses.

Keywords Tricyclic antipsychotics · Pharmacophore · Distance-based features · Quantitative structure-pharmacophore-affinity relationships (QSPAR)

1 Introduction

Psychoses, complex delusions, or hallucinations are also called schizophrenia. Tricyclic antidepressants and tricyclic antipsychotics such as dibenzodiazepine, dibenzothiazepine, and thienobenzodiazepine compounds show very good effects against schizophrenia [1]. Antipsychotics are given along with mood elevators such as lithium/valproate, as a first-line treatment for manic and mixed episodes associated with bipolar disorder [2].

The atypical antipsychotics such as risperidone, lurasidone, aripiprazole, ziprasidone, amisulpride, lumateperone, cariprazine, zotepine, clozapine, loxapine, olanzapine,

A. K. Saxena, S. Nandi (eds.), *Global Trends in Health, Technology and Management*, https://doi.org/10.1007/978-3-031-75457-9_19

and quetiapine have also been produced good efficacy in the treatment of schizophrenia and manic depression as a monotherapy, whereas only olanzapine and quetiapine have been proven as broad-spectrum against manic, mixed, and depressive prophylactic treatments [3–6].

The chlorpromazine does not have 5HT2A receptor blocking action and also it produces extrapyramidal side effects but the second-generation atypical antipsychotics can block D2 as well as 5HT2A receptors and they do not produce extrapyramidal side effects. The zotepine, clozapine, loxapine, olanzapine, and quetiapine are tricyclic antipsychotics with good tolerance and low sedation effects on patients.

Sekhar et al. developed a 3D pharmacophore on 12 atypical antipsychotics having a mode of binding inhibition of D2 and 5HT2A receptors to predict the important structural features that are common to highly active compounds using HipHop in the CATALYST program [7]. To quest the molecular mechanism of tricyclic antipsychotics' binding affinity, a quantitative structure-affinity relationship should be developed.

The quantitative relationship between atypical antipsychotic structures and affinity has not been yet explored. Several TCA derivatives were taken into pharmacophore properties generation for the current study. Then, QSARs on pharmacophoric distant-related descriptors were developed. It can be done by pharmacophore property generation followed by quantitative correlation with the affinity to design congeneric dibenzothiazepine, dibenzodiazepine, and thienobenzodiazepine having potent D2 as well as 5HT2A receptors binding affinity with least side effects and resistance issues.

2 Materials and Methods

2.1 Biological Activity Data

Biological activity data of 26 tricyclic antipsychotic compounds having dibenzodiazepine, dibenzothiazepine, and thienobenzodiazepine scaffolds were obtained from published literature [1, 8, 9]. These above analogues exert their effects by blocking the 5HT2A and D2 receptors (Table 1).

2.2 Computation of Structure and Pharmacophore

The 2D module of the ChemDraw Ultra 8.0 software was used to draw the two-dimensional structure of 26 compounds. The drawn 2D structures were converted into three-dimensional (3D) structures followed by structural optimization to make the chemical structure energetically stable utilizing the MM2 force field and a 0.001 dielectric constant utilizing Chem3D Ultra [10]. These optimized conformations were browsed into BIOVIA software for the generation of pharmacophore and calculation of distance-based features (Fig. 1) that would be taken into consideration for the QSAR modeling.

The term pharmacophore refers to an assembly of 3D features required to enable the best supramolecular interaction with a particular biological target. These 3D features

Table 1 Biological activity data of 26 tricyclic antipsychotics

Compound no.	Structure	Compound no.	Structure
1	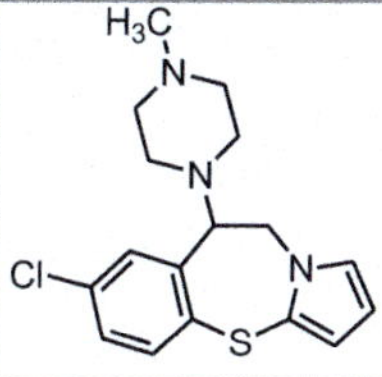	2	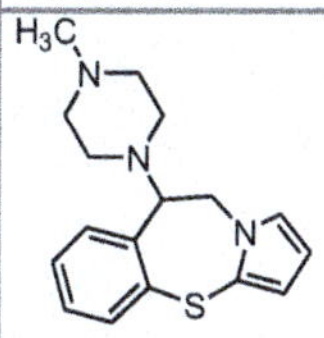
3	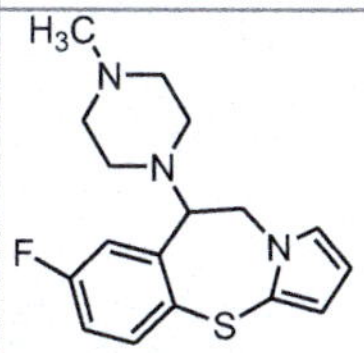	4	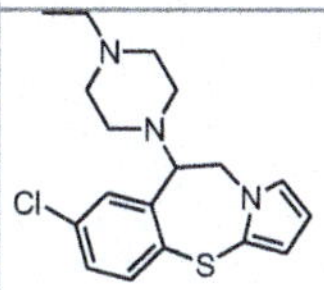
5	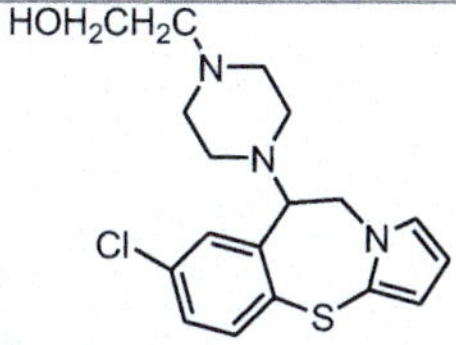	6	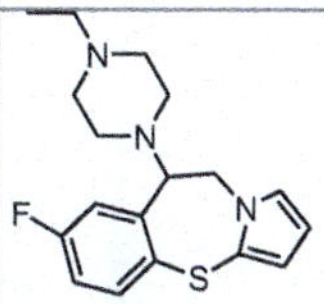
7	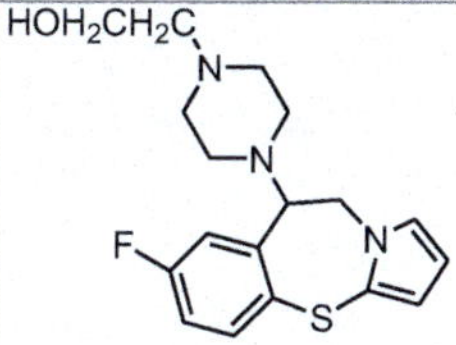	8	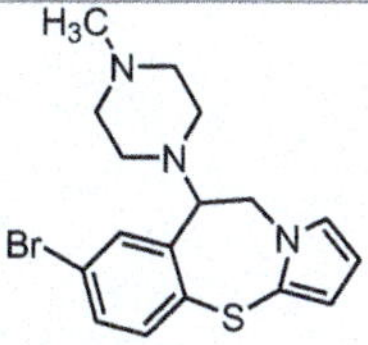
9	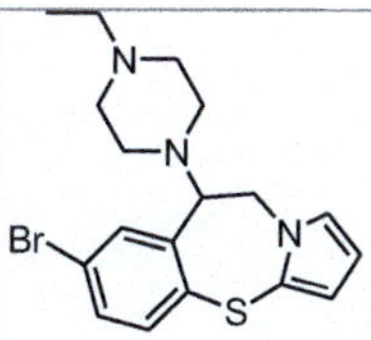	10	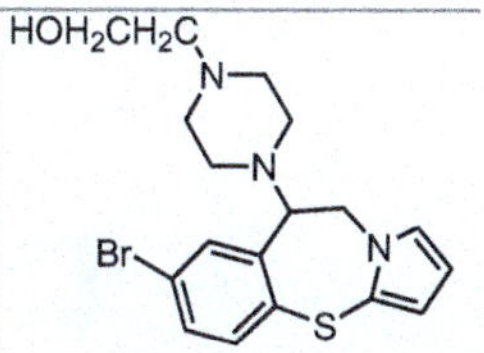
11	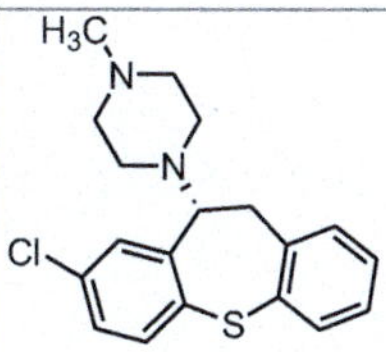	12	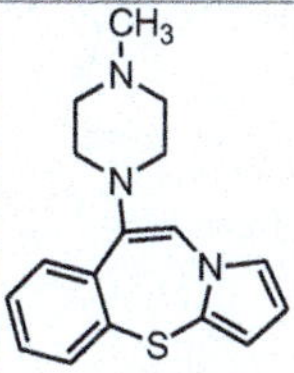
13	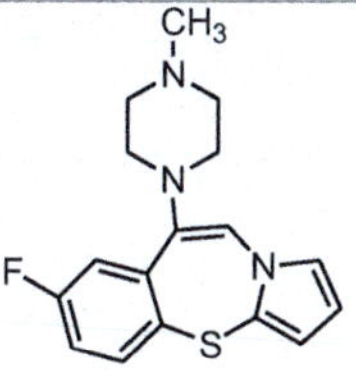	14	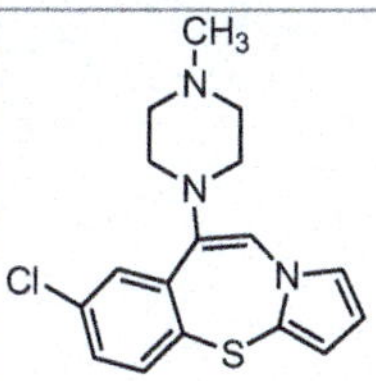

(continued)

Table 1 (continued)

Compound no.	Structure	Compound no.	Structure
15	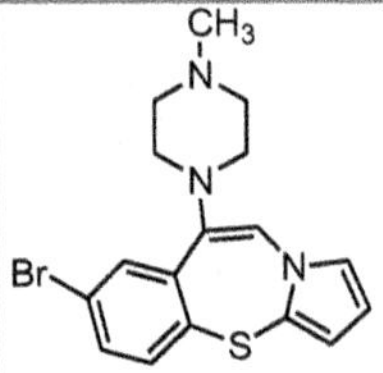	16	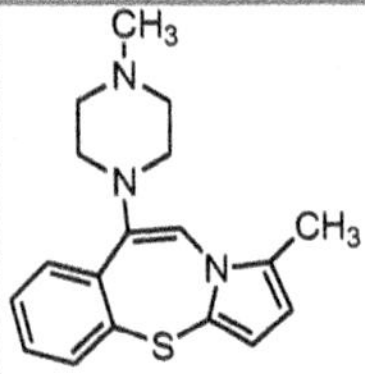
17	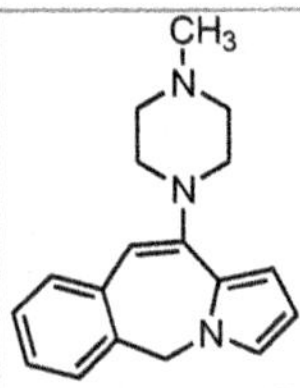	18	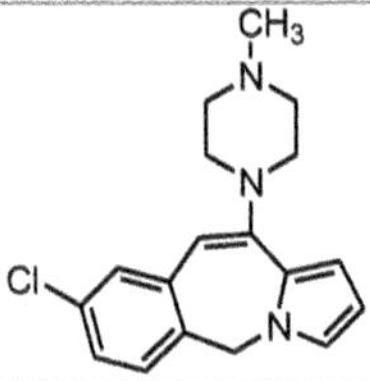
19	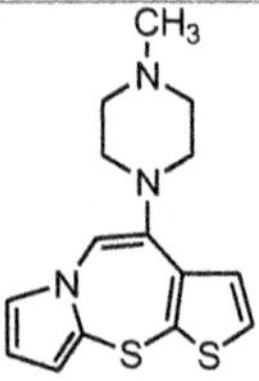	20	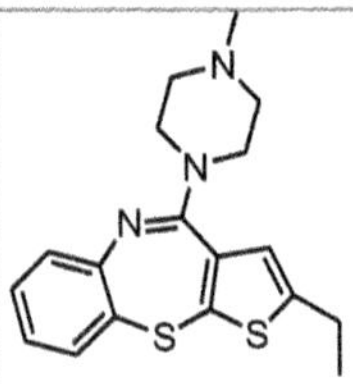
21	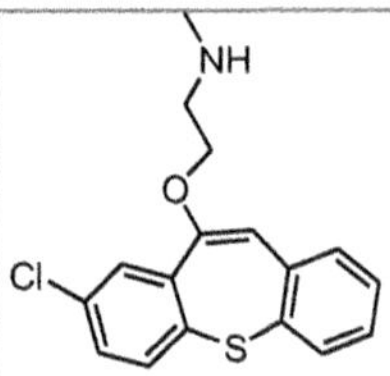	22	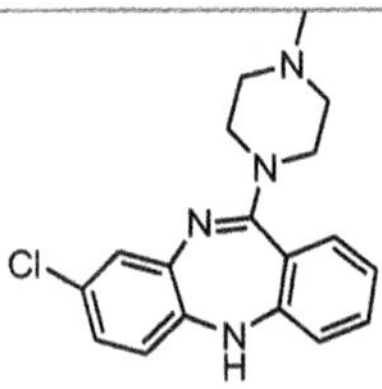
23	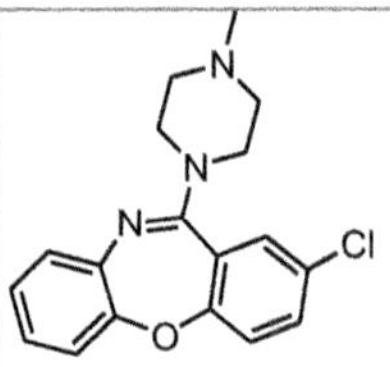	24	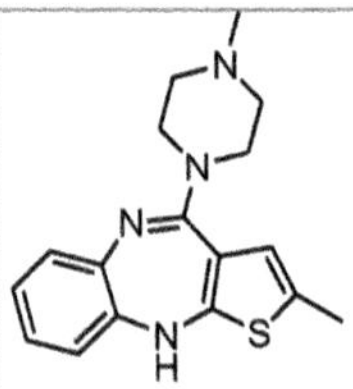
25	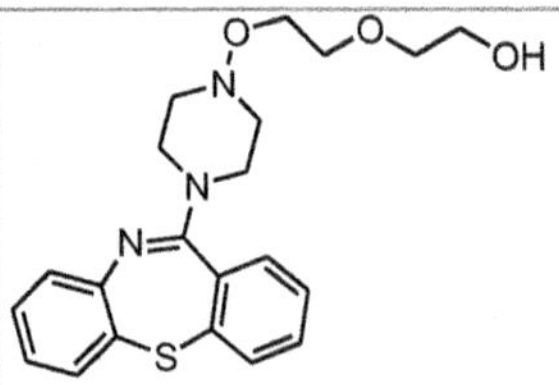	26	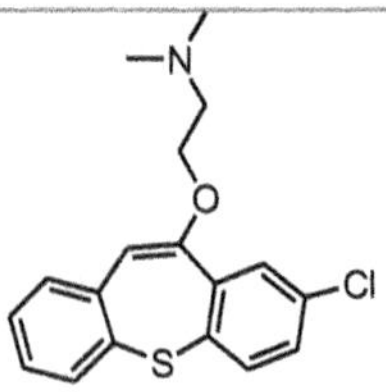

are common to all bioactive compounds. It is useful to determine the number of the 3D characteristics when conforming to the variable spatial zone that binds and shared by all molecules that are active but not active compounds [11, 12].

As a result, the 3D pharmacophore is predicated on the idea that all of the structurally different compounds can attach to the biological target similarly. The six built-in pharmacophoric properties such as hydrogen bond acceptor and donor, hydrophobicity, aromatic rings, positive ionization properties (cation), and negatively ionizing characteristics (anion) were used to define the chemical characteristics among the highly active ligands that could help non-covalent connecting the ligand to its target receptor [13]. Therefore, the pharmacophore has been generated as a common sub-structure of all active compounds having affinities toward the same targets [14].

Pharmacophore features of the 26 tricyclic antipsychotic compounds having dibenzodiazepine, dibenzothiazepine, and thienobenzodiazepine scaffolds were generated. The following features were predicted as significant (Fig. 2).

In this model, vertex A and B represent aromatic hydrophobicity, and mid C and vertex D represent positive ionizations. Now, these vertices have been connected to make topograph where the AB, BD, DA, AC, BC, and CD represent the significant

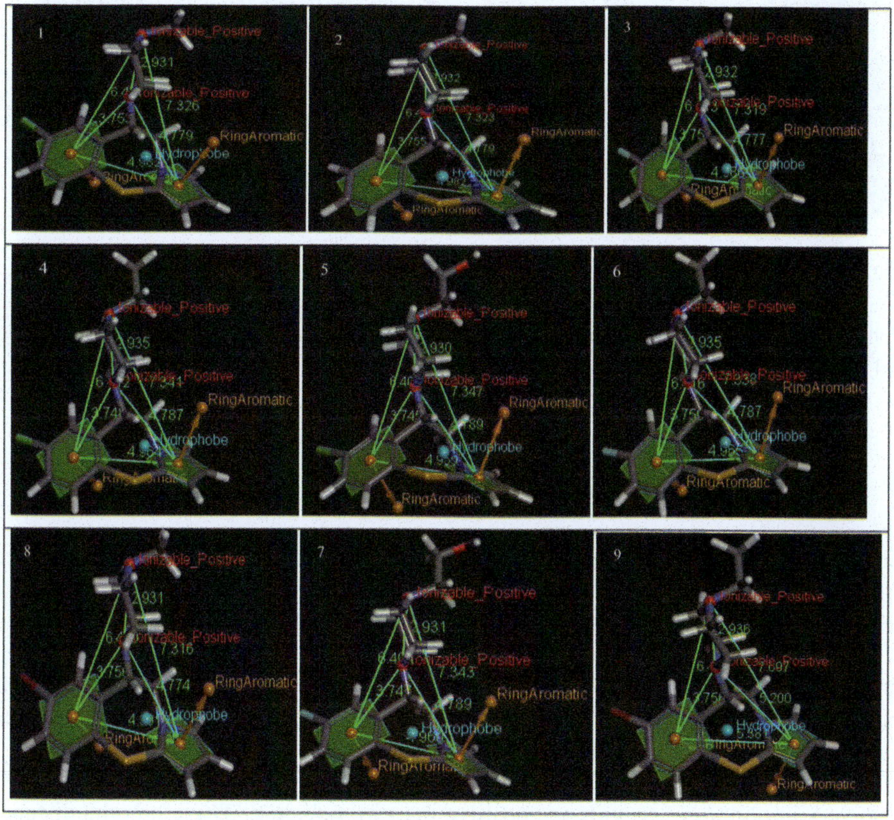

Fig. 1 Pharmacophore models of 26 compounds

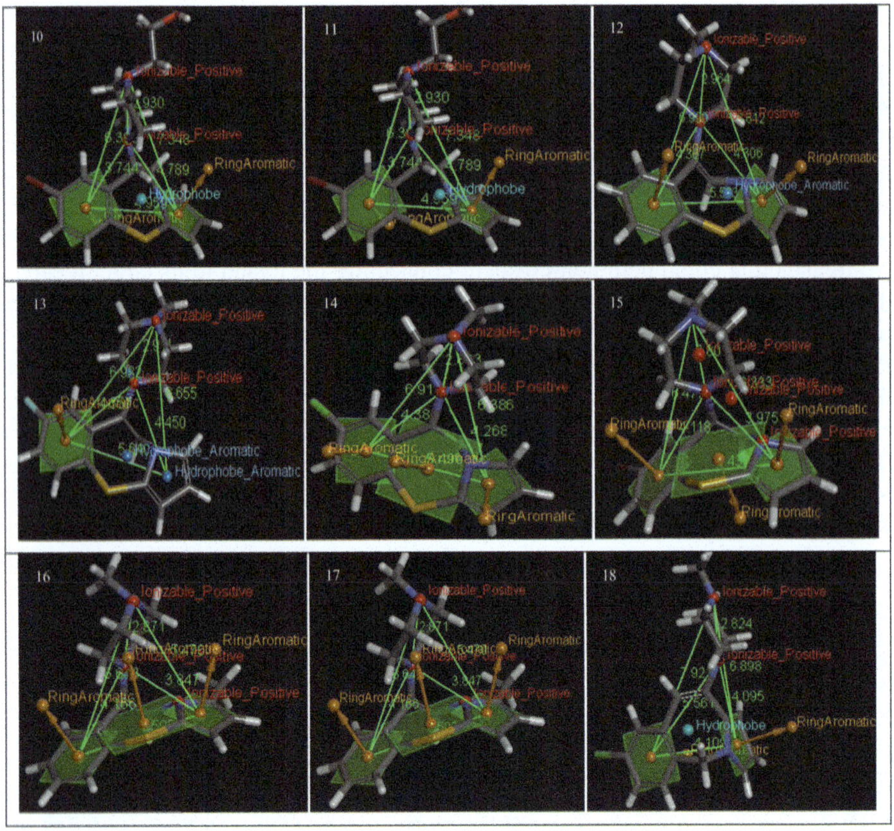

Fig. 1 (continued)

distance-based pharmacophore features responsible for the target affinity and biological activity. These distances are calculated in Å.

3 Results and Discussion

To design the ligand-based active compounds, pharmacophore shape can be a crucial strategy. For the structure-activity relationship correlations of well-known antagonists, several pharmacophore models for TCAs have been developed (Table 2).

Two models have been developed (Figs. 3 and 4). Fig. 3 is based on the five standard drugs such as clozapine, loxapine, olanzapine, quetiapine, and zotepine. The distance range between two aromatic rings (AB = 4.583–7.231 Å) and the distances between each aromatic ring and the terminal basic amine nitrogen are calculated as BD = 4.913–6.429 Å and DA =7.831–8.098 Å. The midpoint basic amino moiety is also very important to provide the positive ionization feature. Therefore, the distances from this center basic amino moiety C to each of the two aromatic moieties (A and B)

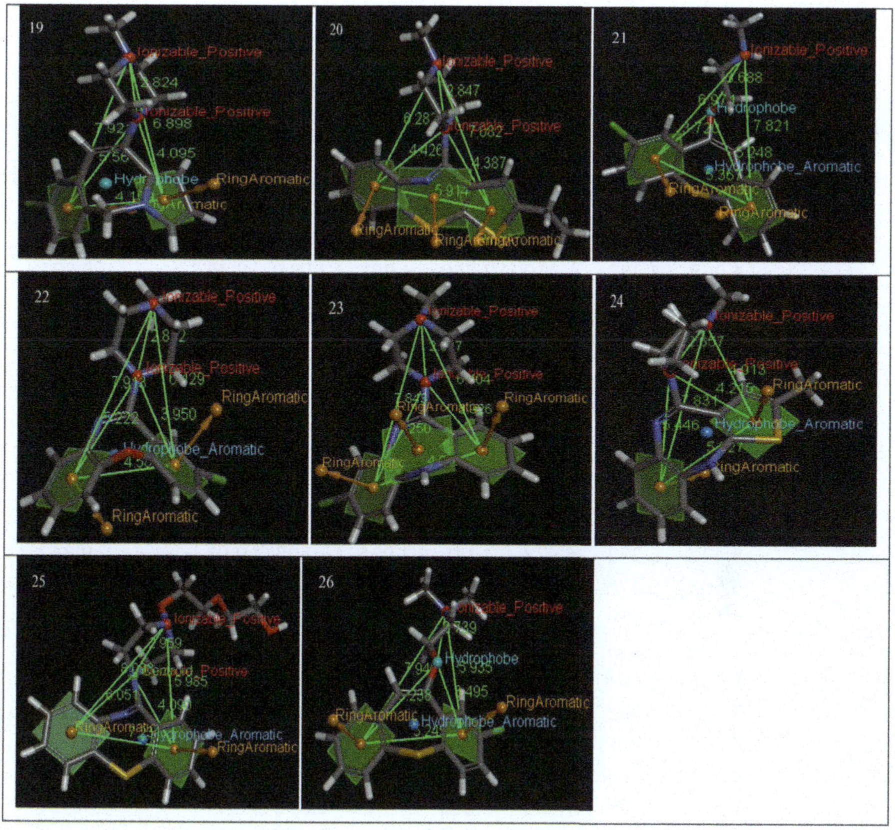

Fig. 1 (continued)

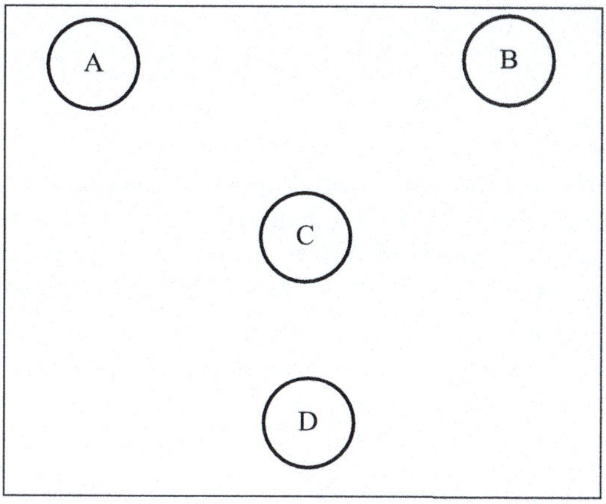

Fig. 2 Predicted pharmacophoric features for 26 compounds

Table 2 The calculated pharmacophore distance-based features (Å) along with 5HT2A and D2 receptors affinity

Sr. No.	AB	BD	DA	AC	BC	CD	pKi(5HT2A)	pKi(D2)
1	4.982	7.326	6.407	3.753	4.779	2.031	8.94	8.69
2	4.982	7.323	6.412	3.755	4.779	2.932	7.60	7.10
3	4.980	7.319	6.413	3.756	4.777	2.932	8.29	7.63
4	4.964	7.341	6.412	3.748	4.787	2.935	8.23	8.51
5	4.959	7.347	6.400	3.745	4.789	2.930	8.06	8.23
6	4.965	7.383	6.410	3.750	4.787	2.935	8.36	8.02
7	4.960	7.343	6.404	3.747	4.789	2.931	7.13	7.66
8	4.991	7.316	6.410	3.756	4.774	2.931	7.92	8.44
9	5.381	7.697	6.427	3.756	5.200	2.936	7.64	8.40
10	4.959	7.348	6.398	3.744	4.789	2.930	7.65	8.30
11	4.959	7.348	6.398	3.744	4.789	2.930	9.64	9.30
12	5.563	6.612	6.990	4.307	4.307	2.964	9.18	7.76
13	5.640	6.355	6.982	4.357	4.450	2.911	9.45	8.07
14	5.490	6.386	6.918	4.380	4.268	2.863	9.46	9.36
15	5.437	6.333	6.477	4.118	3.995	2.807	9.08	9.34
16	5.505	5.470	6.643	4.166	3.847	2.871	8.95	6.89
17	4.110	6.929	7.767	5.574	4.106	2.825	8.13	6.23
18	4.109	6.898	7.921	5.561	4.095	2.824	8.86	7.14
19	5.561	6.741	4.830	3.853	4.251	2.833	8.66	8.07
20	5.914	7.082	6.282	4.426	4.387	2.847	3.70	4.22
21	5.367	7.821	6.974	3.720	5.248	3.688	1.0	6.5
22-clozapine	4.919	6.404	7.843	5.250	3.926	2.907	8.26	6.59
23-loxapine	4.583	6.429	7.913	5.222	3.960	2.892	8.11	7.92
24-olanazapine	5.027	4.913	7.831	5.446	4.215	2.657	8.69	7.46
25-quetiapine	7.231	5.985	8.098	6.051	4.690	2.959	6.99	6.61
26-zotepine	5.249	5.935	7.945	3.238	3.495	3.739	0.69	11

and the terminal amino moiety (D) have been calculated and taken into consideration. These distances are AC = 3.238–6.510, BC =3.495–4.690, and CD = 2.657–3.739. These distance-based pharmacophores are frequently used to characterize the fundamental geometric properties to design the congeneric compounds.

The pharmacophore models of 1–21 congeneric compounds have been generated.

The distances are calculated as AB = 4.109–5.914 Å, BD = 5.470–7.821 Å, DA = 4.830–7.921 Å, AC = 3.720–5.574 Å, BC = 3.847–5.248 Å, CD = 2.031–3.688 Å.

The congeneric compound has been designed by taking standard nuclei such as dibenzodiazepine, dibenzothiazepine, and thienobenzodiazepines. The distance-based pharmacophoric feature would be expected to vary a bit from the standard topograph. Therefore quantitative structure-affinity relationship is carried out to check how the congeneric compound topograph fits along with the standard topograph on the 5HT2A and D2A target affinity.

In the above two topographs, it has been shown that only the distance AC of both test and congeneric compounds is matched while other distances were almost within the

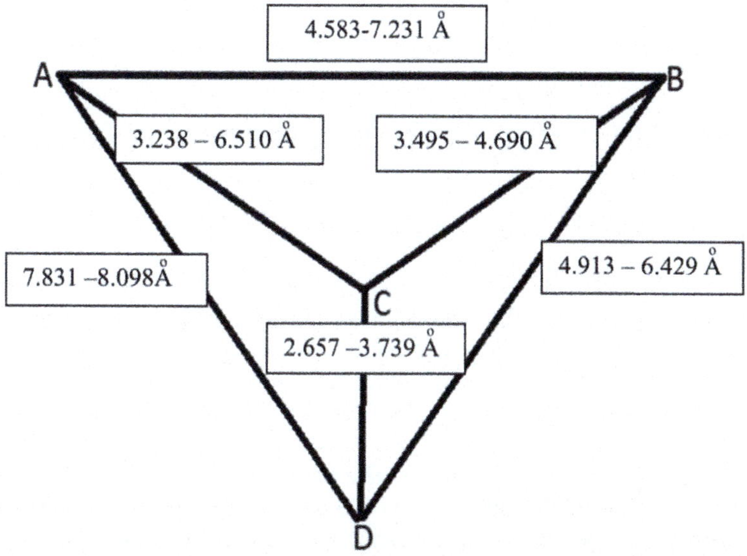

Fig. 3 Topograph for standard drugs (22–26)

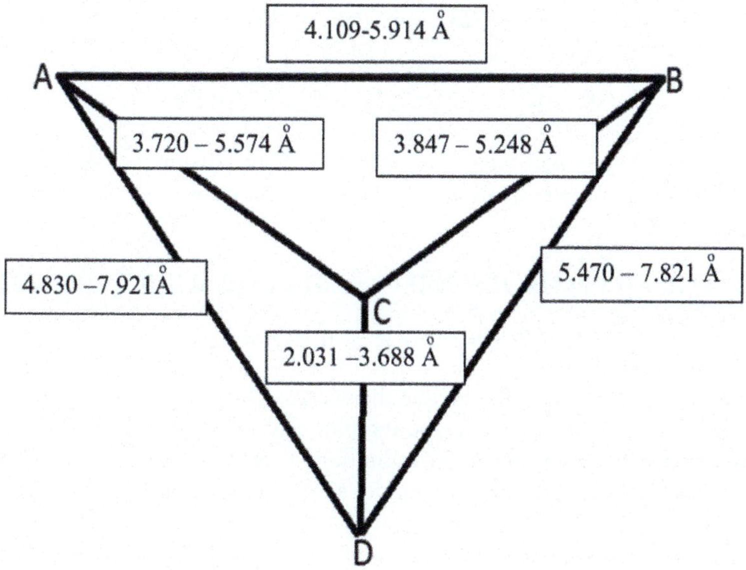

Fig. 4 Topograph developed for congeneric compounds (1–21)

range with a few deviations. If this deviation is significant, it may negatively impact the drug-target affinity and binding. The pharmacophoric feature-based model is a qualitative approach. Therefore, the quantitative structure–pharmacophore affinity relationship has been developed for all compounds (N = 26) taking 5HT2A and D2 binding affinity as dependent variables, whereas the calculated various feature distances were taken as independent variables for the development of QSPAR models (Eqs. 1–4) for the 5HT2A and D2 targets, respectively. The QSPAR models have been developed using MINITAB 14 software [15].

$$\text{pKi}(5\text{HT2A}) = 33.2 - 1.16\,\text{AB} - 1.28\,\text{BD} - 1.18\,\text{DA} + 0.866\,\text{AC} + 1.17\,\text{BC} - 4.03\,\text{CD}$$

$$N = 26, S = 1.645\ R^2 = 0.613, F = 5.03 \qquad \text{(Eq. 1)}$$

In the above model, it was shown that the predictors such as AC and BC have positive contributions, whereas AB, BD, DA, and CD have negative contributions toward 5HT2A inhibition. The model can explain 61.3% of the total data points (R = 0.782 which is the square root of multiple R^2 for regression) with a level of significance (F) of 5.03 and standard error (S) of 1.645. The equation although with a good correlation coefficient of 0.783 was further reduced to make it more statistically significant with the help of a stepwise reduction process. The reduction of one of the parameters (BC) in the form of topological distance and one compound (number 20) with a large residual resulted in Eq. 2 with an improved correlation coefficient (R) of 0.852 and an explained variance of 72.63% with the level of significance (F) of 10.09. The equation was more than 99% significant ($F_{4,\,47(\alpha = 0.001)} = 5.52; F_{(4,47)} = 12.1$)

$$\text{pKi}(5\text{HT2A}) = 29.70 - 0.405\,\text{AB} - 0.406\,\text{BD} - 1.462\,\text{DA} + 1.235\,\text{AC} - 4.20\,\text{CD} \qquad \text{(Eq. 2)}$$

$$N = 25; S = 1.297; R^2 = 0.726; F = 10.09$$

Similarly, another model was developed taking the distances and activity of D2 which resulted in an initial equation (Eq. 3) which had a low correlation coefficient of 0.693.

$$\text{pKi}(\text{D2}) = 17.8 - 0.130\,\text{AB} - 0.530\text{BD} + 0.895\text{DA} - 1.85\text{AC} - 0.201\text{BC} - 1.02\,\text{CD}$$

$$N = 26; S = 1.069, R^2 = 0.48, F = 2.92 \qquad \text{(Eq. 3)}$$

In the above model, it was shown that the decrease in value of distances such as AB, BD, AC, BC, and CD may increase the inhibition of D2 receptors as only one feature (DA) has a positive impact on the D2 inhibition. This model can explain 48% of total variances (R = 0.694) with a level of significance (F) value of 2.92 and standard error (S) of 1.069.

Eq. 3 was reduced using the same technique employed in Eq. 1, wherein a parameter (BC) and two compounds (number 16 and 20) out of which one was the same compound which was removed in developing Eq. 2) and led to Eq. 4.

$$\text{pKi}(\text{D2}) = 20.67 + 0.005\text{AB} - 0.932\text{BD} + 0.481\text{DA} - 1.681\text{AC} - 0.827\text{CD} \qquad \text{(Eq. 4)}$$

$$N = 24; S = 0.715; R^2 = 0.659, F = 6.97 \qquad \text{(Eq. 4)}$$

Eq. 4 was a statistically (>99.99%) significant ($F_{4, 47(\alpha = 0.001)} = 5.52; F_{(4,47)} = 12.1$) equation, which showed good correlation coefficient (R) of 0.812 and an explained variance of 65.94%.

4 Conclusion

The present study is an attempt to develop topographs for TCA which may be valid and fit for both 5HT2A and D2 targets and to assess quantitatively the affinity of the selective TCA compounds toward 5HT2A and D2 targets involved with depression cum psychosis. The topographs with AB, BD, DA, AC, BC, and CD represent the significant distance-based features responsible for the 5HT2A and D2 inhibition. It is very difficult to predict the compounds' affinity followed by intrinsic activity on real in vivo human models; therefore, the developed QSPAR models will help to design potent congeneric compounds.

References

1. Avram, S., Duda-seiman, D., Borcan, F., Wolschann, P.: QSAR–CoMSIA applied to antipsychotic drugs with their dopamine D2 and serotonine 5HT2A membrane receptors. J. Serb. Chem. Soc. **76**(2), 263–281 (2011)
2. Taylor, D., Paton, C., Kapur, S., Taylor, D.: *The Maudsley Prescribing Guidelines in Psychiatry*, 11th edn. Wiley-Blackwell, Chichester, West Sussex, UK (2012)
3. Tripathi, K.D.: Essentials of Medical Pharmacology, 7th edn. Jaypee Brothers Medical Publishers (P) Ltd, New Delhi (2013)
4. Earley, W., Burgess, M.V., Rekeda, L., Dickinson, R., Szatmári, B., Németh, G., McIntyre, R.S., Sachs, G.S., Yatham, L.M.: Cariprazine treatment of bipolar depression: a randomized double-blind placebo-controlled phase 3 study. Am. J. Psychiatry. **176**(6), 439–448 (2019)
5. Tohen, M., Katagiri, H., Fujikoshi, S., Kanba, S.: Efficacy of olanzapine monotherapy in acute bipolar depression: a pooled analysis of controlled studies. J. Affect. Disord. **149**(1–3), 196–201 (2013)
6. Thase, M.E.: Quetiapine monotherapy for bipolar depression. Neuropsychiatr. Dis. Treat. **4**(1), 11–21 (2008)
7. Sekhar, K.G.V.C., Devambhatla, R.K.V., Nagesh, H.N., Rao, V.S.: Pharmacophore hypothesis for atypical antipsychotics. Bull. Korean Chem. Soc. **33**(9), 2930 (2012)
8. Jafari, S., Bouillon, M.E., Huang, X.F., Pyne, S.G., Fernandez-Enright, F.: Novel olanzapine analogues presenting a reduced H1 receptor affinity and retained 5HT2A/D2 binding affinity ratio. BMC Pharmacol. **12**, 8 (2012)
9. Shobo, M., Kondo, Y., Yamada, H., Mihara, T., Yamamoto, N., Katsuoka, M., Harada, K., Ni, K., Matsuoka, N.: Norzotepine, a major metabolite of zotepine, exerts atypical antipsychotic-like and antidepressant-like actions through its potent inhibition of norepinephrine reuptake. J. Pharmacol. Exp. Ther. **333**(3), 772–781 (2010)
10. Mills, N.: ChemDraw ultra 10.0. J. Am. Chem. Soc. **128**(41), 13649–13650 (2006)
11. Leach, A.R., Gillet, V.J., Lewis, R.A., Taylor, R.: Three-dimensional pharmacophore methods in drug discovery. J. Med. Chem. ***53***(2), 539–558 (2009). https://doi.org/10.1021/jm900817u

12. Shoichet, B.K.: Virtual screening of chemical libraries. Nature. **432**(7019), 862–865 (2004)
13. Mason, J.S., Good, A.C., Martin, E.J.: 3-D pharmacophores in drug discovery. Curr. Pharm. Des. **7**(7), 567–597 (2001)
14. Kumar, A., Nandi, S., Saxena, A.K.: Antidepressant drug design on TCAs and Phenoxyphenylpropylamines utilizing QSAR and pharmacophore modeling. Comb. Chem. High Throughput Screen. **25**(3), 451–461 (2022)
15. www.minitab.com

Novel Synthesis of 3,5-Disubstituted-5*H*-Thiazolo[3,2-*α*]Pyrimidine

Shreya Dwivedi[1], Ram Kishore[1] (✉), and Nitin Srivastava[1] (✉)

[1] Department of Chemistry, Amity University Uttar Pradesh Lucknow Campus, Lucknow, India

Abstract Structural similarity between thioamides and 3,4-dihydropyrimidin-2(1*H*)-thiones made us to explore the reaction of 3,4-dihydropyrimidin-2(1*H*)-thiones with doubly electrophilic alkynyl(aryl)iodonium tosylates for synthesis of C2-N3 linked thiazolo[3,2-*a*]pyrimidine derivatives. In the reported research, we have employed alkynyl(aryl)iodonium salts for the ring annulation of 3,4-dihydropy rimidin-2(1*H*)-thiones to give thiazolo[3,2-*a*]pyrimidine derivatives using Triton-B as phase transfer catalyst instead of other bases like K_2CO_3 or Et_3N used earlier.

Keywords Biginelli reaction · Thiazolo[3,2-*a*]pyrimidine · Triton-B · Binary solution

1 Introduction

Multi-component reactions (MCRs) have proved their significance in organic synthesis and medicinal chemistry [1–3]. Biginelli dihydropyrimidine synthesis [4] is one such a prominent MCR that is employed for the synthesis of nitrogen heterocycles through cyclocondensation reaction of ethyl acetoacetate **1**, benzaldehyde **2**, and urea **3** in the presence of an acid (Scheme 1).

The dihydropyrimidinones and their derivatives are versatile heterocyclic compounds with plethora of biological activities like antiviral, antibacterial, adrenergic receptor agonist, mitotic kinesin inhibitor, and calcium channel modulators [5–8]. Dihydropyrimidinone-5-carboxylate core has also been found in biologically active marine alkaloids [9], e.g. batezelladine alkaloids, which are potent HIV gp-120-CD4 inhibitors [10]. Diversified biologically potent molecules are given in Fig. 1.

The methods existing in literature to synthesize thiazolo[3,2-*a*]pyrimidine derivatives mainly employ cyclocondensation of 3,4-dihydropyrimidin-2(1*H*)-thiones with terminal alkynes [11–20]. These processes suffer the constraints of complicated processes, tedious workup, and ultimately low yields. Our group is involved in the cleaner,

A. K. Saxena, S. Nandi (eds.), *Global Trends in Health, Technology and Management*, https://doi.org/10.1007/978-3-031-75457-9_20

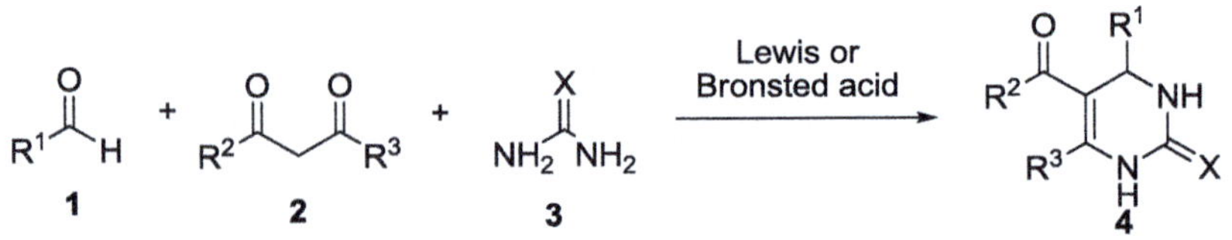

Scheme 1 Biginelli dihydropyrimidine synthesis

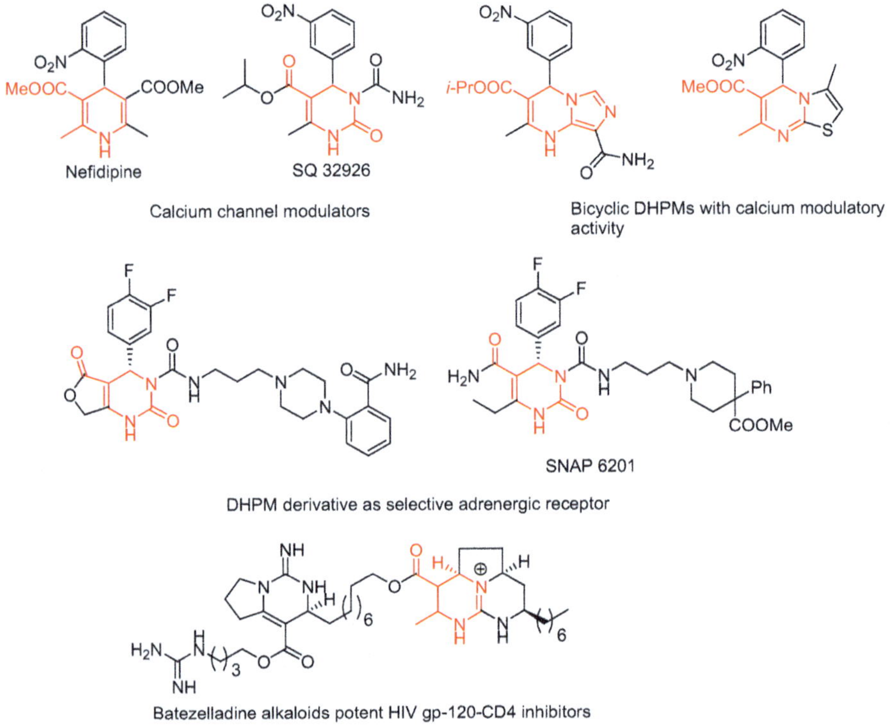

Fig. 1 Structurally diverse biologically potent dihydropyrimidinones

greener, easier process for the synthesis of biologically potent scaffolds with higher yields [21–27]. The earlier reported methods involve the usage of high heat energy, toxic solvent [15–17, 20], lower yield, and enantiomeric complications [17, 20]. In this research paper we have employed alkynyl(aryl)iodonium salts for the ring annulation of 3,4-dihydropyrimidin-2(1*H*)-thiones to give thiazolo[3,2-*a*]pyrimidine derivatives using phase transfer catalyst Triton-B in binary solvent of methyl alcohol and water instead of other bases like K_2CO_3/THF [28] or Et_3N/Toluene [17, 20] used for the synthesis of similar compounds including these compounds. The synthesis was inspired by the structural similarity of thioamides and 3,4-dihydropyrimidin-2(1*H*)-thiones which made us to explore the reaction of 3,4-dihydropyrimidin-2(1*H*)-thiones with doubly electrophilic alkynyl(aryl)iodonium tosylates for the synthesis of C2-N3 linked thiazolo[3,2-*a*]pyrimidine derivatives (Scheme 2).

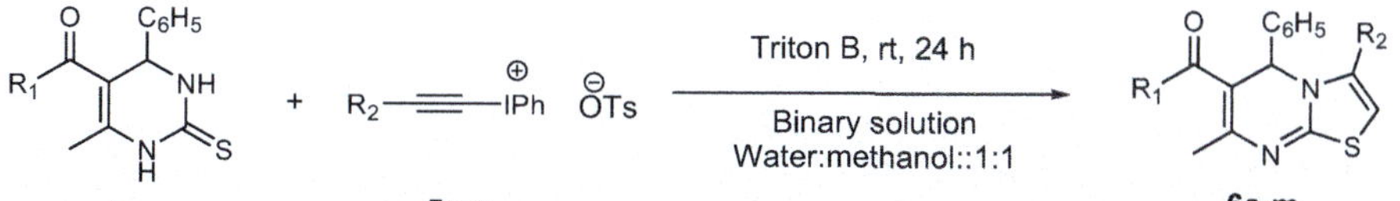

Scheme 2 Synthesis of C2-N3 linked thiazolo[3,2-*a*]pyrimidine derivatives

2 Materials and Methods

2.1 Instruments

All the chemicals used in the reactions were taken from Merck, Alfa-Aesar, Sigma Aldrich, Fluka. Proton NMR and C-13 NMR spectra were recorded on a Bruker Avance II 400 NMR spectrometer at room temperature in the solvent $CDCl_3$ and chemical shifts are reported in ppm on the δ scale and coupling constants, *J*, are in Hz. LCMS spectra were recorded using Shimadzu 2010 (single quadrupole). IR spectra were recorded on a Bruker FTIR-αE. Thin layer chromatography (TLC) was carried out on SiO_2 (silica gel 60F254, Merck), and the spots were located with ultraviolet (UV) light.

2.2 General Procedure for the Synthesis of [Hydroxy(Tosyloxy)Iodo]Benzene from (Diacetoxyiodo)Benzene

A mixture of (diacetoxyiodo)benzene (25 mmol) and *para*-toluene sulfonic acid monohydrate (25.55 mmol) was mixed well in mortar for 1 h, resulting in release of acetic acid and wetting of the reaction mixture. In the next step this solid residue was cleaned with diethyl ether (125 mL) and dried in rotatory evaporator to give the white solid product as [hydroxy(tosyloxy)iodo]benzene, which has been used for the preparation of alkynyl(phenyl)iodonium tosylate without further purification and characterization (Scheme 3) [29].

2.3 Synthesis of Alkynyl(Phenyl)Iodonium Tosylate from Terminal Alkyne and [Hydroxy(Tosyloxy)Iodo]Benzene (5a-c)

A mixture of appropriate alkyne (10 mmol), [hydroxy(tosyloxy)iodo]-benzene (3 mmol), and toluene (70 mL) was stirred and refluxed for 1 h. After cooling the solution was cleaned using 100 mL of H_2O, dried over $MgSO_4$, and concentrated to a volume of 10 mL. This concentrated solution was warmed at water bath, up to 80 °C, and added with 40 mL of diethyl ether with constant stirring which resulted in separation of alkynyl(phenyl)iodonium tosylate crystals on cooling. Keeping the mixture undisturbed for 1 day at 0 °C resulted in isolation of the product. The product formed was washed with diethyl ether and dried in air and has % yield of 70% (Scheme 4) [30].

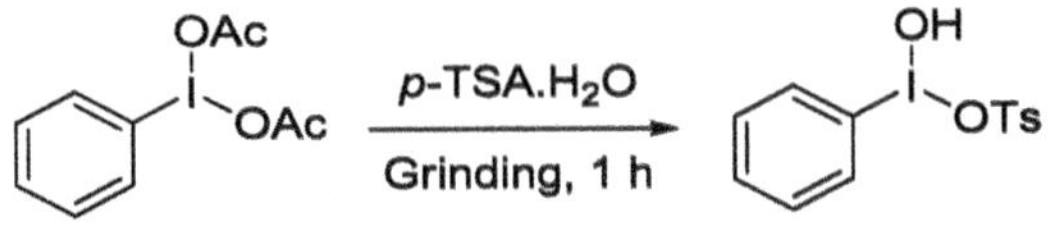

Scheme 3 [Hydroxy(tosyloxy)iodo]benzene from diacetoxyiodobenzene

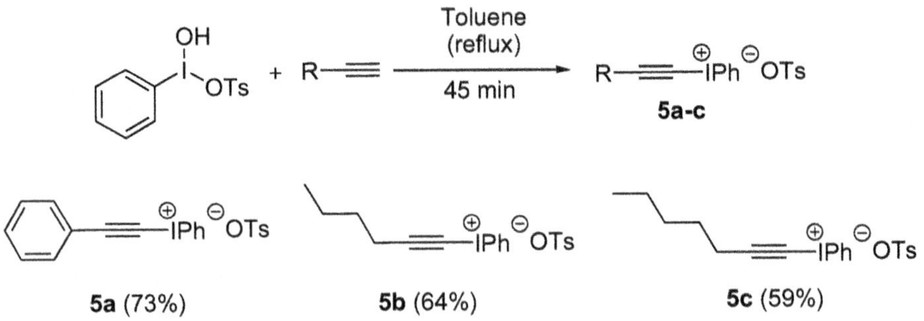

Scheme 4 Preparation of alkynyl(phenyl)iodonium tosylate

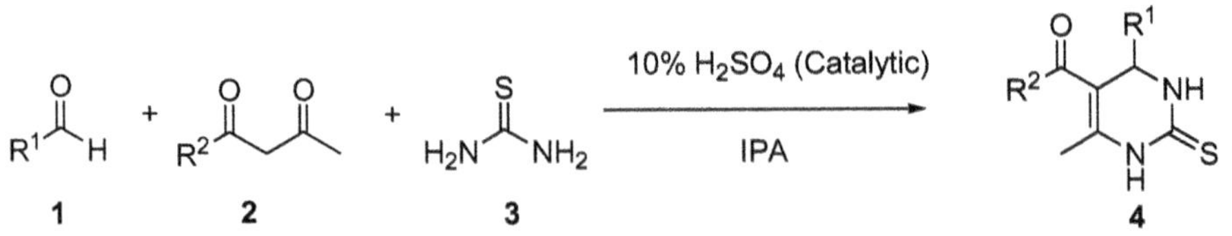

Scheme 5 Synthesis of 3,4-dihydropyrimidin-2(1*H*)-thiones using aromatic aldehyde, 1,3-dicarbonyl compound, and thiourea

2.4 Protocol for the Preparation of 4-Substituted-3,4-Dihydropyrimidin-2(1*H*)-Thione (4a-J)

20 mmols of aromatic aldehyde **1**, 20 mmols of acetyl acetone or ethyl acetoacetate **2**, 22 mols of thiourea **3**, and 2 mL of 10% aq. H_2SO_4 in 50 mL of EtOH were stirred together at room temperature for 24 h (Scheme 5). Progress of the reaction was observed by TLC. After completion, the reaction mixture was poured into beaker having ice cubes, and consequent filtration of the solid compound formed. Recrystallization of the Biginelli compound was done in isopropanol. The products were characterized by the comparison of melting points with the literature value (Table 1).

2.5 General Procedure for the Synthesis of Thiazolo[3,2-*a*]Pyrimidine Derivatives (6a-M) Starting from 4-Aryl-3,4-Dihydropyrimidin-2(1*H*)-Thiones and Alkynyl(Aryl)Iodonium Tosylates

2 mmol of 4-aryl-3,4-dihydropyrimidin-2(1*H*)-thiones **4** was taken in 10 mL of binary solution of methanol in water (1:1) and added with 6 mmols of Triton-B 4.8 mmols of alkynyl(aryl)iodonium tosylates **5**. The reaction mixture was stirred for 24 h at room

Table 1 Synthesis of 3,4-dihydropyrimidin-2(1*H*)-thiones using aromatic aldehyde, 1,3-dicarbonyl compound, and thiourea

Entry	R^1	R^2	Yield (%)[a]	Mp (°C)	Lit. Mp (°C)
4a	C_6H_5	CH_3CH_2O	91	206–208	202–204 [31]
4b	$4\text{-}CH_3OC_6H_4$	CH_3CH_2O	87	138–140	137–139 [32]
4c	$3\text{-}CH_3OC_6H_4$	CH_3CH_2O	93	150–152	150–152 [33]
4d	$3,4\text{-}(CH_3O)_2C_6H_3$	CH_3CH_2O	86	164–166	165–166 [34]
4e	$2,5\text{-}(CH_3O)_2C_6H_3$	CH_3CH_2O	79	184–186	Not available
4f	$4\text{-}CH_3C_6H_4$	CH_3CH_2O	85	190–192	192–194 [31]
4 g	$4\text{-}ClC_6H_4$	CH_3CH_2O	91	181–182	184–185 [35]
4 h	C_6H_5	CH_3	83	219–221	220–222 [36]
4i	$4\text{-}CH_3C_6H_4$	CH_3	76	190–192	Not available
4j	$3\text{-}CH_3OC_6H_4$	CH_3	73	210–212	Not available

[a] Isolated yield

temperature and monitored for its progress by TLC. The product formed was extracted with ethyl acetate (3 × 20 mL) after addition of 10 mL of water on completion of the reaction. The ethyl acetate layer was cleaned with 3 N HCl (4 × 10 mL), while the HCl layer was washed with ethyl acetate (10 mL). The separated acidic layer was neutralized with $NaHCO_3$ (Saturated) and now the product was extracted with ethyl acetate (4 × 10 mL). The ethyl acetate layer was cleaned with 10% NaCl, dried over anhydrous Na_2SO_4, and concentrated using a rotatory evaporator to give crude product, which was further purified with the help of column chromatography as thiazolo[3,2-*a*]pyrimidine derivatives **6** in the form of yellow solid or semi-solid product.

2.6 Characterization of the Products (6a-m)

All the products **6a-m** were characterized by IR, ^{1}H NMR, ^{13}C NMR, and LCMS analysis. LCMS spectrum of products **6a-m** shows $(M + H)^+$ value, which is in good agreement with the annulation product formation. The absence of NH stretch in IR spectrum is a strong evidence for the formation of annulations product. ^{1}H NMR spectrum shows absence of NH protons. The singlet in between δ 6–6.25 shows the presence of proton at C2 position. Also, the number of aliphatic and aromatic protons matches exactly with the structure. The C3 carbon shows peak in the range δ 139–140. Also, the number of peaks matches with the number of non-equivalent carbon atoms.

Ethyl 7-methyl-3,5-diphenyl-5*H*-thiazolo[3,2-*a*]pyrimidine-6-carboxylate (6a): Solid (Yellow), Yield: 88% (301 mg), mp = 110–112 °C; ^{1}H NMR: δ 1.20 (t, 3H, $J = 7.12$ Hz), 2.50 (s, 3H), 3.98 (q, 2H, $J = 7.2$ Hz), 5.98 (s, 1H), 6.20 (s, 1H), 6.81 (d, 2H, $J = 7.32$ Hz), 7.10 (t, 2H, $J = 7.3$ Hz), 7.20 (d, 3H, $J = 8.32$), 7.40 (t, 2H, $J = 7.36$ Hz), 7.50 (t, 1H, $J = 7.34$ Hz); ^{13}C NMR: δ 166.63, 166.20, 141.95, 139.85, 129.73, 129.39, 128.96, 128.87, 128.67, 128.22, 126.56, 103.37, 101.28, 59.86, 57.99, 29.71, 23.19, 14.25; IR (cm^{-1}): 698, 994, 1078, 1225, 1362, 1455, 1567, 1673, 1737, 2917, 2970; MS (ESI) m/z $(M + H)^+$ Calculated for $C_{22}H_{20}N_2O_2S$: 376.12. Found: 377.05.

Ethyl 5-(4-methoxyphenyl)-7-methyl-3-phenyl-5*H*-thiazolo[3,2-*a*]pyrimidine-6-carboxylate (6b): Solid (Yellow), Yield: 67% (333 mg), mp = 122–124 °C; ^{1}H NMR: δ 1.09 (t, *J* = 7.10 Hz, 3H), 4.12 (s, 3H), 3.61 (s, 3H), 3.97 (q, *J* = 7.12 Hz, 2H), 6.05 (s, 1H), 6.09 (s, 1H), 6.50 (d, 2H, *J* = 8.72 Hz, 2H), 6.60 (d, 2H, *J* = 8.72 Hz) 7.10 (d, 2H, *J* = 6.96 Hz), 6.28–6.39 (m, 3H); ^{13}C NMR: δ 166.81, 166.29, 159.24, 155.29, 139.81, 134.64, 129.63, 128.92, 128.67, 127.93, 113.42, 102.89, 101.39, 59.77, 57.37, 55.13, 23.43, 14.28; IR (cm^{-1}): 3005, 2988, 1685, 1482, 1371, 1317, 1275, 1258, 1176, 1032, 763, 749, 694, 600; MS (ESI) m/z (M + H)$^+$ Calculated for $C_{23}H_{22}N_2O_3S$: 406.14. Found: 407.08.

Ethyl 5-(3-methoxyphenyl)-7-methyl-3-phenyl-5*H*-thiazolo[3,2-*a*]pyrimidine-6-carboxylate (6c): Solid (Yellow), Yield: 73% (329 mg), mp = 116–118 °C; ^{1}H NMR: δ 1.10 (t, 3H, *J* = 7.08 Hz), 2.40 (s, 3H), 3.508 (s, 3H), 4.02 (q, 2H, *J* = 7.04 Hz), 6.10 (s, 1H), 6.16 (s, 1H), 6.40 (d, 1H, *J* = 7.56 Hz), 6.60 (d, 1H, *J* = 8.04 Hz), 6.12 (t, 1H, *J* = 7.92 Hz), 7.10 (d, 2H, *J* = 7.24 Hz), 7.30–7.39 (m, 4H); ^{13}C NMR: δ 166.75, 166.45, 159.18, 155.73, 143.54, 139.73, 129.61, 129.31, 128.91, 128.69, 118.87, 113.96, 111.61, 102.95, 100.98, 59.80, 57.78, 54.95, 23.51, 14.31; IR (cm^{-1}): 605, 694, 749, 765, 1001, 1077, 1226, 1252, 1313, 1371, 1483, 1585, 1598, 1689, 2837, 2981; MS (ESI) m/z (M + H)$^+$ Calculated for $C_{23}H_{22}N_2O_3S$: 406.14. Found: 407.05.

Ethyl 7-methyl-3-phenyl-5-(*p*-tolyl)-5*H*-thiazolo[3,2-*a*]pyrimidine-6-carboxylate (6d): Solid, Yield: 70% (273 mg), mp = 130–132 °C; ^{1}H NMR: δ 1.20 (t, 3H, *J* = 7.12 Hz), 2.18 (s, 3H), 2.50 (s, 3H), 4.10 (q, 2H, *J* = 7.08 Hz), 6.08 (s, 1H), 6.20 (s, 1H), 6.65 (d, 2H, *J* = 8.08 Hz), 6.90 (d, 2H, *J* = 7.92 Hz), 7.20 (d, 2H, *J* = 6.90 Hz), 7.50 (t, 2H, 7.24 Hz), 7.50 (t, 1H, *J* = 6.66 Hz); ^{13}C NMR: δ 166.80, 166.47, 155.62, 139.80, 139.30, 137.79, 129.62, 129.60, 128.90, 128.86, 128.65, 126.44, 102.74, 101.29, 59.74, 57.56, 23.55, 21.12, 14.28; IR (cm^{-1}): 598, 656, 697, 788, 998, 1072, 1176, 1227, 1312, 1362, 1469, 1573, 1680, 1737, 2924, 3090; MS (ESI) m/z (M + H)$^+$ Calculated for $C_{23}H_{22}N_2O_2S$: 390.14. Found: 390.98.

Ethyl 5-(3,4-dimethoxyphenyl)-7-methyl-3-phenyl-5*H*-thiazolo[3,2-*a*] pyrimidine-6-carboxylate (6e): Solid (Yellow), Yield: 82% (357 mg), mp = 136–138 °C; ^{1}H NMR: δ 1.20 (t, 3H, *J* = 7.12 Hz), 2.51 (s, 3H), 3.48 (s, 3H), 3.80 (s, 3H), 4.10 (q, 2H, *J* = 8.56 Hz), 6.09 (d, 2H, *J* = 7.76 Hz), 6.21 (s, 1H), 6.50 (d, 1H, *J* = 8.32 Hz), 6.60 (d, 1H, *J* = 8.76 Hz), 7.20 (d, 2H, *J* = 6.56 Hz), 7.51 (m, 3H); ^{13}C NMR: δ 166.87, 166.32, 155.49, 148.55, 148.01, 139.74, 134.99, 129.87, 129.50, 128.93, 128.70, 119.09, 110.77, 109.66, 102.84, 101.27, 59.77, 57.64, 55.70, 55.39, 23.58, 14.31; IR (cm^{-1}): 610, 658, 697, 739, 809, 1023, 1092, 1159, 1231, 1365, 1442, 1486, 1578, 1661, 1737, 2970; MS (ESI) m/z (M + H)$^+$ Calculated for $C_{24}H_{24}N_2O_4S$: 436.15. Found: 437.10.

Ethyl 5-(4-chlorophenyl)-7-methyl-3-phenyl-5*H*-thiazolo[3,2-*a*] pyrimidine-6-carboxylate (6f): Solid (Yellow), Yield: 73% (299 mg), mp = 128–130 °C; ^{1}H NMR: δ 1.20 (s, 3H), 2.48 (s, 3H), 4.10 (q, 2H, *J* = 8.78 Hz), 6.20 (s, 1H), 6.24 (s, 1H), 6.70 (d, 2H, *J* = 8.48 Hz), 7.10 (d, 2H, *J* = 8.48 Hz), 7.20 (d, 2H, *J* = 7.0 Hz), 7.40 (t, 2H, *J* = 7.3 Hz), 7.50 (t, 1H, *J* = 7.36 Hz); ^{13}C NMR: δ 166.61, 166.29, 140.60, 139.64, 133.92, 129.83, 129.32, 128.92, 128.79, 128.40, 128.02, 103.20, 100.95, 59.94, 57.36,

29.72, 22.48, 14.28; IR (cm^{-1}): 598, 654, 697, 770, 824, 996, 1014, 1078, 1178, 1225, 1312, 1369, 1475, 1575, 1698, 1737, 2925, 2970; MS (ESI) m/z $(M + H)^+$ Calculated for $C_{22}H_{19}ClN_2O_2S$: 410.09. Found: 411.05.

Ethyl 5-(2,5-dimethoxyphenyl)-7-methyl-3-phenyl-5*H*-thiazolo[3,2-*a*] pyrimidine-6-carboxylate (6 g): Solid (Yellow), Yield: 79% (344 mg), mp = 144–146 °C; ^{1}H NMR: δ 1.09 (t, 3H, *J* = 12.72 Hz), 2.40 (s, 3H), 3.29 (s, 3H), 3.48 (s, 3H), 4.10 (q, 2H, *J* = 9.57 Hz), 6.10 (s, 1H), 6.40 (d, 1H, *J* = 2.96 Hz), 6.50 (s, 1H), 6.60 (d, 1H, *J* = 8.92 Hz,), 6.59 (d, 1H, *J* = 8.88 Hz), 7.10 (d, 2H, *J* = 7.08 Hz), 7.40 (m, 3H); ^{13}C NMR: δ 166.87, 166.82, 155.57, 152.76, 150.93, 139.99, 130.79, 130.10, 129.12, 128.75, 128.29, 114.72, 114.65, 111.31, 102.04, 99.96, 59.44, 55.50, 55.40, 54.60, 23.52, 14.12; IR (cm^{-1}): 573, 606, 698, 716, 794, 922, 996, 1014, 1079, 1217, 1263, 1318, 1365, 1454, 1571, 1665, 1737, 2970; MS (ESI) m/z $(M + H)^+$ Calculated for $C_{24}H_{24}N_2O_4S$: 436.15. Found: 437.10.

Ethyl 7-methyl-3-pentyl-5-phenyl-5*H*-thiazolo[3,2-*a*]pyrimidine-6-carboxylate (6 h): Solid (Yellow), Yield: 66% (244 mg), mp = 102–104 °C; ^{1}H NMR: δ 0.90 (t, 3H, *J* = 7.64 Hz), 1.49–1.39 (m, 8H), 1.40–1.49 (m, 1H), 1.52–1.61 (m, 1H), 2.20–2.31 (m, 1H), 3.41 (s, 3H), 4.20 (q, 2H, *J* = 7.46 Hz), 6.01 (s, 1H), 6.21 (s, 1H), 7.22–7.30 (m, 5H); ^{13}C NMR: δ 166.98, 166.92, 156.03, 143.02, 139.90, 128.73, 128.18, 126.28, 100.01, 99.56, 59.85, 57.50, 31.10, 27.34, 26.60, 23.85, 22.29, 14.45, 13.93; IR (cm^{-1}): 669, 696, 756, 831, 1007, 1207, 1242, 1321, 1369, 1474, 1591, 1665, 1693, 1737, 2926, 2955; MS (ESI) m/z $(M + H)^+$ Calculated for $C_{21}H_{26}N_2O_2S$: 370.17. Found: 371.00.

Ethyl 5-(4-chlorophenyl)-7-methyl-3-pentyl-5*H*-thiazolo[3,2-*a*]pyrimidine-6-carboxylate (6i): Solid (Yellow), Yield: 72% (290 mg), mp = 110–112 °C; ^{1}H NMR: δ 0.90 (t, 3H, *J* = 7.0 Hz), 1.27–1.32 (m, 7H), 1.41–1.46 (m, 1H), 1.60–1.65 (m, 1H), 2.18–2.26 (m, 1H), 2.32–2.40 (m, 4H), 4.30 (q, 2H, *J* = 6.04 Hz), 6.01 (s, 1H), 6.20 (s, 1H), 7.30–7.38 (m, 4H); ^{13}C NMR: δ 166.80, 156.01, 141.41, 139.68, 134.02, 128.94, 127.70, 99.92, 99.79, 59.99, 56.94, 31.10, 29.71, 27.34, 26.55, 23.84, 22.29, 14.45, 13.90; IR (cm^{-1}): 588, 636, 729, 813, 973, 1014, 1064, 1205, 1369, 1455, 1562, 1677, 1737, 2864, 2928; MS (ESI) m/z $(M + H)^+$ Calculated for $C_{21}H_{25}ClN_2O_2S$: 404.13. Found: 405.05.

1-(7-Methyl-3-phenyl-5-(*p*-tolyl)-5*H*-thiazolo[3,2-*a*]pyrimidin-6-yl)ethanone (6j): Solid (Yellow), Yield: 74% (266 mg), mp = 126–128 °C; ^{1}H NMR: δ 2.19 (s, 3H), 2.29 (s, 3H), 2.50 (s, 3H), 6.31 (s, 1H), 6.36 (s, 1H), 6.71 (d, 2H, *J* = 8.04 Hz), 6.90 (d, 2H, *J* = 7.92 Hz), 7.20 (d, 2H, *J* = 6.96 Hz), 7.39 (t, 2H, *J* = 7.32 Hz), 7.50 (d, 1H, *J* = 7.24 Hz); ^{13}C NMR: δ 194.76, 140.35, 138.90, 137.83, 129.95, 128.98, 126.40, 112.46, 103.59, 57.20, 31.95, 31.47, 29.72, 29.39, 24.76, 22.72, 21.10, 14.16; IR (cm^{-1}): 596, 696, 822, 941, 989, 1062, 1177, 1228, 1311, 1369, 1462, 1611, 1737, 2848, 2916; MS (ESI) m/z $(M + H)^+$ Calculated for $C_{22}H_{20}N_2OS$: 360.13. Found: 361.05.

1-(3-Butyl-7-methyl-5-(*p*-tolyl)-5*H*-thiazolo[3,2-*a*]pyrimidin-6-yl)ethanone (6 k): Solid (Yellow), Yield: 69% (234 mg), mp = 106–108 °C; ^{1}H NMR: δ 0.90 (t, 3H, *J* = 7.26 Hz), 1.34 (q, 2H, *J* = 7.29 Hz), 1.38–1.45 (m, 1H), 1.49–1.56 (m, 1H), 2.30 (s, 3H), 2.32–2.36 (m, 1H), 2.40 (s, 3H), 2.42 (s, 3H), 2.46–2.48 (m, 1H), 6.10 (s, 1H),

6.40 (s, 1H), 7.12 (d, 2H, $J = 7.96$ Hz), 7.21 (d, 2H, $J = 8.12$ Hz); ^{13}C NMR: δ 194.24, 166.73, 140.76, 139.70, 137.85, 129.37, 126.16, 112.04, 100.52, 56.40, 31.71, 29.71, 28.29, 27.06, 25.08, 22.09, 21.12, 13.71; IR (cm^{-1}): 665, 694, 733, 759, 847, 957, 1040, 1079, 1211, 1314, 1454, 1586, 1597, 2930, 2956; MS (ESI) m/z $(M + H)^+$ Calculated for $C_{20}H_{24}N_2OS$: 340.16. Found: 341.00.

1-(3-Butyl-5-(3-methoxyphenyl)-7-methyl-5*H*-thiazolo[3,2-*a*]pyrimidin-6-yl) ethanone (6 l): Solid (Yellow), Yield: 76% (270 mg), mp = 114–116 °C; ^1H NMR: δ 0.91 (t, 3H, $J = 7.26$ Hz), 1.45 (m, 2H), 1.60 (m, 2H), 2.29 (m, 2H), 1.71 (s, 3H), 2.42 (s, 3H), 3.80 (s, 3H), 4.60 (s, 1H), 6.10 (s, 1H), 6.50 (s, 1H), 6.81 (d, 1H, $J = 7.4$ Hz), 6.78 (d, 1H, $J = 7.76$ Hz), 7.21 (t, 1H, $J = 7.86$ Hz); ^{13}C NMR: δ 194.27, 166.98, 159.72, 156.07, 144.01, 140.70, 129.72, 118.32, 113.31, 111.79, 111.63, 100.42, 56.27, 55.18, 31.67, 28.89, 27.01, 25.21, 22.07, 13.70; IR (cm^{-1}): 663, 695, 734, 747, 849, 958, 1041, 1080, 1212, 1315, 1369, 1455, 1586, 1737, 2930, 2956; MS (ESI) m/z $(M + H)^+$ Calculated for $C_{20}H_{24}N_2O_2S$: 356.16. Found: 357.00.

1-(3-Butyl-7-methyl-5-phenyl-5*H*-thiazolo[3,2-*a*]pyrimidin-6-yl)ethanone (6 m): Solid (Yellow), Yield: 79% (257 mg), mp = 126–128 °C; ^1H NMR: δ 0.90 (t, 3H, $J = 7.24$ Hz), 1.29 (m, 2H), 1.39 (m, 1H), 1.48 (m, 1H), 2.30 (m, 1H), 2.40 (s, 6H), 2.46 (m, 1H), 5.99 (s, 1H), 6.51 (s, 1H), 7.31 (m, 5H); ^{13}C NMR: δ 194.21, 167.02, 156.04, 142.61, 140.65, 128.67, 127.95, 126.16, 111.80, 100.35, 56.46, 31.67, 28.87, 27.01, 25.27, 22.04, 13.67; IR (cm^{-1}): 612, 659, 695, 719, 827, 957, 1083, 1211, 1315, 1369, 1458, 1552, 1607, 2929, 2956; MS (ESI) m/z $(M + H)^+$ Calculated for $C_{19}H_{22}N_2OS$: 326.15. Found: 326.90.

3 Results and Discussion

Structural similarity between thioamides and 3,4-dihydropyrimidin-2(1*H*)-thiones made us to explore the reaction of 3,4-dihydropyrimidin-2(1*H*)-thiones with doubly electrophilic alkynyl(aryl)iodonium tosylates for the synthesis of C2-N3 linked thiazolo[3,2-*a*]pyrimidine derivatives. Intermediate alkynyl(aryl)iodonium tosylates were synthesized from various terminal alkynes and hydroxy(tosyloxy)iodobenzene with the help of existing methods (Scheme 6) [24].

For the desired synthesis, we took reaction of 4-phenyl-3,4-dihydropyrimidin-2(1*H*)-thiones **4a** with phenyl(phenylethyny1)iodonium tosylate **5a** as model reaction using mild base Triton-B in various solvents like CH_2Cl_2, $CHCl_3$, THF, acetonitrile, and binary solution of water and methyl alcohol in the ratio of 1:1 as shown in Scheme 7

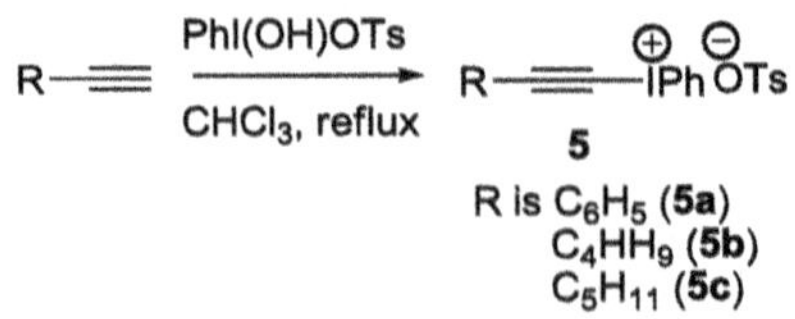

Scheme 6 Preparation of alkynyl(aryl)iodonium tosylates

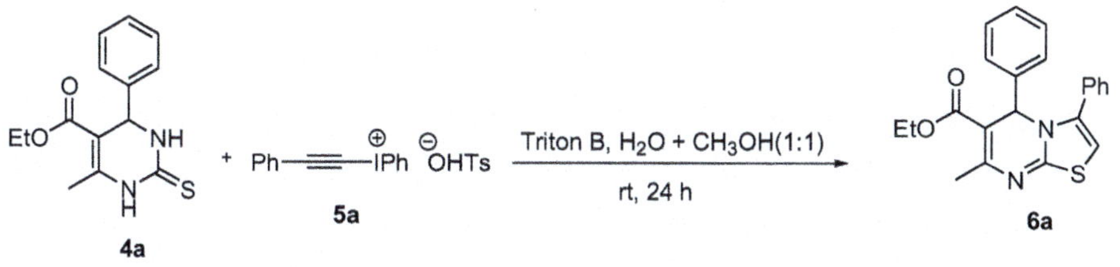

Scheme 7 Synthesis of thiazolo[3,2-*a*] pyrimidine derivatives using alkynyl(aryl)iodonium tosylates and 3,4-dihydropyrimidin-2(1*H*)-thiones

Table 2 Optimization of reaction conditions in terms of solvent

Entry	**5a**	Solvent	Yield of **6a** (%)[a,b]
1	1 mmol	$CHCl_3$	45
2	1 mmol	CH_2Cl_2	42
3	1 mmol	THF	56
4	1 mmol	DMSO	51
5	1 mmol	$H_2O:CH_3OH$:: 1:1	80

[a]Isolated yield
[b]Reaction conditions: **4a** (1 mmol), **5a** and Triton-B (2.5 mmol)

Table 3 Time and yield of the reaction with different catalyst

S.N	Name of phase transfer catalyst	Time in h	Yield %
1	Amberlite IRA 400 (basic resin)	48	30
2	TMAB	56	42
3	**Triton-B**	**24**	**80**
4	TBAI	50	48
5	TBAHS	60	50
6	TBAC	45	45
7	No catalyst	–	No reaction

and given in Table 2. It was found that the formation of **6a** took place in all the solvents. The optimized yield 80% of **6a** was obtained using 1.4 equivalents of **5a** in binary solution of water and methyl alcohol after stirring at room temperature for 24 h (Table 2, entry 5).

After the optimization of the solvent, we carried out reaction in different bases and yield of the **6a** was examined in the presence of various phase transfer catalysts such as Amberlite IRA 400 (basic resin), tetramethyl ammonium bromide (TMAB), Triton-B, tetrabutyl ammonium iodide (TBAI), tetrabutyl ammonium hydrogen sulfate (TBAHS), tetrabutyl ammonium carbonate (TBAC). The analysis of the yield and time led to the conclusion that Triton-B is most appropriate in obtaining high yield in less time. The reaction when carried out in absence of the catalyst did not occur, so it necessitates the use of catalyst in the reaction (Table 3).

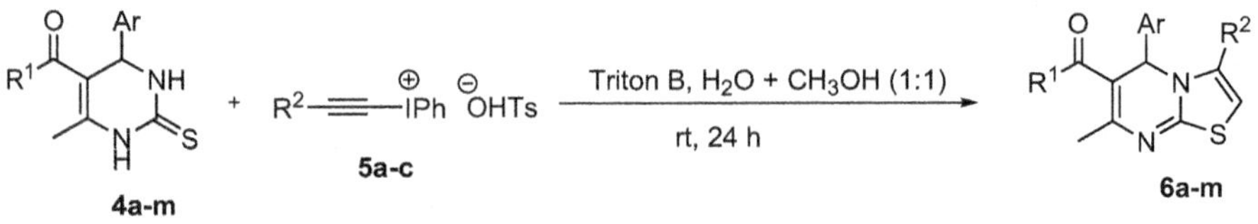

Scheme 8 Synthesis of thiazolo[3,2-*a*] pyrimidine derivatives

Once the reaction was optimized for the solvent and the phase transfer catalyst, the efficiency of the synthesis was further explored through cyclocondensation of 4-aryl-3,4-dihydropyrimidin-2(1*H*)-thiones **4a-m** with different alkynyl(aryl)iodonium tosylates **5a-c** (Scheme 8). It was found that the electron-donating and withdrawing substituents at C-4 aryl group of 3,4-dihydropyrimidin-2(1*H*)-thiones have less effect on the yields of products. The presence of phenyl group on alkynyl(aryl)iodonium tosylates (Table 4, entries 6**a-6f** and 6**j**) as well as alkyl substituents (Table 4, entries 6 **g-6i** and **6 l-6 m**) reacted very well to provide the cyclization products in good to high yields.

Plausible Mechanism The formation of thiazolo[3,2-*a*]pyrimidine derivatives may be represented as shown in Scheme 9. The nucleophilicity of sulfur in 3,4-dihydropyrimid in-2(1*H*)-thiones **4** makes it to attack electrophilic iodine and give **7** which and subsequent Claisen rearrangement results in the formation of vinyl iodonium ylide **8** [37]. Now reductive elimination of iodobenzene from **8** led to the formation of alkylidene carbene **9** as intermediate which finally undergoes cyclization to form **6**.

4 Conclusions

The method reported in the manuscript reports the metal-free, novel, easier, and efficient method for the synthesis of thiazolo[3,2-*a*]pyrimidine through cyclocondensation of 3,4-dihydropyrimidin-2(1*H*)-thione with alkynyl(aryl)iodonium tosylates. This reaction developed significant use of alkynyl(aryl)iodonium tosylates as alkynyl cation and significantly improved the synthesis of thiazolo[3,2-*a*]pyrimidine derivatives.

Table 4 Thiazolo[3,2-*a*] pyrimidine derivatives synthesized[a]

Entry	Structure	Time (h)	% Yield
6a	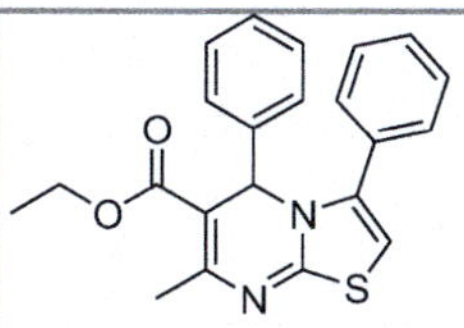	24	81
6b	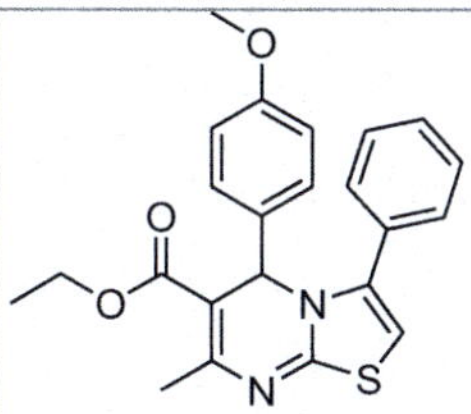	24	79
6c	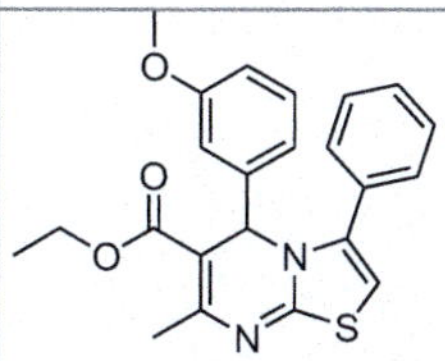	24	76
6d	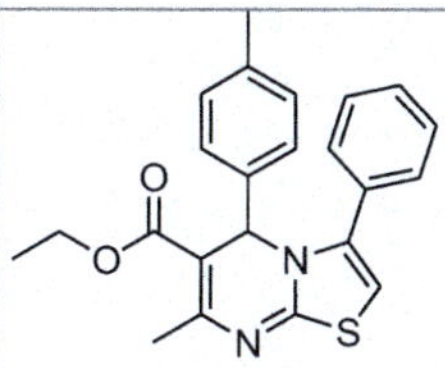	24	78
6e	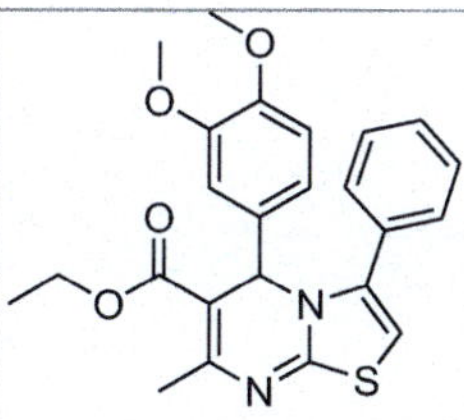	24	80
6f	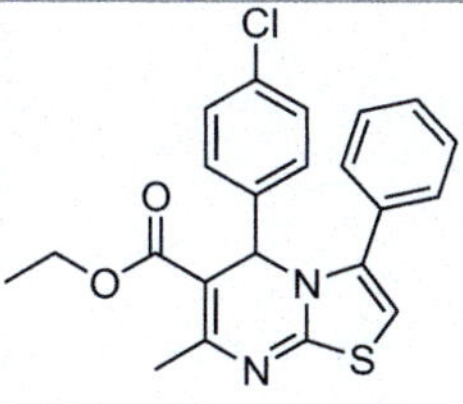	24	71
6 g	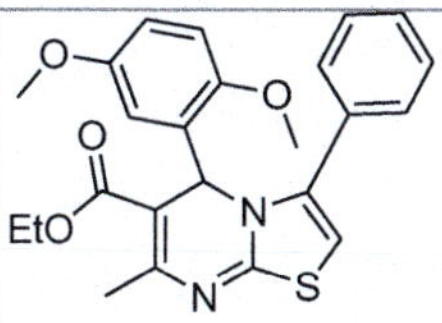	24	77

(continued)

Table 4 (continued)

Entry	Structure	Time (h)	% Yield
6 h	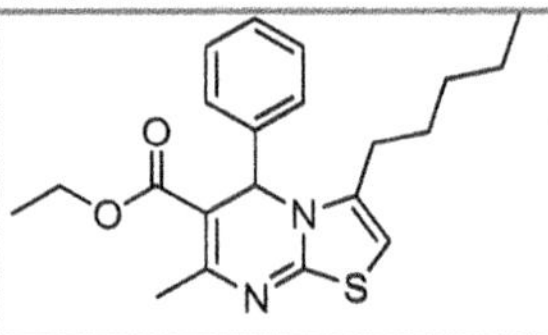	24	73
6i	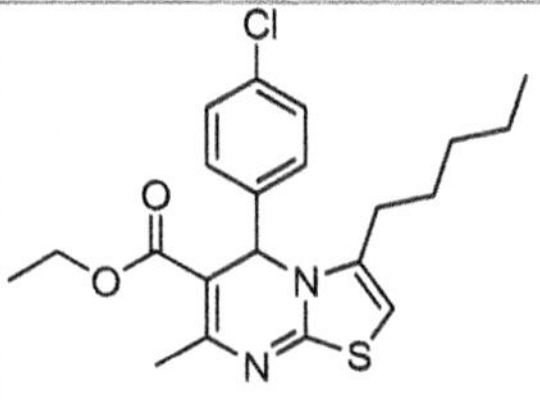	24	78
6j	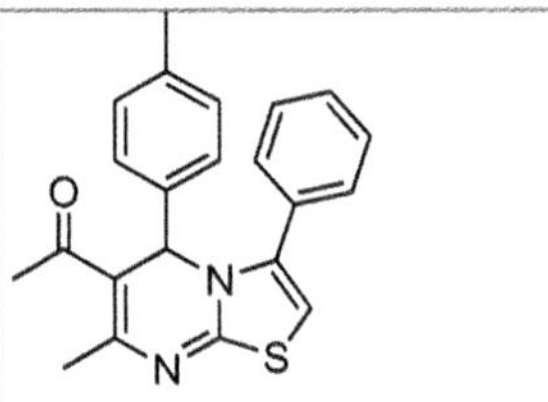	24	71
6 k	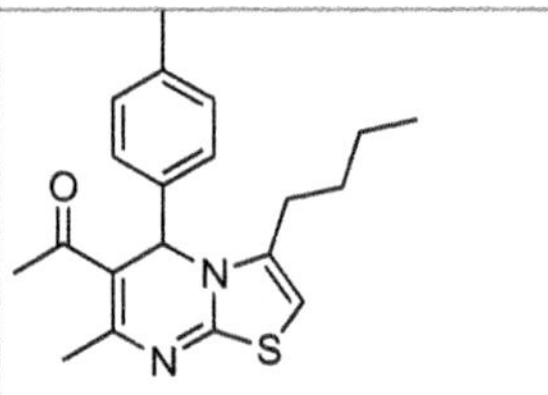	24	74
6 l	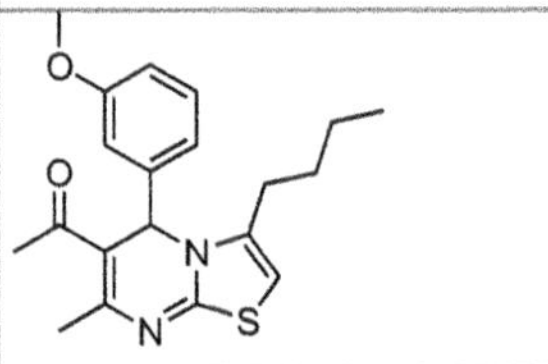	24	69
6 m	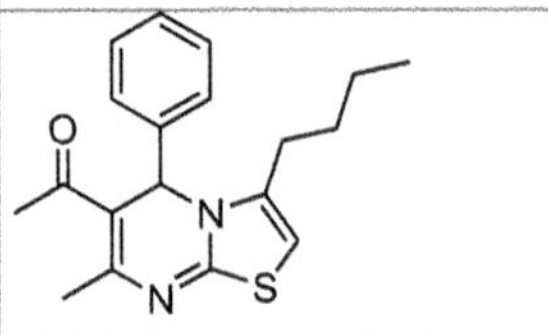	24	67

[a] The compounds from 6a-m have also been synthesized and reported earlier by different method [28]

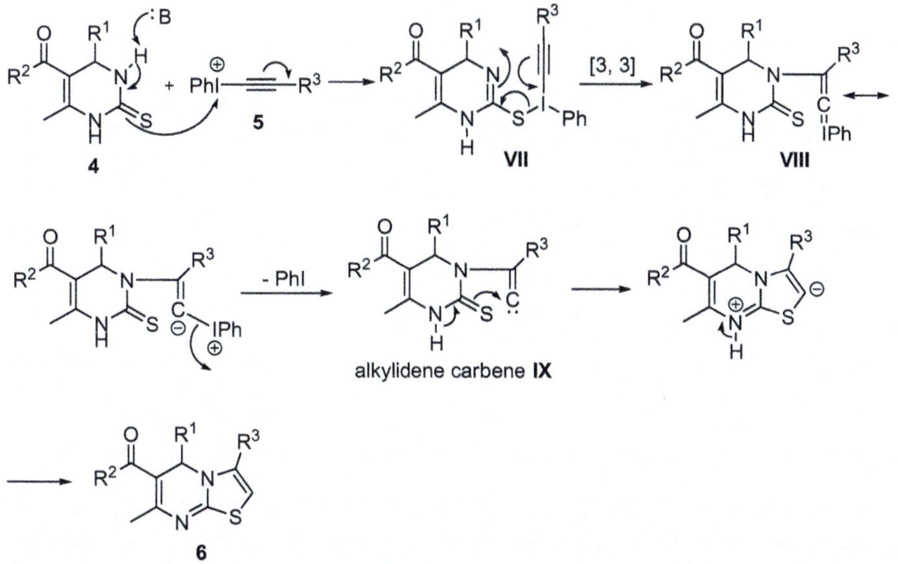

Scheme 9 Mechanism of 3,5-disubstituted-5*H*-thiazolo[3,2-*a*]pyrimidine formation

Conflict of Interest The authors declare no potential conflict of interest.

References

1. Graziano, G., Stefanachi, A., Contino, M., Prieto-Díaz, R., Ligresti, A., Kumar, P., Scilimati, A., Sotelo, E., Leonetti, F.: Multicomponent reaction-assisted drug discovery: a time- and cost-effective green approach speeding up identification and optimization of anticancer drugs. Int. J. Mol. Sci. **24**(7), 6581 (2023). https://doi.org/10.3390/ijms24076581
2. Jinni, L., Li, Y., Liu, N., Huang, N., Wang, L.: D.: A new type of heterogeneous catalysis strategy for organic reactions: Ugi-3CR catalyzed by highly stable MOFs with exposed carboxyl groups. Org. Chem. Front. **9**(22), 6179–6186 (2022). https://doi.org/10.1039/D2QO01257J
3. Grosjean, S., Pachón-Angona, I., Dawra, M., Refouvelet, B., Ismaili, L.: Multicomponent reactions as a privileged tool for multitarget-directed ligand strategies in Alzheimer's disease therapy. Future Med. Chem. **14**(21), 1583–1606 (2022). https://doi.org/10.4155/fmc-2022-0170
4. Bhat, M.A., Al-Omar, M.A., Naglah, A.M., Al-Dhfyan, A.: Biginelli synthesis of novel dihydropyrimidinone derivatives containing phthalimide moiety. J. Chem. **1–10**, 4284628 (2020). https://doi.org/10.1155/2020/4284628
5. Castro, J.M., Silva, A.C.A., Ritter, M., da Silva, A.F., Gonçalves, C.L., Dos Santos, P.R., Borja, L.S., de Pereira, C.M.P., da Silva, N.P.: Dihydropyrimidinones against multiresistant bacteria. Front. Microbiol. **18**(13), 743213 (2022). https://doi.org/10.3389/fmicb.2022
6. Kumarasamy, D., Roy, B.G., Rocha-Pereira, J., Neyts, J., Nanjappan, S., Maity, S., Mookerjee, M., Naesens, L.: Synthesis and in vitro antiviral evaluation of 4-substituted

3,4-dihydropyrimidinones. Bioorg. Med. Chem. Lett. **27**(2), 139–142 (2017). https://doi.org/10.1016/j.bmcl.2016.12.010
7. Prasad, T., Mahapatra, A., Sharma, T., Sahoo, C.R., Padhy, R.N.: Dihydropyrimidinones as potent anticancer agents: insight into the structure–activity relationship. Arch. Pharm. **356**(6), 2200664 (2023). https://doi.org/10.1002/ardp.202200664
8. Kaur, R., Chaudhary, S., Kumar, K., Gupta, M.K., Rawal, R.K.: Recent synthetic and medicinal perspectives of dihydropyrimidinones: a review. Eur. J. Med. Chem. **132**, 108–134 (2017). https://doi.org/10.1016/j.ejmech.2017.03.025
9. Ismaili, L., Nadaradjane, A., Nicod, L., Guyon, C., Xicluna, A., Robert, J.F., Refouvelet, B.: Synthesis and antioxidant activity evaluation of new hexahydropyrimido [5, 4-c]quinoline-2, 5-diones and 2-thioxohexahydropyrimido [5, 4-c] quinoline-5-ones obtained by Biginelli reaction in two steps. Eur. J. Med. Chem. **43**(6), 1270–1275 (2008)
10. Fustero, S., Catalan, S., Aceña, J.L., del Pozo, C.: A new strategy for the synthesis of fluorinated 3, 4-dihydropyrimidinones. J. Fluor. Chem. **130**(12), 1145–1150 (2009)
11. Sekhar, T., Thriveni, P., Venkateswarlu, A., Daveedu, T., Peddanna, K.: Sri Bhashyam Sainath, one-pot synthesis of thiazolo[3,2-a]pyrimidine derivatives, their cytotoxic evaluation and molecular docking studies. Spectrochim. Acta A Mol. Biomol. Spectrosc. **231**(118056), 118056 (2020). https://doi.org/10.1016/j.saa.2020.118056
12. Jadeja, J., Savant, M.: Synthesis of thiazolo[3,2-a]pyrimidine molecules, in vitro cytotoxic evaluation and molecular docking studies. J. Iran Chem. Soc. **20**, 1491–1502 (2023). https://doi.org/10.1007/s13738-023-02772-3
13. Fares, M., McCosker, P.M., Alsherbiny, M.A., Willis, A.C., Clark, T., Neyts, J., Jochmans, D., Keller, P.A.: Regioselective convergent synthesis of 2-arylidene thiazolo[3,2-a]pyrimidines as potential anti-chikungunya agents. RSC Adv. **10**, 5191–5195 (2020). https://doi.org/10.1039/D0RA00257G
14. Kulakov, I.V., Nurkenov, O.A., Turdybekov, D.M., et al.: Synthesis of thiazolo[3,2-a]pyrimidines based on 4-aryl-substituted 3,4-dihydro-pyrimidine(1H)-2-thiones and the crystal structure of ethyl 5-(2,4-dimethoxyphenyl)-7-methyl-3-oxo-3,5-dihydro-2H-thiazolo-[3,2-a]pyrimidine-6-carboxylate. Chem. Heterocycl. Comp. **45**, 856–859 (2009). https://doi.org/10.1007/s10593-009-0346-z
15. Mohamadpour, F.: Visible-light-induced radical condensation cyclization to synthesize 3,4-dihydropyrimidin-2-(1H)-ones/thiones using photoexcited Na2 eosin Y as a direct hydrogen atom transfer (HAT) catalyst. ACS Omega. **7**(10), 8429–8436 (2022). https://doi.org/10.1021/acsomega.1c05808
16. Slimi, H., Moussaoui, Y., Salem, R.B.: Synthesis of 3,4-dihydropyrimidin-2(1H)-ones/thiones via Biginelli reaction promoted by bismuth(III)nitrate or PPh3 without solvent. Arab. J. Chem. **9**, S510–S514 (2016). https://doi.org/10.1016/j.arabjc.2011.06.010
17. Singh, S., Schober, A., Gebinoga, M., Grob, G.A.: Convenient method for synthesis of thiazolo[3,2-a]pyrimidine derivatives in a one-pot procedure. Tetrahedron Lett. **52**, 3814 (2011). https://doi.org/10.1016/j.tetlet.2011.05.067
18. Agarkov, A.S., Shiryaev, A.K., Solovieva, S.E., et al.: Synthesis, chemical properties, and application of 2-substituted thiazolo[3,2-a]pyrimidine derivatives. Russ. J. Org. Chem. **59**, 337–364 (2023). https://doi.org/10.1134/S1070428023030016
19. Abu-Hashem, A.A., Youssef, M.M., Hussein, H.A.R.: Synthesis, antioxidant, antitumor activities of some new thiazolopyrimidines, pyrrolothiazolopyrimidines and triazolopyrrolothiazolopyrimidines derivatives. J. Chin. Chem. Soc. **58**, 41 (2011). https://doi.org/10.1002/jccs.201190056

20. Xiao, D., Han, L., Sun, Q., Chen, Q., Gong, N., Lv, Y., Suzenet, F., Guillaumet, G., Cheng, T., Li, R.: Copper-mediated synthesis of N-fused heterocycles via Csp–S coupling reaction and 5-endo-digcyclization sequence RSC Adv. **2**, 5054 (2012)
21. Srivastava, N., Singh, S., Vishnoi, R.K., Chaturvedi, D., Kishore, R., Rai, S., Chaturvedi, A.K.: Clean and green synthesis of organic isothiocyanates using zinc peroxide as desulfurizing agent and their antimicrobial activity. Russ. J. Org. Chem. **59**(12), 2205–2213 (2023). https://doi.org/10.1134/S1070428023120199
22. Srivastava, N.: Efficient TBAI-CS2 promoted synthesis of substituted hydrazinecarbodithiolates. Org. Prep. Proced. Int. (OPPI). **54**(6), 569–577 (2022). https://doi.org/10.1080/00304948.2022.2111171
23. Srivastava, N., Chaturvedi, D., Kishore, R., Kaur, M., Kaur, R.: Triton-B/CS2 mediated novel synthesis of substituted thioureas: a novel class of anti-cancer agents. Indian J. Chem. 'B'. **61**, 870–877 (2022). https://doi.org/10.56042/ijc.v61i8.65053
24. Srivastava, N., Kishore, R.: Cleaner and greener synthesis of 3H-benzothiazole-2-thione and its derivatives. J Sulfur Chem. **42**(1), 29–39 (2020). https://doi.org/10.1080/17415993.2020.1803321
25. Srivastava, N.: Tetrabutylammonium iodide/I_2Mediated convenient and green synthesis of substituted organic isothiocyanates. Org. Prep. Proced. Int. **53**(6), 562–570 (2021). https://doi.org/10.1080/00304948.2021.1975487
26. Srivastava, N., Rathore, B.K., Vishnoi, R.K., Bajpai, S.: A simple and convenient method for synthesis of artemisinin dimer through aliphatic nitro compounds. Asian J. Chem. **33**(1), 49–52 (2020). https://doi.org/10.14233/ajchem.2021.22918
27. Srivastava, N., Saxena, M.: Manisha Shukla, novel synthesis of 5-Oxo-2-thioxo-2,5-dihydro-3-thiophenecarboxylate derivative in non-aqueous medium. Asian J. Chem. **31**(1), 176–180 (2019). https://doi.org/10.14233/ajchem.2019.21622
28. Shelke, A.V., Bhong, B.Y., Karade, N.N.: New synthesis of 3,5-disubstituted-5H-thiazolo[3,2-a]pyrimidine via ring annulation of 3,4-dihydropyrimidin-2(1H)-thione using alkynyl(aryl) iodonium salts. Tetrahedron Lett. **54**(6), 600–603 (2013). https://doi.org/10.1016/j.tetlet.2012.11.098
29. Yusubov, M.S., Wirth, T.: Solvent-free reactions with hypervalent iodine Ragents. Org. Lett. **7**, 519 (2005). https://doi.org/10.1021/ol047363e
30. Margida, A.J., Koser, G.F.: Exchange of carbon ligands at iodine in iodonium salts. A direct synthesis of aryl(2-furyl)iodonium tosylates from aryl(tert-butylethynyl)iodonium tosylates. J. Org. Chem. **49**, 4703 (1984). https://doi.org/10.1021/jo00198a023
31. Nan-Yan, F., Yao-Feng, Y., Zhong, C., Shan-Wei, W., Ji-Tao, W., Peppe, C.: Indium(III) bromide-catalyzed preparation of dihydropyrimidinones: improved protocol conditions for the Biginelli reaction. Tetrahedron. **58**, 4801 (2002). https://doi.org/10.1016/S0040-4020(02)00455-6
32. Lu, J., Bai, Y., Wang, Z., Yang, B., Ma, H.: One-pot synthesis of 3,4-dihydropyrimidin-2(1H)-ones using lanthanum chloride as a catalyst. Tetrahedron Lett. **41**, 9075 (2000). https://doi.org/10.1016/S0040-4039(00)01645-2
33. Ranu, B.C., Hajra, A., Jana, U.: Indium(III) chloride-catalyzed one-pot synthesis of dihydropyrimidinones by a three-component coupling of 1,3-dicarbonyl compounds, aldehydes, and urea: an improved procedure for the Biginelli reaction. J. Org. Chem. **65**, 6270 (2000). https://doi.org/10.1021/jo000711f
34. Rana, K., Kaur, B., Kumar, B.: Synthesis and antihypertensive activity of some dihydropyrimidines Indian. J. Chem. Sect. B. **43**, 1553 (2004)

35. Xue, S., Shen, Y.C., Li, Y.L., Shen, X.M., Guo, Q.X.: Synthesis of 4-aryl-3,4-dihydropyrimidinones using microwave-assisted solventless Biginelli reaction. Chin. J. Chem. **20**, 385 (2002)
36. Wang, L., Qian, C., Tian, H., Ma, Y.: Lanthanide triflate catalyzed one-pot synthesis of dihydropyrimidin-2(1H)-thiones by a three-component of 1,3-dicarbonyl compounds, aldehydes, and thiourea using a solvent-free Biginelli condensation. Synth. Commun. **33**, 1459 (2003). https://doi.org/10.1081/SCC-120018755
37. Wipf, P., Venkatraman, V.: A new thiazole synthesis by cyclocondensation of thioamides and alkynyl(aryl)iodonium reagents. J. Org. Chem. **61**, 8004 (1996). https://doi.org/10.1021/jo961681c

Insights into the Structure and Function of ClpP from the *Mycobacterium* Genus

Smriti Bhardwaj[1] and Kuldeep K. Roy[1] (✉)

[1] School of Health Sciences and Technology, UPES, Dehradun, Uttarakhand, India

Abstract Tuberculosis (TB) remains a major global health concern, affecting morbidity and mortality rates significantly across the globe. Major obstacles to global TB control efforts are posed by the emergence of drug-resistant TB, including multidrug-resistant TB (MDR-TB) and extensively drug-resistant TB (XDR-TB). Therefore, new antibiotics against novel target(s) are urgently needed. Caseinolytic peptidase P (ClpP) represents a novel prospective target for hunting antibiotics against TB. ClpP is a serine protease that aids in the quality control of proteins in the bacterial cell. It has been reported that changes in the ClpP complex have an impact on the virulence and survival of different bacterial pathogens. Understanding the structure of the mycobacterial ClpP can facilitate the design of novel antibiotics against it. The hetero-tetradecameric form of ClpP represents the active form. N-blocked dipeptides (ligands) are necessary for the activation of the ClpP complex leading to proteolysis of misfolded proteins. The sequence comparison of Mtb and Msm ClpP revealed few residues in Mtb ClpP that were different from Msm ClpP. These residues possess different properties, and their interaction with the catalytic triads and the activator might play an important role in the pathogenicity of Mtb. Through comparison of apo- and ligand-bound ClpP structures, we observed that upon the ligand binding, the complex opens up its pores to perform its proteolytic function. A major shift of handle domain was observed which indicated that there might be a major shift in handle domain during the transit from inactive to active form. In addition, this shift of handle domain might play a very important role in the proteolysis as such shift causes the widening of the pore to carry out the protein degradation process. The insights gained in the present study may be useful in the structure-based drug design of novel anti-TB agents.

Keywords Antibiotics · Pathogens · ClpP · Drug-resistant · Virulence · Serine protease

A. K. Saxena, S. Nandi (eds.), *Global Trends in Health, Technology and Management*, https://doi.org/10.1007/978-3-031-75457-9_21

1 Introduction

Tuberculosis (TB) is the highly contagious disease caused by *Mycobacterium tuberculosis* (Mtb). TB remains a major global health concern, affecting morbidity and mortality rates significantly globally [1]. TB affects the individuals disproportionally those who are marginalized, such as people who are homeless, immunocompromised, people with high alcohol consumption and drug addicts [2]. TB mostly affects the countries with low income. In the 30 countries affected by TB, around 87% of the population with TB were from low- or middle-income families. The urban slum people had a threefold increase in TB causing bacillus concentration in sputum samples compared to rural people [3]. Also, approximately 12% of the population with TB are children under 15 years of age with 1.2 million of TB cases per year [4]. Post the meeting of UN HLM (High-Level Meeting) in 2018, which aimed to end TB by 2030, 7 million people died from TB in the past 5 years, which is a profound tragedy and a reminder of the urgency of the discovery of novel anti-TB agents with better efficacy or the one requiring a shorter duration of treatment [5]. The only vaccine available for the disease is BCG (Bacille Calmette-Guérin), which protects only infants and young children but not adolescents and adults from the severe form of TB [6]. M72/AS01E (M72), a "might be novel vaccine" which is in its last-phase clinical trial, might have the potential to become the first TB vaccine in the last century, if effective [7]. Rapid profiling of resistant cases, its safe, effective, and shortened treatment with regimens might lead to decrease in the incidence rates, prevalence rates, morbidity, and mortality caused by the resistant forms of mycobacteria globally [8].

Mycobacterium tuberculosis is a rod-shaped bacterium, which has a property of lipid-rich cell wall. Mycobacteria retain the carbol fuchsin stain even in the presence of acidic alcohol. TB is a highly contagious disease, transmitted through the aerosolized droplets containing the bacillus responsible. When a TB patient speaks, coughs, or sneezes, they release the air droplets containing Mtb, which are then inhaled by an uninfected person in the proximity. The bacillus enters and resides within the air spaces within lungs, and afterwards, it either replicates or remains in the dormant form. The replication takes place within the pulmonary site of infection, at first, and then spread to the lymph nodes of lungs (pulmonary infection) and then disseminates to other body parts, replicates, and infects (extra-pulmonary infection) [9]. When Mtb enters the alveoli of lungs, macrophage activates and performs phagocytosis of the bacillus. Mtb develops an army of factors that dodge the macrophage defense mechanism to escape its killing and replicate within this immune cell. Manipulation of intracellular macrophages signaling also effects the cytokine environment, which modifies the potency of immune cell response setting the bacillus tolerance by the host and favoring the Mtb survival intracellularly. During primary infection, progressive immune response establishes, and macrophages and other immune cells accumulate that either inhibit the Mtb proliferation or kill macrophages to form caseous necrotic granulomas that facilitates its growth [10, 11]. In these granulomas, Mtb can persist for several years. According to Global Tuberculosis Report 2023 by World Health Organization (WHO), recent estimation of TB burden documents that approximately 10.6 million people in the year 2022 developed the disease and 1.6 million people died. The emergence of drug-resistant forms, including multidrug-resistant TB (MDR-TB), extensively drug-resistant TB (XDR-TB) and total drug-resistant TB (TDR-TB), is a major challenge across the globe in the effective treatment of TB. MDR-TB strains are resistant to Isoniazid and Rifampicin, while the XDR-TB strains are also resistant to the fluoroquinolones as well as one of the injectable second-line drugs (Capreomycin, Amikacin, or Kanamycin). TDR-TB strains are resistant toward all the available antibiotics against Mtb [12]. Overall, tackling drug-resistant TB requires a multifaceted approach involving new drug development, search for novel drug target, drug repurposing, synergistic combination of drugs, and others. In order to discover novel anti-tubercular agents, a variety of drug targets have been explored against Mtb virulence, including InhA (Enoyl acyl carrier protein reductase) [13], LipY [14], MmpL3 (mycobacterial membrane protein large 3) [15], ATP synthase [16, 17], DprE1 (decaprenylphosphoryl-β-D-ribose 2′-epimerase) [18], and QcrB [19, 20].

Living cells contain many proteins; some of them can be intrinsically misfolded or defective because of stress. Therefore, proteolysis of undesired or defective proteins plays an important role in maintaining homeostasis, within cells, and survival of the bacteria. "Proteins to be degraded" are arrested, unfolded, and degraded into smaller fragments by the serine protease, ClpP (Caseinolytic peptidase P), that is present in the cytoplasm itself [21]. The bacterial ClpP complex is a double-ring energy-dependent peptidase consisting of 14-monomeric subunits, which in its active form exists with two discrete ClpP1 and ClpP2 heptamer [22] stacked upon each other, thus forming a hollow pore inside, where proteolysis occurs [23]. ClpP complex is a peculiar drug target in which both hyperactivation or inhibition leads to the lethal effect in bacteria. The complex consists of two subunits, namely, proteolytic and regulatory [24]. The proteolytic subunit, ClpP, is asymmetric in its functional form and comprised of two subunits, ClpP1 and ClpP2, which together perform proteolysis of degraded or misfolded substrates, thereby, it helps bacteria to maintain the good quality of proteins. The regulatory subunit, ClpX or ClpC1, are the AAA+ (ATPases associated with diverse cellular activities) chaperones that are required for substrate delivery. In the absence of these chaperones, ClpP complex remains in the closed state [25]. Research demonstrates that ClpP2 is required for the binding with ClpX or ClpC1, whereas ClpP1 alone is responsible for substrate degradation, growth, and infection in Mtb [26]. The complex together is essential for protein degradation and hence it has been demonstrated that if either of the subunits is degraded, it leads to the death of the bacteria [27]. In addition, N-blocked dipeptides are necessary for the activation of these complexes from homo-tetradecamer (inactive) to hetero-tetradecameric form (active). The activator binds to the complex and interacts with the head and handle domains and widens up the pore, where proteolysis occurs [28]. Here, we aimed to explore and compare the sequence and structure of ClpP complex from *Mycobacterium tuberculosis* (Mtb), a pathogenic form, and *Mycobacterium smegmatis* (Msm), a non-pathogenic form. It was to understand the reason behind virulence of Mtb and non-virulence of Msm. Our aim was to also understand the role of N-blocked dipeptides whose binding leads to activation of the ClpP complex through analyses of the crystal structures of apo- and N-blocked peptide-bound ClpP complexes. The insights gained here are expected to guide the structure-based drug design of novel anti-TB agents.

2 Materials and Methods

2.1 Protein Sequences and Structures

The three-dimensional (3D) coordinates of the Mtb ClpP1P2 complex in apo form (PDB ID: 4U0H; resolution = 3.25 Å) [22], Mtb ClpP1P2 bound with agonist ZIL, i.e. N-[(benzyloxy)carbonyl]-L-isoleucyl-L-leucine (PDB ID: 4U0G; resolution = 3.20 Å) [22], and Msm ClpP1 (PDB ID: 6PBU; resolution = 2 Å) [29] were downloaded from the RCSB Protein Data Bank (PDB) (www.rcsb.org) [30]. All the protein sequences were obtained from Universal Protein Resource (UniProt) database [31]. The UniProt IDs of these proteins are P9WPC5 (Mtb ClpP1), P9WPC3 (Mtb ClpP2), A0A8B4R245 (Msm ClpP1), and A0A8B4R2Y7 (Msm ClpP2).

2.2 Protein Sequence Analysis

Multiple sequence alignment (MSA) of ClpP subunits was done using the CLUSTAL Omega server [32]. Mtb ClpP1 (UniProt ID: P9WPC5), Mtb ClpP2 (UniProt ID: P9WPC3), Msm ClpP1 (UniProt ID: A0A8B4R245), and Msm ClpP2 (UniProt ID: A0A8B4R2Y7) were subjected to multiple sequence alignments.

2.3 Protein and Ligand Preparation

The missing loops in the original crystal structures, downloaded from the RCSB PDB, were modeled with the protein structure homology-modeling approach using the Swiss-model server [33]. Subsequently, addition of hydrogens, assignment of bond orders, and the removal of water molecules and other unwanted residues from the protein structures were carried out using the Python Molecular Viewer (PMV) and/or PyMol v2.5.4 tools [34].

2.4 Protein–Ligand Interaction

The protein–ligand interaction-based study and its visualization were carried out using the free versions of visualizers, namely PyMol v2.5.4 [34] or Biovia [35]. Using these tools, the visualization and comparative evaluation of the protein-ligand bound complexes were carried out. The residues within 4 Å were selected and the polar interactions between the ligand and protein were also studied. These tools were also used to generate graphics that helped to explain protein structure, their arrangement of catalytic triads, the ligand binding protein pocket, interaction of amino acids, and shift of the handle domain after activator binding.

3 Results

3.1 Comparative Sequence Analysis of *Mycobacterium tuberculosis* and *Mycobacterium smegmatis*

It was established that the ClpP complex contributes to pathogenicity and survival of Mtb [24, 36, 37], while it contributes to the survival of non-pathogenic Msm [38]. Comprehending the fundamentals of ClpP-mediated survival and pathogenicity is crucial. Thus, our goal was to thoroughly examine the published crystal structures of ClpP from Mtb and Msm in order to comprehend the role of ClpP in the pathogenicity of Mtb. In order to identify sequence variations at various positions, we used multiple sequence alignments to compare the sequences of ClpP1 and ClpP2 subunit from Mtb and Msm. In contrast to the Msm ClpP1 subunit, the Mtb ClpP1 subunit had three extra amino acids with total 200 amino acids. In contrast, Msm ClpP2 had 218 amino acids, which was four amino acids more than Mtb ClpP2 (214 amino acids). Table 1 provides the sequence information of various subunits of ClpP complexes from Mtb and Msm. Table 2 provides the percentage identity matrix of ClpP1 and ClpP2 subunits from Msm and Mtb. ClpP1 and ClpP2 subunits of both the microorganisms were approximately 7–9% of non-identical amino acids. In addition, within the hetero-tetradecamer, 50–52% amino acids of their sequences were non-identical or different.

Furthermore, it was found that there were few similar amino acids in the subunits of ClpP complexes from Mtb and Msm, which may account for the subunits' different functions. We found that there was 91.28% sequence identity between Mtb ClpP1 with Msm ClpP1, while there was 93.40% sequence identity between Mtb ClpP2 with Msm ClpP2. This prompted us to concentrate on the differences in key amino acids in the ClpP complexes of the two species of mycobacteria. The residues that were present in Mtb but absent in Msm might be crucial in the pathogenicity of the organism since the amino acids within the ClpP complex differ in structure, function, and interactions with other amino acids.

Sequence comparison of Mtb ClpP1 with Mtb ClpP2: ClpP1 and ClpP2 homologues correspond to the NCBI entries for Mtb ClpP1 (UniProt ID: P9WPC5) and Mtb ClpP2 (UniProt ID: P9WPC3). As shown in Fig. 1a, the MSA revealed that 48.26% of the sequences of Mtb ClpP1 and Mtb ClpP2 were identical. There were 85 identical, 41 highly similar, and 20 less similar amino acids in ClpP1 and ClpP2 from *M. tuberculosis*. The remaining amino acids were different. This demonstrated that the majority of the amino acids in the two Mtb ClpP subunits, ClpP1 and ClpP2, were distinct from one another. Since each amino acid has a unique structure and set of properties, they may

Table 1 Summary of the source, sequence ID, and sequence lengths of ClpP complexes

Source	Subunit	UniProt sequence ID	Sequence length
Mycobacterium tuberculosis	ClpP1	P9WPC5	200
Mycobacterium tuberculosis	ClpP2	P9WPC3	214
Mycobacterium smegmatis	ClpP1	A0A8B4R245	203
Mycobacterium smegmatis	ClpP2	A0A8B4R2Y7	218

Table 2 Percentage identity matrix of ClpP1 and ClpP2 subunit of Msm and Mtb

		Mtb		Msm	
Source	Subunit	ClpP1	ClpP2	ClpP1	ClpP2
Mtb	ClpP1	100%	48.26%	91.28%	43.72%
	ClpP2	48.26%	100%	43.37%	93.40%
Msm	ClpP1	91.28%	43.37%	100%	47.78%
	ClpP2	43.72%	93.40%	47.78%	100%

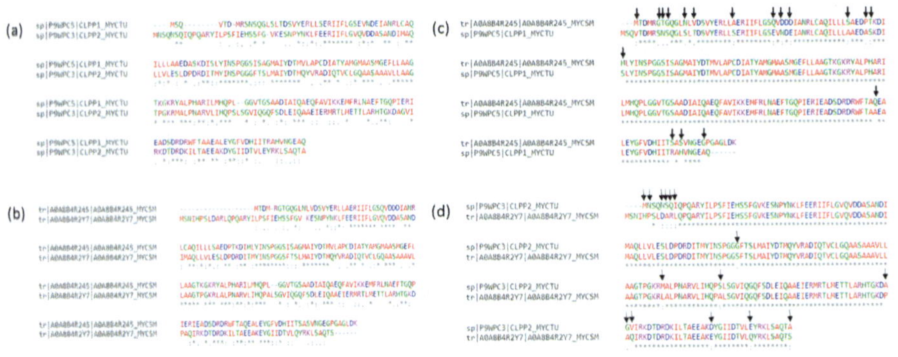

Fig. 1 Multiple sequence alignment of various subunits from the ClpP complex by Clustal Omega: (**a**) ClpP1 and ClpP2 subunits from Mtb ClpP complex, (**b**) ClpP1 and ClpP2 subunits from Msm ClpP complex, (**c**) ClpP1 from Msm and Mtb ClpP complexes, and (**d**) ClpP2 from Msm and Mtb ClpP complexes. ↓ Non-identical or similar amino acids; * Identical amino acid; **:** Highly similar amino acid; . Less similar amino acid

interact differently. Their varying activity in the process of protein breakdown may be due to the differences in the residues of the two subunits of ClpP complex.

Sequence comparison of Msm ClpP1 with Msm ClpP2: ClpP1 and ClpP2 homologues correspond to the NCBI entries for Msm ClpP1 (UniProt ID: A0A8B4R245) and Msm ClpP2 (UniProt ID: A0A8B4R2Y7). The MSA showed that Msm ClpP1 and Msm ClpP2 shared 47.78% sequence identity as shown in Fig. 1b. There were 87 identical, 38 highly similar, and 19 less similar amino acids in ClpP1 and ClpP2 in Msm when aligned. The rest of the amino acids were different. Here, the two subunits, ClpP1 and ClpP2, in Msm, having different activities might be due to differences in the structure and properties of the amino acids that might show different interactions.

Sequence comparison of Msm ClpP1 with Mtb ClpP1: The ClpP1 subunits from Mtb and Msm were then aligned to analyze the identical and varying sequences. ClpP1 homologues correspond to the NCBI entries for Msm ClpP1 (UniProt ID: A0A8B4R245) and Mtb ClpP1 (UniProt ID: P9WPC5). The MSA result showed that sequences were 91.28% identical as shown in Fig. 1c. This showed that around 8.72% residues were either similar or non-identical with respect to each other. Seven residues were highly similar, whereas six residues were less similar. Since the non-identical residues' characteristics and interactions with various amino acids may contribute to the pathogenicity of Mtb, we concentrated on them. The residues that were very different in both the species were HIS-58|SER-61, GLN-175|ALA-178, SER-189|ARG-192, SER-

191|HIS-194, GLY-196|ALA-199, and PRO-197|GLN-200. As far as characteristics go, Histidine and Arginine are positively charged amino acids, while Serine, Glutamine, and Proline are polar and uncharged. Alanine and Glycine are non-polar amino acids. Thus, one possible explanation for the virulence observed in Mtb rather than Msm could be the variation in these amino acids within the ClpP complex.

Sequence comparison of Msm ClpP2 with Mtb ClpP2: After observing differences in many amino acids between ClpP1 of Mtb and Msm, we carried out MSA of the ClpP2 subunit of both organisms. ClpP2 homologues correspond to the NCBI entries for Msm ClpP2 (UniProt ID: A0A8B4R2Y7) and Mtb ClpP2 (UniProt ID: P9WPC3). The MSA result revealed that sequences were 93.40% identical as shown in Fig. 1d. Here, nine amino acids were highly similar, whereas two amino acids were less similar. In fact, 6.6% amino acids were either similar or non-identical with respect to each other. The non-identical amino acids were MET-1|HIS-5, ASN-2|PRO-6, GLN-4|LEU-8, ALA-176|PRO-180, and VAL-178|GLN-182. Methionine, Leucine, Valine, and Alanine are non-polar in nature, whereas Histidine is a positively charged residue. Asparagine, Glutamine, and Proline are polar and uncharged amino acids. The virulence of Mtb may therefore be attributed to certain amino acids (MET|HIS, GLN|LEU, ALA|PRO, and VAL|GLN) that were distinct in Mtb and Msm.

3.2 Structural Analysis of Mycobacterial ClpP Complex

Through multiple sequence alignment of the ClpP subunits from Mtb and Msm, we were able to identify residues of amino acids that were identical, similar, or different. Here, our main concern was to understand if distinct amino acids affected the complex's structure in either of the two mycobacterial species. In order to conduct the analysis of ClpP crystal structures, we superimposed the inactive form of one of the ClpP subunits, i.e., the homo-tetradecameric form (ClpP1P1) of Msm and Mtb. Remarkably, as Fig. 2 illustrates, we observed a conformational variation in one of the β-strands, composed of TRP-174, PHE-175, THR-176 in Mtb, and TRP-171, PHE-172, THR-173 in Msm. Furthermore, no other variations were observed.

We conducted a thorough analysis to determine whether the amino acids in the mentioned β-strand interacted with the activator or with the catalytic triad of ClpP1 subunit. Table 3 provides a summary of the amino acids within 4 Å of catalytic triad in Mtb and Msm. We found that TRP-174 from Mtb and TRP-171 from Msm were located within 4 Å of the ClpP1 subunit's catalytic triad (Table 3). We then performed an interaction-based study between the catalytic triad and amino acid residues in the presence of ZIL (activator) to determine whether or not the amino acids in the β-sheet with varying conformations were present in the activated state of the ClpP complex. We found that catalytic triad had TRP-174 within its 4 Å in case of active ClpP1 in Mtb (Table 4). Furthermore, it was noted that within 4 Å of the catalytic triad, only TRP in the β-sheet of Mtb and Msm in unbound form and of Mtb in bound form was present, but it was not involved in direct contact or interaction. TRP-174 in the β-sheet may be involved in the indirect interaction with amino acids composing catalytic triads; it might be necessary for the proteolysis process by ClpP complex. Although it may or may not contribute to virulence of Mtb, given that it was present in the 4 Å of the catalytic triad in both bound

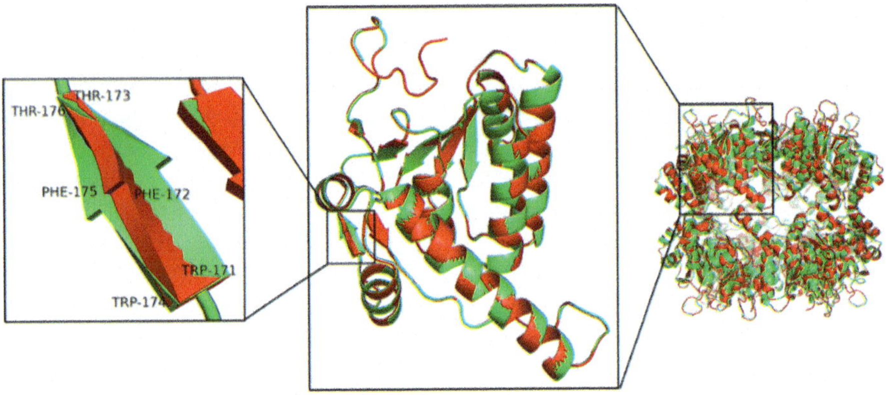

Fig. 2 Superimposition of the ClpP1 subunit of Mtb (UniProt ID: P9WPC5) (Green cartoon) over Msm (UniProt ID: A0A8B4R245) (Red cartoon). The figure represents the monomeric form of ClpP1 revealing the different conformation of one of the β-strands (residues labeled)

Table 3 Summary of the amino acids within 4 Å of catalytic triad in Mtb and Msm

Mycobacterium tuberculosis	*Mycobacterium smegmatis*
GLY-68	GLY-65
GLY-69	GLY-66
ALA-97	ALA-94
MET-99	MET-96
GLY-100	GLY-97
GLU-101	GLU-98
PHE-102	PHE-99
LEU-121	LEU-118
MET-122	MET-119
GLN-124	GLN-121
PRO-125	PRO-122
ASP-168	ASP-165
SER-169	SER-166
ASP-170	ASP-167
ARG-171	ARG-168
ARG-173	ARG-170
TRP-174	TRP-171

and unbound forms in Mtb. In addition, none of the aforementioned studies included PHE-175 or THR-176 from the β-sheet, indicating that these two amino acids were not directly involved in the interaction needed for the activation of the ClpP complex to initiate proteolysis. We then further investigated the amino acids that were present within 4 Å of activator (ZIL in this case) and their interaction with the amino acids as illustrated in Fig. 3a. We aimed to observe if the activator interacted in any way with the three amino acids that were involved in β-sheet exhibiting different conformation.

Table 4 Representation of the amino acids that were interacting with catalytic triad of unbound Mtb and with catalytic triad in presence of N-blocked dipeptide that was bound Mtb ClpP complex

PDB ID: 2CBY (UNBOUND)	PDB ID: 4U0G (BOUND)
GLY-68	[b]PRO-67
GLY-69	GLY-68
ALA-97	GLY-69
[a]MET-99	ALA-97
GLY-100	[a]MSE-99
GLU-101	GLY-100
PHE-102	GLU-101
LEU-121	PHE-102
[a]MET-122	LEU-121
GLN-124	[a]MSE-122
PRO-125	GLN-124
[b]ASP-168	PRO-125
SER-169	[b]LEU-126
[b]ASP-170	SER-169
ARG-171	ARG-171
ARG-173	ARG-173
TRP-174	TRP-174

[a]Amino acids were present as Selenomethionine in the active complex although these amino acids were Methionine in the inactive complex

[b]Amino acids that showed interaction only in one form that was either bound or unbound

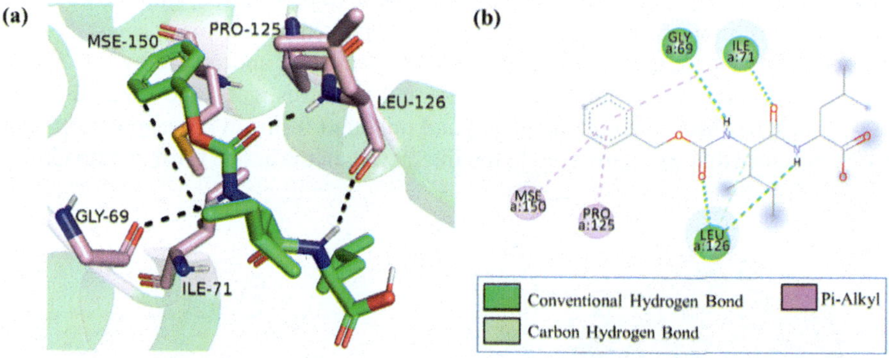

Fig. 3 (**a**) 3-D depiction of interaction of the N-blocked dipeptide, ZIL with the ClpP1 subunit of Mtb (PDB ID: 4U0G). (**b**) 2-D interaction of ZIL, showing different types of bonds with Mtb ClpP1

ZIL was surrounded by GLY-69, SER-70, ILE-71, SER-98, MSE-99, PRO-125, LEU-126, GLY-127, GLY-128, ILE-146, and MSE-150 within its 4 Å. ZIL showed polar interactions with GLY-69, ILE-71, and LEU-126. The various types of interactions of ZIL within 4 Å of the ClpP1 subunit are depicted in Fig. 3b. Nevertheless, none of the aforementioned amino acids from the β-sheet were found interacting with the activator. Therefore, it was determined that one of the three amino acids, i.e. TRP-174 was present within 4 Å of catalytic triad of both active and inactive ClpP1 complex in Mtb and in unbound form of ClpP1 in Msm but none of the three amino acids were present within 4 Å of the activator (ZIL). This implies that rather than interacting directly with the activator, this residue may engage indirectly with the catalytic triad. In addition, the three amino acids, TRP-174, PHE-175, and THR-176 in Mtb, and TRP-171, PHE-172, and THR-173 in Msm, were not those that were found different when MSA of ClpP1 was performed (Fig. 1). Since there was no crystal structure of the active ClpP1 of Msm, it can be hypothesized that even in Msm ClpP1 the presence of amino acids within the catalytic triad, within 4 Å of the activator, and their interaction pattern might be similar to Mtb ClpP1.

3.3 Analysis of Catalytic Triad

The serine protease ClpP complex contains catalytic triad composed of SER-98, HIS-123, and ASP-172 in Mtb, while it contains SER-95, HIS-120, and ASP-169 in Msm. We examined the catalytic triad of ClpP1 subunits from both mycobacterial species by superimposing their ClpP1 subunits, and the results are illustrated in Fig. 4. Our analysis revealed the identical amino acids.

As no difference was observed in the catalytic triads of both species, therefore, in order to look for any differences in amino acids within 4 Å that might differ, we performed interaction-based study of catalytic triads with the amino acids within 4 Å of ClpP1 of Mtb and Msm. Results are depicted in Fig. 5. The catalytic triad of the ClpP1 subunit was discovered to have identical amino acids within its 4 Å in both species in ClpP1's inactive state; the only difference was a three-position shift in numbering (Table 3). This implied that in unbound state of ClpP1 from both the species, the amino acids of their catalytic triad were the same. With the goal of understanding the virulence causing amino acid *via* its property or interaction, we investigated the interaction of the catalytic triad of Mtb in both ZIL-bound and unbound forms in order to search for the critical residues that were involved in the interaction in the active form of Mtb ClpP but

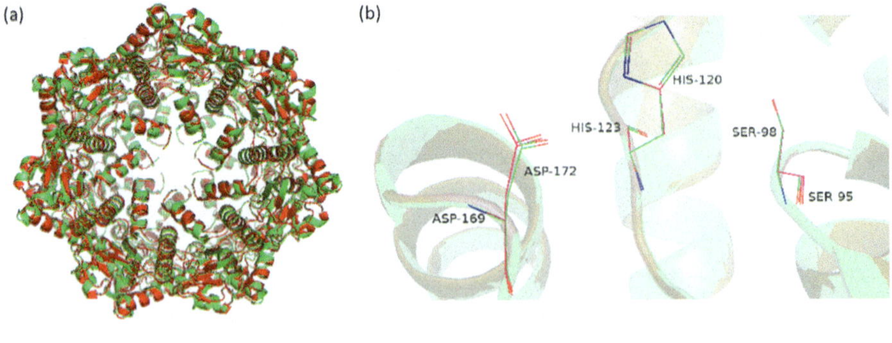

Fig. 4 (**a**) Superimposition of ClpP1 subunits from Msm (Green) (PDB ID: 6PBU) and Mtb (Red) (PDB ID: 2CBY). (**b**) Superimposition of Chain A of Msm ClpP1 (Green) (PDB ID: 6PBU) having catalytic triad as SER-95, HIS-120, and ASP-169 over the Chain A of Mtb ClpP1 (Red) (PDB ID: 2CBY) having catalytic triad as SER-98, HIS-123, and ASP-172

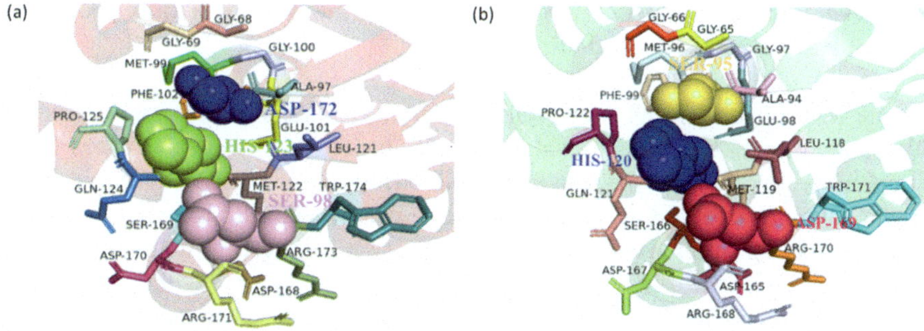

Fig. 5 (**a**) Depiction of Mtb ClpP1 (PDB ID: 2CBY) (Red cartoon); Interacting amino acid (stick representation) with the catalytic triad (SER-98, HIS-123, and ASP-172) shown in sphere representation. (**b**) Depiction of Msm ClpP1 (PDB ID: 6PBU) (Green cartoon); Interacting amino acid (stick representation) with the catalytic triad (SER-95, HIS-120, and ASP-169) shown in sphere representation

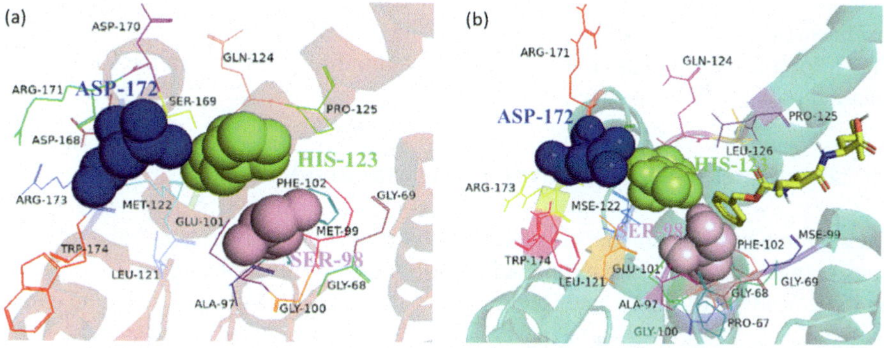

Fig. 6 (**a**) Chain A monomer of ClpP1 (Red cartoon) of the proteolytic complex ClpP1P2 from Mtb (PDB ID: 2CBY) showing interaction of catalytic triads (sphere representation) with residues (lines representation) within 4 Å. (**b**) Chain A monomer of ClpP1 (Cyan cartoon) of the proteolytic complex ClpP1P2 from Mtb (PDB ID: 4U0G) showing interaction of ZIL (Yellow stick representation) with residues (lines representation) within 4 Å

not in the inactive form. Consequently, the interaction-based study was done on the ZIL-bound Mtb ClpP complexes. Figure 6 depicts the protein–ligand interactions observed in these studied complexes. In unbound form of Mtb ClpP, ASP-168 and ASP-170 were present within 4 Å of catalytic triad, whereas these two residues were not present within 4 Å in the ZIL-bound form of Mtb ClpP. Interestingly, PRO-67 and LEU-126 within 4 Å of catalytic triad in case of ZIL-bound form were not present in the inactive or unbound form of ClpP. Moreover, MET-99 and MET-122 were replaced by MSE-99 and MSE-122 in bound state. Table 4 provides a summary of the interactions of catalytic triad with amino acids within its 4 Å in the ZIL-bound (active) and unbound (inactive) forms of Mtb ClpP complexes.

3.4 Interaction of Catalytic Triads of Mtb with Amino Acids within 4 Å with respect to N-Blocked Dipeptide

Aspartate was found within 4 Å of catalytic triad in the inactive (unbound) state of the ClpP1 subunit from Mtb, which is a negatively charged amino acid, whereas Proline and Leucine were present in the active (ZIL-bound) form of the ClpP1 complex. Proline is a polar and uncharged amino acid and Leucine is non-polar in nature. Changes in the amino acid having different structure and property might lead to distinct interaction with the other amino acids in the active ClpP1. This might play a role in maintaining the active state of the complex *via* its interaction. Since the two amino acids, PRO-67 and LEU-126, were found in the active form of the ClpP1 subunit but not in the inactive state within 4 Å of the catalytic triad, we examined which domain within the ClpP complex these amino acids were located. PRO-67 is a component of head domain of ClpP complex, whereas LEU-126 belongs to handle domain [39]. The activator interacts with the head and handle domains of the complex. Here, it was observed that activator did not show any direct interaction with PRO-67 and LEU-126 (Fig. 3a), but they were present within 4 Å of catalytic triad in active form of ClpP complex. These two residues, PRO-67 and LEU-126, were thus identified as the key amino acids in the ClpP1 complex for its function and may have an indirect function in the substrate degradation process.

3.5 Polar Interaction of Catalytic Triads with Residues

The research of polar contacts is preferred since it provides hydrogen–bonding interactions between the protein's amino acids [40]. Figure 7 illustrates the polar contacts of catalytic triads with amino acids within 4 Å of both the ZIL-bound (active) and unbound

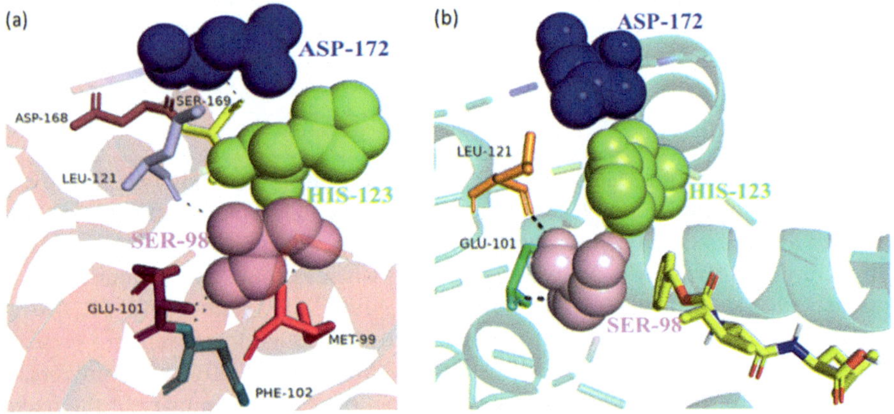

Fig. 7 Depiction of ClpP1 subunit of Mtb amino acids (stick representation) that has polar interaction within 4 Å with the catalytic triad (SER-98-HIS-123-ASP-172) shown in spheres: (**a**) PDB ID: 2CBY in red cartoon (unbound form). (**b**) PDB ID: 4U0G in cyan cartoon (bound form), ZIL (Yellow; stick representation)

(inactive) forms of Mtb ClpP1. As shown in Table 5, we observed that the unbound form of ClpP exhibited polar interactions with five amino acids, while the bound form only displayed interactions with two amino acids (Fig. 7). The two residues that were identified in the ZIL-bound form—a negatively charged amino acid, GLU-101 and a non-polar amino acid with an aliphatic group, LEU-121—were found to exhibit polar interaction with the catalytic triad of the Mtb ClpP1 subunit in both the unbound and ZIL-bound states. These two amino acids may be crucial in stabilizing the ClpP complex's active state in Mtb through interactions with ZIL (activator).

In addition, we have analyzed the conformational difference (movement or shift) of catalytic triad amino acids in both ZIL-bound and unbound forms of Mtb ClpP1. Figure 8 provides a representation of these conformation shifts or differences of amino acids composing the catalytic triad. A few of the catalytic triad amino acids, particularly HIS-123, exhibited a major conformation shifts in the presence of activator (ZIL). The catalytic triads were seen disordered in inactive state and we observed their major shift in the active form in the presence of activator. This conformational shift appears to be crucial for the catalytic triad to be in an ordered state so that the ClpP complex could perform the proteolysis of substrates, e.g., misfolded proteins.

As shown in Fig. 9, we have also examined the shift between the monomers of the ClpP1 heptamers of Mtb in both their active and inactive states. This helped us to understand the movement of the ClpP1 complex during its activation. Here, we have analyzed those amino acids on the outer or peripheral side in the unbound ClpP1, which showed a C_α-C_α distance (distal amino acids) of 69 Å between two handle domains, whereas the ZIL-bound ClpP1 showed a distance of 65.2 Å. In addition, the adjacent handle domain showed a major difference of 1.3 Å from inactive to active forms. This indicated that, in comparison with the inactive form, the peripheral residues of the monomer comprising ClpP1 heptamer move toward each other in the presence of activator during ClpP complex activation. The distance between the inner (axial) amino acid residues of unbound ClpP1 was found to be 23.8 Å, while it was 27.25 Å in the ZIL-bound ClpP1. Additionally, the two adjacent helices were at a distance of 2.1 Å. Therefore, the amino acids in the axial wall of ClpP1 separate out from each other in the presence of activator and migrate from inside to the outside. This led to the conclusion that the peripheral helices of heptamer ClpP1, in the presence of activator (ZIL), move from outward to inward direction, while, on the other hand, the axial wall moves from inward to outward direction. Furthermore, the amino acids in the peripheral helices exhibited a slight shrinkage pattern, whereas the axial wall of the heptamer ClpP1 separated slightly. As a result, this motion opens up the inner pore of the ClpP complex, which is necessary for the recognition of the protein and thus, necessary for the proteolysis to happen.

Table 5 Summary of the amino acids engaged in polar interaction with catalytic triad of the ZIL-bound and unbound forms of the Mtb ClpP complex

PDB ID: 2CBY (UNBOUND)	PDB ID: 4U0G (BOUND)
MET-99	GLU-101
GLU-101	LEU-121
PHE-102	–
LEU-121	–
ASP-168	–
SER-169	–

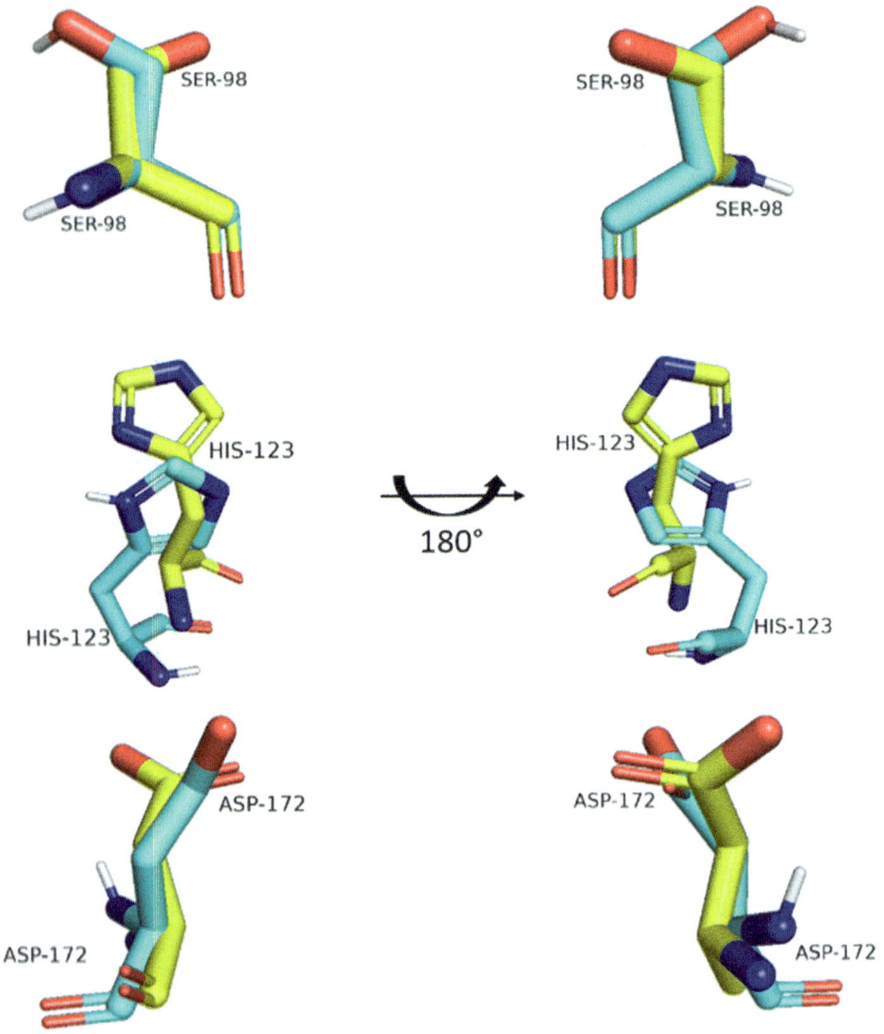

Fig. 8 Representation of shifting of catalytic triad amino acids of Mtb ClpP1 in inactive or unbound state (SER-98, HIS-123, ASP-172—stick representation; Yellow carbon) (PDB ID: 2CBY) and ZIL-bound or active state (SER-98, HIS-123, and ASP-172—stick representation; Cyan carbon) (PDB ID: 4U0G)

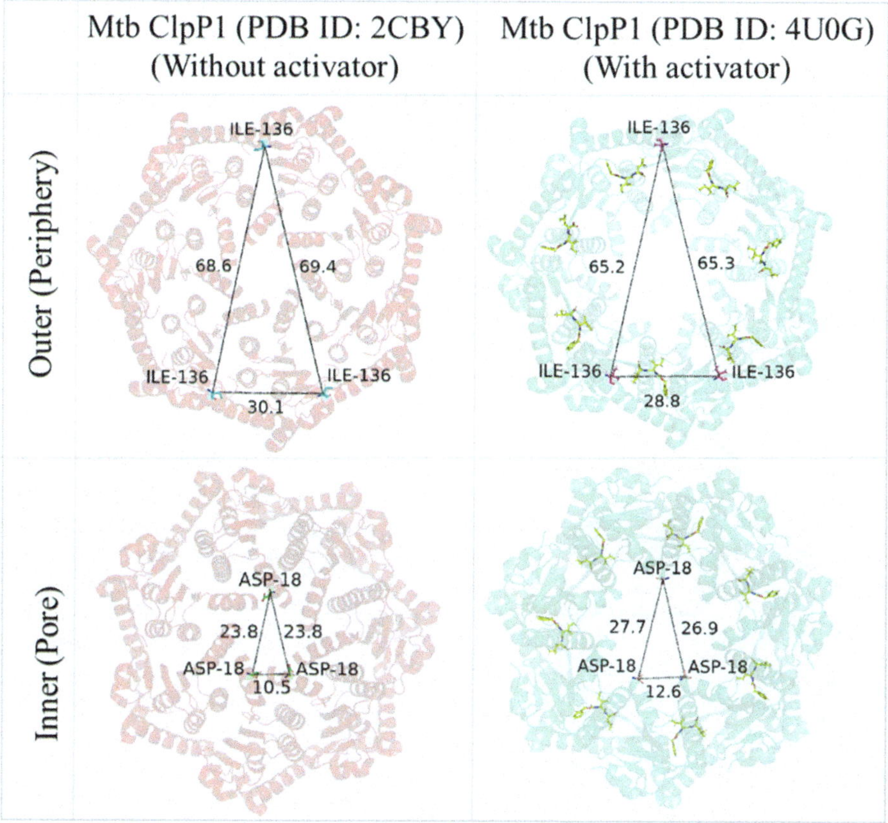

Fig. 9 Representation of C_α-C_α distance of outer (peripheral) and inner (axial) amino acids of the unbound Mtb ClpP1 (Red) (PDB ID: 2CBY) and ZIL-bound Mtb (Cyan) (PDB ID: 4U0G); ZIL is represented in yellow stick. ILE-136 in handle domain is represented in stick (Cyan) and stick (Magenta) in the closed or inactive and active states, respectively. ASP-18 is represented in stick (Green) and stick (Magenta) in the closed or inactive and active states, respectively

3.6 Visualization of N-Blocked Dipeptide Binding Site

The binding of N-blocked dipeptide to the S1 pocket of the ClpP complex is necessary for its activation; it involves the conversion of the homo-tetradecamer form into the hetero-tetradecamer form [28]. Therefore, we visualized the S1 pocket where the N-blocked dipeptide binds to, as shown in Fig. 10. In the Mtb ClpP1 monomeric unit, dipeptide interacted with the amino acids in the head and the handle domains. Next, we examined if the presence of the activator caused any conformational changes in the ClpP complex to analyze the mechanism of pore widening in the ClpP1 subunit to perform proteolysis.

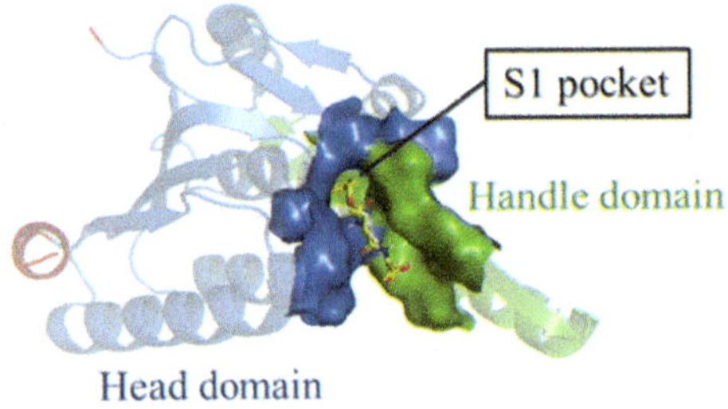

Fig. 10 Chain A monomer of Mtb ClpP1 (PDB ID: 4U0G) showing binding of the dipeptide agonist, ZIL (stick representation; Yellow carbons) in the S1 pocket (Blue and green surface representation)

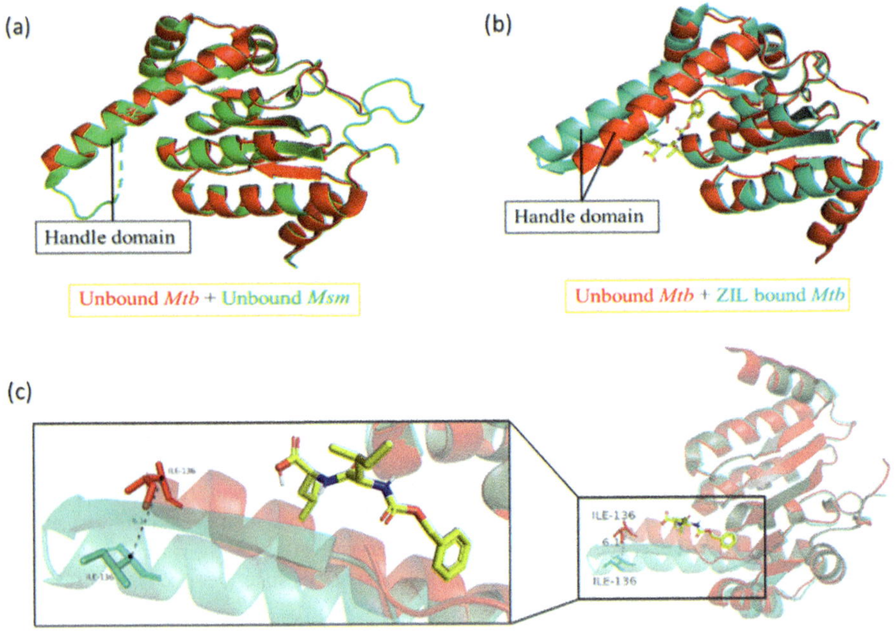

Fig. 11 (**a**) Superimposition of Chain A monomer of Mtb (Red) (PDB ID: 2CBY) over Msm (Green) (PDB ID: 6PBU) showing no visual difference. (**b**) Superimposition of Chain A monomer of Mtb (Red) (PDB ID: 2CBY) over Mtb (Cyan) (PDB ID: 4U0G) bound to ZIL (Yellow; stick representation). (**c**) Shifting of handle domain of Chain A monomer of Mtb (Red) (PDB ID: 2CBY) over Mtb (Cyan) (PDB ID: 4U0G) bound to ZIL (Yellow; stick representation) showing approx. 6 Å shift. ILE-136 is represented in stick form

3.7 Conformational Change in ClpP after Binding of ZIL (Activator)

We examined the structure of apo (unbound) forms of ClpP complexes from Msm and Mtb and activator bound Mtb ClpP complex. Through superimposition, we observed no visual differences except the aforementioned β-sheet in apo or unbound form of Mtb ClpP1 and Msm ClpP1, as shown in Fig. 11a. However, in ZIL-bound form, significant displacement of handle domain toward upward region after binding of the dipeptide (activator) was observed (Fig. 11b), which led to the formation of ordered form of catalytic triads necessary for substrate recognition and subsequent proteolysis. There was a shift of the handle domain by approximately 6.1 Å as shown in Fig. 11c. The mechanism behind the handle domain shift following activator binding or in the presence of an activator is still unknown and needs further investigation.

4 Discussions

ClpP is a serine protease that helps in the survival and pathogenicity of the organism. It maintains proteostasis within the cell by degrading the misfolded /refolded proteins that form within the cell, when stressed. It consists of regulatory and proteolytic subunits. The regulatory subunits are the AAA+ chaperones that show substrate specificity to the complex and assist in the substrate degradation. The proteolytic subunits are the two heptamers stacked upon each other, present either in homo-tetradecamer (inactive—ClpP1P1 or ClpP2P2) or hetero-tetradecamer (active—ClpP1P2) forms. In the presence of N-blocked dipeptides, the homo-tetradecameric form (inactive) of ClpP complex dissociates and re-associates to form hetero-tetradecameric form (active), which subsequently performs proteolysis. The inactivation or hyperactivation of the complex has been reported to stop proteolysis and kill the bacterial strains. Also, ClpP1 is the subunit that is responsible for the proteolytic activity and ClpP2 is responsible for chaperone binding. Therefore, we focused majorly on the ClpP1 subunit of both microorganisms.

We have performed sequence analysis and comparative structural analysis of the two mycobacterial strains, that is *Mycobacterium tuberculosis*, a pathogenic strain, and *Mycobacterium smegmatis*, a non-pathogenic strain. The aim was to understand the role of ClpP complex in the virulence and survival of the mycobacterial species. Upon the sequence analysis, we have found a few residues that were present in Mtb ClpP complex but not in Msm ClpP complex. The amino acids such as SER, ALA, ARG, HIS, ALA, and GLN were present in Mtb ClpP1 but not in Msm ClpP1. Similarly, amino acids such as HIS, PRO, LEU, PRO, GLN were present in Mtb ClpP2 but not in Msm ClpP2. The structure and property of different amino acids can contribute to different interactions. Meanwhile, these interactions can lead to virulence in one species and not in other species. Histidine and Arginine are positively charged amino acids, whereas Serine, Asparagine, Glutamine, and Proline are polar and uncharged amino acids. Leucine and Alanine are non-polar amino acids. It may be suggested here that these amino acids might contribute to the pathogenicity of Mtb due to their different property. Then, we did the interaction-based study, where we analyzed the interaction of the catalytic triads with amino acids within 4 Å in both Mtb and Msm ClpP1. We analyzed that catalytic triads interacted with the same amino acids in both the mycobacterial species. N-blocked dipeptides, e.g., ZIL are essential for the activation of the complex [28]; therefore, we analyzed the interaction of catalytic triads with the residues present within 4 Å of ClpP bound with ZIL. We surprisingly found that in the unbound (inactive) form, catalytic triad was seen interacting with MET-99, GLU-101, PHE-102, LEU-121,

ASP-168, and SER-169 but, by contrast, in the ZIL-bound (active) ClpP, the catalytic triad was seen interacting only with GLU-101 and LEU-121. Therefore, these two amino acids might play an important role in the complex activation.

Furthermore, comparative analysis of the crystal structures of ClpP complexes was performed to search for any structural difference in the proteolytic complex. When the inactive ClpP1 complexes from Mtb and Msm were superimposed and compared, we found that there was a conformational difference in three amino acids in the β-sheet, namely, TRP-174, PHE-175, and THR-176 of Mtb ClpP1, and TRP-171, PHE-172, and THR-173 for Msm ClpP1. Then, we performed the comparative structure analysis of the active and inactive ClpP1 subunits. There was a major shift in the handle domain observed, which might contribute to the arrangement of the disordered catalytic triad. The shifting of handle domain, post the binding of activator, is apparently vital for the function of ClpP complex as it widens the pore to degrade the protein substrate.

5 Conclusions

Tuberculosis (TB) is one of the deadly bacteria causing global threat to humans. Unfortunately, because of prolonged duration of the medicine, poverty, catastrophic cost of the drugs, and other factors, the resistant forms of Mtb have emerged to be a global threat. The serine protease ClpP helps bacteria to survive under unfavorable conditions by regulating protein homeostasis within the cell. The complex contributes to both the survival and virulence of most of the organisms including Mtb. The comparison of the sequences and structures of the ClpP from Mtb and Msm provided the understanding of the contribution of ClpP protease in the virulence of Mtb. N-blocked dipeptides contribute to complex's activation. Upon its binding, the complex opens up its pores to perform its proteolytic function. Our investigations have revealed that there were few residues in Mtb ClpP that were different when compared to Msm ClpP. These residues possess different properties and their interaction with the catalytic triads and the activator might play an important role in the pathogenicity of Mtb. We have observed that there is a major shift of handle domain during the transit from inactive to active form. In addition, this shift of handle domain might play a very important role in the proteolysis as this shift causes the widening of the pore to carry out the protein degradation process. The insights gained in the present study may be useful in further understanding and structure-based drug design of novel anti-TB agents.

Acknowledgments The authors would like to thank the Science and Engineering Research Board (SERB), New Delhi, for providing funding. S.B. is also thankful to UPES, Dehradun, for providing doctoral fellowship. The authors would like to acknowledge UPES, Dehradun, for providing other necessary infrastructural support.

References

1. Litvinjenko, S., Magwood, O., Wu, S., Wei, X.: Burden of tuberculosis among vulnerable populations worldwide: an overview of systematic reviews. Lancet Infect. Dis. **23**, 1395 (2023)
2. Public, T.L., H.: Taking tuberculosis out of the shadows. Lancet Public Health. **8**, e247 (2023)
3. Trajman, A.: The social drivers of tuberculosis, reconfirmed. Lancet Infect. Dis. **24**, 5 (2023)
4. Olbrich, L., Verghese, V.P., Franckling-Smith, Z., Sabi, I., Ntinginya, N.E., Mfinanga, A., Banze, D., Viegas, S., Khosa, C., Semphere, R., Nliwasa, M., McHugh, T.D., Larsson, L., Razid, A., Song, R., Corbett, E.L., Nabeta, P., Trollip, A., Graham, S.M., Hoelscher, M., Geldmacher, C., Zar, H.J., Michael, J.S., Heinrich, N.: Diagnostic accuracy of a three-gene *Mycobacterium tuberculosis* host response cartridge using fingerstick blood for childhood tuberculosis: a multicentre prospective study in low-income and middle-income countries. Lancet Infect. Dis. **24**, 140 (2023)
5. Reid, M., Agbassi, Y.J.P., Arinaminpathy, N., Bercasio, A., Bhargava, A., Bhargava, M., Bloom, A., Cattamanchi, A., Chaisson, R., Chin, D., Churchyard, G., Cox, H., Denkinger, C.M., Ditiu, L., Dowdy, D., Dybul, M., Fauci, A., Fedaku, E., Gidado, M., Harrington, M., Hauser, J., Heitkamp, P., Herbert, N., Herna Sari, A., Hopewell, P., Kendall, E., Khan, A., Kim, A., Koek, I., Kondratyuk, S., Krishnan, N., Ku, C.C., Lessem, E., McConnell, E.V., Nahid, P., Oliver, M., Pai, M., Raviglione, M., Ryckman, T., Schäferhoff, M., Silva, S., Small, P., Stallworthy, G., Temesgen, Z., van Weezenbeek, K., Vassall, A., Velásquez, G.E., Venkatesan, N., Yamey, G., Zimmerman, A., Jamison, D., Swaminathan, S., Goosby, E.: Scientific advances and the end of tuberculosis: a report from the lancet commission on tuberculosis. Lancet. **402**, 1473 (2023)
6. Respiratory, T.L., M.: Time to invest in tuberculosis research and disease control. Lancet Respir. Med. **11**, 297 (2023)
7. The, L.: Tuberculosis: a different way of doing things. Lancet. **402**, 937 (2023)
8. Naidoo, K., Perumal, R.: Advances in tuberculosis control during the past decade. Lancet Respir. Med. **11**, 311 (2023)
9. Glickman, M.S., Jacobs Jr., W.R.: Microbial pathogenesis of *Mycobacterium tuberculosis*: Dawn of a discipline. Cell. **104**, 477 (2001)
10. Queval, C.J., Brosch, R., Simeone, R.: The macrophage: a disputed fortress in the battle against mycobacterium tuberculosis. Front. Microbiol. **8**, 2284 (2017)
11. Mahamed, D., Boulle, M., Ganga, Y., Mc Arthur, C., Skroch, S., Oom, L., Catinas, O., Pillay, K., Naicker, M., Rampersad, S., Mathonsi, C., Hunter, J., Wong, E.B., Suleman, M., Sreejit, G., Pym, A.S., Lustig, G., Sigal, A.: intracellular growth of *Mycobacterium tuberculosis* after macrophage cell death leads to serial killing of host cells. elife. **6**, e22028 (2017)
12. Palomino, J.C., Martin, A.: Drug resistance mechanisms in mycobacterium tuberculosis. Antibiotics (Basel). **3**, 317 (2014)
13. Marrakchi, H., Lanéelle, G., Quémard, A.K.: InhA, a target of the antituberculous drug isoniazid, is involved in a mycobacterial fatty acid elongation system, FAS-II. Microbiology (Reading). **146**(Pt 2), 289 (2000)
14. Saxena, A.K., Roy, K.K., Singh, S., Vishnoi, S.P., Kumar, A., Kashyap, V.K., Kremer, L., Srivastava, R., Srivastava, B.S.: Identification and characterisation of small-molecule inhibitors of Rv3097c-encoded lipase (LipY) of *Mycobacterium tuberculosis* that selectively inhibit growth of bacilli in hypoxia. Int. J. Antimicrob. Agents. **42**, 27 (2013)
15. Bolla, J.R.: Targeting MMPL3 for anti-tuberculosis drug development. Biochem. Soc. Trans. **48**, 1463 (2020)

16. Roy, K.K., Wani, M.A.: Emerging opportunities of exploiting mycobacterial electron transport chain pathway for drug-resistant tuberculosis drug discovery. Expert Opin. Drug Discov. **15**, 231 (2020)
17. Khan, S.R., Singh, S., Roy, K.K., Akhtar, M.S., Saxena, A.K., Krishnan, M.Y.: Biological evaluation of novel substituted chloroquinolines targeting mycobacterial ATP synthase. Int. J. Antimicrob. Agents. **41**, 41 (2013)
18. Amado, P.S.M., Woodley, C., Cristiano, M.L.S., O'Neill, P.M.: Recent advances of dpre1 inhibitors against *Mycobacterium tuberculosis*: computational analysis of physicochemical and admet properties. ACS Omega. **7**, 40659 (2022)
19. Ray, B., Roy, K.K.: Deciphering insights into the binding mechanism and plasticity of telacebec with *M. Tuberculosis* cytochrome bcc-aa3 supercomplex through an unbiased molecular dynamics simulation, free-energy analysis, and dft study. J. Biomol. Struct. Dyn. **1** (2023)
20. Dey, R., Nandi, S., Samadder, A., Saxena, A., Saxena, A.: K.: exploring the potential inhibition of candidate drug molecules for clinical investigation based on their docking or crystallographic analyses against *M. tuberculosis* enzyme targets. Curr. Top. Med. Chem. **20**, 2662 (2020)
21. Moreno-Cinos, C., Goossens, K., Salado, I.G., Van Der Veken, P., De Winter, H., Augustyns, K.: ClpP protease, a promising antimicrobial target. Int. J. Mol. Sci. **20**, 2232 (2019)
22. Schmitz, K.R., Carney, D.W., Sello, J.K., Sauer, R.T.: Crystal structure of *Mycobacterium tuberculosis* ClpP1P2 suggests a model for peptidase activation by AAA+ partner binding and substrate delivery. Proc. Natl. Acad. Sci. USA. **111**, E4587 (2014)
23. Personne, Y., Brown, A.C., Schuessler, D.L., Parish, T.: *Mycobacterium tuberculosis* ClpP proteases are co-transcribed but exhibit different substrate specificities. PLoS One. **8**, e60228 (2013)
24. Ollinger, J., O'Malley, T., Kesicki, E.A., Odingo, J., Parish, T.: Validation of the essential ClpP protease in *Mycobacterium tuberculosis* as a novel drug target. J. Bacteriol. **194**, 663 (2012)
25. Zheng, D., Xu, Y., Yuan, G., Wu, X., Li, Q.: Bacterial ClpP protease is a potential target for methyl gallate. Front. Microbiol. **11**, 598692 (2020)
26. d'Andrea, F.B., Poulton, N.C., Froom, R., Tam, K., Campbell, E.A., Rock, J.M.: The essential *M. tuberculosis* Clp protease is functionally asymmetric in vivo. Sci. Adv. **8**, eabn7943 (2022)
27. Raju, R.M., Unnikrishnan, M., Rubin, D.H., Krishnamoorthy, V., Kandror, O., Akopian, T.N., Goldberg, A.L., Rubin, E.J.: *Mycobacterium tuberculosis* ClpP1 and ClpP2 function together in protein degradation and are required for viability in vitro and during infection. PLoS Pathog. **8**, e1002511 (2012)
28. Bhardwaj, S., Roy, K.K.: ClpP peptidase as a plausible target for the discovery of novel antibiotics. Curr. Drug Targets. **25**, 108 (2023)
29. Nagpal, J., Paxman, J.J., Zammit, J.E., Thomas, A.A., Truscott, K.N., Heras, B., Dougan, D.A.: Molecular and structural insights into an asymmetric proteolytic complex (ClpP1P2) from *Mycobacterium smegmatis*. Sci. Rep. **9**, 18019 (2019)
30. Berman, H.M., Westbrook, J., Feng, Z., Gilliland, G., Bhat, T.N., Weissig, H., Shindyalov, I.N., Bourne, P.E.: The protein data bank. Nucleic Acids Res. **28**, 235 (2000)
31. Coudert, E., Gehant, S., de Castro, E., Pozzato, M., Baratin, D., Neto, T., Sigrist, C.J.A., Redaschi, N., Bridge, A.: Annotation of biologically relevant ligands in UniProtKB using ChEBI. Bioinformatics. **39**, btac793 (2023)
32. Sievers, F., Wilm, A., Dineen, D., Gibson, T.J., Karplus, K., Li, W., Lopez, R., McWilliam, H., Remmert, M., Söding, J., Thompson, J.D., Higgins, D.G.: Fast, scalable generation of high-quality protein multiple sequence alignments using clustal omega. Mol. Syst. Biol. **7**, 539 (2011)
33. Schwede, T., Kopp, J., Guex, N., Peitsch, M.C.: Swiss-model: an automated protein homology-modeling server. Nucleic Acids Res. **31**, 3381 (2003)

34. The pymol molecular graphics system (Version v2.5.4). NY: Schrödinger LLC
35. Biovia. San Diego: Dassault Systèmes. (2021)
36. Illigmann, A., Thoma, Y., Pan, S., Reinhardt, L., Brötz-Oesterhelt, H.: Contribution of the clp protease to bacterial survival and mitochondrial homoeostasis. Microb. Physiol. **31**, 260 (2021)
37. Vahidi, S., Ripstein, Z.A., Juravsky, J.B., Rennella, E., Goldberg, A.L., Mittermaier, A.K., Rubinstein, J.L., Kay, L.E.: An allosteric switch regulates *Mycobacterium tuberculosis* ClpP1P2 protease function as established by cryo-EM and methyl-TROSY NMR. Proc. Natl. Acad. Sci. USA. **117**, 5895 (2020)
38. Hecht, N., Becher, M., Korman, M., Vishkautzan, M., Gur, E.: Inter- and intramolecular regulation of protein depupylation in *Mycobacterium smegmatis*. FEBS J. **287**, 4389 (2020)
39. Ingvarsson, H., Maté, M.J., Högbom, M., Portnoï, D., Benaroudj, N., Alzari, P.M., Ortiz-Lombardía, M., Unge, T.: Insights into the inter-ring plasticity of caseinolytic proteases from the x-ray structure of *Mycobacterium tuberculosis* clpp1. Acta Crystallogr. D Biol. Crystallogr. **63**, 249 (2007)
40. Luscombe, N.M., Laskowski, R.A., Thornton, J.M.: Amino acid–base interactions: a three-dimensional analysis of protein–DNA interactions at an atomic level. Nucleic Acids Res. **29**, 2860 (2001)

Novel Synthesis of 1-arylnaphtho[1,2-*d*]isoxazole Through Oxidation of 1-amidoalkyl-2-Naphthols Using MnO_2

Vanshika Gupta[1], Ram Kishore[1] (✉), and Nitin Srivastava[1] (✉)

[1] Department of Chemistry, Amity University Uttar Pradesh Lucknow Campus, Lucknow, India

Abstract Three-component condensation of β-naphthol, urea, and aldehyde was carried out for the synthesis of 1-amidoalkyl-2-naphthols under acidic conditions. These synthesized 1-amidoalkyl-2-naphthols on reaction with MnO_2 unusually led to the synthesis of 1-arylnaphtho[1,2-*d*] isoxazoles. The reaction signifies efficiency of MnO_2 in forming N-O bond in organic synthesis.

Keywords MnO_2 · Phenolic oxidation · 1,2-benzisoxazole · 1-amidoalkyl-2-naphthol · 2-hydroxyaryl ketoximes

1 Introduction

The indole and 1,2-benzisoxazole show isosterism which proves their similarity in binding to the biologically significant enzymes [1], making them useful as carrier for pharmacophoric moieties in the drug development. The various pharmaceuticals and therapeutics containing 1,2-benzisoxazoles as a sub-structural unit are, thus, known to exhibit plethora of biological activities like antimicrobial [2], antithrombotic [3], anticancer [4], anti-HIV [5], antitumor [6], etc. Methylene-bridged benzisoxazolyl imidazo[2,1-*b*][1,3,4]thiadiazoles were synthesized from benzisoxazolyl-3-acetic acid and thiosemicarbazide. These compounds displayed very good antibacterial and antifungal activity [2] as shown in Fig. 1.

Alkenyldiarylmethanes (ADAMs) with a benzo[*d*]isoxazole ring are an effective bioisosteric replacement of the metabolically labile methyl ester moiety and the compounds exhibited anti-HIV-1 activity with EC50 values in the 20–40 nanomolar range [5]. Heat shock protein 90 (Hsp90) is a molecular chaperone that is responsible for activating many signaling proteins and is a promising target in tumor biology. Benzisoxazole derivatives have been reported as a potent inhibitor of Chaperone heat shock protein 90 (HSP 90) [6].

A. K. Saxena, S. Nandi (eds.), *Global Trends in Health, Technology and Management*, https://doi.org/10.1007/978-3-031-75457-9_22

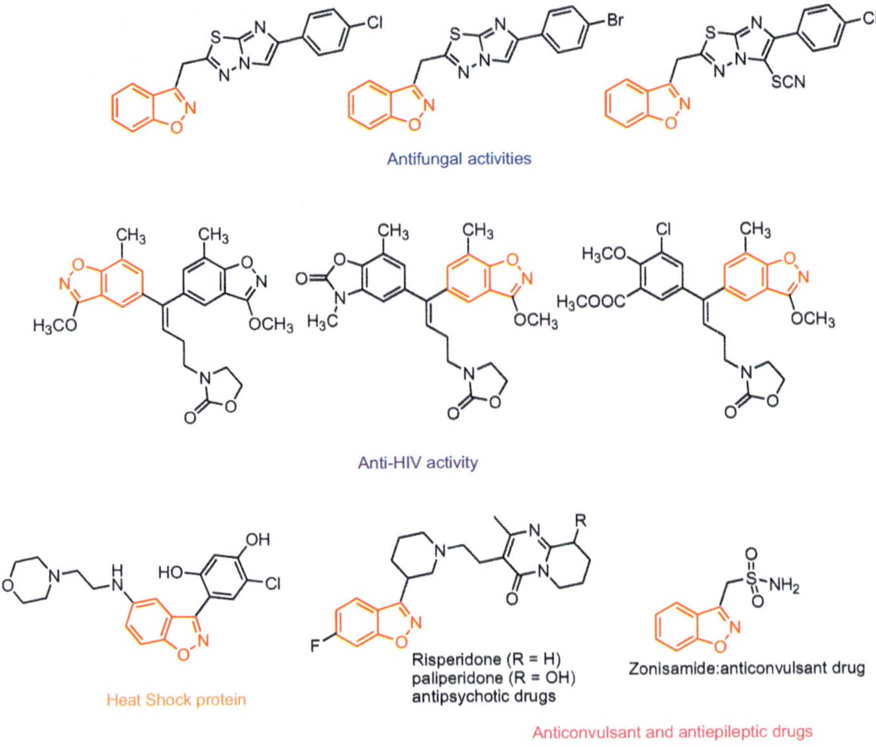

Fig. 1 Structurally diverse biologically potent isoxazolyl derivatives

Many biologically active compounds, zonisamide, are widely prescribed as an antiepileptic drug. Zonisamide is also available in the market as an anticonvulsant drug [7, 8]. Risperidone is a potent antipsychotic drug which is mainly used to treat schizophrenia (including adolescent schizophrenia), schizoaffective disorder, the mixed and manic states associated with bipolar disorder, and irritability in people with autism [9, 10]. Paliperidone, also known as 9-hydroxyrisperidone, is a dopamine antagonist of the atypical antipsychotic class of medications [11, 12].

The diverse applications of the isoxazoles make it necessary to explore novel, easier, and more efficient methods for synthesizing of these compounds. The synthesis of 1,2-benzisoxazole derivatives is relatively a challenging task [4, 13–25], which may be made easier exploring the experimental synthesis.

Analogous to the structure of 2-hydroxy-1-naphthaldehyde oxime, the 1-amidoalkyl-2-naphthol **4** is a typical phenolic compound with a primary amide type moiety at benzylic position. We were, therefore, interested to find out the fate of this multi-functional phenolic molecule with the reactions of hypervalent iodine reagents to extend our previous work for the synthesis of biologically potent scaffolds using cleaner greener and easier methods [26–35]. We envisaged that the reaction of 1-amidoalkyl-2-naphthols **4** with MnO_2 may be employed for phenolic oxidation followed by the trapping of nucleophilic oxygen in the synthesis of spiro compound. We found that

1-amidoalkyl-2-naphthol **4** upon reaction with MnO_2 yielded naphtho[1,2-*d*]isoxazole **5**. The easy and simple synthesis of 1-amidoalkyl-2-naphthol in the reported method gives way to an easy and efficient synthetic protocol for naphtho[1,2-*d*]isoxazole **5** employing oxidant MnO_2 (Scheme 1).

2 Materials and Methods

2.1 General Experimental Remarks

Unless otherwise stated, chemicals and reagents were used as received. The progress of reaction was monitored using TLC in UV chamber. The pure product was obtained using column chromatography where column was filled with silica gel and eluted with mixture of ethyl acetate and *n*-hexane (1,3). ^{1}H and ^{13}C NMR spectra were recorded on a Bruker Advance NMR Spectrometer using DMSO-d6/$CDCl_3$ as solvent and TMS as standard. Infrared (IR) spectra ranging from 4000 to 200 cm^{-1} were obtained using a Bomem MB-104-FTIR spectrophotometer through the neat technique. LCMS of the synthesized compounds were done on Agilent LCMS instrument, and the obtained values agreed with the calculated ones.

2.2 Synthesis of 1-Amidoalkyl-2-napthol (4a-l)

A mixture of appropriate aromatic aldehyde (5 mmol), *β*-naphthol (5 mmol), urea (6 mmol), and trifluoroacetic acid (0.1 mmol) in 1,2-dichloroethane (5 mL) at room temperature was stirred for 8 h. The progress of the reaction was monitored by TLC. After completion, the reaction mixture was filtered, and the precipitate washed with H_2O. The products formed in crude form were purified by recrystallization from EtOH: H_2O (1:3) and the pure products **4a-l** were obtained in 64–89% yields (Table 1).

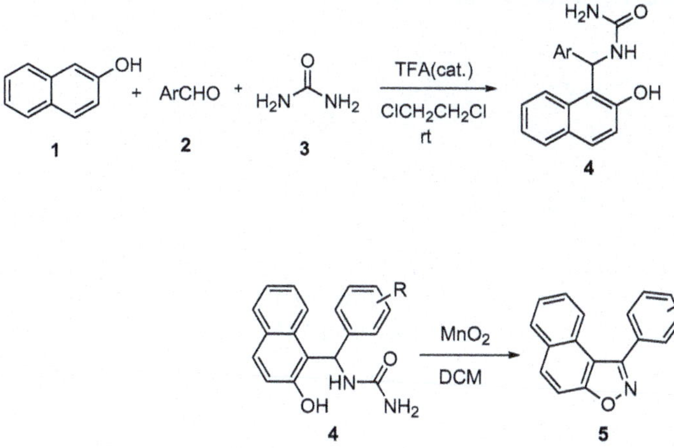

Scheme 1 Synthesis of naphtho[1,2-*d*]isoxazole using MnO_2

Table 1 Three-component condensation of aldehyde, urea, and β-naphthol for synthesis of 1-amidoalkyl-2-napthol

Entry	Ar	Mp (°C)	Isolated yield (%) on recrystallization
4a	C_6H_5	170–172	88
4b	$2\text{-}CH_3C_6H_4$	114–118	71
4c	$3\text{-}CH_3{}^\circ C_6H_4$	240–242	73
4d	$4\text{-}CH_3C_6H_4$	112–114	81
4e	$3\text{-}FC_6H_4$	130–132	78
4f	$2\text{-}ClC_6H_4$	150–152	82
4g	$4\text{-}ClC_6H_4$	164–166	89
4h	$2\text{-}BrC_6H_4$	134–136	67
4i	$3\text{-}BrC_6H_4$	162–164	76
4j	$4\text{-}BrC_6H_4$	168–170	87
4k	$2,3\text{-}Cl_2C_6H_3$	128–130	64
4l	$2,4\text{-}Cl_2C_6H_3$	136–138	67

2.3 General Experimental Procedure for the Formation of 1-arylnaphtho[1,2-*d*]Isoxazole (5a-l)

A clean, dry RB flask was charged with 1-amidoalkyl-2-naphthols **4a-l** (2 mmol) in dichloromethane (15 mL), and 3 mmol MnO_2 and continued stirring for 2 h. at room temperature. The progress of reaction was monitored by TLC. After completion of the reaction, solvent was evaporated under reduced pressure and the product was purified by column chromatography using ethyl acetate (1/100) as eluent to afford white solid product **5a-l**.

2.4 Characterization of the Products

1-((2-Hydroxynaphthalen-1-yl)(phenyl)methyl)urea (4a):

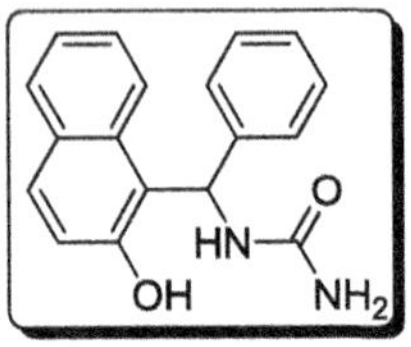

Nature: White solid, Yield: 88% (892 mg); Mp: 170–172 °C; IR (cm^{-1}): 3466, 3346, 3218, 1651, 1630, 1524, 1435, 1353, 1267, 1125, 1034, 1010, 934, 877, 848, 814, 741.

1-((2-Hydroxynaphthalen-1-yl)(*o*-tolyl)methyl)urea (4b):

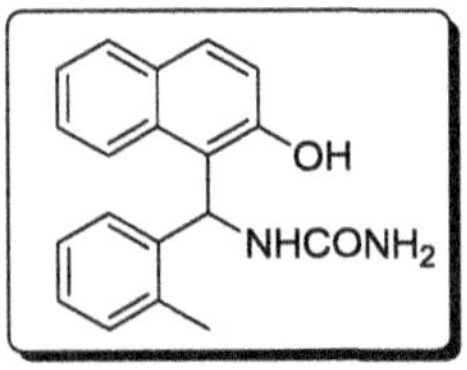

Nature: White solid, Yield: 71% (754 mg); Mp: 114–118 °C; IR (cm^{-1}): 3460, 3398, 3332, 1626, 1524, 1435, 1353, 1267, 1125, 1034, 1010, 934, 877, 848, 814, 741.

1-((2-Hydroxynaphthalen-1-yl)(3-methoxyphenyl)methyl)urea (4c):

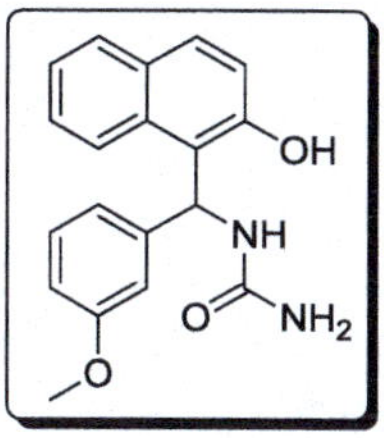

Nature: White solid, Yield: 73% (816 mg); Mp: 240–242 °C; IR (cm^{-1}): 3460, 3398, 3332, 1658, 1626, 1530, 1485, 1436, 1355, 1265, 1157, 1051, 818, 783, 744, 695, 627.

1-((2-Hydroxynaphthalen-1-yl)(*p*-tolyl)methyl)urea (4d):

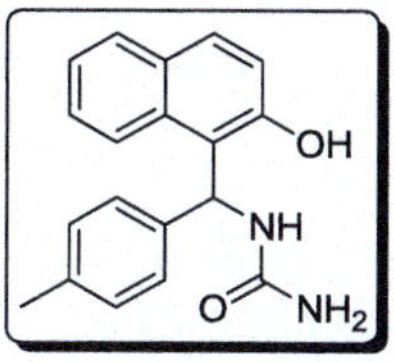

Nature: White solid, Yield: 81% (860 mg); Mp: 112–114 °C. IR (cm^{-1}): 3465, 3351, 3213, 1651, 1524, 1435, 1353, 1267, 1124, 1034, 1010, 934, 877, 848, 814, 771, 741, 615.

1-((3-Fluorophenyl)(2-hydroxynaphthalen-1-yl)methyl)urea (4e):

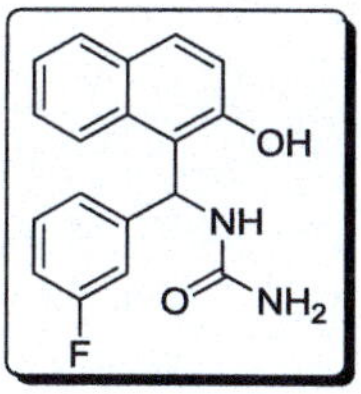

Nature: White solid, Yield: 78% (839 mg); Mp: 130–132 °C; IR (cm^{-1}): 3481, 3345, 3216, 1664, 1625, 1530, 1482, 1437, 1356, 1269, 1163, 1085, 1035, 1017, 959, 858, 822, 789, 769, 688, 617.

1-((2-Chlorophenyl)(2-hydroxynaphthalen-1-yl)methyl)urea (4f):

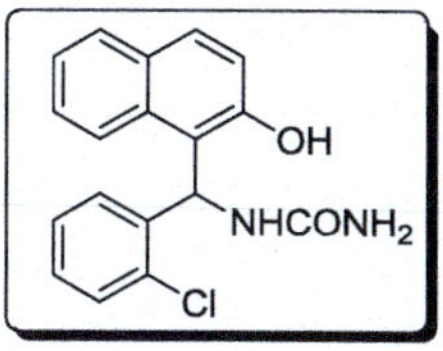

Nature: White solid, Yield: 82% (928 mg); Mp: 150–152 °C; IR (cm^{-1}): 3470, 3350, 3186, 1650, 1626, 1505, 1435, 1354, 1269, 1124, 1034, 1009, 935, 814, 751.

1-((4-Chlorophenyl)(2-hydroxynaphthalen-1-yl)methyl)urea (4g):

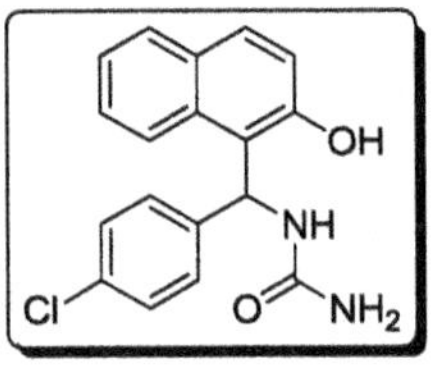

Nature: White solid, Yield: 89% (1.007 g); Mp: 164–166 °C; IR (cm^{-1}): 3495, 3392, 1641, 1626, 1516, 1438, 1360, 1273, 1120, 1069, 1013, 878, 817, 748, 651.

1-((2-Bromophenyl)(2-hydroxynaphthalen-1-yl)methyl)urea (4h):

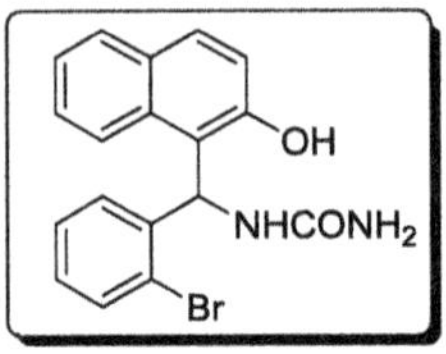

Nature: White solid, Yield: 67% (860 mg); Mp: 134–136 °C; IR (cm^{-1}): 3336, 3212, 1651, 1615, 1510, 1435, 1353, 1271, 1121, 1024, 1009, 937, 817, 740, 687, 579.

1-((3-Bromophenyl)(2-hydroxynaphthalen-1-yl)methyl)urea (4i):

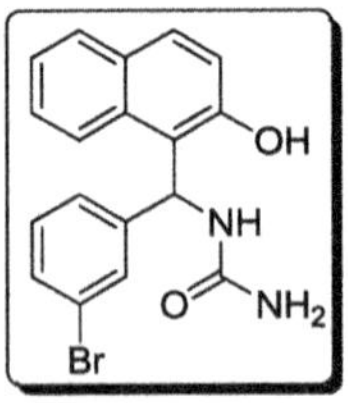

Nature: White solid, Yield: 76% (976 mg); Mp: 162–164 °C; IR (cm^{-1}): 3490, 3446, 3387, 1644, 1588, 1515, 1469, 1353, 1255, 1123, 1068, 1010, 948, 813, 747, 681, 652, 620.

1-((4-Bromophenyl)(2-hydroxynaphthalen-1-yl)methyl)urea (4j):

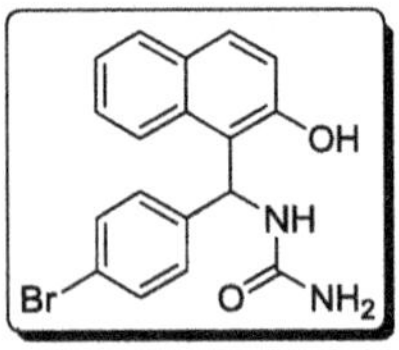

Nature: White solid, Yield: 87% (1.143 g); Mp: 168–170 °C; IR (cm^{-1}): 3496, 3392, 1644, 1626, 1589, 1526, 1438, 1360, 1272, 1122, 1009, 878, 852, 816, 749, 647, 631.

1-((2,3-Dichlorophenyl)(2-hydroxynaphthalen-1-yl)methyl)urea (4k):

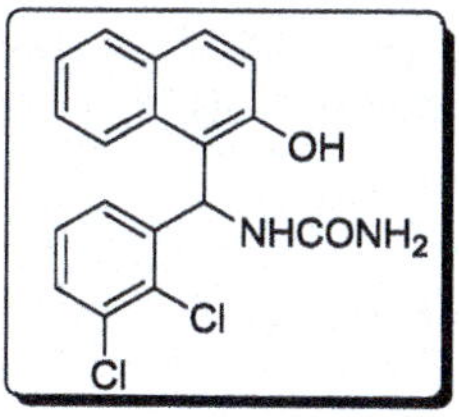

Nature: White solid, Yield: 64% (800 mg); Mp: 128–130 °C; IR (cm^{-1}): 3439, 3321, 1645, 1606, 1515, 1436, 1416, 1377, 1269, 1183, 1146, 1010, 939, 816, 783, 740, 685, 606.

1-((2,4-Dichlorophenyl)(2-hydroxynaphthalen-1-yl)methyl)urea (4l):

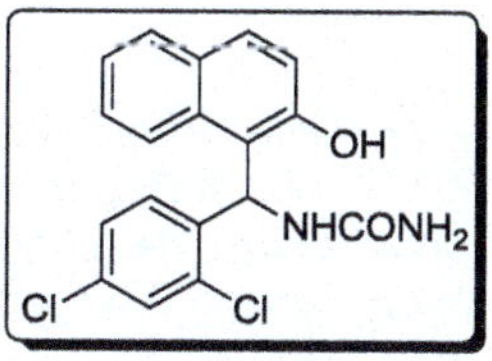

Nature: White solid, Yield: 67% (837 mg); Mp: 136–138 °C; IR (cm^{-1}): 3426, 3342, 3176, 1657, 1599, 1536, 1469, 1436, 1358, 1268, 1235, 1139, 1102, 1011, 942, 849, 794, 743, 589.

1-Phenylnaphtho[1,2-*d*]isoxazole (5a):

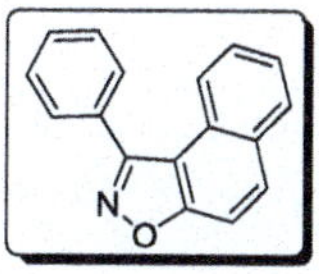

Nature: White solid, Yield: 72% (353 mg); Mp: 172–174 °C; ^{1}H NMR ($CDCl_3$): δ 7.42–7.48 (m, 4H), 7.59 (t, 1H, J = 7.8 Hz), 7.65 (d, 1H, J = 7.8 Hz), 7.71 (d, 1H, J = 7.8 Hz), 7.88 (d, 1H, J = 7.8 Hz), 8.23–8.28 (m, 2H), 8.52 (d, 1H, J = 8 Hz); ^{13}C NMR 100 MHz ($CDCl_3$): δ 109.85, 121.30, 124.40, 125.05, 126.58, 127.00, 128.37, 128.53, 127.91, 128.60, 128.94, 130.96, 131.25, 137.60, 141.34, 148.06, 162.36; IR (cm^{-1}): 3046, 2921, 2850, 1733, 1635, 1586, 1547, 1484, 1443, 1370, 1235, 1086, 1003, 927, 876, 865, 815, 758, 740, 704, 679; MS (ESI) m/z $(M + H)^+$ Calculated for $C_{17}H_{11}NO$: 245.08. Found: 245.95.

1-(*o*-Tolyl)naphtho[1,2-*d*]isoxazole (5b):

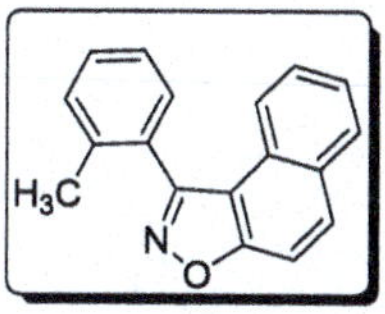

Nature: White solid, Yield: 67% (347 mg); Mp: 78–80 °C; ^{1}H NMR(400 MHz): δ 2.90 (s, 3H), 7.29–7.40 (m, 3H), 7.49 (t, 1H, *J* = 7.8 Hz), 7.70 (t, 1H, *J* = 7.8 Hz), 7.80 (d, 1H, *J* = 8.0 Hz), 7.80 (d, 1H, *J* = 8 Hz), 7.89 (d, 1H, *J* = 8 Hz), 8.19 (d, 1H, *J* = 8 Hz), 8.58 (d, 1H, *J* = 8 Hz,); ^{13}C NMR: δ 162.69, 147.58, 138.62, 137.57, 131.84, 131.18, 130.56, 129.70, 128.58, 126.94, 126.72, 126.55, 126.10, 125.90, 125.32, 122.34, 110.85, 22.35; IR (cm^{-1}): 561, 627, 702, 725, 752, 773, 802, 873, 922, 1008, 1027, 1084, 1203, 1231, 1271, 1371, 1471, 1544, 2849, 2919; MS (ESI) *m/z* $(M + H)^+$ Calculated for $C_{18}H_{13}NO$: 259.10. Found: 260.00.

1-(3-Methoxyphenyl)naphtho[1,2-*d*]isoxazole (5c):

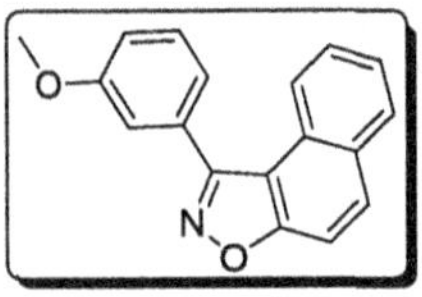

Nature: White solid, Yield: 70% (385 mg); Mp: 126–128 °C; ^{1}H NMR ($CDCl_3$): δ 3.90 (s, 3H), 6.98 (d, *J* = 7.8 Hz, 1H), 7.40 (t, 1H, *J* = 7.8 Hz), 7.48 (t, *J* = 8 Hz, 1H), 7.72 (2H, t, *J* = 7.8 Hz), 7.68 (d, 1H, *J* = 7.8 Hz), 7.81 (s, 1H), 7.90–7.99 (m, 2H), 7.60 (d, 1H, *J* = 8 Hz); ^{13}C NMR 100 MHz ($CDCl_3$): δ 162.20, 159.96, 148.02, 137.60, 131.24, 130.00, 128.70, 128.62, 126.99, 126.59, 126.07, 125.39, 122.32, 119.85, 117.71, 111.79, 110.83, 55.51; IR (cm^{-1}): 642, 724, 739, 761, 815, 877, 928, 1016, 1052, 1088, 1175, 1236, 1271, 1373, 1489, 1557, 2851, 2919; MS (ESI) *m/z* $(M + H)^+$ Calculated for $C_{18}H_{13}NO_2$: 275.09. Found: 276.00.

1-*p*-Tolylnaphtho[1,2-*d*]isoxazole (5d):

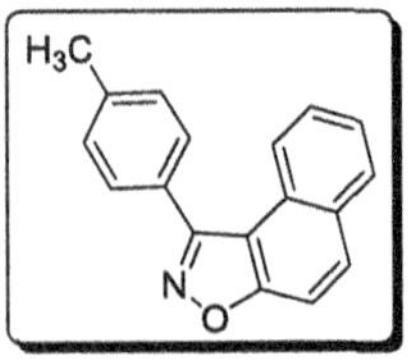

Nature: White solid, Yield: 57% (296 mg); Mp: 146–148 °C; ^{1}H NMR ($CDCl_3$): δ 2.39 (s, 3H), 7.29 (d, 2H, *J* = 7.8 Hz), 7.49 (t, 1H, *J* = 7.8 Hz), 7.58 (t, 1H, *J* = 7.8 Hz,), 7.68 (d, *J* = 7.6 Hz, 1H), 7.80 (d, *J* = 7.6 Hz, 1H), 8.00 (d, *J* = 7.6 Hz, 1H), 8.19 (d, 2H, *J* = 8 Hz), 8.60 (d, 1H, *J* = 8 Hz,); ^{13}C NMR 100 MHz ($CDCl_3$): δ 162.65, 147.92, 144.48, 141.57, 137.64, 131.22, 130.25, 129.67, 129.21, 128.58, 127.34, 126.91, 126.53, 125.76, 125.31, 124.78, 122.29, 110.82, 21.77; IR (cm^{-1}): 554, 702, 725, 752, 773, 867, 922, 1008, 1027, 1084, 1163, 1203, 1231, 1231, 1271, 1371, 1455, 1488, 1544, 2849, 2919; MS (ESI) *m/z* $(M + H)^+$ Calculated for $C_{18}H_{13}NO$: 259.10. Found: 260.00.

1-(3-Fluorophenyl)naphtho[1,2-*d*]isoxazole (5e):

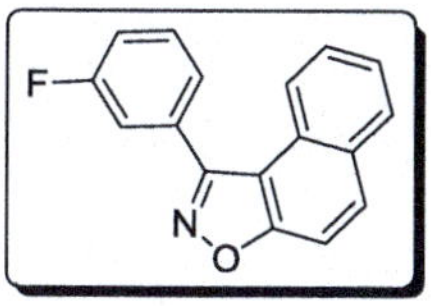

Nature: White solid, Yield: 68% (358 mg); Mp: 144–148 °C; ^{1}H NMR ($CDCl_3$): δ 7.38–7.60 (m, 2H), 7.70–7.80 (m, 2H), 7.82 (d, 1H, $J = 8$ Hz), 7.85–8.12 (m, 2H), 8.18 (d, 1H, $J = 8$ Hz), 8.60 (d, 1H, $J = 8$ Hz); ^{13}C NMR 100 MHz ($CDCl_3$): δ 161.76, 148.12, 137.52, 131.27, 130.57, 128.62, 127.14, 126.58, 126.48, 125.54, 123.01, 122.25, 118.13, 117.91, 114.37, 114.13, 110.78; IR (cm^{-1}): 571, 626, 662, 724, 745, 760, 791, 819, 859, 884, 947, 1008, 1061, 1087, 1202, 1248, 1300, 1372, 1446, 1473, 1553, 1592, 1615, 2922, 3050, 3069;MS (ESI) *m/z* (M + H)$^+$ Calculated for $C_{17}H_{10}FNO$: 263.07. Found: 263.95.

1-(2-Chlorophenyl)naphtho[1,2-*d*]isoxazole (5f):

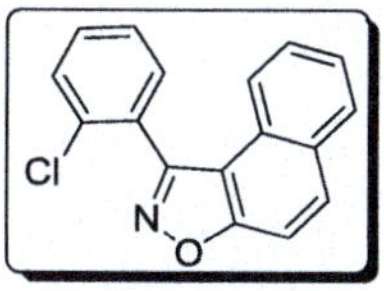

Nature: White solid, Yield: 75% (418 mg); Mp: 96–98 °C; ^{1}H NMR ($CDCl_3$): δ 7.38–7.50 (m, 2H), 7.54–7.64 (m, 2H), 7.70 (t, 2H, $J = 8$ Hz), 7.80 (d, 1H, $J = 8$ Hz), 7.90 (d, 1H, $J = 8$ Hz), 8.02 (d, 1H, $J = 8$ Hz), 8.30–8.38 (m, 1H), 8.61 (d, 1H, $J = 8$ Hz); ^{13}C NMR 100 MHz ($CDCl_3$): δ 160.31, 148.16, 137.18, 133.29, 131.84, 131.64, 131.36, 131.24, 128.61, 127.16, 126.98, 126.64, 126.58, 125.54, 122.33, 110.96; IR (cm^{-1}): 695, 727, 753, 800, 842, 878, 924, 1005, 1049, 1084, 1113, 1179, 1235, 1270, 1370, 1398, 1475, 1543, 1601, 1633, 2848, 2917; MS (ESI) *m/z* (M + H)$^+$ Calculated for $C_{17}H_{10}ClNO$: 279.05. Found: 279.85.

1-(4-Chlorophenyl)naphtho[1,2-*d*]isoxazole (5g):

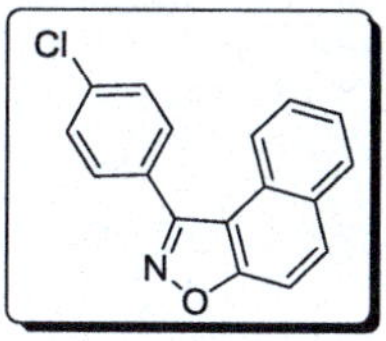

Nature: White solid, Yield: 59% (330 mg); Mp: 184–186 °C; ^{1}H NMR (DMSO-d6): δ 7.60 (t, 1H, $J = 8$ Hz), 7.70 (d, 2H, $J = 8$ Hz), 7.82 (t, 1H, $J = 8$ Hz), 7.92 (d, 1H, $J = 8$ Hz), 7.94 (d, 1H, $J = 8$ Hz), 8.10 (d, 1H, $J = 8$ Hz), 8.30 (d, 2H, $J = 8$ Hz), 8.50 (d, 1H, $J = 8$ Hz); ^{13}C NMR 100 MHz (DMSO-d6): δ 160.68, 147.66, 136.77, 136.40, 130.83, 129.21, 128.62, 128.45, 127.05, 126.34, 125.77, 125.50, 125.41, 121.53, 110.91; IR (cm^{-1}): 2920, 2851, 1705, 1596, 1471, 1453, 1369, 1269, 1234, 1203, 1083,

1048, 1003, 878, 832, 798, 742, 725, 693, 632. MS (ESI) *m/z* $(M + H)^+$ Calculated for $C_{17}H_{10}ClNO$: 279.05. Found: 279.95.

1-(2-Bromophenyl)naphtho[1,2-*d*]isoxazole (5h):

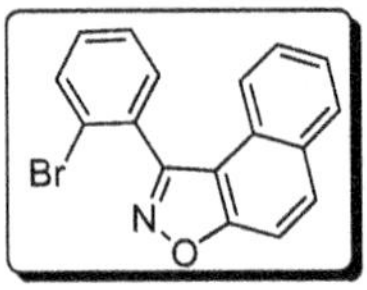

Nature: White solid, Yield: 63% (407 mg); Mp: 104–106 °C; ^{1}H NMR ($CDCl_3$): δ 7.40 (t, 1H, *J* = 8 Hz), 7.50 (t, 1H, *J* = 8 Hz), 7.60 (t, 1H, *J* = 8 Hz), 7.70 (t, 1H, *J* = 8 Hz), 7.80 (t, 2H, *J* = 8 Hz), 7.92 (d, 1H, *J* = 8 Hz), 8.01 (d, 1H, *J* = 8 Hz), 8.20 (d, 1H, J = 8 Hz), 8.72 (d, 1H, J = 8 Hz); ^{13}C NMR 100 MHz ($CDCl_3$): δ 160.82, 148.16, 137.14, 134.65, 132.17, 131.73, 131.25, 128.75, 128.60, 127.49, 126.69127, 126.57, 125.53, 122.34, 121.82, 110.96, 0.17; IR (cm^{-1}): 642, 687, 726, 755, 802, 931, 1003, 1022, 1076, 1095, 1231, 1315, 1373, 1455, 1538, 2852, 2921;

MS (ESI) m/z $(M + H)^+$ Calculated for $C_{17}H_{10}BrNO$: 323.85. Found: 323.85.

1-(3-Bromophenyl)naphtho[1,2-*d*]isoxazole (5i):

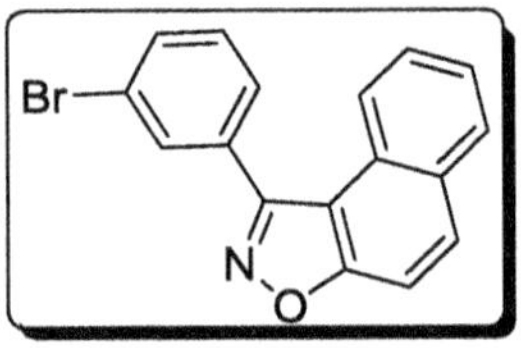

Nature: White solid, Yield: 73% (472 mg); Mp: 180–182 °C; ^{1}H NMR 400 MHz (DMSO-d6): δ 7.54–7.62 (m, 2H), 7.69–7.77 (m, 2H), 7.89 (d, 1H, *J* = 8 Hz), 7.95 (d, 1H, *J* = 8 Hz), 8.07 (d, 1H, *J* = 8 Hz), 8.26 (d, 1H, *J* = 8 Hz), 8.41 (s, 1H), 8.49 (d, 1H, *J* = 8 Hz); ^{13}C NMR 100 MHz (DMSO-d6): δ 160.13, 147.74, 136.70, 133.86, 131.05, 130.84, 129.29, 128.82, 128.61, 127.10, 126.57, 125.79, 125.62, 125.46, 122.37, 121.55, 110.91; IR (cm^{-1}): 651, 673, 719, 755, 768, 788, 816, 878, 933, 1005, 1049, 1096, 1230, 1369, 1470, 1544, 2850, 2920; MS (ESI) *m/z* $(M + H)^+$ Calculated for $C_{17}H_{10}BrNO$: 323.85. Found: 323.85.

1-(4-Bromophenyl)naphtho[1,2-*d*]isoxazole (5j):

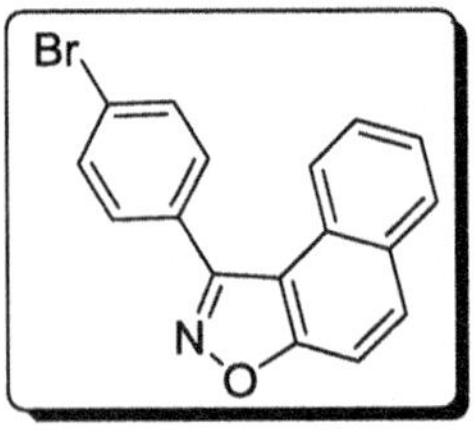

Nature: White solid, Yield: 61% (394 mg); Mp: 194–198 °C; ^{1}H NMR ($CDCl_3$): δ 7.60 (t, 1H, *J* = 8 Hz), 7.70–7.82 (m, 4H), 7.91 (d, 1H, *J* = 8 Hz), 7.98 (d, 1H, *J* = 8 Hz), 8.20 (d, 2H, *J* = 12 Hz), 8.56 (d, 1H, *J* = 8 Hz); ^{13}C NMR 100 MHz ($CDCl_3$): δ 161.42,

148.11, 137.56, 132.23, 131.29, 128.76, 128.63, 127.13, 126.45, 125.71, 125.53, 122.24, 110.77; IR (cm^{-1}): 2919, 1595, 1475, 1372, 1270, 1235, 1202, 1087, 1071, 1048, 1004, 877, 834, 806, 745, 723, 682.

1-(2,3-Dichlorophenyl)naphtho[1,2-*d*]isoxazole (5k):

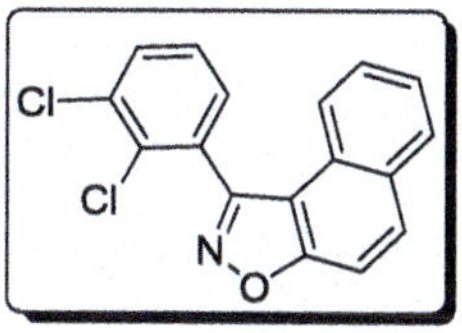

Nature: White solid, Yield: 67% (291 mg); Mp: 17–172 °C; ^{1}H NMR (DMSO-d6): δ 7.60 (t, 1H, J = 8 Hz), 7.70 (t, 1H, J = 8 Hz), 7.76 (t, 1H, J = 8 Hz), 7.90 (d, 1H, J = 8 Hz), 7.96 (d, 1H, J = 8.9 Hz), 8.02 (d, 1H, J = 8.9 Hz), 8.10 (d, 1H, J = 8 Hz), 8.20 (d, 1H, J = 8 Hz), 8.50 (d, 1H, J = 8 Hz); ^{13}C NMR 100 MHz (DMSO-d6): δ 158.83, 147.66, 136.30, 133.91, 132.58, 130.79, 130.45, 130.21, 128.30, 128.08, 128.05, 127.17, 126.90, 125.86, 125.50, 121.55, 110.91; IR (cm^{-1}): 3056, 1662, 1558, 1516, 1442, 1313, 1187, 1155, 1122, 1064, 1003, 1008, 941, 882, 801, 779, 742, 726, 701, 632, 563; MS (ESI) *m/z* $(M + H)^+$ Calculated for $C_{17}H_9Cl_2NO$: 313.01. Found: 313.95.

1-(2,4-Dichlorophenyl)naphtho[1,2-*d*]isoxazole (5l):

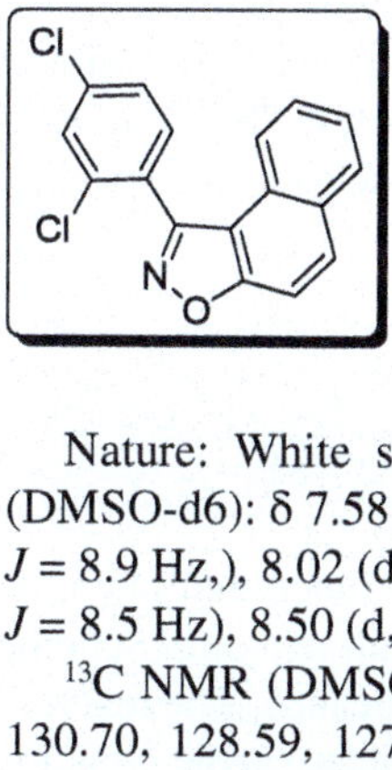

Nature: White solid, Yield: 74% (321 mg); Mp: 162–164 °C; ^{1}H NMR MHz (DMSO-d6): δ 7.58 (t, 2H, J = 8 Hz), 7.67 (d, 1H, J = 8 Hz), 7.78 (s, 1H), 7.90 (d, 1H, J = 8.9 Hz,), 8.02 (d, 1H, J = 8.8 Hz), 8.10 (d, 1H, J = 8 Hz), 8.15 (s, 1H), 8.30 (d, 1H, J = 8.5 Hz), 8.50 (d, 1H, J = 8 Hz);

^{13}C NMR (DMSO-d6): δ 158.62, 147.57, 136.44, 136.35, 133.01, 132.55, 130.79, 130.70, 128.59, 127.62, 127.13, 126.79, 125.82, 125.48, 124.46, 121.55, 110.86; IR (cm^{-1}): 2964, 1680, 1653, 1558, 1506, 1456, 1378, 1314, 1209, 1186, 1096, 1033, 1007, 944, 864, 826, 797, 762, 729, 693, 639; MS (ESI) *m/z* $(M + H)^+$ Calculated for $C_{17}H_9Cl_2NO$: 313.01. Found: 313.90.

3 Results and Discussion

In this protocol the condensation of β-naphthol, urea, and aldehyde led to the formation of starting materials 1-amidoalkyl-2-naphthols **4a–l**. The model reaction was carried between 1 mmol of 1-((2-hydroxynaphthalen-1-yl)(phenyl)methyl)urea **4a** and 1.2 mmol of MnO_2. Screening of different solvents such as CH_3Cl, CH_2Cl_2, $ClCH_2CH_2Cl$,

CH_3CN, and methanol was attempted for the reaction of **4a** with different equivalents of (MnO_2) as an oxidant. The reaction was optimized in dichloromethane as solvent and 1.5 equiv. of the oxidant with respect to the substrate as shown in Table 2. The ^1H NMR analysis of **5a** shows the presence of eleven aromatic protons around δ_H 7.17–8.53 ppm. The ^{13}C NMR of **5a** indicates the presence of oxygen-aryl carbon (δ_c 148.06 ppm) and imine carbon C=N (δ_c 162.36 ppm). The LCMS analysis of **5a** indicated a peak at $m/z = 245.95$ (M + 1) which is in confirmation with the structure. Oxidation of **4a** took place in CH_3OH a nucleophilic solvent using 1.2 equivalent of MnO_2. However, the formation of **5a** again took place without formation of any dearomatization product as well as the Hoffmann rearrangement product.

Now different 1-amidoalkyl-2-naphthols **4a-l** were employed for the synthesis of 1-arylnaphtho[1,2-*d*]isoxazole **5a-l** in appreciable yields as shown in Table 3.

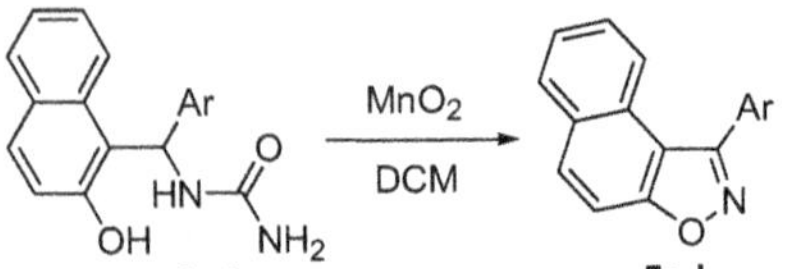

4a-l **5a-l**

Mechanism The formation of the product may have followed the following mechanism. The higher the nucleophilicity of phenolic oxygen in **4,** the higher is the association for formation of product. The additive compound **6** deprotonates to give **6b** and undergoes protonation and then gives out basic manganese oxide to yield **7b.** Compound **7b** deprotonates to give **8**. Compound **8** further releases formamide to give final product 5, which is an isoxazole.

4 Conclusions

To conclude we may say that 1-amidoalkyl-2-naphthol is a multi-functional phenolic molecule with the speculation of oxidation with hypervalent manganese oxide (Scheme 2). Therefore, its reaction study with hypervalent MnO_2 is of significance. The reaction of 1-amidoalkyl-2-naphthols with MnO_2 led to form naphtho-fused isoxazole which is

Table 2 Optimization of oxidation of **4a**

Entry	Solvent	Equiv. of (MnO_2)	Reaction time (h)	% Yield
1	DCM	1.0	4	70
2	**DCM**	**1.5**	**2**	**78**
3	DCM	1.5	4	68
4	1,2-dicholoroetane	1.2	4	59
5	1,2-dicholoroetane	1.5	4	69
6	Chloroform	1.2	6	60
7	Methane carbonitrile	1.2	6	48
8	Methyl alcohol	1.2	8	61

Table 3 Synthesis of 1-arylnaphtho[1,2-*d*]isoxazole from 1-amidoalkyl-2-naphthol using MnO_2 in DCM

Entry	Product	Time (h)	% Yield	Ref
5a	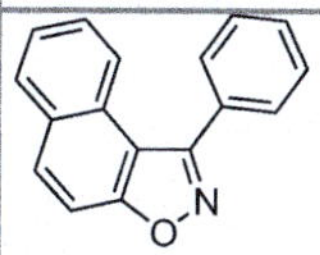	8	72	17
5b	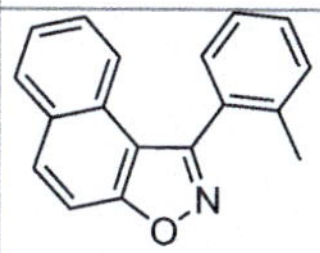	8	67	17
5c	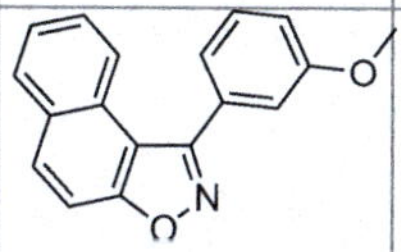	8	70	17
5d	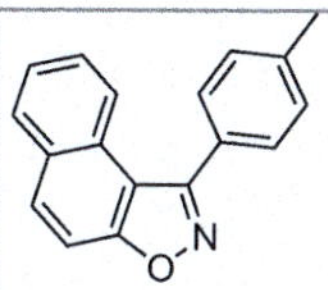	8	57	17
5e	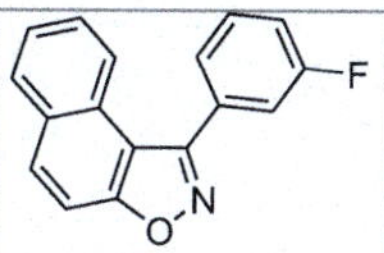	8	68	17
5f	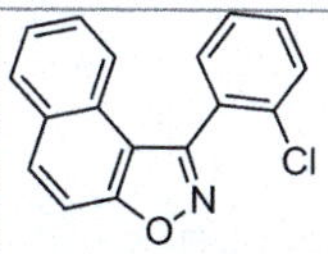	8	75	18
5g	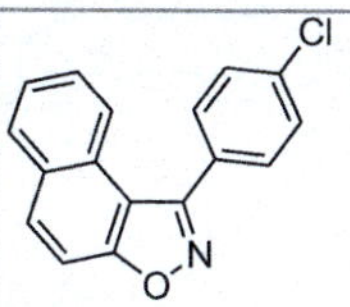	8	59	17
5h	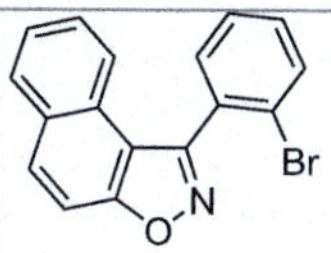	8	63	17
5i	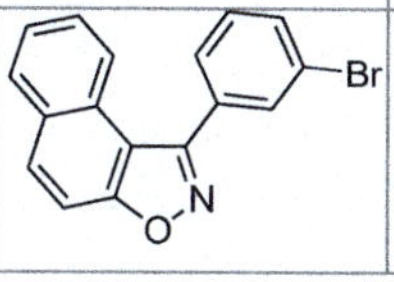	8	73	17

(continued)

Table 3 (continued)

Entry	Product	Time (h)	% Yield	Ref
5j	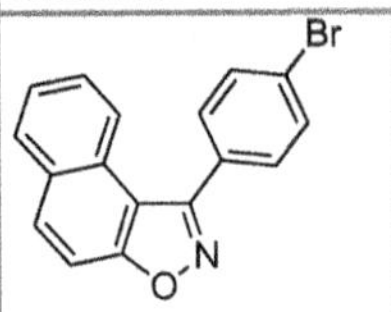	8	61	17
5k	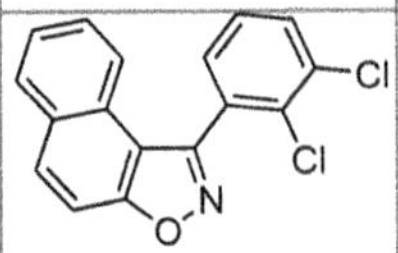	8	67	17
5l	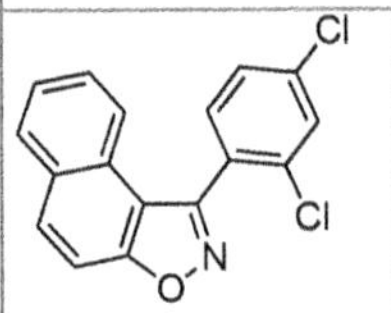	8	74	17

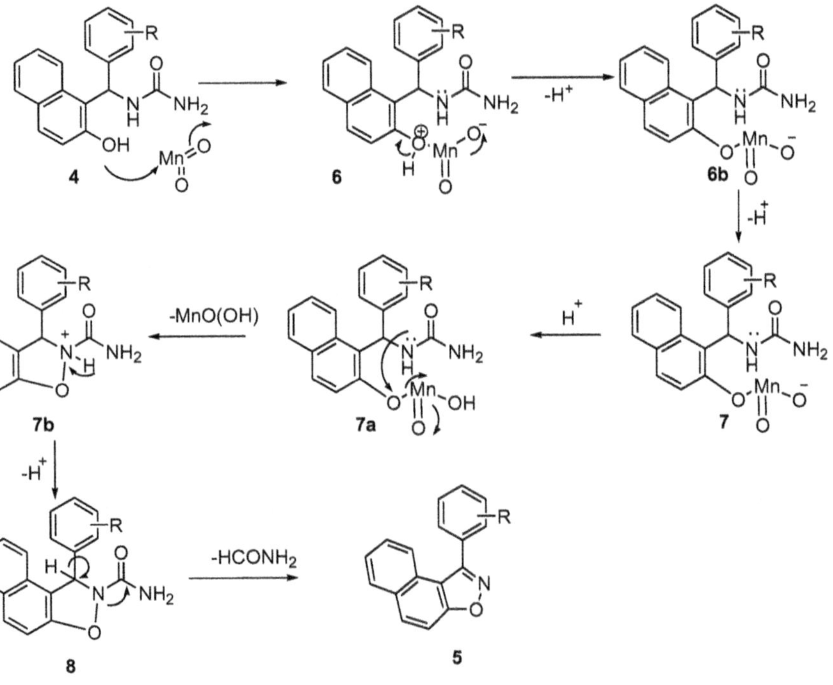

Scheme 2 Plausible mechanism for the formation of 1-arylnaphtho[1,2-*d*]isoxazole

known as a sub-structural unit of many biologically important scaffolds. It is the first example of the reaction of relatively "new phenol" 1-amidoalkyl-2-naphthols with MnO_2. The product formation took place through the unusual pathway involving crucial N-O bond formation induced using mild reaction conditions.

Conflict of Interest The authors declare no potential conflict of interest.

References

1. Arava, V.R., Siripalli, U.B.R., Nadkarni, V., Chinnapillai, R.: Novel base catalyzed rearrangement of sultone oximes to 1,2-benzisoxazole-3-methane sulfonate derivatives. Beilstein J. Org. Chem. **3**, 20 (2007). https://doi.org/10.1186/1860-5397-3-20
2. Lamani, R.S., Shetty, N.S., Kamble, R.R., Khazi, I.A.: Synthesis and antimicrobial studies of novel methylene bridged benzisoxazolyl imidazo[2,1-b][1,3,4]thiadiazole derivatives. Eur. J. Med. Chem. **44**(7), 2828–2833 (2009). https://doi.org/10.1016/j.ejmech.2008.12.019
3. Nuhrich, A., Varache-Lembege, M., Renard, P., Devaux, G.: Synthesis and inhibitory effects on platelet aggregation of 3-(2-thienyl)- and 3-(1-imidazolyl)-1,2-benzisoxazole derivatives. Eur. J. Med. Chem. **29**, 75 (1994). https://doi.org/10.1016/0223-5234(94)90129-5
4. Jain, M., Kwon, C.H.: 1,2-Benzisoxazole phosphorodiamidates as novel anticancer prodrugs requiring bioreductive activation. J. Med. Chem. **46**, 5428 (2003). https://doi.org/10.1021/jm020581y
5. Deng, B.L., Zhao, Y., Hartman, T.L., Watson, K., Buckheit Jr., R.W., Pannecouque, C., Clercq, E.D., Cushman, M.: Synthesis of alkenyldiarylmethanes (ADAMs) containing benzo[d]isoxazole and oxazolidin-2-one rings, a new series of potent non-nucleoside HIV-1 reverse transcriptase inhibitors. Eur. J. Med. Chem. **44**(3), 1210–1214 (2009). https://doi.org/10.1016/j.ejmech.2008.09.013
6. Gopalsamy, A., Shi, M., Goals, J., Vogan, E., Jacob, J., Johnson, M., Lee, F., Nilakantan, R., Petersen, R., Svenson, K., Chopra, R., Tam, M.S., Wen, Y., Ellingboe, J., Arndt, K., Boschelli, F.: Discovery of benzisoxazoles as potent inhibitors of chaperone heat shock protein 90. J. Med. Chem. **51**(3), 373–375 (2008). https://doi.org/10.1021/jm701385c
7. Andreas, S.B.: Zonisamide in treatment of epilepsy. Expert Opin. Pharmaco. **11**, 115 (2010). https://doi.org/10.1517/14656560903468728
8. Wroe, S.J.: Zonisamide. Treatment of Epilepsy, 3rd edn, p. 713. Wiley (2009)
9. Swainston, H.T., Goa, K.L.: Long-acting risperidone: a review of its use in schizophrenia. CNS Drugs. **18**, 113 (2004)
10. Deeks, E.D.: Risperidone long-acting injection. Drugs. **70**, 1001–1012 (2010). https://doi.org/10.2165/11204480-000000000-00000
11. Solanki, P.V., Uppelli, S.B., Pandit, B.S., Mathad, V.T.: An improved and efficient process for the production of highly pure Paliperidone, a psychotropic agent, via DBU catalyzed N-alkylation. ACS Sustain. Chem. Eng. **1**, 243 (2013). https://doi.org/10.1021/sc3000916
12. Yang, L.P.H., Plosker, G.L.: Paliperidone extended release. CNS Drugs. **21**, 417–425 (2007). https://doi.org/10.2165/00023210-200721050-00005
13. Kalkote, U.R., Goswami, D.D.: New synthesis of 1,2-benzisoxazole derivatives. Aust. J. Chem. **30**, 1847 (1977)
14. Udd, S., Jokela, R., Franzen, R., Tois, J.: Copper-catalyzed cyclization of Z-oximes into 3-methyl-1,2-benzisoxazoles. Tetrahedron Lett. **51**, 1030–1033 (2010). https://doi.org/10.1016/j.tetlet.2009.12.070
15. Spiteri, C., Mason, C., Zhang, F., Ritson, D.J., Sharma, P., Keeling, S., Moses, J.E.: An efficient entry to 1,2-benzisoxazoles via1,3-dipolar cycloaddition of in situ generated nitrile oxides and benzyne. Org. Biomol. Chem. **8**, 2537 (2010). https://doi.org/10.1039/B927235F

16. Iranpoor, N., Firouzabadi, H., Nowrouzi, N.: A novel method for the highly efficient synthesis of 1,2-benzisoxazoles under neutral conditions using the Ph_3P/DDQ system. Tetrahedron Lett. **47**, 8247 (2006). https://doi.org/10.1016/j.tetlet.2006.09.120
17. Inamoto, K., Katsuno, M., Yoshino, T., Arai, Y., Hiroya, K., Sakamoto, T.: Synthesis of 3-substituted indazoles and benzoisoxazoles via Pd-catalyzed cyclization reactions: application to the synthesis of nigellicine. Tetrahedron Lett. **63**, 2695 (2007). https://doi.org/10.1016/j.tet.2007.01.010
18. Chen, C., Cui, S.: Iterative assembly of nitrile oxides and ynamides: synthesis of isoxazoles and pyrroles. J. Org. Chem. **84**(18), 12157–12164 (2019). https://doi.org/10.1021/acs.joc.9b01430
19. Molander, G.A., Cavalcanti, L.N.: Synthesis of trifluoromethylated isoxazolidines: 1,3-dipolar cycloaddition of nitrosoarenes, (trifluoromethyl)diazomethane, and alkenes. Org. Lett. **15**(12), 3166–3169 (2013). https://doi.org/10.1021/ol401402d
20. Shelke, A., Bhong, B., Karade, N.: Oxidation of 1-amidoalkyl-2-naphthols using (diacetoxyiodo)benzene: unusual formation of 1-arylnaphtho[1,2-d]isoxazoles. Synthesis. **46**(6), 752–756 (2014). https://doi.org/10.1055/s-0033-1338590
21. Stokker, G.: Preparation of 1,2-benzisoxazoles from salicylaldoximes via trichloroacetyl isocyanate. J. Org. Chem. **48**, 2613 (1983). https://doi.org/10.1021/jo00163a042
22. Roman, G., Comanita, E., Comanita, B.: Synthesis and reactivity of Mannich bases. Part 15: synthesis of 3-(2-(1-pyrazolyl)ethyl)-1,2-benzisoxazoles. Tetrahedron. **58**, 1617 (2002). https://doi.org/10.1016/S0040-4020(02)00027-3
23. Dubrovskiy, A.V., Larock, R.C.: Synthesis of benzisoxazoles by the [3 + 2] cycloaddition of *in situ* generated nitrile oxides and arynes. Org. Lett. **12**, 1180 (2010). https://doi.org/10.1021/ol902921s
24. Crossley, J.A., Browne, D.L.: Cycloaddition of benzynes and nitrile oxides: synthesis of benzisoxazoles. Tetrahedron Lett. **51**, 2271 (2010). https://doi.org/10.1016/j.tetlet.2010.02.103
25. Kobayashi, E., Togo, H.: Facile one-pot transformation of primary alcohols into 3-aryl- and 3-alkyl-isoxazoles and -pyrazoles. Synthesis. **51**(19), 3723–3735 (2019). https://doi.org/10.1055/s-0039-1690102
26. Srivastava, N., Saxena, M.: Novel synthesis of symmetrical dialkyl/diaryl trithiocarbonates in non-aqueous medium at room temperature using CS_2. CS_2CO_3 and alkyl/aryl halide. Rasayan J. Chem. **12**(1), 333–337 (2019). https://doi.org/10.31788/RJC.2019.1215038
27. Srivastava, N.: Efficient TBAI-CS2 promoted synthesis of substituted hydrazinecarbodithiolates org. Prep. Proced. Int. **54**(6), 569–577 (2022). https://doi.org/10.1080/00304948.2022.2111171
28. Srivastava, N., Chaturvedi, D., Kishore, R., Kaur, M., Kaur, R.: Triton-B/CS2 mediated novel synthesis of substituted thioureas: a novel class of anti-cancer agents. Indian J. Chem. 'B'. **61**, 870–877 (2022). https://doi.org/10.56042/ijc.v61i8.65053
29. Srivastava, N., Kishore, R.: Cleaner and greener synthesis of 3H-benzothiazole-2-thione and its derivatives. J Sulfur Chem. **42**(1), 29–39 (2020). https://doi.org/10.1080/17415993.2020.1803321
30. Srivastava, N.: Tetrabutylammonium iodide/I_2 mediated convenient and green synthesis of substituted organic isothiocyanates. Org. Prep. Proced. Int. **53**(6), 562–570 (2021). https://doi.org/10.1080/00304948.2021.1975487
31. Srivastava, N., Rathore, B.K., Vishnoi, R.K., Bajpai, S.: A simple and convenient method for synthesis of artemisinin dimer through aliphatic nitro compounds. Asian J. Chem. **33**(1), 49–52 (2020). https://doi.org/10.14233/ajchem.2021.22918

32. Kishore, R., Kamboj, M., Shukla, M., Chaturvedi, D., Srivastava, N.: Novel strategy for synthesis of cyclic dithiocarbamates catalyzed by triton-B. Asian J. Chem. **31**(5), 1091–1094 (2019). https://doi.org/10.14233/ajchem.2019.21830
33. Srivastava, N., Saxena, M.: Manisha Shukla, novel synthesis of 5-Oxo-2-thioxo-2,5-dihydro-3-thiophenecarboxylate derivative in non-aqueous medium. Asian J. Chem. **31**(1), 176–180 (2019). https://doi.org/10.14233/ajchem.2019.21622
34. Tyagi, R., Singh, K., Srivastava, N., Sagar, R.: Recent advances in carbohydrate based gelators. Mater. Adv. **4**, 3929–3950 (2023). https://doi.org/10.1039/D3MA00321C
35. Srivastava, N., Saxena, A.K.: Caspase-3 activators as anticancer agents. Curr. Protein Pept. Sci. **24**(10), 783–804 (2023). https://doi.org/10.2174/1389203724666230227115305

Invited lectures (Abstracts)

Green- and Click-Chemistry: An Opportunity for Simplification and Innovation Toward Natural and Nature-Inspired Synthetic Small Molecules of Biological and Commercial Importance

Arun K. Sinha
aksinha08@rediffmail.com
Former Pro-Vice Chancellor, Ranchi University, Ranchi, Jharkhand, India& Former Chief Scientist and Professor (AcSIR), C.S.I.R-C.D.R.I., Lucknow, U.P., India.

The concept of green chemistry has a more than 25-year history of invention and innovation of creating new transformations, new products, new manufacturing processes that perform better with less waste and less cost while being sustainable and safe for humans and the environment. Similarly, click chemistry has become known as a fast, selective, wide in scope, efficient, reliable, and simple to perform for the synthesis of novel and bioactive compounds in high yield with good atom economy and desired functionalities. Thus, the range of products and processes invented and improved by green chemistry and click chemistry touch virtually every industry sector from medicine to agriculture to energy to plastics to nanotechnology to electronics to polymer to natural products. Small molecules of natural origin including polyphenolics and thiophenol-based bioactive compounds have drawn great interest from the scientific community as they are associated with wide range of biological activities including anticancer, antimalarial, antibacterial, antifungal, anti-inflammatory, etc. However, exploration of these natural bioactive molecules is severely hindered by their insufficient percentage in their natural resources, difficult isolation procedure, limiting trials for wider applications besides their tedious multistep synthesis involving protection–deprotection strategy. The shortcomings of the prevalent methodologies have provided a fresh stimulus to develop new strategies based on "Green" and Click Chemistry. Our groups from noticeable time have been working on such green methodologies for extraction and purification of natural molecules as well as design and synthesis of various polyphenolic and thiophenolic-based bioactive molecules like FEMA-GRAS approved 4-vinylphenols, stilbenoids (e.g., resveratrol, pterostilbene), hybrid molecules (e.g., salvianolic acid F), and 3-sulphenylated indole derivatives with their biological evaluation. Moreover, the details of microwave- and ionic liquid-promoted step-economical synthesis of such natural and non-natural molecules of biological and commercial importance will be discussed during presentation.

References:

K. Sinha* et al., *Angew. Chem. Int. Ed.* (**2012**)51, 12250.
K. Sinha* et al. *Angew. Chem. Int. Ed.* (**2012**) *51, 2636.*

A. K. Saxena, S. Nandi (eds.), *Global Trends in Health, Technology and Management*, https://doi.org/10.1007/978-3-031-75457-9

A.K. Sinha*et al., *J. Med. Chem.* (**2012**) 55, 297.
A.K. Sinha* et al., *Angew. Chem. Int. Ed.* (**2015**) 54, 828.
K. Sinha* et al., *J. Org. Chem* (**2019**) 84, 2660.
K. Sinha* et al., *Chem. Asian J.* (**2022**) *17, e202200619.*
K. Sinha* et al., *Adv. Synth. Catal.* (**2023**) 365, 4163–416

Technology Life Cycle—Concept to Commissioning

Alok Kumar Saxena
Technology Consultant

The ultimate aspiration of every scientist or inventor is to witness their creation playing a pivotal role in the lives of the masses, impacting the environment positively, and serving not only the nation's interests but also global well-being. The journey of innovation begins with conceptualization, traverses numerous stages, and culminates in the development of a technology that can seamlessly integrate into the fabric of humanity's needs. However, for a technology to truly make a difference, it must also be commercially viable.

This presentation delves into the intricate process that any technology undergoes before achieving widespread commercial adoption. The journey involves a combination of time- consuming and investment-intensive steps, collectively shaping the characteristic "S" curve associated with the maturation of any process. Throughout the presentation, these steps on the "S" curve are elucidated, shedding light on the inherent challenges associated with technology scale-up and development, as well as emphasizing the crucial aspect of long-term sustainability.

Identification of Targeted Therapeutics Against Oral Squamous Cell Carcinoma

Anshuman Dixit
Institute of Life Sciences, Bhubaneswar

Oral cancer with about 80000 cases/year is a big concern in India. Majority of the oral cancer cases are oral squamous cell carcinoma (OSCC). Consumption of alcohol, chewing tobacco, areca nut, and smoking are significant risk factors associated with oral cancer. Major problems associated with oral cancer treatment are unavailability of targeted therapies and development of chemoresistance. To address the challenge, we devised a bioinformatics pipeline using gene expression analysis, Cox proportional hazard regression, and machine learning to identify genes that can stratify the OSCC patients by risk profile (high-low). We also analyzed the identified genes (BOP1, CCNA2, CKS2, PLAU, and SERPINE1) for their involvement in other cancers as well as in chemoresistance in multiple cancer cell lines.

It is pertinent to note there are no drugs available against these biomarkers to date. Therefore, a comprehensive in silico drug design strategy assimilating homology modeling, extensive molecular dynamics (MD) simulation, and ensemble molecular docking has been applied to identify potential compounds against identified targets, and potential molecules have been identified. We hope that this study will help in deciphering potential genes having roles in chemoresistance and a significant impact on OS. It will also result in the identification of new targeted therapeutics against OSCC.

Pyrazolo[4,3-c]Pyridine Sulfonamides as Carbonic Anhydrase Inhibitors: Synthesis, Biological and in Silico Studies

Andrea Angeli[1,2], Victor Kartsev[3], AnthiPetrou[4], Boris Lichitsky[5], Athina Geronikaki[4,*], and Claudiu T. Supuran[1,*]

[1] NeuroFarba Department, Sezione di ScienzeFarmaceutiche, UniversitàdegliStudi di Firenze, Via Ugo Schiff 6, 50019 Sesto Fiorentino, Italy;

[2] Centre of Advanced Research in Bionanoconjugates and Biopolymers, PetruPoni Institute of Macromolecular Chemistry, AleeaGrigoreGhica-Voda, no. 41A, 700487 Iasi, Romania;

[3] InterBioScreen, 142432 Chernogolovka, Russia

[4] Department of Pharmacy, School of Health, Aristotle University of Thessaloniki, 54124 Thessaloniki, Greece;

[5] Zelinsky Institute of Organic Chemistry, Leninsky Prospect, 119991 Moscow, Russia;

Carbonic anhydrases (CAs, EC 4.2.1.1) catalyze the essential reaction of CO_2 hydration in all living organisms, being actively involved in the regulation of a plethora of patho-/physiological conditions. A series of chromene-based sulfonamides were synthesized and tested as possible CA inhibitors. On the other hand, in microorganisms, the β- and γ- classes are expressed in addition to the α- class, showing substantial structural differences to the human isoforms. In this scenario, not only human but also bacterial CAs are of particular interest as new antibacterial agents with an alternative mechanism of action for fighting the emerging problem of extensive drug resistance afflicting most countries worldwide. Pyrazolo[4,3-c]pyridine sulfonamides were synthesized using methods of organic chemistry. Their inhibitory activity assessed against the cytosolic human isoforms hCA I and hCA II, the transmembrane hCA IX and XII, and β- and γ- CAs from three different bacterial strains was evaluated by a stopped-flow CO_2 hydrase assay. Several of the investigated derivatives showed interesting inhibition activity toward the cytosolic associate isoforms hCA I and hCA II, as well as the 3β- and 3γ-CAs. Furthermore, computational procedures were used to investigate the binding mode of this class of compounds within the active site of hCA IX. Four compounds (**1f**, **1g**, **1h,** and **1k**) were more potent than AAZ against hCA I. Furthermore, compound **1f** also showed better activity than AAZ against the hCA II isoform. Moreover, ten compounds out of eleven appeared to be very potent against the γ-CA from *E. coli,* with a Ki much lower than that of the reference drug. Most of the compounds showed better activity than AAZ against hCA I as well as the γ-CA from *E. coli* and the β-CA from *Burkholderia pseudomallei (BpsCAβ).* Compounds **1f** and **1k** showed a good selectivity index against hCA I and hCA XII, while **1b** was selective against all 3β-CA isoforms from *E. coli*, *BpsCA*, and *VhCA* and all 3γ-CA isoforms from *E. coli*, *BpsCA*, and *PgiCA*.

Quantum Computing Machine Learning for Drug Discovery

Atul Tripathi
atul.tripathi2007@gmail.com
Ex AI and Big Data Consultant, Prime Minister's Office, National Security Council Secretariat

Quantum computing and machine learning are poised to revolutionize drug discovery, offering novel approaches to address the complexities of identifying and developing new therapeutics. Quantum computing harnesses the principles of quantum mechanics to perform computations exponentially faster than classical computers, opening avenues for simulating molecular interactions and predicting drug-target binding with unprecedented precision. At its core, quantum computing leverages the principles of quantum mechanics to process information in ways that far exceed the capabilities of classical computers. In the realm of drug discovery, this translates into the ability to simulate and analyze molecular structures and interactions with unprecedented accuracy and efficiency. By harnessing the unique properties of quantum systems, such as superposition and entanglement, quantum computers can explore vast chemical spaces, predict molecular properties, and elucidate intricate biochemical pathways that underpin disease mechanisms. Machine learning complements this by analyzing vast datasets of molecular structures, biological pathways, and clinical data to uncover patterns and relationships that guide drug discovery efforts. One of the most promising applications of quantum computing and machine learning in drug discovery is the optimization of molecular structures. Quantum algorithms can efficiently explore the vast chemical space to identify lead compounds with desirable properties, while machine learning models can predict the bioactivity and pharmacokinetic profiles of these compounds, accelerating the process of drug candidate selection. Other exciting areas where Quantum Machine Learning can enable are Molecular Property Prediction, Molecular Docking and Binding Affinity Prediction, Drug Repurposing and Multi-Modal Data Integration, Adverse Drug Reaction Prediction, and Drug Safety Assessment and more.

Navigating the Expected and Unexpected Twists and Turns of Lead Optimization: The Discovery of Clinical Candidate for the Treatment of Parkinson Disease

Diwan S Rawat
Senior Professor of Chemistry, University of Delhi
Vice Chancellor, Kumaun University, Nainital-263601, Uttarakhand

In order to address the issue of drug resistance and improve the ADME properties of a drug molecule concept of molecular hybridization was put forward wherein two or more distinct pharmacophores are covalently linked into a single molecule. This approach may lead to a molecule with improved efficacy and may solve the problem of drug resistance and reduce the undesired side effects [1, 2]. The development of such molecular frameworks with synthetic selectivity and economic viability is still a challenging task for the pharmaceutical industry. Drugs developed through this approach can be used for the cure of infectious diseases where treatment is limited to few drugs and the known drugs have limitations such as toxicity, pharmacokinetics, pharmacodynamic and drug resistance. The benefit of using molecular hybrids is to activate different or the same targets by a single molecule, thereby increasing therapeutic efficacy and improving bioavailability. The molecular hybridization approach has resulted in many drug candidates with improved activity profile and some of these compounds are in clinical trials. We have utilized this concept in designing antimalarial molecules and many molecules with aminoquinoline and pyrimidine pharmacophore showed low nanomolar activity. Later a massive multi-institutional collaboration was started and over 700 new molecules were studied for Nurr1 activation, a potential target for Parkinson disease model and identified 15 hits out of which 3 compounds have cleared pre-clinical trials and technology has been transferred to NURRON pharmaceuticals for further development [3–10]. These molecules activate the Nurr1 enzyme which is essential for the survival of the dopamine neurons, stops the aggregation of α-synuclein protein in the brain, and promotes autophagy. Systematic studies demonstrated that these compounds can cure the Parkinson-induced mice model at 5 mg/kg body weight without any toxicity and recently phase I clinical trials of one of the molecules have begun.

References:

Meunier, B. *Acc. Chem. Res.* **2008**, *41*, 2008.
S. S. Shikha, M. Sharma, P.M.S. Chauhan, *Drug News & Perspective*, **2010**, *23*, 632.
S. Manohar, D. S. Rawat, *ACS Med. Chem. Lett.* **2012,** *3*, 555.
Beena, K. K. Raj, S. M. Siddiqui, A. Azam, D. S. Rawat, *Chem. Med. Chem.* **2014**, *9*, 2439.
A. Thakur, S. Manohar, V. Kumar, S. V. Malhotra, D. S. Rawat, *Med. Chem. Commun.* **2014**, *5*, 576.
M. Tripathi, D. Taylor, S. I. Khan, B. L. Tekwani, P. Ponnan, T. Velpandian, U. Das, S. Rawat, *ACS Med. Chem. Lett.* **2019**, *10*, 714.
S. S. Maurya, A. Bahuguna, S. I. Khan, D. Kumar, R. Kholiya, D. S. Rawat, *Eur. J. Med. Chem.* **2019,** *162*, 277.
A. Bahuguna, S. Rawat, D. S. Rawat, *Med. Res. Rev.* **2021**, *41*, 2565.
W. Kim, M. Tripathi, C. Kim, S. Vardhineni, Y. Cha, S. K. Kandi, M. Feitosa, R. Kholiya, E. Sah, A. Thakur, Y. Kim, S. Ko, K. Bhatia, S. Manohar, Y.-B. Kong, G. Sindhu, Y.-S. Kim, B. Cohen, D. S. Rawat, K.-S. Kim, *Nature Communications*, **2023**, 14, 4283.
D. S. Rawat, Sunny Manohar, Ummadisetty Chinna Rajesh, Deepak Kumar, Anuj Thakur, Mohit Tripathi, Panyala Linga Reddy, Shamseer Kulangara Kandi, Satyapavan Vardhineni, Kwang-Soo, and Chun-Hyung Kim, Amino-quinoline based

hybrids and uses thereof. Pub no: US 2017/0209441 A1; EP Application No. 13758678; PCT/US2013/28329; WO2013134047 A3, PCT/US2013/028329 (2013). US 11,026,943 B2 (2021).

Way2drug: Efficient Prediction of Biological Activities from Global to Local Scales

Dmitry S. Druzhilovskiy, Dmitry A. Filimonov, Anastasia V. Rudik, Polina I. Savosina, Vladimir V. Poroikov.

Department for Bioinformatics, Institute of Biomedical Chemistry, Moscow, 119121, Russian Federation, dmitry.druzhilovsky@ibmc.msk.ru

Global chemical space is extremely vast and finding a molecule with the required pharmacotherapeutic property is a formidable challenge. Starting with the analysis of big chemical-biological data obtained *in silico, in vitro, in vivo,* and *in clinico* it is necessary to finish with one active pharmaceutical ingredient possessing the needed safety and efficacy. Combining information extracted from the curated datasets of active/inactive compounds available via World Wide Web with computational AI/ML tools, investigators are surfing from global to local scales in pharmaceutical R&D, enabling faster and more efficient development of new therapeutic remedies.

Way2Drug (https://www.way2drug.com) is a rapidly expanding web portal focused on integrating in-demand in silico tools for drug discovery. Way2Drug currently hosts services for predicting the biological activities (PHARMA), toxicity (TOX), metabolism (META), and physicochemical characteristics (ADME) of drug-like molecules. All tools are freely available for non-commercial academic research.

Way2Drug is evolving as the basis for development of computational platform for efficient analysis and interpretation of the extensive biomedical and clinical data, comparative analysis of information extracted from these data to differentiate the normal and pathological states, obtaining new knowledge to identify potential pharmacological targets and biomarkers, designing potential pharmacological substances with the required properties, determining the optimal approach to therapy taking into account the individuality of patients.

The study is performed in the framework of the Program for Basic Research in the Russian Federation for a long-term period (2021–2030) (No. 122030100170-5).

The Male Pill: Are We There Yet?

Gunda Georg
Regents Professor of Medicinal Chemistry, University of Minnesota, USA

To provide couples with additional safe and reversible options for contraception, the development of non-hormonal contraceptives for both men and women is highly desirable to assist with family planning and reducing unintended pregnancies. The talk will provide an overview of the current state of hormonal and non-hormonal male contraceptive drug discovery and development with an emphasis on the discovery and development of retinoic acid receptor alpha antagonists, including preclinical efficacy studies of our preclinical candidate YCT529 in mouse and non-human primates. YCT529 entered a Phase 1a clinical trial in December 2023.

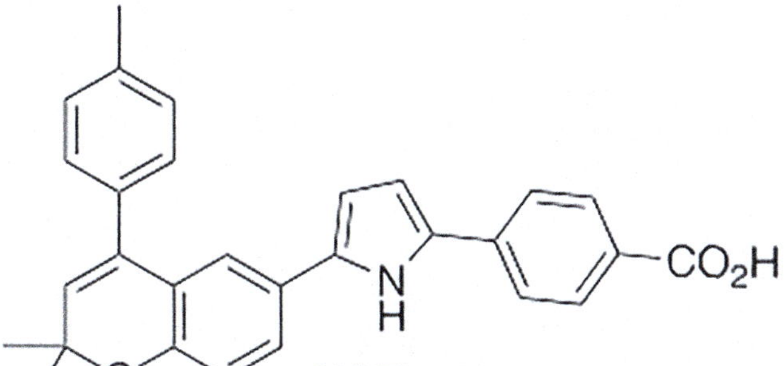

YCT529

	α	β	γ
IC_{50} (nM)	9.7	> 3500	>35000

no agonist activity

Development of Natural Template-Based Novel Diagnostic and Theranostic Agents and Their Formulation Development for Alzheimer's Disease

Gyan Modi*[1], Saroj Kumar[2], Saripella Srikrishna[3], Sarika Gupta[4], Geeta Rai[5], Vegi GM Naidu[6], Gauri Shankar[1], Himanshu Rai[1], and Gourav Singh[1]

gpmodi.phe@itbhu.ac.in

[1]Department of Pharmaceutical Engineering & Technology, Indian Institute of Technology (Banaras Hindu University), Varanasi, 221005, India

[2]Department of Biophysics, All India Institute of Medical Sciences, New Delhi-110029, India

[3]Department of Biochemistry, Banaras Hindu University, Varanasi-201005, India

[4]National Institute of Immunology, New Delhi-110029, India

[5]Department of Molecular and human genetics, Institute of Sciences, Banaras Hindu University, Varanasi-201005, India

[6]Department of Pharmacology & Toxicology, National Institute of Pharmaceutical Education and Research (NIPER)-Guwahati-781032, India

Alzheimer's disease (AD) is a brain-related neurological disorder characterized by gradual loss in memory along with other peripheral and central symptoms. The currently available treatments for AD provide only symptomatic relief without addressing the pathological hallmarks of the disease, therefore, neurodegeneration continues with these therapies[1]. Natural products have gained enormous interest in preventing or diagnosis of AD. Intriguingly, these natural products also suffer from several drawbacks and are not druggable[2]. To address the issues associated with natural products, we developed ferulic acid (FA) and rivastigmine analogs as therapeutic and theranostic agents for AD.

Naturally inspired novel molecules were designed, synthesized, and characterized following the art of medicinal chemistry. The enzyme inhibition studies were carried out with modified Ellman method[3]. SH-SY5Y cells were utilized to evaluate the neuroprotection ability of the lead molecules. The *in-vivo* studies were carried out in various AD mice, drosophila models, and in AD autopsy samples.

The lead molecules were found to be potent inhibitors of AChE/BChE (IC_{50}= 0.79±0.1nM, and 3.97±0.8nM for BChE and AChE, respectively). The lead molecules exhibited promising antioxidant, $A\beta_{1-42}$ and tau aggregation modulation, NLRP3 inhibition, and metal chelation property. Further, the lead molecules were efficacious in the AD drosophila model, *in vivo* and *ex vivo* in scopolamine-induced AD models.

The development of another series of diagnostic and theranostic probes is also inspired on nature product. The lead probe molecules have shown promising and selective Aβ aggregation detection ability in different AD models including transgenic AD mice model and AD patient autopsy samples. The unique ocular imaging pattern in the AD Drosophila model strongly suggests that probes hold promise as a dependable indicator for rapid, noninvasive assessment of new therapeutic modulators or inhibitors in AD.

We are developing the exosome-based formulation of the lead identified in our laboratory.

Acknowledgments
GM is thankful to the Science and Engineering Research Board under Core Research Grant (SERB-CRG/2018/003490), the Indian Council of Medical Research (ICMR/EMR/2019-3088), (ICMR/ EMR/2021-10363). GM also thanks Dr. Anita Mahadevan, Coordinator, Human Brain Bank (NIMHANS), for providing the autopsy tissue samples from clinically diagnosed AD patients.

Strategy to Combat Persistent Tuberculosis: Entrapment of Flow of Substrate in Non-Essential Pathway

Indira Ghosh, (former professor) School of Computational & Integrative Sciences, Jawaharlal Nehru University, New Delhi, India

The post genomic era has provided the leap into the hypothesis-based biology in the field of the "omes" & "omics" world. Designing of drugs is no longer confined to the domains of physiology & medicinal chemistry. The process has changed other way, for biologists it is of utmost importance to know the cause & mechanism of disease, relevance of balancing the flow of substrate & inhibitor, etc. in the metabolism of systems. But the importance of reorganizing our thinking process is mostly reflected in the merger & acquisition of Biotechnology, Bioinformatics, and Chemoinformatics companies by/with large pharmaceutical companies recently. The development in the field of molecular modeling and simulation has played the leading role in the drug design technology in pharmaceutical industry. Starting from Genomics, alignment of genes of interest amongst the species, three-dimensional structure prediction from the knowledge of sequence of amino acids, prediction of active sites, searching for lead compounds using database and structure-based design, prediction of binding efficiency prior to the experiment and lead optimization using physico-chemical properties of the compounds are a few to cite. One such case study, using in silico attempt towards the designing from known anti- tuberculosis chemicals, a set of novel inhibitors and finding the drug target (s) using pathway simulation and analysis [1–3], will be discussed in brief. Recent news has warned the medical community that war against pathogenic bacteria is yet not over, repeated breaking of different drug resistance in different range of pathogens has been observed in patients especially in developing & underdeveloped countries. Many researchers, probably the next to cancer are experimentally using multi-prong attack on the understanding the mechanism and developing new strategy of inhibitors for combating this phenomena, resurging XDR, MDR, PDR. The extensive discussion to define these classes is available in literature [4–7]. Research work engaging clinicians, epidemiologists, biologists, microbiologists, and others in this field is emerging to highlight the complexity of the problem. However, many chemicals have been developed to combat the MDR/XDR Tb, yet they have not been fully successful for a long time. Tuberculii bacteria have shown to have multiple spontaneous mutations at a predictable rate and independent of different drug administration. Such phenomena are prevalent to many bacterial & more common to viral disease which leads researchers to develop multiple target drug designing. Present discussion will open up the future of drug design using the understanding of disease, its inter-relationship within biochemical pathway & effect of protein–protein interaction identified as target for combating persistence stage in bacteria.

References:

Developing an Antituberculosis compounds database & data mining in the search of a motif responsible for the activity of a diverse class of Antituberculosis agents, Om Prakash & Indira Ghosh, J. Chem. Inf. Model., 46, 17–23, 2006.

Kinetic modeling of tricarboxylic acid cycle and glyoxylate bypass in Mycobacterium tuberculosis, and its application to assessment of drug targets, Vivek Kumar Singh and Indira Ghosh, Theoretical Biology and Medical Modelling, BMC Journal. 3; 3:27. 2006.

Determination of Phosphorylation Sites for NADP-specific Isocitrate dehydrogenase from Mycobacterium tuberculosis, Rithvik Vinekar &Indira Ghosh in J. Biomol. Struct. Dyn., Volume 26, Issue Number 6, p 663–895, 2009.

Multidrug-resistant, extensively drug-resistant and pan drug-resistant bacteria: an international expert proposal for interim standard definitions for acquired resistance, a.-P. Magiorakos, et al., Clin. Microbiol. Infect. July 2011, https://doi.org/10.1111/j.1469-0691.2011.03570.x.

Treatment of Multidrug-Resistant Tuberculosis, Michael D. Iseman, New Engl J Med 1993; 329:784–791 September 9, 1993.

Multidrug-Resistant Tuberculosis and Culture Conversion with Bedaquiline, Andreas H. Diacon etal. New Engl J Med 371; 8nejm.org August 21, 2014.
Drug-resistance mechanisms and tuberculosis drugs, Claudio U Köser etal, Lancet 2015 January 24; 385(9965): 305–307. doi:https://doi.org/10.1016/S0140-6736(14)62450-8.

Big and Small Data in Computational Modeling of Toxicological Endpoints

Marjan Vračko
Kemijski inštitut/National Institute of Chemistry, Slovenia

In the era of traditional toxicology a lot of data have been collected, which are compiled in different databases (also called "small data"). On this basis many Quantitative Structure-Activity Relationship (QSAR) models have been developed with intention to expand the knowledge about toxicity to larger chemical space. On the other side, we have a large depository of "big data," which refers to high-throughput screening results and -omics results. It is a particular challenge to compress and integrate the "big and small" data into a single model.

The presentation is oriented to the computational modeling of different toxicological properties considering "big and small" data. Focus is on representations of "big data" (genomic and proteomic data) and the application of Counter-Propagation Artificial Neural Networks (CPANN). CPANN represent one of the basic algorithms of Artificial Intelligence (AI). In the presentation we introduce their architecture and the training algorithm [1]. An accent is placed on their ability for clustering and classification.

In the talk several studies are introduced. Some studies have focused on the binding affinity to androgen nuclear receptors, evaluated with respect to carcinogenic potency data and the CPANN analysis of binding affinities to estrogen, androgen, and glucocorticoid receptors. The clustering of bisphenols with respect to carcinogenic potency database and the developmental toxicity database has been performed with CPANN. Furthermore, several chemometrical studies have been performed on proteomic data using the CPANN, genetic algorithm, and the Principal Component Analysis (PCA). The aims are to study the correlation between proteomic data and different in vitro endpoints and to get insight into mechanisms of activity [2]. As example of analysis of genomic data we present the clustering of Zika and Covid virus data regarding the geographical origin of viruses [3].

References:

M. Novič, Quantitative structure activity/toxicity relationship through neural networks for drug discovery or regulatory use. Curr. Top. Medicinal Chem. 2023, 23, 2792–2804
M. Vračko, S.C. Basak & F. Witzmann, Chemometrical analysis of proteomics data obtained from three cell types treated with multiwalled carbon nanotubes and TiO2 nanobelts. SAR&QSAR Environ. Res. 2018, 29, 567–577

M. Vračko, S. C. Basak, T Dey, A. Nandy, Cluster analysis of coronavirus sequences using computational sequence descriptors: with applications to SARS, MERS and SARS-CoV-2 (CoVID-19). Curr. Comput. Aided Drug Des. 2021, 17, 936–945

Conquering the Neutrophil-A Tale of Novel Neutrophil Serine Protease Inhibitors

Om Prakash
Department of Biochemistry and Molecular Biophysics, Kansas State University, Manhattan, KS 66506, USA

The neutrophils are responsible for our body's innate response against *Staphylococcus aureus* infections and the main driver of inflammatory processes in human diseases. Recently, our collaborator, Dr. Brian Geisbrecht's lab has discovered family of *S. aureus* proteins known as extracellular adherence proteins (Eaps). These proteins target immune proteases found exclusively in human neutrophils. The Eaps and their smaller homologues EapH1 and EapH2 are selective inhibitors of neutrophil serine proteases (NSPs), including cathepsin-G (CG) and neutrophil elastase (NE). To better understand the structure basis for inhibition of these NSPs by EapH1 and EapH2, we have carried out NMR chemical shift perturbation (CSP) studies in solution. Our CSP analysis showed that both EapH1 and EapH2 inhibit CG through a similar binding mode, while EapH1 and EapH2 use distinct binding mode to inhibit NE. Our results define a new mechanism of simultaneous inhibition of two serine proteases by a single Eap protein. The understanding how *S. aureus* blocks neutrophil function may lead us to treating Staphylococcal infections and to develop novel anti-inflammatory inhibitory therapeutics.

Keywords
- Neutrophil Serine Protease (NSP)
- Chemical Shift Perturbation (CSP)
- Extracellular Adherence Proteins (Eaps)
- Structure-Activity Relationship (SAR)

The Importance of 3D Thinking in Anti-Bacterial Agent Discovery

Prasad V. Bharatam
Department of Medicinal Chemistry, National Institute of Pharmaceutical Education and Research, SAS Nagar, Mohali, Punjab, India

3D thinking is very important in drug discovery. 3D thinking helped us in finding nitreones as drugs. Nitreones are cationic but not very reactive. This is an advantage. Though nitreones carry two lone pairs of electrons at the central cationic nitrogen, these two lone pairs do not participate in any interactions (including hydrogen bonds). Though nitreones carry two coordination bonds between carbene units and the central nitrogen, these bonds do not participate in any reactions. Hence, nitreones are safe for drug discovery. Metformin, chlorhexidine, famotidine, etc. drugs carry nitreone character. We have been working on the computational and synthetic aspects of nitreones for the past 15 years. We found that a few antidiabetic, anti-malarial, and anti-leishmanial agents exhibit the nitreone character. Recent synthetic and in vitro efforts lead to the discovery of antibiotic nitreones.

Development of Biased KOR Agonist for Managing Chronic Pain

Sanjay Batra, s_batra@cdri.res.in, batra_san@yahoo.co.uk
Medicinal and Process Chemistry Division, CSIR-Central Drug Research Institute, Sector 10, Jankipuram Extension, Sitapur Road, Lucknow-226031 (U.P.), INDIA

Pain, both acute and chronic, affects millions of people across the globe. Whereas neuropathic pain is usually chronic arising from progressive nerve disease, traumatic nerve injury or infection, non- neuropathic pain (nociceptive pain) is typically due to somatic injury or illness. Opioid analgesics have been long employed as effective method for treating chronic pain but have limitation of serious adverse effects, viz. sedation, dizziness, nausea, vomiting, constipation, physical dependence, and respiratory depression. Therefore, many efforts to discover novel compounds targeting known pain targets and emerging targets for treating pain are being pursued.[1]

The β-carboline subunit is widely represented in natural sources and bioactive agents displaying activity against different diseases. Mitragynine (**1**), a β-carboline alkaloid which is the major constituent of *M. speciosa* leaves, has opium-like analgesic effect and antitussive properties comparable to codeine but does not produce respiratory depression.[2]Experimental evidence suggests that mitragynine binds preferably to μ-opioid receptor and κ-opioid receptor (KOR). Another β-carboline derivative 6-chloro-8-(glycinyl)- amino-β-carboline (**2**) is known to display potent inhibitory activity of nitric oxide implicated as a key factor in the pathogenesis of neuropathic pain.[3]The β-carboline derivative (**3**) is reported to be a potent analgesic that acts via selectively inhibiting the GABA uptake.[4] During our work on exploring the potential of fused-β-carbolines for different bioactivities, we have discovered several new β-carboline derivatives (**4,5**) which display potent KOR agonist activity and display better pain relieving activity than duloxetine, the present standard of care. The results pertaining to our study will be presented and discussed.

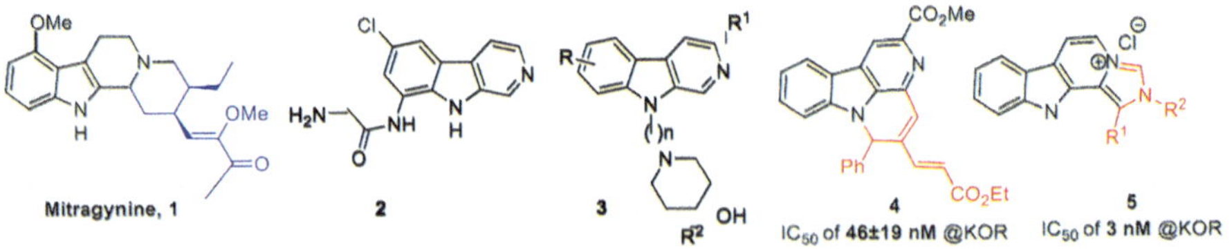

References:

S. Obeng, T. Hiranita, F. León, L. R. McMahon, C. R. McCurdy J. Med. Chem. 2021, 64, 6523.2 a) Jansen, K. L. R.; Prast, C. J. J. Ethnopharmacol. 1988, 23, 115; b) Takayama, Chem. *Pharm. Bull.* **2004**, *52*, 916.3. Grodzki, A. C. G.; Poola, B.; Pasupuleti, N.; Nantz, M. H.; Lein, P. J.; Gorin, F. *J. Pharmacol. Exp. Ther.***2015**, *352*, 438.4. Huunsfeldt, P. O.; Søkilde, B.; Lundbeck, J. M., **WO2001032176**; *Chem. Abstr.* 134:340495

a) Yadav, V. D.; Kumar, L.; Kumari, P.; Kumar, S.; Singh, M.; Siddiqi, M. I.; Yadav, P. N.; Batra, S. ChemMedChem **2021**, 16, 1917; b) P. N. Yadav, **Sanjay Batra** et al. Patent application 202211036633 on 24-06-2022***

Best oral presentation award (Abstract)

Docking-Based Experimental Approach Examining the Effectiveness of Nano-formulations as Diabetes Therapeutics

Rishita Dey, Debojyoti Tarafdar, Sisir Nandi, and Asmita Samadder

Cytogenetics and Molecular Biology Lab., University of Kalyani, Kalyani, Nadia, 741235, India

Department of Pharmaceutical Chemistry, Global Institute of Pharmaceutical Education and Research Kashipur-244713, India

Department of Chemistry, Chanchal College (Affiliated to University of Gour Banga), Malda-732123, India

Diabetes mellitus is a chronic, metabolic disorder that needs to be addressed globally. Diabetes aids in the onset of several associated disorders, including neuropathy, nephropathy, cardiomyopathy, pulmonary arterial disease, etc. Patients develop a reliance on regular medication or insulin dosage, which triggers these related issues. A theoretical validation using computational docking study-based prediction utilizing phyto-based nutraceuticals offers an edge over conventional treatment methods in preventing such situations. The present study explores the potential benefits of plant-based compounds as replacements or alternatives to synthetic pharmaceuticals to improve therapeutic effectiveness while minimizing side effects. Further, to minimize the problems of solubility and dose of phytocompound during internalization, nanotechnology-based drug delivery was explored. Pelargonidin, a phyto-based nutraceutical belonging to the anthocyanidin group known for its therapeutic properties when encapsulated in a biodegradable poly-lactide-co-glycolic-acid (PLGA) polymer, demonstrated increased activity, exceeding that of its non-encapsulated equivalents. Studies using AFM, TEM, and DLS were conducted to determine its physicochemical properties. Therefore, nano-mediated delivery of drugs could improve the solubility, dissolution, retention, and stability of drugs, thereby extending their pharmacological effect. Hence, computational molecular docking studies of drug–protein interactions emerge as a new field aimed at selecting molecules that act to modulate biotargets, thereby establishing the concept of drug delivery with "target-specificity" for combating a variety of diseases including diabetes and its complications.

A. K. Saxena, S. Nandi (eds.), *Global Trends in Health, Technology and Management*, https://doi.org/10.1007/978-3-031-75457-9

Keywords
- Diabetes
- Molecular docking
- Nanotechnology
- Pelargonidin
- PLGA

Poster presentations (Abstracts)

Optimizing Solubility Through Co-crystal Technology: A Rigorous Investigation into Tablets Loaded with Azelnidipine Co-crystals

Aparna Joshi[1], joshi.aparna0306@gmail.com, and G. Gnana Rajan[2]
[1]School of Pharmaceutical Sciences, SGRR University, Dehradun, Uttarakhand, India
[2]Department of Pharmaceutics, SGRR University, Dehradun, Uttarakhand, India.

In the pursuit of enhancing the therapeutic efficacy of Azelnidipine, a widely prescribed antihypertensive agent with limited aqueous solubility, this research delves into the innovative realm of co-crystal technology. The formulation phase incorporates co-crystal strategies designed to optimize the Azelnidipine solubility profile. Three co-formers were employed for co-crystal formulation by wet grinding and solvent evaporation methods. Best co-crystals were found with co-former saccharine by solvent evaporation methods with the help of different analytical characterization. Azelnidipine-loaded co-crystals using saccharin as a former were compressed in tablet dosage form using the wet granulation method. The evaluation process encompasses a spectrum of pharmaceutical attributes, including dissolution behavior, and stability, ensuring a holistic understanding of the developed dosage form. In-vitro bioequivalent studies form a pivotal component of this investigation, elucidating the comparative pharmacokinetics and bioavailability of the optimized Azelnidipine co-crystal tablet formulation against the conventional dosage form. The findings of this research promise to contribute significantly to the field of drug delivery, offering a comprehensive exploration of co-crystal technology's role in optimizing solubility and bioavailability.

Keywords
- Azelnidipine
- Co-crystals
- Bioequivalence
- Solubility enhancement
- Co-former

A. K. Saxena, S. Nandi (eds.), *Global Trends in Health, Technology and Management*, https://doi.org/10.1007/978-3-031-75457-9

Natural Product and Their Derivatives: Promising Caspase-3 Activators and Anticancer Agents

Nitin Srivastava[1] and A.K Saxena[2*]
[1]Department of Chemistry, Amity University Uttar Pradesh Lucknow Campus-226028, India
[2]Department of Pharmaceutical Chemistry, GIPER, Kashipur-244713, India

Cancer is one of the major causes of global morbidity and mortality. The development of various targets found that caspase-3 is an important target whose activation and cleavage may induce apoptosis and ultimately rid from cancer cells. Approaching nature for the treatment of cancer is the most demanding for sustainable development. Apart from many synthetic small molecules naturally occurring molecules have also been reported as caspase-3 activators that induce apoptosis in cancer cells. The work presented here will review the published natural products exhibiting caspase-3 activation or the cleavage to provide some insight into the structural features responsible for caspase-3 activation leading to anticancer activity. This review will help in the development of some novel lead molecules for designing novel molecules of improved cancer therapeutic. The review encompasses the established natural drugs, extracts exhibiting caspase-3 activation and anticancer activity which may be useful for development of lead molecules and optimization of anticancer activity. It may facilitate drug development in a sustainable way.

Keywords
- Cancer
- Caspase-3
- Apoptosis
- Natural products

Modeling of the Cyclic Guanidines as 5-Ht5a Receptor Ligands Using Chemometric Tools

Manju Choudhary[1], Brij K. Sharma[2], Sisir Nandi[3], and Pradeep Pilania[1*], pradeep.pilania@gmail.com
[1]Department of Chemistry, Government Science College, Sikar, 332001(Rajasthan) India
[2]Government College, Talera, Bundi, 323021 (Rajasthan) India
[3]Department of Pharmaceutical Chemistry, Global Institute of Pharmaceutical Education & Research (GIPER), Kashipur, 244713 (Uttarakhand) India

The 5-HT5A receptor binding affinities of the cyclic guanidine derivatives have been quantitatively analyzed in terms of Dragon descriptors. The derived QSAR models have provided rationales to explain the binding affinity of titled compounds. The associations of atomic mass to path length 3 of the Moran autocorrelation (MATS3m), van der Waals volume to path length 3 and 4 of the Moran autocorrelations (MATS3v and MATS4v), polarizability to the highest eigenvalue n.1 of the Burden matrix (BEHp1) and emergence of descriptors Me (mean atomic Sanderson electronegativity scaled on carbon atom) and JGI4 (fourth-order mean topological charge index) have shown the prevalence of atomic properties and charge content to explain the binding affinity. PLS analysis has also confirmed the dominance of information content of the CP-MLR identified descriptors. The derived models and participating descriptors in them have suggested that the substituents of guanidine moiety have sufficient scope for further modification.

Keywords
- QSAR
- Cyclic guanidines
- 5-HT5A ligands
- Binding affinity
- PLS
- CP-MLR

Exploring the Potential of Herbal Nanosponges in Combating the Skin Infections

Priyanka Tomar[1], prianka07@gmail.com, and G. Gnana Rajan[2]
[1]School of Pharmaceutical Sciences, SGRR University, Dehradun, Uttarakhand, India
[2]Department of Pharmaceutics, SGRR University, Dehradun, Uttarakhand, India

Topical drug delivery system faced many problems like poor permeability, skin irritation, allergic reactions. Different types of nanoparticles of plant medicine are gaining a lot of attention currently but they have certain disadvantages like drug loading and toxicity. This research unfolds the potential of nanosponges to conquer these problems. They solubilize poorly water-soluble drug and provide prolonged release. They can load both hydrophilic and hydrophobic drug molecules because of their inner hydrophobic cavities and external hydrophilic branching. Nanosponges are tiny sponges having a size of about a virus and can easily penetrate through skin. The outer surface is typically porous, allowing controlled release of drug. Nanosponge drug delivery system has been emerging as one of the most promising fields in drug delivery. Thus, nanosponges loaded herbal formulation not only able to reduce the conventional treatment related to issues but also may serve as future drug delivery carriers for skin care products which improves the patient compliance.

Keywords
- Topical drug delivery
- Nanosponges
- Increased penetration
- Improved patient compliance

Comparative Studies of Some Common FDA-Approved Drugs Against Human and Fungal Protein

Sarita[1*], sarita7@s.amity.edu; bbk.sarita@gmail.com, Neha Mathur[1], Pooja Chawla[2], Sudip Roy[3], Aiswarya Pawar[4], and Mahima Kori[5]

[1]Amity Institute of Pharmacy, Amity University Uttar Pradesh, Lucknow, 226010, Uttar Pradesh, India

[2]Dr.Pooja Chawla ISF College of Pharmacy, GT Road Moga, 142001, Punjab

[3,4,5]Prescience Insilco Private Limited72/12, Whitefield, Nallurhalli, Bengaluru, Karnataka, 560066

The pharmacotherapy that is now available for the management of fungal infections is restricted to medications that fall within the azole, allylamine, echinocandin, etc. categories. Since the fungus strain grows resistance to various treatment classes, we have employed a drug repurposing strategy in this case, investigating currently licensed medications and utilizing computational methods to validate novel therapeutic outcomes.

The new antifungal potential of some FDA-approved drugs and comparative studies validating the results against human protein as well as fungal protein using *in silico* tools have been explored.

The selected ligand SDF files were obtained from pubchem.ncbi.nlm.nih.gov. and protein of interest data were obtained from protein bank. The docking was done using the Prescience in silico Solution Suite version V 2.1.0. The application X-ESS was used for PRinS3 and MD simulations were performed against two proteins: human lanosterol 14-alpha demethylase PDB ID:3LD6 and fungal protein lanosterol 14-alpha-demethylase PDB ID:5V5Z for comparative studies.

In this investigation, the effects of lansoprazole, phenytoin, and atorvastatin were examined. The outcomes were compared with those of commonly used modern medications, such as itraconazole and ketoconazole. By comparing all three drugs against standard medications, the MD simulation was also used to evaluate the stability of the ligands and protein complexes. The average RSMD, COM distance, and interaction energies indicated that all of the chosen drugs formed stable complexes against the proteins that were tested, allowing for further evaluation in *in vitro* & *in vivo* antifungal efficacy studies.

In silico studies showed that all three drugs showed potential activity against as a potential antifungal drug and are used as an alternative drug against the common fungal strain.

Keywords

- Docking
- Drug repurposing
- Resistance
- Minimum inhibitory concentration

Digitalization in Healthcare

Anshika Bathla, anshii0601@gmail.com
Six Sigma Institute of Technology and Science

The digital revolution has ushered in a new era of possibilities in healthcare, transforming the way we access, deliver, and manage medical services. This poster explores the multifaceted impact of digitalization on healthcare systems, highlighting key trends, challenges, and opportunities. One significant trend is the adoption of electronic health records (EHRs) and health information systems (HIS), which streamline data management, improve patient information accessibility, and enhance care coordination among providers. Additionally, telemedicine and virtual care platforms have emerged as vital tools, facilitating remote consultations, monitoring, and follow-ups, thus expanding access to healthcare services, especially in underserved areas. Moreover, the integration of artificial intelligence (AI) and machine learning algorithms is revolutionizing diagnostics, treatment planning, and personalized medicine, leading to more accurate and efficient healthcare delivery. Wearable devices, mobile health apps, and remote monitoring technologies empower patients to actively participate in their health management, promoting preventive care and wellness interventions. However, digitalization in healthcare also presents challenges such as data privacy concerns, interoperability issues, and the digital divide among populations. Addressing these challenges requires collaborative efforts from policymakers, healthcare providers, technology developers, and regulatory bodies to ensure equitable access, data security, and regulatory compliance. In conclusion, digitalization holds immense promise for improving healthcare efficiency, accessibility, and outcomes. By embracing innovative technologies and fostering a culture of digital health literacy, we can unlock the full potential of digitalization to create a more connected, patient-centric healthcare ecosystem.

Virtual Screening, ADMET Profiling, Molecular Docking Approaches to Search for Potent Selective Synthetic Inhibitors Against Kra_{sg12d}

Divya Pandey, divya.pandey84@gmail.com and Kuldeep K. Roy*, kuldeep.roy@ddn.upes.ac.in, kuldeepkroy@gmail.com
Department of Pharmaceutical Sciences, School of Health Sciences and Technology, UPES, Dehradun, 248007, Uttarakhand, India

Cancer, a complex and multifaceted group of diseases, is a leading cause of death worldwide. Cancer is characterized by uncontrolled cell growth and the potential to invade surrounding tissues. There are multiple mechanisms reported for cancer pathophysiology. Notably, 25% of cancer cases are due to mutations in KRAS (Kirsten rat sarcoma viral oncogene homolog) protein that result in cancerous cell proliferation and metastasis (Kessler et al. 2019). Recently, the Food and Drug Administration (FDA) has granted fast-track approval to Adagrasib and Sotorasib for use in adult patients with locally progressed or metastatic KRASG12C-mutated nonsmall cell lung cancer (NSCLC) (Lanman et al. 2020; Li et al. 2022). In this study, using an *in silico* approach comprised of tool validation, molecular docking, virtual screening and ADMET profiling using the KRAS G12Dcrystal structure and commercial database were investigated. We screened 7,00,000 compounds from ZINC database. The top-ranked 50 compounds identified from the ZINC database were subjected to ADMET profiling. Molecular docking and ADMET results highlighted a higher affinity of four hit compounds toward KRASG12D in comparison with the reference inhibitor, BI-2852. Overall, the present research provides strong support for further in vitro testing of these newly identified KRASG12D inhibitors.

Keywords
- KRAS
- In silico
- Virtual screening
- Molecular docking

Hypertension: A Review

Einab Ansari, and Shivani Lodha
Faculty of Pharmacy-Veer Madho Singh Bhandari Uttarakhand Technical University, Dehradun, India

Objective: The aim of this review is to gather the current information about the various drug therapies and diagnostic evaluations given to control the disease. **Background:** Hypertension (HTN or HT), also known as high blood pressure (HBP), is a long-term medical condition in which the blood pressure in the arteries is persistently elevated. Blood pressure is defined as the force of blood pushing against the walls of blood vessels. Blood pressure is measured through an apparatus called sphygmomanometer. Over a billion people around the world have hypertension or high blood pressure. Hypertension is represented by two numbers—The top number is the **systolic blood pressure** which is the arterial pressure when the heart is contracting and the lower number is **diastolic blood pressure** which is the arterial pressure when the heart is relaxing or refilling. The blood pressure which is considered normal is 120/80 mm Hg.

Symptoms: In some patients, the symptoms will develop like severe headache, blurred vision, dizziness, nausea, vomiting, fatigue, confusion, chest pain, shortness of breath.

Diagnostic evaluations: History collection and physical examination (an appropriate measurement of BP), chest x-ray, ECG, lab tests including blood cell count, blood chemistry (potassium, sodium, creatinine, fasting glucose, total cholesterol, and HDL cholesterol).

Management of hypertension: (A) The lifestyle modification measures mainly include—Weight reduction, DASH diet (dietary approaches to stop hypertension), dietary sodium reduction, reduce alcohol, exercise, stress management.

B) The pharmacological therapy includes—Various groups of drugs are used for the treatment of hypertension, collectively these drugs are called as antihypertensive drugs, which includes—**Diuretics:** E.g., Chlorothiazide, Furosemide, **Beta blockers:** E.g., Atenolol, Propranolol, **Alpha blockers:** E.g., Prazosin, **Vasodilators:** E.g., Nitroglycerin, Sodium nitroprusside, **ACE Inhibitors:** E.g., Captopril, Ramipril, **Calcium channel blockers:** E.g., Amlodipine, Verapamil.

Conclusion: High blood pressure causes stroke by causing a break in blood vessels and bleeding in the brain, may lead to impaired vision and blindness, congestive heart failure, and can cause the kidneys to fail.

Optimizing Chemotherapy Outcomes: Investigating the Effects of Fasting and Low-Calorie Diet on Cancer Patients' Health and Treatment Response

Harshita[1] and Shikha Saxena[1*], Shikha.saxena@ddn.upes.ac.in
Department of Food and Nutrition, School of Health Sciences and Technology, UPES, Dehradun, Uttarakhand, India

Chemotherapy, a common cancer treatment, often causes distressing side effects that impact patients' health and treatment effectiveness. This study explores whether dietary changes, like fasting or eating fewer calories, can improve chemotherapy outcomes and reduce side effects in cancer patients. Using a randomized controlled trial, a diverse group of cancer patients undergoing chemotherapy will be studied. A comparison will be done between patients fasting and patients on a normal diet. The impact will be validated with the help of various biomarkers and complications associated with chemotherapy. By closely examining the treatment responses, and how patients feel, the study aims to understand how these diets affect cancer treatment. Previous studies suggest that fasting and low-calorie diets might help chemotherapy work better by making cancer cells more sensitive to treatment while protecting healthy cells. However, it's not clear how these diets work in real-life situations. The study will be carefully planned, ensuring ethical standards and patient safety. The study intends to uncover how fasting and low-calorie diets impact cancer treatment and patient well-being. The proposed research could change how we approach cancer treatment. If these diets are proven to be beneficial, they could be incorporated to foster cancer treatment and care plans. The outcomes of the study will be crucial in improving treatment outcomes and quality of life for patients. This study also sets the stage for further investigation into the role of nutrition in cancer treatment, potentially leading to more effective therapies in the future.

Keywords
- Low-Calorie Diet
- Fasting
- Patient well-being
- Dietary Change

Oncolytic Viral Therapy: A Promising Approach in Cancer Treatment

Javed Saifi, jsaifi051@gmail.com and Sisir Nandi
Global Institute of Pharmaceutical Education & Research (GIPER), Kashipur-244713, Uttarakhand, India

Oncolytic viral therapy has emerged as a promising and innovative approach in the field of cancer treatment. This therapeutic strategy exploits the selective replication and oncolysis of viruses in tumor cells, while sparing the surrounding healthy tissues. The unique ability of oncolytic viruses to not only directly kill cancer cells but also induce an immune response against tumor antigens has garnered significant attention. In this abstract, we provide an overview of the principles, historical evidence, mechanisms, and potential applications of oncolytic viral therapy. We discuss the different types of oncolytic viruses and their specific mechanisms of action and application of OVs in clinical field. Furthermore, we highlight the challenges and limitations associated with this therapy, such as host immune response and potential side effects. Finally, we discuss future directions to enhance the efficacy and optimize the clinical utility of oncolytic viral therapy, including combination strategies with other treatment modalities. The growing body of evidence supports the potential of oncolytic viral therapy as a valuable addition to the armamentarium of cancer therapy, offering new hope for improved patient outcomes and ultimately, the possibility of eradicating cancer.

Striking a Balance: Unraveling the Relationship Among Corporate Social Responsibility, Sustainability, and Organizational Performance in the Pharmaceutical Industry

Kanika Saxena, kanika1984gupta@gmail.com
Amity Business School, Amity University, Lucknow, India

Purpose—This paper aims to study the inter-relationship among corporate social responsibility (CSR), sustainability and organizational performance in the pharmaceutical sector.

Methodology—In-depth analysis of organizational performance, CSR, and sustainability activities of selected 8 pharmaceutical companies was undertaken for a period of consecutive 5 years from 2016 till 2020. This included data collection for the organizational performance variables net income, annual sales, and stock market performance parameters (market capitalization, price-to-earnings ratio, and market price per share) for each year. A cumulative score based on the number of activities performed under the CSR and sustainability agenda was calculated for each selected company for each year. This followed a linear regression analysis to identify whether there is an existence of a relationship among the three entities.

Results—Both sustainability and CSR positively influence organizational performance in terms of annual sales and net income (before and after tax). However, market capitalization and market price per share were negatively influenced by both CSR and sustainability. Another interesting finding showed that price-to-earnings ratios are not influenced by both CSR and sustainability.

Conclusion—The new insights obtained in the study have further provided real-world evidence to support that both CSR and sustainability improve long-term profits and can be considered investments into the future in the pharmaceutical sector. It advocates investing in activities related to CSR and sustainability initiatives to improve organizational performance which will not only reap profit in the long term but also do good to the society.

Keywords
- Corporate social responsibility
- Sustainability
- Pharmaceutical sector
- Organizational performance
- Sustainability development goals

Telemedicine and Remote Healthcare

Komal Bisht, komalbisht541@gmail.com
Six Sigma Institute of Technology and Science

Particularly in recent years due to the COVID-19 pandemic, **telemedicine and remote healthcare** has experienced tremendous growth and transformation. An extensive summary of the most recent developments and potential future paths in telemedicine and remote healthcare delivery are the goals of this review essay. The study investigated various important aspects such as improved patient engagement, cost-effective delivery models, remote monitoring and management of chronic conditions, availability of emergency and urgent care services, ease of access to specialized consultations and second opinions, integration with electronic health records (EHR), and changes in regulations and reimbursement.

We address the effects of telemedicine on the provision of healthcare, emphasizing its capacity to raise system efficiency, lower healthcare inequities, and improve patient outcomes. We also look at the potential and problems that come with the use of telemedicine, including as technological obstacles, legal issues, privacy issues, and the need for workforce development. This review clarifies the revolutionary significance of telemedicine and remote healthcare in changing the face of healthcare delivery globally through a thorough analysis of recent literature and case studies. We also note new developments in telemedicine, such as the use of blockchain, wearable technology, and artificial intelligence, and the growth of telemedicine services in niche markets including mental health, geriatrics, and rural healthcare.

Overall, this analysis highlights how important telemedicine and remote healthcare are to enhancing patient-centered care, increasing access to high-quality medical services, and meeting the changing global healthcare needs of society. We highlight the significance of ongoing innovation and cooperation to achieve the full potential of telemedicine in influencing the future of healthcare delivery in our concluding remarks, which include insights into the implications for researchers, policymakers, and stakeholders in the healthcare industry.

Keywords
- Remote healthcare
- Telemedicine services
- Healthcare delivery
- EHR
- Potential
- Revolutionary
- Blockchain
- Healthcare inequities
- Technological obstacles

Insights into the Mechanism of Action of Specific Phytoconstituents from the *Euphorbiaceae Family, Derived from Docking Analysis*

Jamal Mahvish*, mahvishjamaal@gmail.com and Singh Mhaveer
School of Pharmaceutical Sciences, IFTM University, Moradabad-244102

Diabetes mellitus (DM) represents a significant metabolic syndrome, affecting approximately 2.8% of the global population with metabolic disorders, a figure projected to escalate to 4.4% by 2030. Among the array of drugs available in the market, including sulfonylureas, biguanides, α-glucosidase inhibitors, thiazolidinediones, and non-sulfonylurea secretagogues, mechanisms such as enhancement of insulin sensitivity, secretion, insulin supplementation, and stimulation of glucose uptake are employed. However, these conventional drugs are often linked to various side effects such as hepatic failure, weight gain, tachycardia, and hypothyroidism. In contrast, herbal remedies offer a promising alternative, characterized by safety, accessibility, fewer side effects, and cost-effectiveness compared to synthetic drugs. Euphorbiaceae, a vast family of dicotyledons boasting 317 genera and 8000 species, yields phytoconstituents like flavonoids, terpenoids, and tannins, which have been utilized in treating ailments ranging from viral infections to skin diseases. While numerous plants within this family have exhibited in vivo antidiabetic properties, the precise mechanism of action of their phytoconstituents remains elusive and warrants further investigation. To address this, docking studies were conducted on the 3D structures of four proteins implicated in type-2 DM, namely 11-β hydroxysteroid dehydrogenase type 1, glutamine: fructose-6-phosphate amidotransferase, protein-tyrosine phosphatase 1B, and mono-ADP ribosyltransferase sirtuin-6, employing 32 known phytoconstituents from this family. The study's findings, based on docking energies and key amino acid interactions, provide insights into the potential mechanisms of action of these phytoconstituents.

Keywords
- Diabetes mellitus
- Euphorbiaceae
- Flavonoids
- Tannins
- Docking

Development and Characterization of Nanosuspension of an Anticancer Drug to Enhance Therapeutic Efficacy in Gastric Cancer

Mayank Kumar Khede*[1], mayank.khedeadcst@gmail.com, Bhabani Shankar Nayak[2], Biswaranjan Mohanty[3], Harekrishna Roy[4]

[1]Ph.D.Research Scholar, BPUT NCR, Institute of Pharmacy and Technology, Salipur, Cuttack, 754202, Odisha, India

[2]Faculty in Pharmacy, KIIT School of Pharmacy, KIMS, KIIT Deemed to be University, Bhubaneswar-751024, Odisha, India

[3]Faculty in Pharmacy, Institute of Pharmacy and Technology, Salipur, Cuttack, 754202, Odisha, India

[4]Faculty in Pharmacy, Nirmala College of Pharmacy, Atmakur Village, Mangalagiri Mandal, Guntur, 522503,Andhra Pradesh, India

Nanosuspensions, distinguished by their reduced particle size and enhanced drug bioavailability, represent a promising avenue to achieve targeted drug delivery in cancer therapeutics. The complicated nature of gastric cancer necessitates innovative drug delivery approaches to enhance therapeutic efficacy. In this study, we present a comprehensive investigation into the development, formulation, and characterization of a nanosuspension formulated with capecitabine, an anticancer drug, to optimize targeted therapy for gastric cancer. The study focuses on tailoring critical physicochemical parameters, including particle size, zeta potential, drug encapsulation efficiency, and drug release kinetics, to improve drug solubility, absorption, and overall therapeutic efficacy. The capecitabine nanosuspension is designed with the objective of achieving enhanced targeted drug delivery to the gastric tumor site, thereby potentially improving treatment outcomes. The advantages of this approach are discussed, highlighting the potential for increased drug efficacy and the possibility of reducing dosing frequency. The ultimate goal is to bring this novel drug delivery system into clinical settings, providing a promising avenue for more effective and targeted therapy in gastric cancer.

Keywords
- Capecitabine
- Nanosuspension
- Anticancer drugs
- Gastric cancer
- Targeted therapy

Impact of Ashwagandha on Insomnia

Shantanu Tyagi and Sisir Nandi

Insomnia is a sleep disorder that makes it difficult to fall asleep or stay asleep. Symptoms include having trouble falling asleep, stress and anxiety, mental health conditions which are depression, schizophrenia, etc. Ashwagandha which is an Indian herbal medicine is effective to cure sleeping disorder (insomnia). Its scientific name is Withania somnifera, withanolides are chemical components of ashwagandha show effective results to cure insomnia by acting on neurotransmitter (GABA).

Role of *Bidens pilosa* L. in Health Management

Shivani Lodha
Faculty of Pharmacy, Veer Madho Singh Bhandari Uttarakhand Technical University, Dehradun, India

Biden pilosa is a perennial herb that widely grows all around us. It is commonly known as blackjack, beggar's tick, and farmer's friend, etc. Although it is easily available, it is one of the most neglected medicinal plants in India having tremendous health benefits. It would not be incorrect to say that it is a kind of plant with the potential to cure various diseases in single plant. In many parts of the world it has been traditionally used as a source of food and medicine. In Vietnam, during the Vietnam War, soldiers adopted the herb as a vegetable, which led to it being known as the "soldier vegetable." It is especially important in eastern Africa, where it is known as *michicha*. In traditional Chinese medicine, this plant is considered a medicinal herb, called xianfengcao (Chinese). In traditional Bafumbira medicine, this plant is applied on a fresh wound and is known as inyabalasanya. Extracts from *Bidens pilosa* are used in Southern Africa for malaria. Also there are various scientific and research-based studies on *B. pilosa* that shows its anti-allergy, antihypertensive and smooth muscle relaxant, anti-cancerogenic, antidiabetic, anti-inflammatory, antimicrobial, antioxidant, antidiabetic, immunomodulatory, antitumoral, antimicrobial properties, etc. All these activities are due to the presence of compounds known as polyacetylenes, flavonoids, chalconeokanin, ethyl caffeate, hydroxycinnamic acid, etc. As *Biden pilosa* has already been screened for its potent antidiabetic activity, and considering that a majority of elderly population in India are suffering from diabetes— which can cause associated disease in long term—it could become a choice of drug to treat the diabetes and its associated diseases with a single drug.

Insights into the Structure and Function of Caseinolytic Peptidase P from the *Mycobacterium* Genus

Smriti Bhardwaj, smriti.explores@gmail.com and Kuldeep K. Roy*, kuldeep.roy@ddn.upes.ac.in

School of Health Sciences and Technology, UPES, Dehradun, 248007, Uttarakhand, India

Antibiotics have shown to be of crucial relevance in preventing diseases brought on by dangerous bacteria. However, the irrational antibiotic usage and other confounding factors have led to the emergence of drug-resistant strains. Pathogenic strains of *Mycobacterium* genus have always been a threat to humans. Because of the development of resistance toward the available antibiotics, new antibiotics against novel druggable target(s) are urgently needed. One of the prospective targets for the development of novel antibiotics for tuberculosis is the Caseinolytic peptidase P (ClpP), particularly the pathogenic forms of mycobacteria that infect and wreak havoc on people (1, 2). Understanding the crystal structures of the mycobacterial ClpP can facilitate the design of novel antibiotics against it. ClpP is a serine protease that aids in the quality control of proteins in the bacterial cell. It has been reported that changes to the ClpP complex have an impact on the virulence and survival of different bacterial pathogens (3). The heterotetradecameric form of ClpP represents the active form. N-blocked dipeptides are necessary for the activation of the ClpP complex leading to proteolysis of misfolded proteins (4). The goal of the current research was to understand the ClpP structure and function, its sequence conservation analysis among mycobacteria, both pathogenic and non-pathogenic, and designing a strategy for the identification of new potential drug that can efficiently kill the drug-resistant mycobacteria.

Keywords
- Antibiotics
- Pathogens
- ClpP
- Drug-resistant
- Virulence
- Serine protease

References:

- Pan S, Malik IT, Thomy D, Henrichfreise B, Sass P. The functional ClpXP protease of Chlamydia trachomatis requires distinct clpP genes from separate genetic loci. Sci. Rep. 2019;9(1):14129

De SS, Khambete MP, Degani MS. Oxadiazole scaffolds in anti-tuberculosis drug discovery. Bioorg. Med. Chem. Lett. 2019;29(16):1999–2007

Raju RM, Unnikrishnan M, Rubin DH, Krishnamoorthy V, Kandror O, Akopian TN, et al. Mycobacterium tuberculosis ClpP1 and ClpP2 function together in protein degradation and are required for viability in vitro and during infection. PLoS Pathog. 2012;8(2):e1002511

Bhardwaj S, Roy KK. ClpP peptidase as a plausible target for the discovery of novel antibiotics. Curr. Drug Targets 2023

Wearable Devices in Health Monitoring

Tushali Khanna, Motikhanna57@gmail.com

Bachelor of pharmacy, Six Sigma Institute of Technology and Science

Wearable gadgets have evolved quickly, becoming essential components in the field of health monitoring. This abstract investigates the changing landscape of wearable technology, focusing on its transformative impact on personal health and well-being. As these devices grow more common, both consumers and healthcare providers must grasp their different types, benefits, and limitations.

This study classifies wearable devices into three types: fitness trackers, smart watches, and medical-grade wearables. Each type is examined for its distinct features and functionalities, offering a thorough review of the many tools available for health monitoring. The taxonomy offered explains the rising span of wearable technology, ranging from lifestyle wearables to those built for specialized medical uses.

The advantages of wearable health monitoring are numerous and go beyond basic data collection. Real-time monitoring, tailored information, and the promotion of preventative health practices are some of the primary benefits. This abstract dives into the positive effects of wearables, proving their ability to raise health awareness, promote physical activity, and aid in the early diagnosis and management of health conditions.

Wearable gadgets have significant promise, but they also present issues that must be carefully considered. Concerns about privacy, data security, and the trustworthiness of health data collected by wearables are addressed. Furthermore, difficulties such as user adherence, technological limits, and the possibility of information overload are investigated. Recognizing these limitations is critical for developing a balanced perspective on the use of wearables in health monitoring.

This abstract provides a thorough overview of wearable technologies, including their various varieties, benefits, and problems.

Keywords
- Wearable gadgets
- Fitness tracker
- Watch
- Data monitoring

Novel Synthesis of 1-Arylnaphtho[1,2-*d*]Isoxazole Through Oxidation of 1-Amidoalkyl-2-Naphthols Using (Diacetoxyiodo) Benzene

Vanshika Gupta, Ram Kishore, and Nitin Srivastava*, nitinsriv5@gmail.com
Department of Chemistry Amity University Uttar Pradesh Lucknow Campus, 226028, India

A series of 1-amidoalkyl-2-naphthols were synthesized via three-component condensation of β-naphthol, aldehyde, and urea under acidic conditions. The subsequent reactions of 1-amidoalkyl-2-naphthols with (diacetoxyiodo)benzene resulted in the unusual formation of 1-arylnaphtho[1,2-*d*]isoxazoles. This reaction demonstrates a useful application of (diacetoxyiodo)benzene for the formation of N-O bond.

Keywords

- Hypervalent iodine
- Phenolic oxidation
- 1,2-benzisoxazole
- 1-amidoalkyl-2-naphthol
- 2-hydroxyaryl ketoximes

MicroRNA: The Unseen Regulator of Wound Healing

Yashodha Rawat, Tripti Rawat and Shahjad

Department of Pharmacology, Global Institute of Pharmaceutical Education and Research, Kashipur, 244713, India

Small non-coding RNA molecules called microRNAs (miRNAs) control the expression of genes by attaching to the 3′ untranslated regions of their target messenger RNAs (mRNAs) and causing either their destruction or translational suppression. As prospective indicators and therapeutic targets for wound healing, miRNAs are emerging. Here, we give a general summary of the pathogenic process, types, and definition of miRNA. RNA polymerase II or III converts genomic DNA into lengthy primary transcripts (pri-miRNAs), which may contain one or more hairpin forms. The first stage of wound healing, known as hemostasis, tries to control bleeding and avoid infection. At the location of the damage, platelets gather and release several substances, including growth factors, cytokines, and chemokines, which start the inflammatory process. The second stage of wound healing is inflammation, during which immune cells including neutrophils, macrophages, and lymphocytes are drawn to and activated at the site of the injury.

The third stage of wound healing, known as proliferation, entails the development of granulation tissue, re-epithelialization, and angiogenesis. Fibroblasts go into the wound area during this phase, when they create collagen and other extracellular matrix elements that give tissues the structural support they need to heal. Several genes, including PTEN, PDCD4, SPRY1, and TIMP3, have a specialized function in the healing of wounds. MicroRNAs including miR-21, miR-146a, miR-155, miR-181a, miR-223, and miR-let-72 are identified as regulators of the inflammatory response in wound healing and have therapeutic potential. TNF-, IL-6, IL-10, TGF-, NF-B, and NLRP3 are examples of inflammatory mediators whose expression is modulated by these miRNAs. This chapter focuses on the obstacles in miRNA therapy, including factors in miRNA target selection and delivery, as well as the miRNAs that may be possibly targeted to improve skin wound healing.

The Influence of the Global Carbon Markets in Mitigating the Risks of Climate Change with Focus on India

Parikshit Narain*, Kanika Saxena*

Climate change is a real problem facing our generation and the next, with potential impacts threatening our existence. It is a real issue requiring real action on the ground to reduce GHG emissions across the globe. Global economies have recognized the urgency of limiting carbon emissions to fight climate change. This study aims to develop a framework for a digital carbon market to assist in monitoring carbon emissions in India, thereby helping in achieving India's commitment to reduce its carbon emission by 45% each year by 2030 and achieve its target of net zero by 2070.

Initially, a literature review was conducted on the role of the carbon market and current greenhouse gas (GHG) emissions in different regions of the world with an emphasis on India. This included studying mobilizing green finance toward suppliers by allowing buyers to offset their GHG emissions through verified, high-quality, and trusted credits. The market structure had the following components:

1. Mechanisms to create demand for credits.
2. Supplier and buyer on boarding
3. Carbon accounting to measure emissions for offsetting
4. Verification to ensure carbon reductions are credible and follow core carbon principles like additionality.
5. Carbon exchange and trading infrastructure
6. Carbon credit pricing methodology
7. Globally accessible registry
8. Regulatory compliance and reporting
9. Measurement of co-benefits including SDG compliance for positive impact.

The framework developed is based on the carbon market incentivizing emission reductions by assigning a monetary value to carbon emissions, driving investment in cleaner technologies and practices. The framework promotes global cooperation by enabling countries to collaborate on emission reduction efforts and facilitate the efficient allocation of resources by encouraging emission reductions where they are most cost-effective.

Fossil fuel consumption, agriculture, land-use change, and cement production are the dominant anthropogenic sources of CO_2 in the atmosphere. The digital market space for carbon emissions provided us to not only monitor the carbon emissions but also provide opportunities to reduce the emissions by providing monetary incentives.

Dengue: A Review

Vikas Kori, Shivani Lodha

Faculty of Pharmacy- Veer Madho Singh Bhandari Uttarakhand Technical University, Dehradun, India

Dengue fever is a mosquito-borne viral infection prevalent in tropical and subtropical regions. Dengue is an acute viral illness caused by RNA virus of the family Flaviviridae and spread by Aedes mosquitoes. Presenting features may range from asymptomatic fever to dreaded complications such as hemorrhagic fever and shock. This abstract provides an overview of the disease, encompassing its etiology, transmission, clinical manifestations, and epidemiological significance. Emphasizing the increasing global burden, the abstract discusses prevention measures, challenges in diagnosis, and potential avenues for future research to mitigate the impact of dengue fever on public health.

The manufacturer's authorised representative in the EU is Springer Nature Customer Service Centre GmbH, Europaplatz 3, 69115 Heidelberg, Germany. If you have any concerns regarding our products, please contact ProductSafety@springernature.com

Printed and bound by CPI Group (UK) Ltd, Croydon, CR0 4YY
30/09/2026
02225587-0008